Lecture Notes in Physics

Springer-Verlag Berlin Heidelberg GmbH

The Editorial Policy for Proceedings

The series Lecture Notes in Physics reports new developments in physical research and teaching – quickly, informally, and at a high level. The proceedings to be considered for publication in this series should be limited to only a few areas of research, and these should be closely related to each other. The contributions should be of a high standard and should avoid lengthy redraftings of papers already published or about to be published elsewhere. As a whole, the proceedings should aim for a balanced presentation of the theme of the conference including a description of the techniques used and enough motivation for a broad readership. It should not be assumed that the published proceedings must reflect the conference in its entirety. (A listing or abstracts of papers presented at the meeting but not included in the proceedings could be added as an appendix.)

When applying for publication in the series Lecture Notes in Physics the volume's editor(s) should submit sufficient material to enable the series editors and their referees to make a fairly accurate evaluation (e.g. a complete list of speakers and titles of papers to be presented and abstracts). If, based on this information, the proceedings are (tentatively) accepted, the volume's editor(s), whose name(s) will appear on the title pages, should select the papers suitable for publication and have them refereed (as for a journal) when appropriate. As a rule discussions will not be accepted. The series editors and Springer-Verlag will normally not interfere with the detailed editing except in fairly obvious cases or on technical matters.

Final acceptance is expressed by the series editor in charge, in consultation with Springer-Verlag only after receiving the complete manuscript. It might help to send a copy of the authors' manuscripts in advance to the editor in charge to discuss possible revisions with him. As a general rule, the series editor will confirm his tentative acceptance if the final manuscript corresponds to the original concept discussed, if the quality of the contribution meets the requirements of the series, and if the final size of the manuscript does not greatly exceed the number of pages originally agreed upon. The manuscript should be forwarded to Springer-Verlag shortly after the meeting. In cases of extreme delay (more than six months after the conference) the series editors will check once more the timeliness of the papers. Therefore, the volume's editor(s) should establish strict deadlines, or collect the articles during the conference and have them revised on the spot. If a delay is unavoidable, one should encourage the authors to update their contributions if appropriate. The editors of proceedings are strongly advised to inform contributors about these points at an early stage.

The final manuscript should contain a table of contents and an informative introduction accessible also to readers not particularly familiar with the topic of the conference. The contributions should be in English. The volume's editor(s) should check the contributions for the correct use of language. At Springer-Verlag only the prefaces will be checked by a copy-editor for language and style. Grave linguistic or technical shortcomings may lead to the rejection of contributions by the series editors. A conference report should not exceed a total of 500 pages. Keeping the size within this bound should be achieved by a stricter selection of articles and not by imposing an upper limit to the length of the individual papers. Editors receive jointly 30 complimentary copies of their book. They are entitled to purchase further copies of their book at a reduced rate. As a rule no reprints of individual contributions can be supplied. No royalty is paid on Lecture Notes in Physics volumes. Commitment to publish is made by letter of interest rather than by signing a formal contract. Springer-Verlag secures the copyright for each volume.

The Production Process

The books are hardbound, and the publisher will select quality paper appropriate to the needs of the author(s). Publication time is about ten weeks. More than twenty years of experience guarantee authors the best possible service. To reach the goal of rapid publication at a low price the technique of photographic reproduction from a camera-ready manuscript was chosen. This process shifts the main responsibility for the technical quality considerably from the publisher to the authors. We therefore urge all authors and editors of proceedings to observe very carefully the essentials for the preparation of camera-ready manuscripts, which we will supply on request. This applies especially to the quality of figures and halftones submitted for publication. In addition, it might be useful to look at some of the volumes already published. As a special service, we offer free of charge LATEX and TEX macro packages to format the text according to Springer-Verlag's quality requirements. We strongly recommend that you make use of this offer, since the result will be a book of considerably improved technical quality. To avoid mistakes and time-consuming correspondence during the production period the conference editors should request special instructions from the publisher well before the beginning of the conference. Manuscripts not meeting the technical standard of the series will have to be returned for improvement.

For further information please contact Springer-Verlag, Physics Editorial Department II, Tiergartenstrasse 17, D-69121 Heidelberg, Germany

J. J. Brey   J. Marro   J. M. Rubí   M. San Miguel (Eds.)

25 Years
of Non-Equilibrium
Statistical Mechanics

Proceedings of the XIII Sitges Conference,
Held in Sitges, Barcelona, Spain,
13–17 June 1994

 Springer

Editors

J. J. Brey
Física Teórica, Fac. de Física
University of Sevilla, Apdo. 1065
E-41080 Sevilla, Spain

J. Marro
Instituto Carlos I de Física Teórica y Computacional
University of Granada
E-18071 Granada, Spain

J. M. Rubí
Dept. Física Fonamental
University of Barcelona, Diagonal 647
E-08028 Barcelona, Spain

M. San Miguel
Dept. Física
University of Illes Baleares
E-07071 Palma de Mallorca, Spain

Library of Congress Cataloging-in-Publication Data

25 years of non-equilibrium statistical mechanics : proceedings of the
  XIII Sitges Conference, held in Sitges, Barcelona, Spain, 13-17 June
  1994 / J. Brey ... [et al.], eds.
      p.   cm. -- (Lecture notes in physics ; 445)
  The Sitges Conference on Statistical Mechanics.
  Includes bibliographical references.

  1. Statistical mechanics--Congresses.   I. Brey, J., 1946-
II. Sitges Conference on Statistical Mechanics (13th : 1994)
III. Series.
QC174.7.A15 1995
530.1'3--dc20   95-34611
                                                                  CIP

ISBN 978-3-662-14018-5   ISBN 978-3-540-49203-0 (eBook)
DOI 10.1007/978-3-540-49203-0

Typesetting: Camera-ready by the authors
SPIN: 10481070   55/3142-543210 - Printed on acid-free paper

Preface

The XIII Sitges Conference was organized for the occasion of the 25th year of its foundation in 1969 by Prof. L. Garrido and the aim was to give an overview of the previous conferences, which have been held during these years. Special emphasis has been devoted to topics related to non-equilibrium phenomena. It is a privilege to thank all the participants who, with their collaboration and support during these years, have contributed to make these conferences a success. As a homage we have collected at the end of this issue the most significant lectures. In particular, in this commemorative edition we acknowlede the participation of Profs: G. Ahlers, B. Alder, K. Binder, E.G.D. Cohen, J.R. Dorfman, J. Dufty, M. Ernst, M. Feigenbaum, B.U. Felderhof, D. Frenkel, R. Graham, F. Guinea, P. Hänggi, K. Kawasaki, J. Lebowitz, P. Mazur, S. Miracle, G. Parisi, Y. Pomeau, H. Reiss, D. Sherrington, N. van Kampen, J.M.J. van Leeuwen, J. Villain, and others. They certainly contributed to make this event important and to give it a dimension of celebration.

We wish to express our gratitude to institutions which generously provided financial support for the organization of the Conference: DGCyT of the Spanish Government, CIRIT of the Generalitat of Catalunya, the universities of Barcelona and Zaragoza and "La Caixa". We also thank the city of Sitges for allowing us again to use the Palau Maricel as a lecture hall and for creating a very pleasant atmosphere.

We are also very grateful and extend our warmest thanks to all those who collaborated in the organization of this conference particularly to Profs. L.J. Boya and J. Biel for their constant attention as well as to I. Pagonabarraga, C. Miguel, and Drs. A. Pérez and A. Diaz for their priceless help in the infrastructure and in the edition of the lectures.

Barcelona
June 1995   The Organizing Committee

Portrait of Luis Garrido [1]

I think it was in 1959 that I first met Luis Garrido. At a hall of residence in Madrid he spoke before a group of students. [...] both his form of address to us and the fact that he had arranged to meet us, as well as the style of the meeting itself all seemed unusual to me. He was in his late twenties then, and expressed himself with difficulty, in a Spanish full of foreign words. He spoke of the development of Physics, of research and of some things that I was unable to fully understand, that somehow did not fit into what I had previously heard. But he was convincing. [...] his unorthodox manner, his keenness and his power to seduce, qualities which, along with his determination, have made Garrido a uniquely important figure [...]. Firstly in Zaragoza and later - from 1962 onwards - in Barcelona, with the skilful support of Professors Sanchez del Rio and Otero Navascues, he brought new concerns and a distinctly different style to our University. [...] The formation of a study group that was successful in publishing articles in journals of international prestige; the recruitment of students and graduates from different universities; the enthusiasm shown to him by his most talented students; the creation of a convivial atmosphere of scientific activity that surrounded him, and that could be physically felt in his light-filled laboratories; his defense of a preferential order of academic values; and his lack of respect for those conventional norms that hindered his projects -all of this, as well as the personal style of the man, led to this spirit of innovation. [...]

In spite of the difficulties of those years, a stubbornly determined Garrido, was able to use his power of seduction to obtain support. And thanks to that, we can also speak today of Garrido's personal work, and not only of the triggering effect he had on those around him. Today we celebrate the 25th anniversary of the foundation of the School of Statistical Mechanics, one of the most discernible results of Garrido's determination and intuition. For intuition is another of the qualities that I would like to stress in Garrido's character. An intuition not only evident in his role as a scientist and researcher - quite obviously shown in the depth of his intellectual ability - but also to be found in his way of confronting any other problems that lay before him. Here, perhaps, lies the reason for our frequent difficulty in following his line of thought: by intuition he quickly arrived at solutions that others could only reach after a slower, rational thought process.

---

[1] Excerpts from the speech by Prof. Jesus Biel (University of Granada) in homage to Prof. Luis Garrido.

Garrido's intuition in relation to the Sitges School was shown in the first place by his desire to found it and mould it to the character he wanted it to have. It had to be a School. [...] At that time in Spain there were few paths open to the young student wanting an initiation in those subjects, whose content, rather than in Spain itself, was developed outside the country. [...] the Sitges School was opened from the beginning to all Spanish young physicists who could benefit from his generosity and thus take part in its activity. The choice of Statistical Mechanics as the School's general subject was yet another of Garrido's skilful touches. [...] a subject that possessed a highly promising future. The Statistical Mechanics of irreversibility was surrounded by special circumstances. Some twelve years before, the subject had acquired a spectacular impulse with the borrowing of concepts and techniques from other parts of Mathematical Physics; there were therefore many problems already conceived and a wide field open for research. On the other hand, those were also the years which heralded the new era of the theory of phase transitions and critical phenomena. And as another example, remember those advances that were being made, at that time, in the rigorous treatment of the foundations of Statistical Mechanics and its connection with Thermodynamics.

[...] Garrido [...] has always preferred from the beginning to point out what his goals are, and thus, attempt to attract others to join him in his projects. The Sitges School is not the final result of a process of evolution; it is the happy creation of an enthusiast, the magnet which Garrido cleverly devised to orientate in the right direction the scientific trajectories of many young Spaniards and the attention of a fair number of physicists of international renown. For that was the challenge, the goal of the enterprise: not only to obtain the interest of Spain in Statistical Mechanics, but at the same time to facilitate communication between researchers here and innovations from outside and thus open the doors of Europe and America to our postgraduates. All of Garrido's efforts [...] would have been fruitless if he had not been able to set the Sitges School up on a level that allowed for its continuity. And once more his intuition and personal seduction made it possible for him to count on the advice and assistance of figures of great scientific standing. Thanks to this help, the topics which the School concentrated upon were, year in and year out, the most appropriate for the moment. [...] We would not be celebrating today the 25th anniversary of the Sitges School without international cooperation and the altruism of those who let themselves be attracted by what was then no more than an ambitious project [...]. [We wish to] [...] express our thanks to him, and to all those who have helped in its creation. Those who have participated as professors, conference speakers and members, as well as those who have contributed in any way to the foundation, maintenance and diffusion of the School and its subject matter. [...]

The importance of this School and Conference during the "25 Years of Non-Equilibrium Statistical Mechanics" is made evident this week by the attendance of those participants in the "XIII Sitges Conference", which aims to give an overview of the previous conferences, featuring as lecturers distinguished physicists. Most of them have been lecturers of the School, and I would like to extend my thanks to all for their presence in this homage to Luis Garrido and his work. I believe that the "Sitges Conference" should feel itself honoured to count on their collaboration in this enterprise.

[...] there are, undoubtedly, many more [of Luis Garrido's achievements] I could mention. For example, I have not chosen to speak now of another of his works, which has been carried out on a more personal level on his pupils and collaborators, and which will only be familiar to each one of them. And I do not think it is fitting for me to dwell on his scientific work in detail here. However, if I have succeeded in highlighting what Luis Garrido has meant for the development of physics in Spain - and, in particular, that of Statistical Mechanics - then I hope you will allow me to turn to Prof. Garrido and say, not only on my behalf, but on behalf of all of us here today: "Thank you, Luis; thank you very much for everything".

Sitges, June 1994

Prof. J. Biel

Contents

Microscopic Reversibility and Macroscopic Behavior:
  Physical Explanations and Mathematical Derivations .   1

J.L. Lebowitz
1 Introduction   2
2 The Problem of Macroscopic Irreversibility   5
3 Boltzmann's Answer   6
4 The Use of Probability   8
5 Initial Conditions   9
6 Origin of Low-Entropy States   10
7 Irreversibility and Macroscopic Stability   13
8 Boltzmann vs. Gibbs Entropies   17
9 Remarks   18
10 Typical vs. Averaged Behavior   18

Twenty-Five Years of Non-Equilibrium Statistical Mechanics:
  Towards a Better Understanding of Dense Fluids   21

E.G.D. Cohen
1 Introduction   21
2 Divergences, Green-Kubo Relations,
   Long-Time Tails, Vortex Diffusion   22
3 Long-Range Non-Equilibrium Correlations, Neutron Scattering,
   Cage Diffusion, Fast and Slow Sound   28
4 Lattice Gas Cellular Automata, Transport Coefficients,
   and Lyapunov Exponents   37
5 Analogy Atomic Liquids and Concentrated Colloidal Suspensions   39
6 Conclusion   42

Models for Dissipation in Quantum Mechanics 51

N.G. van Kampen
1 The Problem 51
2 First Model: Free Particle 53
3 Second Model: Harmonic Oscillator 54
   Appendix 57

Brownian Motion Theory:
   Properties of Non-Linear Langevin Equations 60

P. Mazur
1 Introduction: Theory and Theories of Brownian Motion 60
2 The Random Force Autocorrelation Function 63
3 When and Why is the Random Force a Gaussian Process? 64
4 The Genuinely Non-Linear Case 67

Non-Equilibrium, Phase Transition, and Related Problems 70

K. Kawasaki
1 Introduction 70
2 Recent Period 72
3 Modern Period 76
4 Dynamics of Two-Dimensional Foams 77
5 Discussion 80

Simulation of the Consistent Boltzmann Equation for
   Hard Spheres and Its Extension to Higher Densities .. 82

F.J. Alexander, A.L. Garcia, and B.J. Alder
1 Introduction 82
2 Non-ideal Gas DSMC 83
3 Dense Gas DSMC 84
4 Computer Simulations 84
5 Efficiency 88
6 Conclusions 88

Over Two Decades of Pattern Formation:
   A Personal Perspective 91

G. Ahlers
1 Introduction 91
2 The 1970's and Before 94

3   The 1980's 101
4   The 1990's 112
5   Outlook 117

Fluctuations in the Steady State 125

R. Graham
1   Introduction 125
2   Potential Plateaus 128
3   Noisy Attractors 130
4   Potential Plateaus and Escape Times Near Bifurcations 131
5   Conclusion 133

Slow Non-Equilibrium Dynamics 135

G. Parisi
1   Introduction 135
2   Local Equilibrium States 136
3   The Non Equilibrium Equations 138
4   Real Glasses 140
5   Conclusions 141

Theory of Glass Transition in Spin Glasses,
   Orientational Glasses and Structural Glasses 143

K. Binder
1   Introduction: Basic Facts and Concepts 143
2   Orientational Glasses 146
3   Towards the Modeling of the Glass Transition
    in Amorphous Polymers 152
4   Concluding Remarks 157

Evolution of Order Parameters in Disordered Spin Systems.
   A Closure Procedure 161

D. Sherrington and A.C.C. Coolen
1   Derivation of Closed Macroscopic Flow Equations 161
2   Results for the Hopfield Model 166
3   Results for the Sherrington-Kirkpatrick Model 169
4   An Exactly Solvable Toy Model 173
5   Summary 176

Elastic Instabilities in Crystal Growth 177

J. Villain, C. Duport, and P. Nozières
1   Introduction 177
2   The Asaro-Tiller-Grinfeld Instability 179
3   The Grinfeld Instability on a Singular Surface 182
4   Dynamics in the Step Flow Regime: Step Bunching Instability   184
5   Other Instabilities and Other Mechanisms 186

**Vapor Phase Nucleation: Molecular Mechanisms
    for Embryo Development** 189

H. Reiss, C.L. Weakliem, and H.M. Ellerby
1   Introduction 189
2   The Fluctuation Cluster 190
3   Characteristics of $W_i(v)$ 191
4   Rate Theory 193
5   Nucleation Rate Based on the MLDM 194
6   Summary 197

Chaos in Lorentz Lattice Gases 199

M.H. Ernst and J.R. Dorfman
1   Introduction 199
2   Chapman-Kolmogorov Equation 200
3   Lorentz Lattice Gas 201
4   Deterministic Map 202
5   Lyapunov Exponents for Random Walks 203
6   Links Between Chaos and Kinetic Theory 205
7   Kinetic Theory of Closed Systems 207
8   Conclusions and Perspectives 208

Dynamics of Reptation 211

J.D. Balkenende, J.A. Leegwater, and J.M.J. van Leeuwen
    Introduction 211
1   The Master Equation 213
2   The Drift Velocity 216
3   Generalizations of the Drift Formula 218
4   Long Polymers 220
5   The Fast Extron Limit 222
6   Discussion 225

Non-Equilibrium Fluctuations in Simple Fluids 227

J. Dufty and M. C. Marchetti
1 Introduction 227
2 Generating Functional 228
3 The Boltzmann Gas 231
4 The Hard Sphere Enskog Fluid 233
5 Practical Applications 235
6 Discussion 238

Long Time Tails in Stress Correlation Functions 240

M.H.J. Hagen, C.P. Lowe, and D. Frenkel
1 Introduction 240
2 Lattice-Boltzmann Model 242
3 Results 245
4 Conclusions 249

New Advances in Laplacian Growth Models 250

F. Guinea, O. Pla, E. Louis, and V. Hakim
1 Introduction 250
2 Morphologies in the Dielectric Breakdown Model 252
3 Fracture by Thermal Stresses 255
4 Conclusions 257

Slow Dynamics and Linear Relaxation 259

B.U. Felderhof
1 Introduction 259
2 Slow Dynamics 260
3 Pole Structure 263
4 Diffusion in a Potential 266
5 Discussion 267

**Driven Tunneling: New Possibilities for Coherent and
Incoherent Quantum Transport** 269

T. Dittrich, P. Hänggi, B. Oelschlägel, and R. Utermann
1 Introduction 269
2 The Model and Its Symmetries 270
3 Driven Tunneling and Localization 271
4 Tunnel Splittings and the Onset of Chaos 273

5   Driven Tunneling with Dissipation 275
6   Summary 277

**From the Random Sequential Adsorption
to the Ballistic Model** 282

P. Schaaf
1   Introduction 282
2   Experimental Results 283
3   Simulation Results 287
4   From the RSA to the Ballistic Model 291
5   Conclusion 294

High-Frequency Viscosity of a Dilute Polydisperse Emulsion   296

R.B. Jones
1   Introduction 296
2   Polydisperse Formalism 297
3   Droplets 299
4   Results and Discussion 300

Mesoscopic Theory of Liquid Crystals 303

W.Muschik, H. Ehrentraut, C. Papenfuss, and S. Blenk
1   Introduction 303
2   Mesoscopic Concept 304
3   Orientational Balances 305
4   Alignment Tensors 308
5   Remark on Constitutive Equations 309

On the Microscopic Theory of Phase Coexistence 312

S. Miracle Solé
1   Introduction 312
2   Gibbs States and Interfaces 312
3   The Surface Sension 314
4   The Step Free Energy 315
5   Facets in the Equilibrium Crystal 317
6   Nucleation and Growing Crystals 318

A Non-Equilibrium Phase Transition Induced
  by Multiplicative Noise 322

C. van den Broeck, J.M.R. Parrondo, and R. Toral
1 Introduction 322
2 Mean Field Model 323
3 Noise-Induced Phase Transition 324
4 Discussion 325

Electric Field Domains in Superlattices: Dynamics 327

L.L. Bonilla, J.A. Cuesta, J. Galán,
F.C. Martínez, and J.M. Molera
1 Introduction 327
2 Steady States 331
3 Phase Diagram and PC Time-Dependent Oscillations 332
4 Final Remarks 336

Molecular Dynamics Simulation of Phase Separation
  and Domain Growth 338

S. Toxvaerd
1 Non-Equilibrium Molecular Dynamics 338
2 MD of Phase Separations in 2 Dimensional Mixtures 340

Finite-Size Effects in the Kardar-Parisi-Zhang Equation .. 344

R. Toral and B. Forrest

Competing Scaling Behaviours in the Late
  Stage of Growth Kinetics 352

A. Coniglio, P. Ruggiero, and M. Zannetti

Stretched-Exponential Relaxation of Transient Electric
  Birefringence in Polymer-Like Reverse Micelles 363

G.J.M. Koper, C. Cavaco, and P. Schurtenberger
1 Introduction 363
2 Theory 364
3 Materials and Methods 366
4 Results and Discussion 367

List of Posters 371

List of Participants 375

Contributions to the Sitges Conferences 379

Microscopic Reversibility and Macroscopic Behavior:
Physical Explanatoins and Mathematical Derivations

Joel L. Lebowitz

Departments of Mathematics and Physics
Rutgers University, New Brunswick, NJ 08903

Abstract: The observed general time-asymmetric behavior of macroscopic systems—embodied in the second law of thermodynamics—arises naturally from time-symmetric microscopic laws due to the great disparity between macro and micro-scales. More specific features of macroscopic evolution depend on the nature of the microscopic dynamics. In particular, short range interactions with good mixing properties lead, for simple systems, to the quantitative description of such evolutions by means of autonomous hydrodynamic equations, e.g. the diffusion equation. These deterministic time-asymmetric equations accurately describe the observed behavior of *individual* macro systems. Derivations using ensembles (or probability distributions) must therefore, to be relevant, hold for almost all members of the ensemble, i.e. occur with probability close to one. Equating observed irreversible macroscopic behavior with the time evolution of ensembles describing systems having only a few degrees of freedom, where no such typicality holds, is misguided and misleading.

> *"The equations of motion in abstract dynamics are perfectly reversible; any solution of these equations remains valid when the time variable t is replaced by $-t$. Physical processes, on the other hand, are irreversible: for example, the friction of solids, conduction of heat, and diffusion. Nevertheless, the principle of dissipation of energy is compatible with a molecular theory in which each particle is subject to the laws of abstract dynamics."*

> W. Thomson, (1874)[1]

1 Introduction

Given the success of the statistical approach, pioneered by James Maxwell, William Thomson (later Lord Kelvin) and made quantitative by Ludwig Boltzmann, in both explaining and predicting the observed behavior of macroscopic systems on the basis of their reversible microscopic dynamics, it is quite surprising that there is still so much confusion about the "problem of irreversibility". I attribute this to the fact that the originality of these ideas made them difficult to grasp. When put into high relief by Boltzmann's precise and elegant form of his famous kinetic equation and H-theorem they became ready targets for attack. The confusion created by these misunderstandings and by the resulting "controversies" between Boltzmann and some of his contemporaries, particularly Ernst Zermelo, has been perpetuated by various authors who either did not understand or did not explain adequately the completely satisfactory resolution of these questions by Boltzmann's responses and later writings. There is really no excuse for this, considering the clarity of the latter. In Erwin Schrödinger's words, "Boltzmann's ideas really give an understanding" of the origin of macroscopic behavior. All claims of inconsistencies (known to me) are in my opinion wrong and I see no need to search for alternate explanations of such behavior—at least on the non-relativistic classical level . I highly recommend some of Boltzmann's works [2], as well as the beautiful 1874 paper of Thomson [1] and the more contemporary references [3-7], for further reading on this subject; see also [8] for more details on the topics discussed here.

Boltzmann's statistical theory of nonequilibrium (time-asymmetric, irreversible) behavior associates *to each microscopic state of a macroscopic system*, be it gas, fluid or solid, a number S_B: the "Boltzmann entropy" of that state [4]. This entropy agrees (up to terms negligible in the size of the system) with the macroscopic thermodynamic entropy of Clausius *when* the system is in equilibrium. It also coincides *then* with the Gibbs entropy S_G, which is defined not for individual microstates but for statistical ensembles or probability distributions (in a way to be described later). The agreement extends to systems in local equilibrium. However, unlike S_G, which does not change in time even for ensembles describing (isolated) systems not in equilibrium, e.g. fluids evolving according to hydrodynamic equations, S_B typically increases in a way which *explains* and describes qualitatively the evolution towards equilibrium of macroscopic systems.

This behavior of S_B is due to the separation between microscopic and macroscopic scales, i.e. the very large number of degrees of freedom involved in the specification of macroscopic properties. It is this separation of scales which enables us to make definite predictions about the evolution of a *typical individual realization* of a macroscopic system, where, after all, we actually observe irreversible behavior. As put succinctly by Maxwell [9] "the second law is drawn from our experience of bodies consisting of an immense number of molecules. ...it is continually being violated, ..., in any sufficiently small group of molecules .... As the number ...is increased ...the probability of a measurable variation ...may be regarded as practically an impossibility". The various ensembles commonly used in statistical mechanics are to be thought of as nothing more than mathematical tools for describing behavior which is practically the same for "almost all" individual macroscopic systems in the ensemble. While these tools can be very useful and some theorems that are proven about them are very beautiful they must not be confused with the real thing going on in a single system. To do that is to commit the scientific equivalent of idolatry, i.e. substituting representative images for reality. Moreover, the time-asymmetric behavior manifested in a single typical evolution of a macroscopic system distinguishes macroscopic irreversibility from the mixing type of evolution of ensembles which are caused by the *chaotic* behavior of systems with but a few degrees of freedom, e.g. two hard spheres in a box. To call the latter irreversible is, therefore, confusing.

The essential qualitative features of macroscopic behavior can be understood on the basis of the incompressible flow in phase space given by Hamilton's equations. They are not dependent on assumptions, such as positivity of Lyapunov exponents, ergodicity, mixing or "equal a priori probabilities," being *strictly* satisfied. Such properties are however important for the quantitative description of the macroscopic evolution which is given, in many cases, by time-asymmetric autonomous equations of hydrodynamic type. These can be derived (rigorously, in some cases) from reversible microscopic dynamics by suitably scaling macro and micro units of space and time and then taking limits in which the ratio of macroscopic to microscopic scales goes to infinity [10]. (These limits express in a mathematical form the physics arising from the very large ratio of macroscopic to microscopic scales.) Using the law of large numbers then shows that these equations describe the behavior of almost all individual systems in the ensemble, not just that of ensemble averages, i.e. the dispersion goes to zero in the scaling limit. Such descriptions also hold, to a high accuracy, when the macro/micro ratio is finite but very large. They are however clearly impossible when the system contains only a few particles.

The existence and form of such hydrodynamic equations depends on the nature of the microscopic dynamics. In particular, instabilities of trajectories induced by chaotic microscopic dynamics play an important role in determining many features of macroscopic evolution. A simple example in which this can be worked out in detail is provided by the Lorentz gas. This consists of a macroscopic number of non-interacting particles moving among a periodic array of fixed convex scatterers arranged in the plane so that there is a maximum distance a particle can travel between collisions. The chaotic nature of the microscopic dynamics, which leads to an approximately isotropic local distribution of velocities, is directly responsible for the existence of a simple autonomous deterministic description, via a diffusion equation, for *typical* macroscopic particle density profiles of this system [10]. Another example is the description via the Boltzmann equation of the density in the six dimensional position and velocity space of a macroscopic dilute system of hard spheres [7], [10]. I use these examples, despite their highly idealized nature, because here all the mathematical *i*'s have been dotted. They thus show *ipso facto*, in a way that should convince even (as Mark Kac put it) an "unreasonable" person, not only that there is no conflict between reversible microscopic and irreversible macroscopic behavior but also that, *for essentially all initial microscopic states consistent with a given nonequilibrium macroscopic state*, the latter follows from the former—in complete accord with Boltzmann's ideas.

Boltzmann's analysis was of course done in terms of classical Newtonian mechanics and I shall use the same framework for this article. The situation is in many ways similar in quantum mechanics where reversible incompressible flow in phase space is replaced by unitary evolution in Hilbert space. In particular I do not believe that quantum measurement is a *new* source of irreversibility. Such assertions in effect "put the cart before the horse". Real measurements on quantum systems are time-asymmetric because they involve, of necessity, systems with very large number of degrees of freedom whose irreversibility can be understood using natural extensions of classical ideas [11], [13].

There are however also some genuinely new features in quantum mechanics relevant to our problem. First, to follow the classical analogy directly one would have to associate a macroscopic state to an arbitrary wave function of the system, which is impossible as is clear from the Schrödinger cat paradigm [12] (or paradox). Second, quantum correlations between separated systems arising from wave function entanglements lead to the impossibility, in general, of assigning a wave function to a subsystem S_1 of a system S in a definite state ψ even at a time when there is no direct interaction between S_1 and the rest of S, and this makes the idealization of an isolated system much more problematical in quantum mechanics than in classical theory. These features of quantum mechanics require careful analysis to see how they affect the irreversibility observed in the real world. An in depth discussion is not only beyond the scope of this article but would also require some new ideas and quite a bit of work which is yet to be done. I refer the reader to references [11–14] for a discussion of some of these questions from many points of view.

I will also, in this article, completely ignore relativity, special or general. The phenomenon we wish to explain, namely the time-asymmetric behavior of spatially localized macroscopic objects, has certainly many aspects which are the same in the relativistic (real) universe as in a (model) non-relativistic one. This means of course that I will not even attempt to touch the deep conceptual questions regarding the nature of space and time itself which have been much discussed recently in connection with reversibility in black hole radiation and evaporation [14]. These are beyond my competence and indeed it seems that their resolution may require new concepts which only time will bring. I will instead focus on the problem of the origin of macroscopic irreversibility in the simplest idealized classical context. The Maxwell-Thomson-Boltzmann resolution of this problem in these models does, in my opinion, carry over essentially unchanged to real systems.

2 The Problem of Macroscopic Irreversibility

Consider a macroscopic system evolving in time, as exemplified by the schematic snapshots of a binary system, say two different colored inks, in the four frames in Figure 1. The different frames in this figure represent pictorially the two local concentrations of the components at different times. *Suppose* we know that the system was isolated during the whole time of picture taking and we are asked to identify the time order in which the snapshots were taken.

The obvious answer, based on experience, is that time increases from 1a to 1d—any other order is clearly absurd. Now it would be very simple and nice if this answer could be shown to follow directly from the microscopic laws of nature. But this is not the case, for the microscopic laws, as we know them, tell a different story: if the sequence going from left to right is permitted by the microscopic laws, so is the one going from right to left.

This is most easily seen in classical mechanics where the complete microscopic state of an isolated classical system of N particles is represented by a point $X = (r_1, v_1, r_2, v_2, \ldots, r_N, v_N)$ in its phase space Γ, r_i and v_i being the position and velocity of the ith particle. Now a snapshot in Fig. 1 clearly does not specify completely the microstate X of the system; rather each picture specifies a coarse grained description of X, which we denote by $M(X)$, the macrostate corresponding to X. For example, if we imagine that the (one liter) box in Fig. 1 is divided into a billion little cubes, then the macrostate M could simply specify (within some tolerance) the fraction of particles of each type in every cube j, $j = 1, \ldots, 10^9$. To each macrostate M there corresponds a very large set of microstates making up a region Γ_M in the phase space Γ. In order to specify properly the region Γ_M we need to know also the total energy E, and any other *macroscopically relevant*, e.g. additive, constants of the motion (also within some tolerance). While this specification of the macroscopic state clearly contains some arbitrariness, this need not concern us unduly here. All the qualitative statements we are going to make about the time evolution of macrostates M are sensibly independent of its precise definition as long as there is a large separation between the macro and microscales.

Let us consider now the time evolution of microstates which underlies that of the macrostates $M(X)$. They are governed by Hamiltonian dynamics which connects a microstate $X(t_0)$ at some time t_0, to the microstate $X(t)$ at any other time t. Let $X(t_0)$ and $X(t_0 + \tau)$, $\tau > 0$, be two such microstates. Reversing (physically or mathematically) all velocities at time $t_0 + \tau$, we obtain a new microstate. If we now follow the evolution for another interval τ we arrive at a microstate at time $t_0 + 2\tau$ which is just the state $X(t_0)$ with all velocities reversed. We shall call RX the microstate obtained from X by velocity reversal, $RX = (r_1, -v_1, r_2, -v_2, \ldots, r_N, -v_N)$.

Returning now to the snapshots shown in the figure it is clear that they would remain unchanged if we reversed the velocities of all the particles; hence if X belongs to Γ_M then also RX belongs to Γ_M. Now we see the problem with our definite assignment of a time order to the snapshots in the figure: that a macrostate M_1 at time t_1 evolves to another macrostate M_2 at time $t_2 = t_1 + \tau$, $\tau > 0$, means that there is a microstate X in Γ_{M_1} which gives rise to a microstate Y at t_2 with Y in Γ_{M_2}. But then RY is also in Γ_{M_2} and following the evolution of RY for a time τ would produce the state RX which would then be in Γ_{M_1}. Hence the snapshots depicting M_a, M_b, M_c and M_d in Fig. 1 could, as far as the laws of mechanics (which we take here to be the laws of nature) go, correspond to a sequence of times going in either direction.

It is thus clear that our judgement of the time order in Fig. 1 is not based on the dynamical laws of evolution alone; these permit either order. Rather it is based on experience: one direction is common and easily arranged, the other is never seen. *But why should this be so?*

3 Boltzmann's Answer

The above question was first raised and the answer developed by theoretical physicists in the second half of the nineteenth century when the applicability of the laws of mechanics to thermal phenomena was established by the experiments of Joule and others. The key people were Maxwell, Thomson and Boltzmann. As already mentioned I find the 1874 article by Thomson an absolutely beautiful exposition containing the full qualitative answer to this problem. This paper is, as far as I know, never referred to by Boltzmann or by latter writers on the subject. It would or should have cleared up many a misunderstanding. I can only hope (but do not really expect) that my article will do better. Still I will try my best to say it again in more modern (but less beautiful) language. The answer can be summarized by a quote from Gibbs which appears (in English) on the flyleaf of Boltzmann's Lectures on Kinetic Theory, Vol. 2, [15] (in German): "In other words, the impossibility of an uncompensated decrease of entropy seems to be reduced to improbability [16]."

This statistical theory can be best understood by associating to each macroscopic state M and *thus to each phase point X giving rise to M*, a "Boltzmann entropy", defined as

$$S_B(M) = k \log |\Gamma_M|, \tag{1}$$

where k is Boltzmann's constant and $|\Gamma_M|$ is the phase space volume associated with the macrostate M, i.e. $|\Gamma_M|$ is the integral of the time invariant Liouville volume element ($\prod_{i=1}^{N} d^3\mathbf{r}_i \, d^3\mathbf{v}_i$) over Γ_M. (S_B is defined up to additive constants, see [4].)

Boltzmann's stroke of genius was to make a direct connection between this microscopically defined function $S_B(M(X))$ and the thermodynamic entropy of Clausius, S_{eq}, which is a macroscopically defined, operationally measurable (up to additive constants), extensive property of macroscopic systems in *equilibrium*. For a system in equilibrium having a given energy E (within some tolerance) volume V and particle number N, Boltzmann showed that

$$S_{eq}(E, V, N) = N s_{eq}(e, n) \simeq S_B(M_{eq}), \quad e = E/N, \ n = N/V, \qquad (2)$$

where $M_{eq}(E, V, N)$ is the equilibrium macrostate (corresponding to M_d in Fig. 1). By the symbol $\simeq$ we mean that for large N, such that the system is really macroscopic, the equality holds up to terms negligible when both sides of equation (2) are divided by N and the additive constant is suitably fixed. It is important that the cells used to define M_{eq} contain many particles, i.e. that the macroscale be very large compared with the microscale.

Having made this identification it is natural to use Equation (1) to also define (macroscopic) entropy for systems not entirely in equilibrium and thus identify increases in such entropy with increases in the volume of the phase space region $\Gamma_{M(X)}$. This identification explains in a natural way the observation, embodied in the second law of thermodynamics, that when a constraint is lifted from an isolated macroscopic system, it evolves toward a state with greater entropy. To see how the explanation works, imagine that there was initially a wall dividing the box in Fig. 1 which is removed at time t_a. The phase space volume available to the system without the wall is fantastically enlarged: If the system in fig. 1 contains 1 mole of fluid in a 1-liter container the volume ratio of the unconstrained region to the constrained one is of order 2^N or $10^{10^{20}}$, roughly the ratio $|\Gamma_{M_d}|/|\Gamma_{M_a}|$. We can then expect that when the constraint is removed the dynamical motion of the phase point X will with very high "probability" move into the newly available regions of phase space, for which $|\Gamma_M|$ is large. This may be expected to continue until $X(t)$ reaches $\Gamma_{M_{eq}}$ corresponding to the system now being in its unconstrained equilibrium state. After that time we can expect to see only small fluctuations from macroscopic equilibrium—typical fluctuations being of order of the square root of the number of particles involved. It should be noted here that an important ingredient in the whole analysis is the constancy in time of the Liouville volume of sets in the phase space Γ. Without this invariance the connection between phase space volume and probability would be impossible or at least very problematic.

Of course, if our isolated system remains isolated forever, Poincaré's Recurrence Theorem tells us that the system phase point $X(t)$ would have to come back very close to its initial value $X(t_a)$, and do so again and again. But these Poincaré recurrence times are so enormous (more or less comparable to the ratio of $|\Gamma_{M_d}|$ to $|\Gamma_{M_a}|$) that when Zermelo brought up this objection to Boltzmann's explanation of the second law, Boltzmann's response [17] was as follows: "Poincaré's theorem, which Zermelo explains at the beginning of his paper, is clearly correct, but his application of it to the theory of heat is not. ...Thus when ...Zermelo concludes, from the theoretical fact that the initial states in a gas must recur—without having calculated how long a time this will take—that the hypotheses of gas theory must be rejected or else fundamentally changed, he is just like a dice player who has calculated that the probability of a sequence of 1000 one's is not zero, and then concludes that his dice must be loaded since he has not yet observed such a sequence!"[a]

---

[a] It is remarkable that in the same paper Boltzmann also wrote "likewise, it is observed that very small particles in a gas execute motions which result from

Thus not only did Boltzmann's great insights give a microscopic interpretation of the mysterious thermodynamic entropy of Clausius; they also gave a natural generalization of entropy to nonequilibrium macrostates M, and with it an explanation of the second law of thermodynamics—the formal expression of the time-asymmetric evolution of macroscopic states occurring in nature.

4 The Use of Probability

Boltzmann's ideas are, as Ruelle [6] says, at the same time simple and rather subtle. They introduce into the "laws of nature" notions of probability, which, certainly at that time, were quite alien to the scientific outlook. Physical laws were supposed to hold without any exceptions, not just almost always and indeed no exceptions were (or are) known to the second law; nor would we expect any, as Richard Feynman [3] rather conservatively says, "in a million years". The reason for this, as recognized by Maxwell, Thomson and Boltzmann, is that, for a macroscopic system, the fraction of microstates for which the evolution leads to macrostates with larger S_B is so close to one (in terms of their Liouville volume) that such behavior is exactly what should be seen to "always" happen. As put by Boltzmann [17], "Maxwell's law of the distribution of velocities among gas molecules is by no means a theorem of ordinary mechanics which can be proved from the equations of motion alone; on the contrary, it can only be proved that it has very high probability, and that for a large number of molecules all other states have by comparison such a small probability that for practical purposes they can be ignored." In present day mathematical language we say that such behavior is *typical*, by which we mean that the set of microstates X in Γ_{M_a} for which it occurs have a volume fraction which goes to 1 as N increases. Thus in Fig. 1 the sequence going from left to right is typical for a phase point in Γ_{M_a} while the one going from right to left has "probability" approaching zero with respect to a uniform distribution in Γ_{M_d}, for N tending towards infinity.

Note that Boltzmann's argument does not really require the assumption that over very long periods of time the macroscopic system should be found in different regions Γ_M, i.e. in different macroscopic states M, for fractions of time *exactly* equal to the ratio of $|\Gamma_M|$ to the total phase space volume specified by its energy. Such behavior, which can be considered as a mild form of Boltzmann's ergodic hypothesis, mild because it is only applied to those regions of the phase space representing macrostates Γ_M, seems very plausible in the absence of constants of the motion which decompose the energy surface into regions with different macroscopic states. It appears even more reasonable when we take into account the lack of perfect isolation in practice which will be discussed later.

---

the fact that the pressure on the surface of the particles may fluctuate". This shows that Boltzmann completely understood the cause of Brownian motion ten years before Einstein's seminal papers on the subject. Surprisingly he never used this phenomenon in his arguments with Ostwald and Mach about the reality of atoms.

Its implication for "small fluctuations" from equilibrium is certainly consistent with observations. (The stronger form of the ergodic hypothesis also seems like a natural assumption for macroscopic systems. It gives a simple derivation for many equilibrium properties of macro systems.)

5 Initial Conditions

Once we accept the statistical explanation of why macroscopic systems evolve in a manner that makes S_B increase with time, there remains the nagging problem (of which Boltzmann was well aware) of what we mean by "with time". Since the microscopic dynamical laws are symmetric, the two directions of the time variable are *a priori* equivalent and thus must remain so *a posteriori* [18]. In particular if a system with a nonuniform macroscopic density profile, such as M_b, at time t_b in Fig. 1 had a microstate that is typical for Γ_{M_b}, then almost surely its macrostate at both times $t_b + \tau$ and $t_b - \tau$ will be like M_c. This is inevitable: Since the phase space region Γ_{M_b} corresponding to M_b at some time t_b is invariant under the transformation $X \rightarrow RX$, it must make the same prediction for $t_b - \tau$ as for $t_b + \tau$. Yet experience shows that the assumption of typicality at time t_b will give the correct behavior only for times $t > t_b$ and not for times $t < t_b$. In particular, given just M_b and M_c, we have no hesitation in ordering M_b before M_c.

If we think further about our ordering of M_b and M_c, we realize that it seems to derive from our assumption that M_b is itself so unlikely that it must have evolved from an initial state of even lower entropy like M_a. From an initial microstate typical of the macrostate M_a, which can be readily created by an experimentalist, we get monotonic behavior of S_B with the time ordering M_a, M_b, M_c and M_d. If, by contrast, the system in Fig. 1 had been completely isolated for a very long time compared with its hydrodynamic relaxation time, then we would expect to always find it in its equilibrium state M_d (with possibly some small fluctuations around it). Presented instead with the four pictures, we would (in this very, very unlikely case) have no basis for assigning an order to them; microscopic reversibility assures that fluctuations from equilibrium are typically symmetric about times at which there is a local minimum of S_B. In the absence of any knowledge about the history of the system before and after the sequence of snapshots presented in Fig. 1, we use our experience to conclude that the low-entropy state M_a must have been an initial prepared state. In the words of Roger Penrose [5]: "The time-asymmetry comes merely from the fact that the system has been *started off* in a very special (i.e. low-entropy) state, and having so started the system, we have watched it evolve in the *future* direction".

The point is that a microstate corresponding to M_b (at time t_b) which comes from M_a (at time t_a) must be *atypical* in some respects of points in Γ_{M_b}. This is so because, by Liouville's theorem, the set Γ_{ab} of all such phase points has a volume $|\Gamma_{ab}| \leq |\Gamma_{M_a}|$ that is *very much smaller* than $|\Gamma_{M_b}|$. This need not however prevent the overwhelming majority of points in Γ_{ab} (with respect to

Liouville measure on Γ_{ab} which is the same as Liouville measure on Γ_a) from having future macrostates like those typical of Γ_b—while still being very special and unrepresentative of Γ_{M_b} as far as their past macrostates are concerned. This sort of behavior is what is explicitly proven by Lanford in his derivation of the Boltzmann equation [7], and is implicit in all derivation of hydrodynamic equations [10]; see also [19]. To see intuitively the origin of such behavior we note that for systems with realistic interactions the domain Γ_{ab} will be so convoluted that it will be "essentially dense" in Γ_b, so that any slight thickening of it will cover all of Γ_{M_b}. It is therefore not unreasonable that their future behavior, as far as macrostates go, will be unaffected by their past history.

(This can be worked out completely for a model macroscopic system in which the (large) N noninteracting atoms are each specified not by $(\mathbf{r}, \mathbf{v})$ but by $\sigma = (\ldots, \sigma_{-2}, \sigma_{-1}; \sigma_0, \sigma_1, \ldots)$, a doubly infinite sequence of zeros and ones (equivalently a point in the unit square). Their discrete time dynamics is that of a shift to the left $(T\sigma)_i = \sigma_{i+1}$ (equivalently the baker's transformation). If we define "velocity reversal" by $(R\sigma)_i = \sigma_{-i-1}$ and the macrostate $M(\sigma)$ by the k values, $M(\sigma) = (\sigma_0 + \sigma_{-1}, \sigma_1 + \sigma_{-2}, \sigma_2 + \sigma_{-3}, \ldots, \sigma_{k-1} + \sigma_{-k})$ then a little thought shows that the future behavior of typical points in $\Gamma_{M_{ab}}$ is indeed as described above.)

5 Origin of Low-Entropy States

The creation of low-entropy initial states poses no problem in laboratory situations such as the one depicted in Fig. 1. Laboratory systems are prepared in states of low Boltzmann entropy by experimentalists who are themselves in low-entropy states. Like other living beings, they are born in such states and maintained there by eating nutritious low-entropy foods, which in turn are produced by plants using low-entropy radiation coming from the Sun. That was already clear to Boltzmann as may be seen from the following quote [20]: "The general struggle for existence of living beings is therefore not a fight for the elements—the elements of all organisms are available in abundance in air, water, and soil—nor for energy, which is plentiful in the form of heat, unfortunately untransformably, in every body. Rather, it is a struggle for entropy that becomes available through the flow of energy from the hot Sun to the cold Earth. To make the fullest use of this energy, the plants spread out the immeasurable areas of their leaves and harness the Sun's energy by a process as yet unexplored, before it sinks down to the temperature level of our Earth, to drive the chemical syntheses of which one has no inkling as yet in our laboratories. The products of this chemical kitchen are the object of the struggles in the animal world".

Note that while these experimentalists have evolved, thanks to this source of low entropy energy, into beings able to prepare systems in particular macrostates with low values of $S_B(M)$, like our state M_a, the total entropy S_B, including the entropy of the experimentalists and that of their environment, must always increase: There are no Maxwell demons. The low entropy of the solar system is also manifested in events in which there is no human participation—so that, for example, if instead of Fig. 1 we are given snapshots of the Shoemaker-Levy comet and Jupiter before and after their collision, then the time direction is again obvious.

We must then ask what is the origin of this low entropy state of the solar system. In trying to answer this question we are led more or less inevitably to cosmological considerations of an initial "state of the universe" having a very small Boltzmann entropy. To again quote Boltzmann [10]: "That in nature the transition from a probable to an improbable state does not take place as often as the converse, can be explained by assuming a very improbable initial state of the entire universe surrounding us. This is a reasonable assumption to make, since it enables us to explain the facts of experience, and one should not expect to be able to deduce it from anything more fundamental". That is, the universe is pictured as having been "created" in an initial microstate X typical of some macrostate M_0 for which $|\Gamma_{M_0}|$ is a very small fraction of the "total available" phase space volume. In Boltzmann's time there was no physical theory of what such an initial state might be and Boltzmann toyed with the idea that it was just a very large, very improbable, fluctuation in an eternal universe which spends most of its time in an equilibrium state. Richard Feynman argues convincingly against such a view [3].

In the current big bang scenario it is reasonable, as Roger Penrose does in [5], to take as initial state the state of the universe just after the big bang. Its macrostate would then be one in which the energy density is approximately spatially uniform. Penrose estimates that if M_f is the macrostate of the final "Big Crunch", having a phase space volume of $|\Gamma_{M_f}|$, then $|\Gamma_{M_f}|/|\Gamma_{M_0}| \approx 10^{10^{123}}$. The high value of $|\Gamma_{M_f}|$ compared with $|\Gamma_{M_0}|$ comes from the vast amount of phase space corresponding to a universe collapsed into a black hole, see Fig. 2.

I do not know whether these initial and final states are reasonable, but in any case one has to agree with Feynman's statement [3] that "it is necessary to add to the physical laws the hypothesis that in the past the universe was more ordered, in the technical sense, than it is today...to make an understanding of the irreversibility." "Technical sense" clearly refers to the initial state of the universe M_0 having a smaller S_B than the present state. Once we accept such an initial macrostate M_0, then the initial microstate can be assumed to be typical of Γ_{M_0}. We can then apply our statistical reasoning to compute the typical evolution of such an initial state, i.e. we can use phase-space-volume arguments to predict the future behavior of macroscopic systems—but not to determine the past. As put by Boltzmann [2], "we do not have to assume a special type of initial condition in order to give a mechanical proof of the second law, if we are willing to accept a statistical viewpoint...if the initial state is chosen at random ...entropy is almost certain to increase."

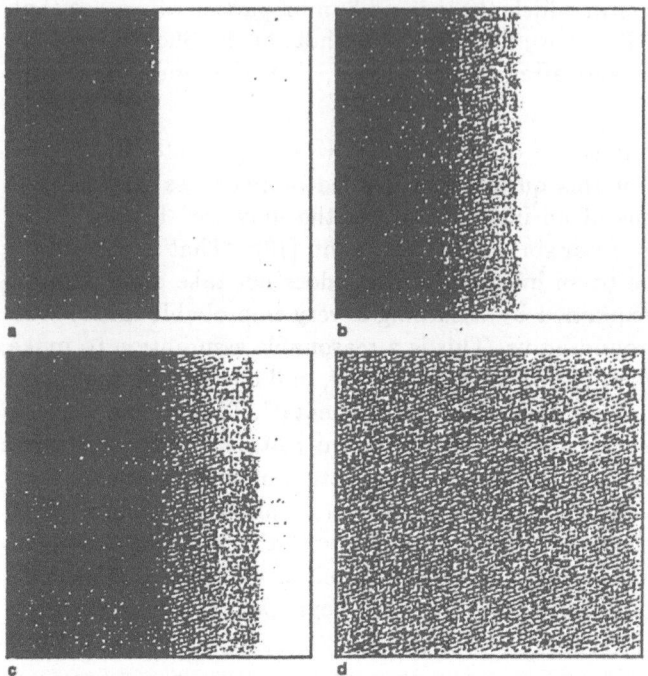

Fig. 1 How would you order this sequence of "snapshots" in time? Each represents a macroscopic state of a system containing, for example, two differently colored fluids.

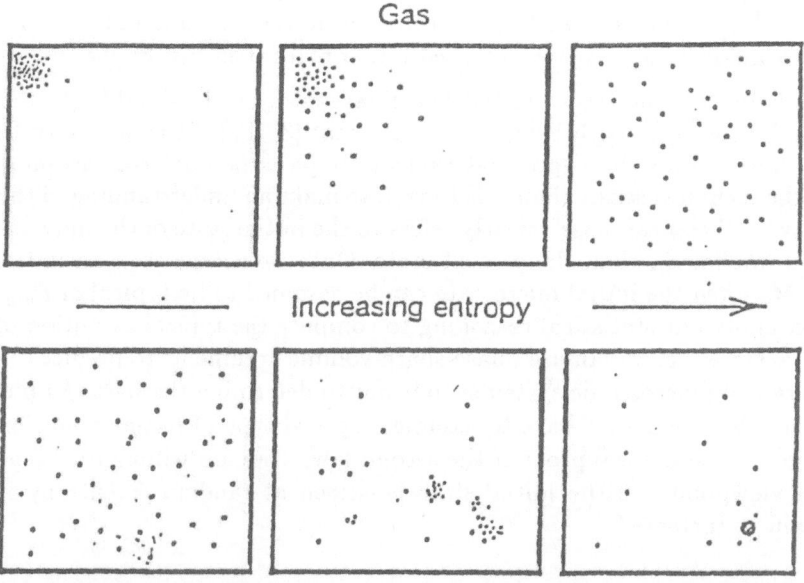

Fig. 2 With a gas in a box, the maximum entropy state (thermal equilibrium) has the gas distributed uniformly; however, with a system of gravitating bodies, entropy can be increased from the uniform state by gravitational clumping leading eventually to a black hole. From Ref. [5].

6 Irreversibility and Macroscopic Stability

Of course mechanics itself doesn't preclude having a microstate X for which $S_B(M(X_t))$ decreases as t increases. An experimentalist could, *in principle*, reverse all velocities of the system in Fig. 1b, and then watch the system unmix itself. It seems however impossible to do so in practice: Even if he/she managed to do a perfect job on the velocity reversal part, as occurs (imperfectly) in spin echo experiments [21] , we would not expect to see the system in Fig. 1 go from M_b to M_a. This would require that *both* the velocity reversal and system isolation be *absolutely perfect*. The reason for requiring such perfection now and not before is that while the macroscopic behavior of a system with microstate Y in the state M_b *coming* from a microstate X typical with respect to Γ_{M_a} is *stable* against perturbations as far as its future is concerned it is very *unstable* as far as its *past* (and thus the future behavior of RY) is concerned (see Figs. 3 and 4).

(I am thinking here primarily of situations like those depicted in Fig. 1 where the macroscopic evolution is described by the stable diffusion equation. However, even in situations, such as that of turbulence, where the forward macroscopic evolution is chaotic, i.e. sensitive to small perturbations, all evolutions will still have increasing Boltzmann entropies in the forward direction. For the isolated evolution of the velocity reversed microstate, however, one has decreasing S_B while the perturbed ones can be expected to have, at least after a very short time, increasing S_B. So even in macroscopically "chaotic" regimes the forward evolution of M is in this sense much more stable than the backward one. Thus in turbulence all forward evolutions are still described by solutions of the same Navier-Stokes equation while the backward macroscopic evolution for a *perfectly isolated* fluid and for an actual one will have no connection with each other.)

The above analysis is based on the very reasonable assumption that almost any perturbation of the microstate Y will tend to make it more typical of its macrostate $M(Y)$, here equal to M_b. The perturbation will thus not interfere with behavior typical of Γ_{M_b}. The forward evolution of the unperturbed RY is on the other hand, by construction, heading towards a smaller phase space volume and is thus untypical of Γ_{M_b}. It therefore requires "perfect aiming" and will very likely be derailed by even small imperfections in the reversal and/or tiny outside influences. After a *very short* time in which S_B decreases the imperfections in the reversal and the "outside" perturbations, such as one coming from a sun flare, a star quake in a distant galaxy (a long time ago) or from a butterfly beating its wings [6], will make it increase again. This is somewhat analogous to those pinball machine type puzzles where one is supposed to get a small metal ball into a particular small region. You have to do things just right to get it in but almost anything you do gets it out into larger regions. For the macroscopic systems we are considering, the disparity between relative sizes of the comparable regions in the phase space is unimaginably larger. In the absence of any "grand conspiracy", the behavior of such systems can therefore be confidently predicted to be in accordance with the second law (except possibly for very short time intervals).

Sensitivity to small perturbations in the entropy decreasing direction is commonly observed in computer simulations of systems with "realistic" interactions where velocity reversal is easy to accomplish but unavoidable roundoff errors play the role of perturbations. It is possible, however, to avoid this effect in simulations by the use of discrete time integer arithmetic. This is clearly illustrated in Figs. 3 and 4. The latter also shows how a small perturbation which has no effect on the forward macro evolution completely destroys the time reversed evolution. This point is very clearly formulated in the 1874 paper of Thomson [1]:

"Dissipation of energy, such as that due to heat conduction in a gas, might be entirely prevented by a suitable arrangement of Maxwell demons, operating in conformity with the conservation of energy and momentum. If no demons are present, the average result of the free motions of the molecules will be to equalize temperature-differences. If we allowed this equalization to proceed for a certain time, and then reversed the motions of all the molecules, we would observe a disequalization. However, if the number of molecules is very large, as it is in a gas, any slight deviation from absolute precision in the reversal will greatly shorten the time during which disequalization occurs. In other words, the probability of occurrence of a distribution of velocities which will lead to disequalization of temperature for any perceptible length of time is very small. Furthermore, if we take account of the fact that no physical system can be completely isolated from its surroundings but is in principle interacting with all other molecules in the universe, and if we believe that the number of these latter molecules is infinite, then we may conclude that it is impossible for temperature-differences to arise spontaneously. A numerical calculation is given to illustrate this conclusion." Thomson goes on to say: "The essence of Joule's discovery is the subjection of physical phenomena to dynamical law. If, then, the motion of every particle of matter in the universe were precisely reversed at any instant, the course of nature would be simply reversed for ever after. The bursting bubble of foam at the foot of a waterfall would reunite and descend into the water; ... Boulders would recover from the mud the materials required to rebuild them into their previous jagged forms, and would become reunited to the mountain peak from which they had formerly broken away. And if also the materialistic hypothesis of life were true, living creatures would grow backwards, with conscious knowledge of the future, but no memory of the past, and would become again unborn. But the real phenomena of life infinitely transcend human science; and speculation regarding consequences of their imagined reversal is utterly unprofitable. Far otherwise, however, is it in respect to the reversal of the motions of matter uninfluenced by life, a very elementary consideration of which leads to the full explanation of the theory of dissipation of energy."

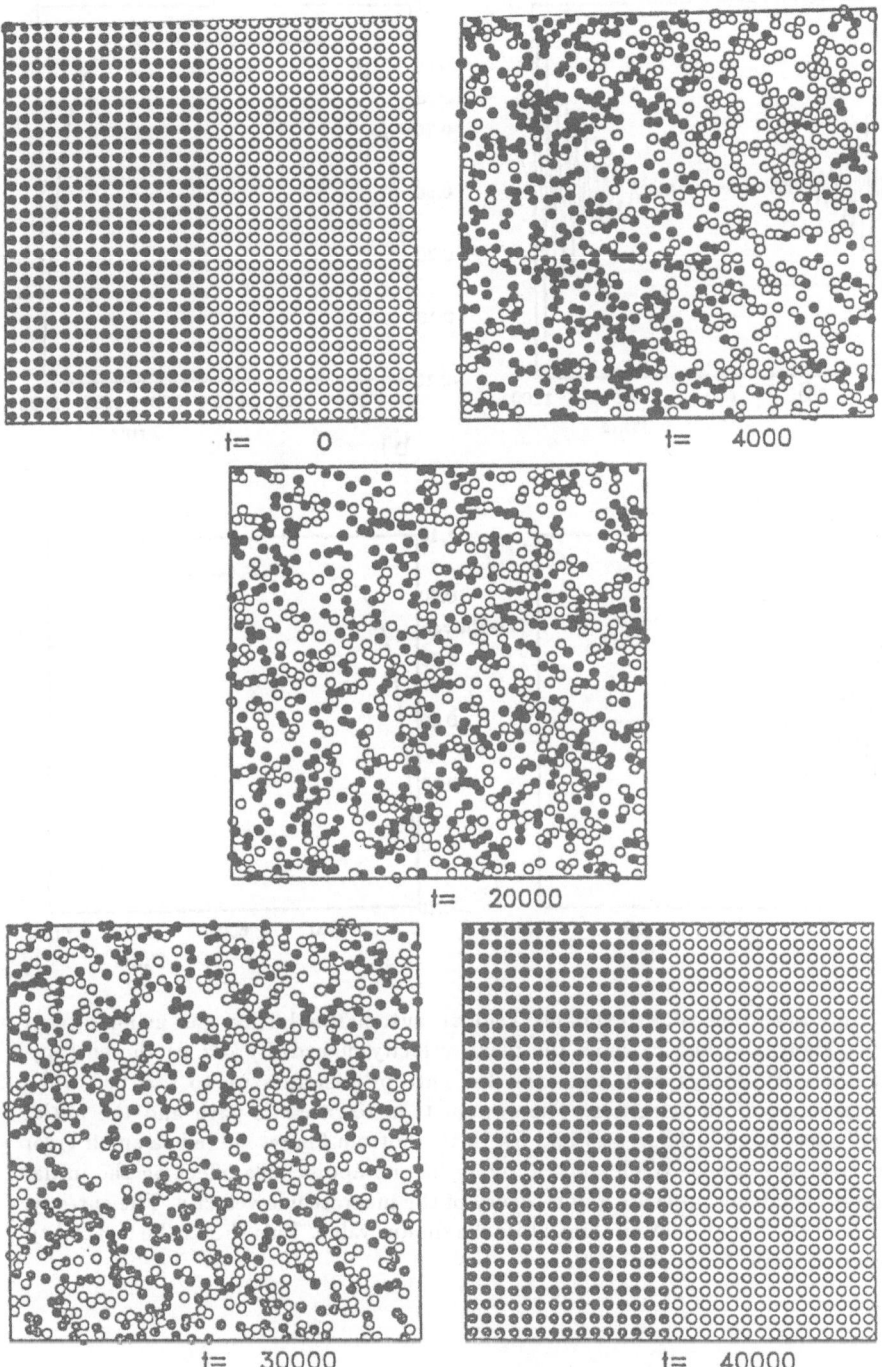

Fig. 3 Time evolution of a system of 900 particles all interacting via the same cutoff Lennard-Jones pair potential using integer arithmetic. Half of the particles are colored white, the other half black. All velocities are reversed at $t = 20,000$. The system then retraces its path and the initial state is fully recovered. From Ref. [22].

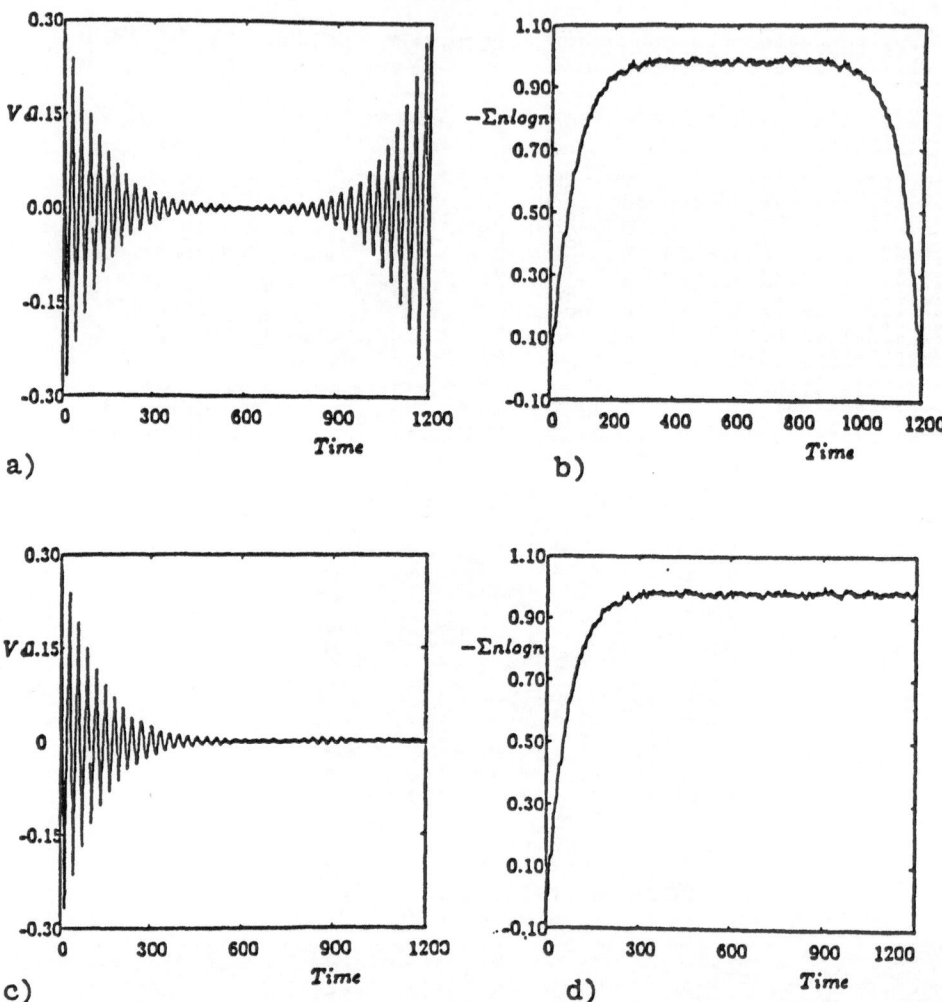

Fig. 4 Time evolution of a reversible cellular automaton lattice gas using integer arithmetic. Figures a) and c) show the mean velocity, figures b) and d) the entropy. The mean velocity decays with time and the entropy increases up to $t = 600$ when there is a reversal of all velocities. The system then retraces its path and the initial state is fully recovered in figures a) and b). In the bottom figures there is a small error in the reversal at $t = 600$. While such an error has no appreciable effect on the initial evaluation it effectively prevents any recovery of the macroscopic velocity. The entropy, on the scale of the figure, just remains at its maximum value. This shows the instability of the reversed path. From Ref. [23].

7 Boltzmann vs. Gibbs Entropies

The Boltzmannian approach, which focuses on the evolution of a particular macroscopic system, is conceptually different from the Gibbsian approach, which focuses primarily on ensembles. This difference shows up strikingly when we compare Boltzmann's entropy—defined in (1) for a microstate X of a macroscopic system—with the more commonly used (and misused) entropy S_G of Gibbs, defined for an ensemble density $\rho(X)$ by

$$S_G(\{\rho\}) = -k \int \rho(X)[\log \rho(X)]dX. \tag{3}$$

Here $\rho(X)dX$ is the probability (obtained some way or other) for the microscopic state of the system to be found in the phase space volume element dX and the integral is over the phase space Γ. Of course if we take $\rho(X)$ to be the generalized microcanonical ensemble associated with a macrostate M,

$$\rho_M(X) \equiv \begin{cases} |\Gamma_M|^{-1}, & \text{if } X \in \Gamma_M \\ 0, & \text{otherwise} \end{cases}, \tag{4}$$

then clearly,

$$S_G(\{\rho_M\}) = k \log |\Gamma_M| = S_B(M). \tag{5}$$

Generalized microcanonical ensembles like $\rho_M(X)$, or their canonical version, are commonly used to describe systems in which the particle density, energy density and momentum density vary slowly on a microscopic scale *and* the system is, in each small macroscopic region, in equilibrium with the prescribed local densities, i.e. when we have local equilibrium [10]. In such cases $S_G(\{\rho_M\})$ and $S_B(M)$ agree with each other, and with the macroscopic hydrodynamic entropy.

Note however that unless the system is in complete equilibrium and there is no further systematic change in M or ρ, the time evolutions of S_B and S_G are *very* different. As is well known, it follows from the fact that the volume of phase space regions remains unchanged under the Hamiltonian time evolution (even though their shape changes greatly) that $S_G(\{\rho\})$ never changes in time as long as X evolves according to the Hamiltonian evolution, i.e. ρ evolves according to the Liouville equation; $S_B(M)$, on the other hand, certainly does change. Thus, if we consider the evolution of the microcanonical ensemble corresponding to the macrostate M_a in Fig. 1a after removal of the constraint, S_G would equal S_B initially but subsequently S_B would increase while S_G would remain constant. S_G therefore does not give any indication that the system is evolving towards equilibrium.

This reflects the fact, discussed earlier, that the microstate $X(t)$ does not remain typical of the local equilibrium state $M(t)$ for $t > 0$. As long as the system remains truly isolated the state $T_t X$ will contain subtle correlations, which are reflected in the complicated shape which an initial region Γ_M takes on in time but which do not affect the future time evolution of M (see the discussion at end of section on Initial Conditions). *Thus the relevant entropy for understanding the time evolution of macroscopic systems is S_B and not S_G.* (Of course if we are willing to do a "course graining" of ρ over cells Γ_M then we are essentially back to dealing with ρ_M, or superpositions of such ρ_M's and we are just defining S_B in a backhanded way.)

8 Remarks

a) The characterization of a macrostate M usually done via density fields in three dimensional space as in Fig. 1 can be extended to mesoscopic descriptions. This is particularly convenient for a *dilute gas* where M can be usefully characterized by the density in the six dimensional position and velocity space of a single molecule. The deterministic macroscopic (or mesoscopic) evolution of this M is *then* given by the Boltzmann equation and $S_B(M)$ coincides with the negative of Boltzmann's famous H-function.

It is important to note however that for systems in which the potential energy is relevant, e.g. non-dilute gases, the H-function does not agree with S_B and $-H$ (but not S_B) will decrease for suitable macroscopic initial conditions. As pointed out by Jaynes [24] this will happen whenever one starts with an initial total energy E and kinetic energy $K = K_0$ such that $K_0 > K_{eq}(E)$, the value that K takes when the system is in equilibrium with energy E. This can be readily seen if the initial macrostate is one in which the spatial density is uniform and the velocity distribution is Maxwellian with the appropriate temperature $T_0 = \frac{2}{3}K_0/kN$. The temperature will then decrease as the system goes to equilibrium and $-H$ which, for a Maxwellian distribution, is proportional to $\log T$ will therefore be smaller in the equilibrium state when $T = T_{eq}(E) < T_0$.
b) Einstein's formula for the probability of fluctuations in an equilibrium system,

$$\text{Probability of } M \quad \sim \quad \exp\{[S(M) - S_{eq}]/k\}$$

is essentially an inversion of formulas (4) and (5). When combined with the observation that the entropy $S_B(M)$ of a macroscopic system, prepared in a specified nonuniform state M, can be computed from macroscopic thermodynamic considerations it yields useful results. In particular when $S_B(M)$ in the exponent is expanded around M_{eq}, and only quadratic terms are kept, we obtain a Gaussian distribution for normal (small) fluctuations from equilibrium. This is one of the main ingredients of Onsager's reciprocity relations [25].

9 Typical vs. Averaged Behavior

I conclude by emphasizing again that having results for typical microstates rather than averages is not just a mathematical nicety but goes to the heart of the problem of understanding the microscopic origin of observed macroscopic behavior — *we neither have nor do we need ensembles when we carry out observations like those illustrated in Fig. 1.* What we do need and can expect to have is typical behavior. Ensembles are merely mathematical tools, useful as long as the dispersion, in the quantities we are interested in, is sufficiently small. This is always the case for properly defined macroscopic variables in equilibrium Gibbs ensembles. The use of such an ensemble as the initial "statistical state" immediately following the lifting of a constraint from a macroscopic system in equilibrium at some time t_0 is also sensible, as long as the evolution of $M(t)$ is, with probability close to one, the same for all systems in the ensemble.

There is no such typicality with respect to ensembles describing the time evolution of a system with only a few degrees of freedom. This is an essential difference (unfortunately frequently overlooked or misunderstood) between the irreversible and the chaotic behavior of Hamiltonian systems. The latter, which can be observed already in systems consisting of only a few particles, will not have a uni-directional time behavior in any particular realization. Thus if we had only a few hard spheres in the box of Fig. 1, we would get plenty of chaotic dynamics and very good ergodic behavior (mixing, K-system, Bernoulli) but we could not tell the time order of any sequence of snapshots.

Finally I note that my discussion has focused exclusively on what is usually referred to as the thermodynamic arrow of time and on its connection with the cosmological initial state. I did not discuss other arrows of time such as the asymmetry between advanced and retarded electromagnetic potentials or "causality". It is my general feeling that these are all manifestations of the low entropy initial state of the universe. I also believe that the violation of time reversal invariance in the weak interactions is not relevant for macroscopic irreversibility.

Acknowledgments: I want to thank Y. Aharonov, G. Eyink, O. Penrose and especially S. Goldstein, H. Spohn and E. Speer for many very useful discussions. I also thank the organizers of this conference for their kind hospitality. Various aspects of this research have been supported in part by the AFOSR and NSF.

References

1   W. Thomson, Proc. of the Royal Soc. of Edinburgh, **8** 325 (1874), reprinted in 1a).

2   a) For a collection of original articles of Boltzmann and others from the second half of the nineteenth century on this subject (all in English) see S.G. Brush, *Kinetic Theory*, Pergamon, Oxford, (1966).
b) For an interesting biography of Boltzmann, which also contains many references, see E. Broda *Ludwig Boltzmann, Man—Physicist—Philosopher*, Ox Bow Press, Woodbridge, Conn (1983); translated from the German.
c) For a historical discussion of Boltzmann and his ideas see also articles by M. Klein, E. Broda, L. Flamm in *The Boltzmann Equation, Theory and Application*, E.G.D. Cohen and W. Thirring, eds., Springer-Verlag, 1973.
d) For a general history of the subject see S.G. Brush, *The Kind of Motion We Call Heat*, Studies in Statistical Mechanics, vol. VI, E.W. Montroll and J.L. Lebowitz, eds. North-Holland, Amsterdam, (1976).

e) G. Gallavotti, Ergodicity, Ensembles, Irreversibility in Boltzmann and Beyond, J. Stat. Phys., to appear, (1995).

3   R. Feynman, *The Character of Physical Law*, MIT P., Cambridge, Mass. (1967), ch.5. R.P Feynman, R. B Leighton, M. Sands, *The Feynman Lectures on Physics*, Addison-Wesley, Reading, Mass. (1963), sections 46–3, 4, 5.

4   O. Penrose, *Foundations of Statistical Mechanics*, Pergamon, Elmsford, N.Y. (1970), ch. 5.

5   R. Penrose, *The Emperor's New Mind*, Oxford U. P., New York (1990), ch. 7.

6   D. Ruelle, *Chance and Chaos*, Princeton U. P., Princeton, N.J. (1991), ch. 17, 18.

7   O. Lanford, Physica A **106**, 70 (1981).

8   J.L. Lebowitz, Physica A **194**, 1 (1993).

9   J.C. Maxwell, *Theory of Heat*, p. 308: "Tait's Thermodynamics", Nature **17**, 257 (1878). Quoted in M. Klein, ref. 1c).

10   H. Spohn, *Large Scale Dynamics of Interacting Particles*, Springer-Verlag, New York (1991). A. De Masi, E. Presutti, *Mathematical Methods for Hydrodynamic Limits*, Lecture Notes in Math 1501, Springer-Verlag, New York (1991). J.L. Lebowitz, E. Presutti, H. Spohn, J. Stat. Phys. **51**, 841 (1988).

11   Y. Aharonov, P.G. Bergmann, J.L. Lebowitz, Phys. Rev. B **134**, 1410 (1964). D.N. Page, Phys. Rev. Lett. **70**, 4034 (1993).

12   J.S. Bell, *Speakable and Unspeakable in Quantum Mechanics*, Cambridge U. P., New York (1987).

13   D. Dürr, S. Goldstein, N. Zanghi, J. Stat. Phys. **67**, 843 (1992).

14   See articles by M. Gell-Mann, J. Hartle, R. Griffiths, D. Page and others in *Physical Origin of Time Asymmetry*, J.J. Halliwell, J. Perez-Mercader and W.H. Zurek, eds., Cambridge University Press, 1994.

15   *Vorlesungen über Gastheorie.* 2 vols. Leipzig: Barth, 1896, 1898. This book has been translated into English by S.G. Brush, *Lectures on Gas Theory*, (London: Cambridge University Press, 1964) .

16   J.W. Gibbs, Connecticut Academy Transactions **3**, 229 (1875), reprinted in *The Scientific Papers*, **1**, 167 (New York, 1961).

17   L. Boltzmann, Ann. der Physik **57**, 773 (1896). Reprinted in 1a).

18   E. Schrödinger, *What is Life? And Other Scientific Essays*, Doubleday Anchor Books, New York (1965), section 6.

19   J.L. Lebowitz and H. Spohn, *Communications on Pure and Applied Mathematics*, XXXVI,595, (1983); see in particular section 6(i).

20   L. Boltzmann (1886) quoted in E. Broda, 1b), p. 79.

21   E.L. Hahn, Phys. Rev. **80**, 580 (1950). See also S. Zhang, B.H. Meier, R.R. Ernst, Phys. Rev. Let.. **69** 2149 (1992).

22   D. Levesque and L. Verlet, J. Stat. Phys. **72**, 519 (1993).

23   B.T. Nadiga, J.E. Broadwell and B. Sturtevant, *Rarefield Gas Dynamics: Theoretical and Computational Techniques*, edited by E.P. Muntz, D.P. Weaver and D.H. Campbell, Vol 118 of *Progress in Astronautics and Aeronautics*, AIAA, Washington, DC, ISBN 0–930403–55–X, 1989.

24   E.T. Jaynes, Phys. Rev. A4, 747 (1971).

25   A. Einstein, Am. Phys. (Leipzig) **22**, 180 (1907); **33**, 1275 (1910). L. Onsager, Phys. Rev. **37**, 405 (1931); **38**, 2265 (1931).

Twenty-five Years of Non-equilibrium Statistical Mechanics: Towards a Better Understanding of Dense Fluids

E. G. D. Cohen
The Rockefeller University
New York, NY 10021, USA

Abstract

A survey is given of some of the new insights that have emerged over the last 25 years in the physics of dense non-equilibrium fluids: atomic liquids as well as concentrated colloidal suspensions. First a brief discussion of some previous developments is given: the non-existence of a density expansion of the transport coefficients; the existence of general expressions for the transport coefficients for a fluid in terms of average current time-auto-correlation functions in equilibrium and the slow decay of the latter due to mode-coupling by vortex-diffusion, a new mechanism of diffusion leading to the long time tails. Next, very long range correlations, also due to mode-coupling, in non-equilibrium fluids subject to a temperature gradient and their experimental detection by light or by microwave scattering are discussed. Also, a reinterpretation of the neutron spectra of dense fluids in equilibrium is given in terms of effective fluid eigenmodes and cage-diffusion, the most important diffusion process in dense fluids. Then some recent work on Lattice Gas Cellular Automata and a new connection between transport coefficients and Lyapunov exponents are briefly mentioned. Finally, a far going analogy between the self-diffusion coefficient and the viscosity of atomic liquids and those of concentrated colloidal suspensions is discussed, based on the similarity of Newtonian and Brownian dynamics on long time scales and that of the cage-diffusion processes in both systems. Very similar expressions for the viscosity of these two fluid systems, based on cage-diffusion, are compared with recent experiments.

I. Introduction

It is not possible for me to give a survey of all developments over the last twenty-five years in our understanding of such a wide field as dense fluids. Therefore, I have had to select those aspects with which I was more or less familiar and was forced to leave out many results which would certainly be of great importance to discuss, but which I am simply not able to do justice to. Thus, for instance, the behavior of liquids or colloids near phase transitions, such as the freezing or glass transition, have not been included. Also, for the conventional interpretation of neutron spectra of atomic liquids in terms of memory kernels, I had to refer to the literature. Instead another, in my opinion, more physical interpretation has been included. It will be clear therefore that this survey is a rather personal overview. I will first briefly discuss some of the developments that preceded those of the last twenty-five years, to set the stage, so to speak, for a discussion of the latter.

II. Divergences, Green-Kubo Relations, Long-Time Tails, Vortex Diffusion

Twenty-five years ago non-equilibrium statistical mechanics was in full swing to work out a number of seminal developments that had taken place in the preceding five years. These developments included the discovery of the non-existence of virial expansions for the transport coefficients in moderately dense gases, the existence of general expressions for the transport coefficients for all densitites - the Green-Kubo relations - and the so called longtime tails. Since over the years many survey papers have appeared, which discuss these developments in great detail, I will confine myself here to a brief discussion of the basic physics of some of them, referring for details to the literature and references therein.

1. <u>Non-existence of virial expansions for the transport coefficients</u>

The discovery that there exist no density or virial expansions for the transport coefficients of moderately dense fluids, grew out of the post-World War II attempts to generalize the Boltzmann equation for a dilute gas systematically to higher densitites as a first attempt to come to a theory of dense fluids. Such an equation would have the form[1]:

$$\frac{\partial f}{\partial t} = -\mathbf{v} \cdot \frac{\partial f}{\partial \mathbf{r}} + J(ff) + K(fff) + L(ffff) + ...$$ (1)

where $f = f(\mathbf{r}, \mathbf{v}, t)$ is the single particle distribution function, giving the average number of gas particles at the position $\mathbf{r}$ with velocity $\mathbf{v}$ at time t. Eq.(1) expresses the rate of change of f as a sum of a streaming term and collision terms, which systematically contain the contributions of two particle ($J(ff)$), three particle ($K(fff)$), four particle ($L(ffff)$) etc. collisions, where the number of f's indicates the number of particles involved in the collisions. The transport coefficients can be obtained from eq.(1) by applying the Chapman-Enskog method of solution to eq.(1).

The Chapman-Enskog solution[2] is an expansion of f around local equilibrium (i.e., around a local Maxwellian velocity distribution function) in powers of the gradients, i.e., the derivatives of the local density $n(\mathbf{r}, t)$ temperature $T(\mathbf{r}, t)$ and velocity $\mathbf{u}(\mathbf{r}, t)$. The linear Navier-Stokes transport coefficients are then obtained by expanding to first order in the gradients; higher order linear (so-called Burnett, super-Burnett, etc.) transport coefficients are obtained when higher order derivatives of $T(\mathbf{r}, t)$ and $\mathbf{u}(\mathbf{r}, t)$ are considered.

This way, a density or virial expansion of the Navier-Stokes transport coefficients is obtained, which reads for the shear viscosity η as:

$$\eta(n, T) = \eta_0(T) + n\eta_1(T) + n^2 n_2(T) + \dots \tag{2}$$

Here n is the number density and T the temperature of the gas; $\eta_0(T)$ the contribution of binary collisions from $J(ff)$, which was already obtained by Chapman and Enskog from the Boltzmann equation; $\eta_1(T)$ contains collisional transfer contributions from $J(ff)$ in addition to ternary collision contributions from $K(fff)$, while $\eta_2(T)$ contains collisional transfer contributions from $K(fff)$ in addition to quaternary collision contributions from $L(ffff)$ etc. The power of n, in front of each virial coefficient $\eta_\ell(T)$, plus two gives the maximum number of particles involved in the collisions that contribute to that coefficient. A closer study of the formal expressions for $K(fff)$ etc. revealed in 1964 that these collision terms diverged[1], the faster the more particles were involved, with the ℓ-particle collision term diverging as $\lim_{L\to\infty}(L/\sigma)^{\ell-(d+1)}$ (exponent zero means a logarithmic divergence) in $d=2$ or 3 dimensions, so that only $K(fff)$ in $d=3$ is finite. Here σ characterizes the size of a particle. The origin of these "kinetic" divergences, as I will call them, is the unrestricted free path each particle in a group of ℓ particles can traverse between successive collisions. This divergence is eliminated by introducing in eq.(1) mean free path cut-offs of these free paths due to collisions with *all* the particles in the gas not just with those in the group, i.e., a virial-or-individual particle expansion of the transport coefficients is not possible. One thinks one knows how to make a rearrangement of eq.(1) leading to such free path cut-offs at a few mean free paths, thus effectively replacing everywhere L/σ by $\ell/\sigma \sim n\sigma^d$, where the mean free path $\ell \sim 1/n\sigma^{d-1}$[3]. This then leads in $d=3$ to an expansion for η, the first few terms of which are:

$$\eta(n, T) = \eta_0(T) + n\eta_1(T) + n^2 \ln n\, \eta_2'(T) + n^2 \eta_2''(T) + \dots \tag{3}$$

where the $n^2 \ln n$ term is due to the logarithmic divergence of $\eta_2(T)$ in eq.(2) in $d = 3$.

However, one does not know what the nature of the general term is and whether only terms of the form $n^r(\ln n)^s(r, s$ integer) occur. For $d= 2$, see section 3.

2. Green-Kubo relations[4]

These are general expressions for the linear transport coefficients occurring in the Navier-Stokes equations of hydrodynamics valid for *all* (fluid) densities. They appear in terms of the average time correlations between (microscopic) current fluctuations in the fluid in equilibrium. They are based on Onsager's regression hypothesis[5] which states that the average decay of a (microscopic) current fluctuation proceeds according to the corresponding macroscopic law, i.e., is determined by the (macroscopic) transport coefficient L[6]. Thus if $j_L(t)$ presents the (microscopic) current at time t, then the conditional average $< j_L(t) >_{[j_L(0)]}= e^{-t/\tau_L}j_L(0)$, where the average on the left hand side (l.h.s.) is taken over all phases (i.e., positions and velocities) of the particles in the fluid, such that the (microscopic) current at time $t = 0$ is $j_L(0)$ as is indicated by $[j_L(0)]$. τ_L is the relaxation time associated with the macroscopic transport coefficient L and to each transport coefficient L belongs a particular (microscopic) current $j_L(t)$ and relaxation time $\tau_L^{[7]}$.

A linear Navier-Stokes transport coefficient L is defined as the proportionality factor between the macroscopic current J and the gradient X as $J = LX$, where $X \sim \partial T/\partial x$ or $\partial u/\partial x$ etc. Onsager's expressions for the transport coefficients can be written in the form[8]:

$$L = \frac{1}{k_B T} \int_0^\infty dt < j_L(t)j_L(0) >_{eq} \tag{4}$$

where k_B is Boltzmann's constant, T the temperature and the average over an equilbrium (e.g. canonical) ensemble. Green[9], Kubo[10] and others using a variety of methods determined, between about 1955 - 1965, the precise expressions for j_L needed to obtain the correct formula for L. Thus for the viscosity one found, for example, the Green-Kubo (G-K) relation:

$$\eta(n, T) = \frac{V}{k_B T} \int_0^\infty dt < P_{xy}(t)P_{xy}(0) >_{eq} \tag{5}$$

where $P_{xy}(t)$ is the xy-component of the microscopic pressure tensor given by:

$$P_{xy}(t)V = \sum_{i=1}^N m\, v_{ix}v_{iy} + \sum_{\substack{i<j \\ 1}}^N r_{ij,x}F_{ij,y} \tag{6}$$

Here m is the mass of a particle, v_{ix}, v_{iy}, the x and y components of the velocity of particle i, respectively, $r_{ij,x}$ the x-component of $r_{ij} = \vec{r}_j - \vec{r}_i$, where r_i is the position of particle i and $F_{ij,y}$ the y-component of $\mathbf{F}_{ij}$, the force on particle i due to particle j of the fluid and

V is the volume of the fluid. The average is taken over an equilibrium (e.g. canonical) ensemble and the thermodynamic limit ($N \to \infty, V \to \infty, N/V = n$) is understood. The G-K relation for η expresses the transport coefficient as a time integral over the equilibrium average of a microscopic current (fluctuation) auto-correlation function. It was shown in 1965[11] that a density expansion of the G-K relations for the transport coefficients in $d = 3$ leads to $0(n)$ in eq.(2) to the same result as obtained from the generalized Boltzmann equation (1) for the non-equilibrium distribution function f on the basis of the Chapman-Enskog method. These density expansions were obtained in the same fashion as for the generalized Boltzmann equation (1) and in general any calculational procedure for eq.(1) (such as the above mentioned rearrangement) can be duplicated for the evaluation of the G-K formulae. As far as density expansions of the transport coefficients are concerned the G-K relations did not bring anything new, but in many other respects, they revolutionized their study: they have been the starting point for a host of other, in particular numerical, calculations of the transport coefficients. It is interesting to remark that at first sceptics either judged their derivation as wrong[12] or their appearance as empty formal relations, with which not much could be done. The same could have been said of Gibbs' relations for the thermodynamic properties of a system in equilibrium in terms of the canonical partition function.

Similar expressions in terms of time integrals over time correlation functions involving three microscopic currents have been obtained for the linear Burnett coefficients etc. These higher order linear transport coefficients occur in higher order hydrodynamic equations, obtained from the Chapman-Enskog solution of eq.(1) as the coefficients of higher order (than the first) derivatives of the local equilibrium parameters: e.g., $T(\mathbf{r}, t)$ or $\mathbf{u}(\mathbf{r}, t)$, in the expressions for the (macroscopic) currents. They lead, to expressions of the macroscopic current J of the form: $J = LX + L_1 X_1 + L_2 X_2 + ...$, where, e.g., $X_1 \sim \partial^2 T / \partial x^2$ etc., $X_2 \sim \partial^3 T / \partial x^3$ etc.

3. Longtime tails - Vortex Diffusion

In numerical studies of the longtime behavior of the velocity auto-correlation function integrand of the G-K relation for the self diffusion coefficient D (where $j_D(t) = v_{1x}(t)$, the x-component of the velocity of a tagged particle 1), Alder and Wainwright[13,14] found in 1968 a short time exponential decay, followed by a slow long time power law decay according to:

$$< v_{1x}(t) v_{1x}(0) >_{eq} = \alpha_D (t_0/t)^{d/2} \tag{7}$$

for $t > 10t_0$ in $d = 2$ and similarly later for $t > 25t_0$ in $d = 3$, i.e., over many mean free times t_0. Since

$$D = \int_0^\infty dt < v_{1x}(t)v_{1x}(0) >_{eq} \tag{8}$$

eqs.(7) and (8) imply that D does not exist in $d = 2$.

The coefficient α_D reads for $d = 2, 3$:

$$\alpha_D = (\frac{1}{4d\pi(\nu + D)t_0})^{d/2}, \tag{9}$$

where $\nu = \eta/nm$ is the kinematic viscosity. α_D contains contributions from two hydrodynamic modes: a viscous mode and a self-diffusion mode, which are long wave length or small wave number $k \to 0$ eigenmodes of the linear hydrodynamic and diffusion equations, respectively. For a fluid of hard disks the mean free time t_0 is given by $t_0 = (\beta m\pi)^{1/2}/2n\sigma\chi$ while for a fluid of hard spheres $t_0 = (\beta m\pi)^{1/2}/4n\sigma^2\chi$, where χ is the radial distribution function $g(r = \sigma)$ at contact. The expression for α_D – and similar expressions for the other transport coefficients – have been obtained both by kinetic theory[15] (with eigenvalues $\nu_E k^2$ and $D_E k^2$, where ν_E and D_E are an (Enskog) approximation[2] to η and D), as well as by a quasi-hydrodynamical evaluation[16] of the l.h.s. of eq.(7) (and of similar equations for other transport coefficients), giving then the full transport coefficients η and D in α_D, as shown in eq.(9). That two transport coefficients appear in the expression for α_D in (9) – or that mode-mode coupling occurs – can be seen as follows. In order to obtain the dominant longtime behavior of the left hand side of eq.(7), one has to couple the microscopic (one particle) current v_{1x} to at least two hydrodynamic modes, whose combined eigenfunctions have the same character as the current v_{1x}, in order to obtain a non-vanishing result. These two hydrodynamic modes are the viscous and the self-diffusion eigenmodes here. Similarly, in the case of the viscosity, the two hydrodynamic modes that couple to $v_{ix}v_{iy}$ (cf.eq.(6)) are either two viscous modes or two sound modes. The transport coefficients ν and D appearing in eq.(9) are therefore directly related to all the eigenvalues of the two hydrodynamic modes involved, i.e. to a contribution $\sim \int_0^{k_0} dk\, k^{d-1}\exp[-(\nu + D)k^2t]$, which leads to the behavior shown on the right hand side (r.h.s.) of eq.(7). Here k_0 is a cut-off wave number of the order of many inverse mean free paths[15]. Equations (7) and (8) imply that D (and similarly the other transport coefficients) will not exist in $d = 2$, due to what one could call "hydrodynamic" divergences. Similar considerations lead for the linear Burnett coefficients to a longtime behavior of their Green-Kubo integrands $\sim (t_0/t)^{d/2-1}$, implying that they do not exist in three dimensions.

Mode-mode coupling was introduced into kinetic theory by Pomeau in 1968[17], showing that the linear transport coefficients did not exist in two dimensions, in spite of the rearrangement of the r.h.s. of eq.(1) to eliminate the above mentioned "kinetic" divergences.

It is interesting to note that simultaneously and independently mode-mode coupling was introduced by Kadanoff and Swift in a discussion of the anomalous divergent behavior of the transport coefficients near a critical point[18].

The appearance of the (macroscopic) transport coefficients in the expressions for the longtime tails (cf.eq.(9)) is an expression of what one could call "the principle of macroscopic dominance", which says that even in evaluating microscopic properties, quantities related to the macroscopic quantities of the system play a dominant role. The reason for this is that the macroscopic quantities are related to the conservation laws, which hold, of course, also on the microscopic level. In the case of the longtime tails the macroscopic quantities that play a dominant role are the hydrodynamic modes, the eigenmodes of the linear hydrodynamic equations, which are directly related to the conservation laws of mass, momentum and energy. They are characterized by the fact that their eigenvalues go to zero when their wave number k goes to zero. I also note that the discovery of the longtime tails was made by studying numerically the integrand of a Green-Kubo relation, a proof of the fruitfulness of their introduction in statistical mechanics.

Before leaving the longtime tails, I want to emphasize their physical origin. I will use the self-diffusion as an example. In kinetic theory they are due to so-called ring-collisions, independent chains of binary collisions that connect e.g. two subsequent binary collisions of the same two particles, thus introducing a (long-range) correlation between the velocities of these same two particles[19]. The two hydrodynamic modes that occur in the mode-mode coupling expression for the longtime tails are associated with these two chains of binary collisions, respectively, where each mode is associated with one chain. They determine the strength of the correlation, relevant for the particular transport coefficient, between an earlier and a later binary collision of the two particles.

In quasi-hydrodynamic language a tagged particle 1 moving with velocity v_1 in the fluid will shed some of its momentum to the surrounding fluid particles. Or, considering this tagged particle as a macroscopic sphere moving through a continuum fluid, the particle will in the long wave length limit excite hydrodynamic viscous, i.e., vortex-like, modes in the surrounding fluid, which couple to the particle's motion, when the particle itself is simultaneously diffusing in the fluid (i.e., being molecular rather than macroscopic)[13c].

Thus the two modes that couple will be a viscous and a (self-)diffusive mode, as stated before. The viscous mode represents the vortex-like motion which returns to the moving particle part of the momentum it shed to the fluid. Since this momentum returns at the back of the particle in the same direction as the particle's velocity, the particle receives an extra kick so to speak, enhancing its diffusion through the fluid, i.e. the longtime tails give, through *vortex diffusion*, a positive contribution to D. Similar results hold for the other transport coefficients. Thus for the viscosity, where one considers e.g. the quantity $v_{ix}v_{iy}$, associated with the motion of each fluid particle i, rather than v_{1x} of one (tagged) particle 1, a coupling of P_{xy} to two viscous modes occurs, representing a sharing of the momentum flux of each fluid particle i with the surrounding particles. Also here some of the shed momentum flux is returned to the particles, such that an enhancement of the total momentum flux associated with the fluid particles occurs, leading to a positive contribution to the viscosity. I remark that the velocity autocorrelation function defined on the l.h.s. of eq.(7), becomes at high densities negative for intermediate times $5t_0 < t < 10t_0$ in d = 2 and $5t_0 < t < 20t_0$ in $d = 3$. This behavior can be understood, on the basis of the cage-diffusion process, discussed below, as due to a mode-coupling of a cage-diffusion and a self-diffusion mode.

The theory of the longtime tails was completed between 1970-1975, ending the work emanating from the seminal developments around 1965. All these developments were relevant for dense fluids and although they occur in principle also in dilute gases, the collision sequences that give rise to them are so rare there, that a Boltzmann binary collision approximation is adequate.

III. Long Range Nonequilibrium Correlations, Neutron Scattering, Cage Diffusion, Fast and Slow Sound

4. Long range nonequilibrium correlations

Around 1980 two new developments were initiated, which led to new insights in dense fluids: one associated with the interpretation of neutron scattering spectra of dense fluids in equibrium, the other with the presence of long range correlations in a fluid in a nonequilibrium stationary state. Since the latter is closely related to the just discussed longtime tails, I will discuss this work first.

5. Long range correlations in a fluid in a nonequilibrium stationary state

A number of different theoretical developments by Procaccia et al[20,21,22,], Kirkpatrick et al[23] and others[24,25], all lead around 1980, to the prediction that in a fluid with a constant temperature gradient, very long range correlations appear between microscopic density fluctuations at different positions in the fluid, which should be observable by light scattering. After preliminary evidence for this by Beyssens et al[26a] the first complete experimental confirmation was made in 1984 by Kiefte, Clouter and Penney[26b], who observed the difference in intensity of the two Brillouin lines, due to the presence of a temperature gradient (cf.fig.1). Later in 1988, Law, Gammon and Sengers[27,28] found very detailed agreement between theory and experiment for the change in intensity of the central or Rayleigh line.

The physics of the origin of the long range correlations is easiest discussed for the Brillouin lines for small temperature gradients, when the intensity of each of the two Brillouin lines changes proportional to the temperature gradient by the same amount in opposite directions, respectively[20,23]:

$$S_{neq}^{\sigma}(\mathbf{k}) = S_{eq}^{\sigma}(k)[1 - \sigma\frac{c}{2\Gamma k^2}\frac{\hat{\mathbf{k}}\cdot\nabla T}{T}] \; (\sigma = \pm 1) \tag{10}$$

Here the two Brillouin lines are distinguished by the index $\sigma = +1$ or -1 and k is determined by the scattering angle θ of the scattered light ($k = 2k_i\sin\theta/2$, where k_i is the wave number of the incoming light). c and Γ are the velocity and the damping of a sound mode of wave number k in the fluid in equilibrium, i.e., related to the imaginary and the real part of the complex eigenvalues of the (hydrodynamic) sound eigenmodes of the fluid:

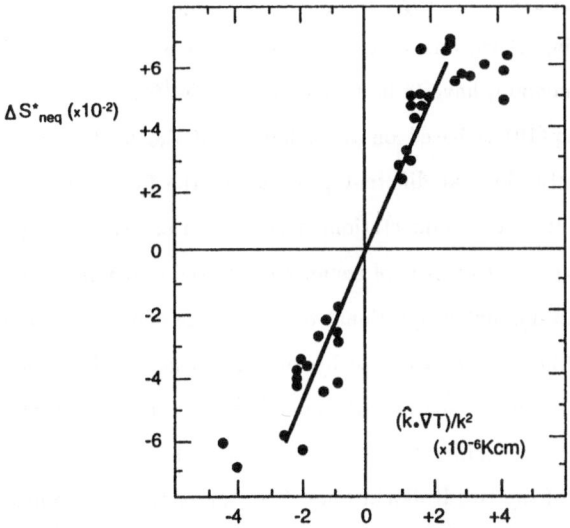

Fig. 1 Normalized intensity difference $\Delta S_{neq}^* = (S_{neq}^- - S_{neq}^+)/(S_{neq}^- + S_{neq}^+)$ of the two Brillouin lines versus $(\hat{\mathbf{k}}\cdot\nabla T)/k^2$ for water with $|\nabla T| = 45$ K cm^{-1} and $T = 307$ K [ref.26b].

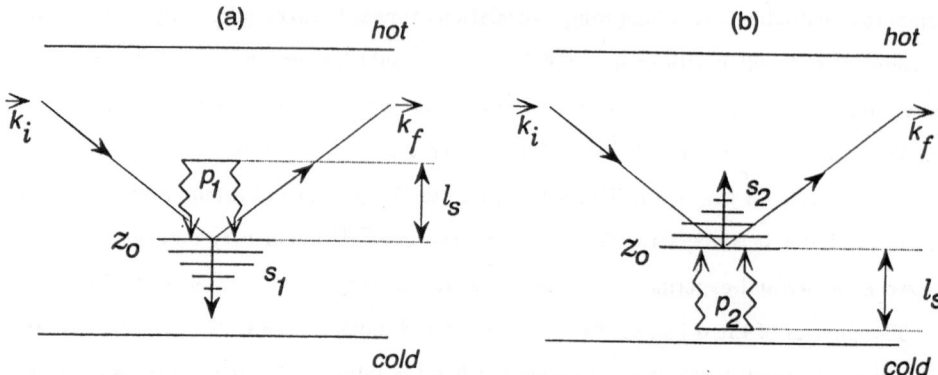

Fig. 2 Incoming light, with wave number k_i is scattered at z_0 to k_f by a sound mode s_1 (down shifted Brillouin line S_{eq}^-) in (a) or s_2 (upshifted Brillouin line S_{eq}^+) in (b), respectively. A correction from the heat flux is due to a pair of sound modes p_1 or p_2, originating at $z_0 + \ell_s$ (a) or $z_0 - \ell_s$ (b), respectively, since a single sound mode does not couple to the heat flux. For, schematically, the heat flux = energy transport = (scalar x vector), while a sound mode = (pressure + longitudinal velocity) = (scalar + vector). Only a pair, i.e., a product of two sound modes gives a (scalar x vector), i.e., a heat flux and couples to the heat flux in the fluid.

$i\sigma c k + \Gamma k^2; \hat{k} = k/k$ with $k = |k|$, respectively. In equilibrium the intensity of the two Brillouin lines $S_{eq}^\sigma(k)(\sigma = \pm 1)$ is the same, i.e. $S_{eq}^{+1} = S_{eq}^{-1}$[36]. At high gradients, wall effects and nonlinear effects, including those due to the bending of sound modes when moving through an inhomogeneous medium, have to be taken into account[29].

The physical interpretation of eq.(10) is based on the influence of the heat flux on the correlations between density fluctuations at different positions in the fluid. In fact, the scattering intensity S_{neq}^σ in the nonequilibrium stationary state at a scattering angle θ is higher (lower) for a Brillouin line due to a *pair* of sound modes originating from the hotter (cooler) side of the fluid by an amount proportional to their combined "mean free paths" $\ell_s = c/2\Gamma k^2$, respectively (cf.fig.2). $\pm \ell_s |\nabla T|$ can be interpreted as the difference in temperature and therefore in amplitude of each pair of sound modes - coming from the hot or cold side, respectively – due to the temperature gradient in the fluid leading to the observed difference in the intensities of the two Brillouin lines. That pairs of sound modes are relevant rather than single ones, is again a mode-mode coupling effect, due to the fact that one needs the product of two sound mode eigenfunctions to couple to the heat flux caused by the temperature gradient, i.e. to have an object of the same character as a heat flux[30]. Similarly, for the Rayleigh line, one needs a pair of a viscous and a heat mode to couple to the heat flux.

The $1/k^2$-dependence of $S_{neq}^\sigma(k)$ in (10) implies a $1/r$-dependence in its Fourier transform, the pair correlation function $G_{neq}(r)$ of two particles in the fluid as a function of their difference in position r, clearly exhibiting the existence of long range correlations in the nonequilibrium fluid. The Rayleigh line has a change in intensity compared to (local) equilibrium given by[22,23,28]:

$$S_{neq}^R(k = S_{eq}^R(k)[1 + c_p T \frac{1}{D_T(\nu + D_T)} \frac{1}{k^4}(\hat{k}_\perp \cdot \frac{\nabla T}{T})^2] \tag{11}$$

where $\hat{k}_\perp$ is a unit vector perpendicular to k, the thermal diffusivity $D_T = \lambda/\rho c_p$, with λ the thermal conductivity and c_p the specific heat at constant pressure per unit mass and $S_{eq}^R(k)$ the (local) equilibrium contribution to the intensity of the Rayleigh line. As mentioned above, there are again mode-mode coupling contributions of two hydrodynamic modes in eq.(11): the viscous and the heat mode, with eigenvalues νk^2 and $D_T k^2$ respectively. Contrary to eq.(10) for the Brillouin lines, eq.(11) is valid for all temperature gradients as long as local equilibrium is maintained. The $1/k^4$-dependence implies even longer ranged correlations than betrayed by the $1/k^2$-dependence of the Brillouin lines. The static structure factor (11) and its extension, the intermediate scattering function $S_{neq}^R(k, t)$, have been extensively checked experimentally by Sengers and collaborators for a fluid layer of about 0.1 cm thickness[27,28]. They confirmed the theoretical predictions and therefore the long range correlations in fluids in a non-equilibrium stationary state within their experimental accuracy of about 1–2%. In fact, for temperature gradients up to 224 K/cm and scattering angles of about 0.5 degree, they measured correlations over a range of about 10^{-3} cm, five orders of magnitude larger than in a fluid in equilibrium at the same density and (average) temperature.

In so far as these long range correlations have been introduced by a (heat) flux (or equivalently a (temperature) gradient) in the fluid, they could be called "flux or field induced long range correlations"; since they are also associated with a "generic scale invariance" (i.e., power law decaying spatial correlations) in the nonequilibrium fluid, they could also be called "generic long range correlations"[28]. It seems a matter of taste, which description one prefers.

While the long range correlations discussed so far can be probed with visible light, there are other long range correlations which extend over the entire fluid layer, so that they can only be explored with microwaves, since the typical height d of the fluid layer is about $d = 0.1\text{cm}$[31]. In this case boundary and gravity effects have to be taken into account, which were neglected in the above mentioned calculations. The mode-mode coupling contributions to the dynamic structure factor $S^R_{neq}(\text{k},\omega)$ can then be many orders of magnitude ($\approx 10^7$) larger than the local equilibrium contribution. As an illustrative example, $S^R_{neq}(\text{k},\omega)$ is sketched in fig. 3 for a stationary state, away from the Rayleigh-Bénard (R-B) instability, when the fluid is heated from above. Then two peaks appear due to two "second sound"-like propagating modes (cf.fig.3)[31,32].

Near the R-B instability the analogue of critical opalescence can be observed in the micro-wave regime in that $S^R_{neq}(\text{k},\omega)$ is a Lorentzian of the form $S^R_{neq}(\text{k},\omega) \sim 1/[\sum(k)^2+\omega^2]$, where the line width goes to zero like $\sum(k)$ and the height goes to infinity like $1/\sum(k)^2$ as the instability point is approached, with $\sum(k)$ the eigenvalue of the critical mode[31].

6. Neutron Scattering Spectra, Effective Eigenmodes, Cage Diffusion

Around 1980 a description of the observed neutron spectra of fluids similar to that of light spectra in terms of three Lorentzian lines was developed by de Schepper and myself[33,34]. Since neutron scattering occurs at wavelengths a thousand times smaller than light scattering, the question arises with what eigenmodes or excitations of the fluid these three lines should be identified.

There is a large literature on the interpretation of neutron spectra based on memory-kernels, which is rather formal and contains quite a few adjustable parameters[35]. I do not feel competent to review this work. I will discuss instead a different approach, based on a generalization to small wavelengths of Landau-Placzek's theory of the Rayleigh-Brillouin triplet observed in light scattering[36]. This implies that the three fluid eigenmodes, corresponding to the three Lorentzians observed in neutron scattering, are the small wave length extensions of the three hydrodynamic modes observed in light scattering: the heat and the two sound modes, which cause this triplet. In fact, if one varies continuously the wavelength from light to neutron size, the Rayleigh-Brillouin triplet transforms continuously into the three neutron spectrum lines, as has been shown in computer simulations of a Lennard-Jones fluid[37] and in experiment[38].

Landau and Placzek's assumption that the (macroscopic) hydrodynamic modes could be used to explain the (microscopic) density fluctuations as observed in light scattering was based in a way on "the principle of macroscopic dominance". The physical justification was that the wavelength of light $\lambda_\ell >> \sigma$, so that at sufficiently high fluid densities there

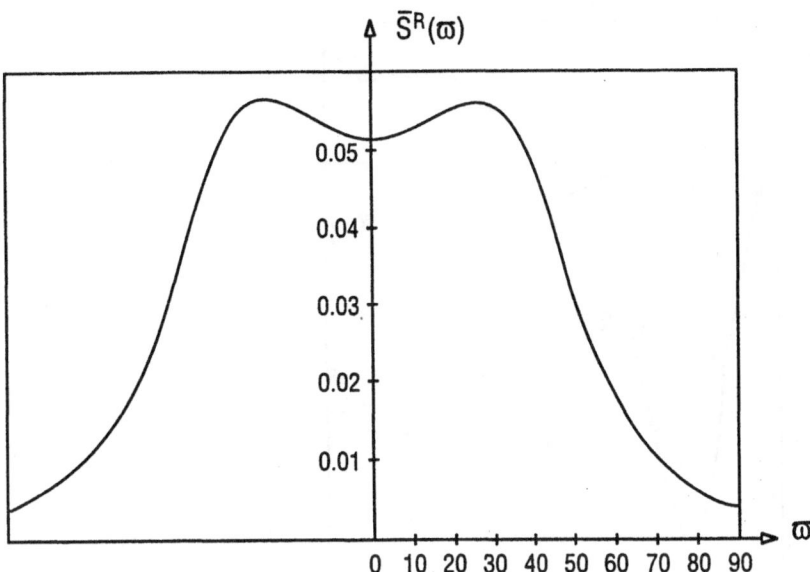

Fig. 3 Reduced nonequilibrium dynamic structure factor $\bar{S}^R(\bar{\omega}) = (\nu/d^2)S_{neq}^R(\omega)/(2\pi)^{-1}$
$\int S_{eq}^R(\omega)d\omega$ as a function of reduced frequency $\bar{\omega} = (d^2/\nu)\omega$, for a Prandtl number 3.6, a
Rayleigh number - 10,000, a reduced horizontal wave number $k_{\parallel}d = 4.5553$ and a vertical
wave number $k_z = 0$. There are two "second sound" peaks at $\bar{\omega} = \pm 28$[31]. To obtain the
actual values of $S(\bar{\omega})$ one has to multiply the ordinates by approximately 10^7 (cf.ref.31,
fig.4a).

are enough particles inside λ_ℓ that their number, momentum and energy is to a very good
approximation conserved and the hydrodynamic equations can be used to describe their
changes in time. Similarly in neutron scattering where the wavelength of a neutron $\lambda_n \approx \sigma$
so that there is only one particle inside λ_n only its number (or mass) is conserved (not
its momentum and energy which are constantly changing due to many collisions with the
surrounding particles) and the (self-)diffusion equation can be used to describe its change
with time. This is another example of the "principle of macroscopic dominance", mentioned before. The central (Lorentz) line in neutron spectra corresponds then to the central
(Rayleigh) line in light spectra, while the other two (Brillouin-like) (Lorentz) lines in the
neutron spectra correspond to two effective eigenmodes of the fluid, which are propagating and damped, like sound modes. For fluids of hard spheres all the (effective) lowest
eigenmodes of the fluid have been approximately determined by kinetic theory on the basis
of a generalized Enskog operator. The agreement with the eigenmodes derived directly
from computer simulations of hard sphere fluids is good (cf.fig.4). For hard sphere fluids

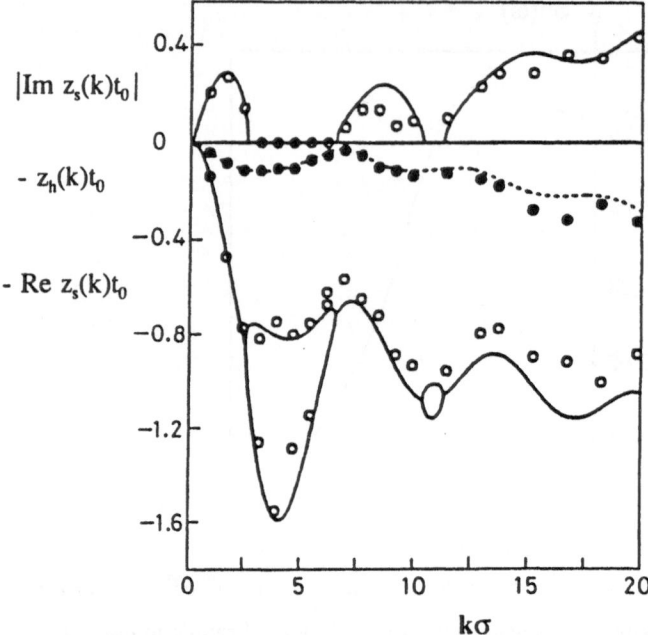

Fig. 4 Reduced (effective) extended hydrodynamic eigenvalues for a hard sphere fluid at a reduced density of $n\sigma^3 = 0.88$ as a function of the reduced wave number $k\sigma$ deduced from computer simulations. The imaginary part $|\mathrm{Im}z_s(k)t_0|$ gives the dispersion curve of the extended sound mode (o), with two clearly visible propagation gaps around $k\sigma \approx 5$ and $k\sigma \approx 10$, respectively. The real part $\mathrm{Re}z_s(k)t_0$ gives the damping of the extended sound mode (o), with two overdamped modes in the propagation gaps, while $z_h(k)t_0$ gives the heat mode (•) with a minimum at $k\sigma \approx 2\pi$. Solid and dashed lines are obtained from kinetic theory[34] with experimental transport coefficients.

the Brillouin-like lines are indeed direct extensions of the (hydrodynamic) sound modes to much smaller wavelengths, just as the Rayleigh-like (central) line is due to such an extension of the (hydrodynamic) heat mode. However, for simple liquids, like liquid Ar[39] or a Lennard-Jones liquid[37], (cf.fig.5) the Brillouin-like modes are not sound-like for large k and this difference is probably due to the importance here of attractive interparticle forces, resulting, amongst others, in a much higher compressibility of real liquids than that of dense hard sphere fluids.

I emphasize that the eigenmodes which are responsible for the neutron (or for that matter also the light) scattering spectra are *effective* eigenmodes of the fluid, which effectively represent the dynamical behavior of the fluid on the molecular scale as it is observed in the

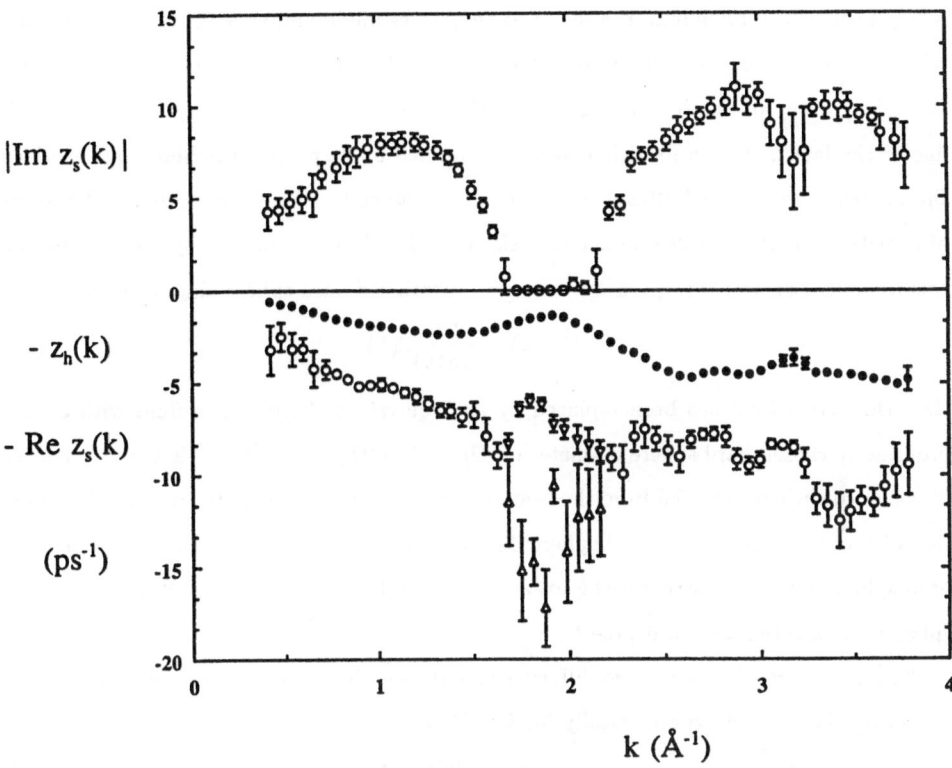

$|Im\ z_s(k)|$

$-\ z_h(k)$

$-\ Re\ z_s(k)$

(ps^{-1})

k (Å^{-1})

Fig. 5 Extended heat mode eigenvalue (•) for gaseous Argon at $T = 120K$ and $p = 20$ bar, deduced from neutron spectra as a function of wavenumber k[38] showing a minimum at $k \approx 2Å^{-1}$. Corresponding imaginary and real parts of the extended soundmode eigenvalue (o), showing a propagation gap around $k \approx 2Å^{-1}$, where two overdamped modes appear (∇ and $\triangle$).

neutron (or light) spectra of the fluid[34,40]. These effective eigenmodes consist, in general, of combinations of real, i.e. extended hydrodynamic as well as kinetic modes[40], where the eigenvalues of the latter do not go to zero for $k \to 0$. They can be determined theoretically either from kinetic theory[40] or from generalized hydrodynamics[37]. For small wavelengths, in the neutron scattering regime, only the extended heat mode can be considered as a real (extended) eigenmode of the fluid also for large k, all the other modes are conglomerates obtained from the superposition of many real eigenmodes. At high fluid densities $n\sigma^3 \geq$ 0.70, the dominant mode is the extended heat mode, which exhibits a strong minimum at $k^* \approx 2\pi/\sigma$, i.e., at $\lambda_n^* \approx \sigma$ (cf.fig.4)[33,34]. For this $k \sim k^*$ the static structure factor exhibits a sharp maximum and the half width at half height $\omega_H(k)$ neutron spectrum shows a very pronounced minimum, the so-called de Gennes minimum, which is therefore directly related to the strong heat mode minimum at $k = k^*$. This implies in turn a very slow (diffusive) decay of density fluctuations in the fluid at this value of k. Physically this is due to the fact that each particle is in a cage formed by its nearest neighbors out of which the particle can escape (diffuse) only with great difficulty. As a consequence, the decay of density fluctuations will proceed very slowly and will be determined by a cage-diffusion coefficient $D_c(k)$, which is given by the above mentioned hard sphere kinetic theory as:

$$D_c^{hs}(k) = \frac{D_B}{\chi S(k)} d(k) \tag{12}$$

Here the actual fluid has been replaced by an "equivalent" hard sphere fluid with an appropriately chosen hard sphere diameter σ. $D_B = 0.214(k_B T/m)^{1/2}/n\sigma^2$ is the low-density diffusion coefficient obtained from the Boltzmann equation, χ is the radial distribution function $g(r)$ of the (equivalent) hard sphere fluid at particle contact, i.e., for $r = \sigma$, $S(k)$ is the hard sphere fluid static structure factor and $d(k) = 1/[1 - j_0(k\sigma) + 2j_2(k\sigma)]$ where j_ℓ is the spherical Bessel function of degree ℓ

The connection between cage-diffusion and the de Gennes minimum is established by observing that indeed experimentally for $k \approx k^*$ (cf.fig.6):

$$\omega_H(k) = D_c^{hs}(k)k^2 \tag{13}$$

clearly indicating the (cage) diffusive origin of the de Gennes minimum.

The extended sound modes show at sufficiently high densities a sound propagation ga near $k \approx k^*$, where (non-propagating) overdamped modes appear (cf.fig.4). This has also been observed in neutron spectra (cf.fig.5)[39].

7. Fast and Slow Sound

An interesting application of the kinetic method to determine the effective eigenmodes of a fluid has been the prediction of fast sound in both low and high density binary fluid mixtures[41] . In this case a short wavelength highly damped but fast sound-like mode was predicted and observed to propagate exclusively in the light component of the mixture. Later a similar slow sound-like component was observed to propagate exclusively in the heavy compound of the mixture. The modes were observed in the light and neutron scattering spectra of appropriately chosen liquid[42] and gaseous[43] binary mixtures, respectively. The fast (slow) sound velocities were higher (lower) – by about a factor two – than those of the hydrodynamic sound velocity of the mixture and very close to that of the pure light (heavy) fluid obtained by removing the heavy (light) particles from the mixture[41].

I now turn to more recent developments, starting around 1985.

IV. Lattice Gas Cellular Automata, Transport Coefficients and Lyapunov Exponents

8. Lattice Gas Cellular Automata

Around 1986, Frisch, Hasslacher and Pomeau[44] and independently Wolfram[45] introduced Lattice Gas Cellular Automata (LGCA) to mimic fluid flow using supercomputers. They considered point particles moving in discrete time steps along the bonds of a regular (e.g. triangular) lattice from site to site. Upon collision of two or more particles at a lattice site the particles scatter according to certain collision rules. Based on the universality of the applicability of the hydrodynamic equations, the hope was that, in spite of the schematic discrete representation of what was in reality a motion of particles in continuous position and velocity spaces, the LLGCA would give a sufficiently accurate and fast representation of fluid motion to compete with conventional methods like wind-tunnels or numerical solutions of the non-linear Navier-Stokes equations. Although this was initially borne out in that flow pictures based on the particle trajectories on the lattice were obtained that – at least to the eye – looked very much like those found in real fluids, more detailed studies revealed many technical difficulties – such as large local fluctuations in the hydrodynamic fields[46] – which have made LGCA not competitive so far with conventional methods, at least for the study of chaotic flow. For spinodal decomposition, phase-separation dynamics and multiphase flow through porous media[47] they appear to provide new insights. Also chemical reactions have been modeled on LGCA[48].

A special case of LGCA: Lorentz Lattice Gas Cellular Automata (LLGCA) have been introduced by Binder and Ernst[49,50]. These LLGCA can be considered as mimicking the diffusion of a particle (or a group of particles) through a set of scatterers (or obstacles) on the lattice sites, from which they scatter according to certain scattering rules. For probabilistic scattering rules, Ernst and van Velzen[50] have successfully adapted a discrete, linear version of the modern kinetic theory, mentioned in section 1, to describe the diffusion process on regular lattices. LLGCA with deterministic scattering rules were introduced by Ruijgrok and myself[51]. Here a variety of abnormal diffusive behavior was found, ranging from super-diffusion to no-diffusion, when all particles are trapped[52]. A number of theorems have recently been proved for these deterministic LLGCA[53].

9. Transport Coefficients and Lyapunov Exponents

In the second half of the 1980's Hoover and Posch studied numerically the properties of a variety of systems from the point of view of dynamical system theory, e.g., determining their Lyapunov spectra. They studied systems in an equilibrium as well as in a nonequilibrium stationary state[54]. In the latter case they used effective equations of motion for the particles in a sheared or heat conducting fluid where the presence of a velocity or temperature gradient in the fluid as well as a thermostat were incorporated into the equations of motion through body forces on each individual fluid particle. This constitutes in the spirit of linear response theory a "mechanization", i.e. a putting into a purely dynamical framework, of what really are statistical forces. This method of numerical study of many particle systems in a nonequilibrium state was called Non Equilibrium Molecular Dynamics (NEMD) and was pioneered from about 1975 – 1985 by Hoover, Evans and Morriss and others[55,56]. For the special case of viscous flow, the SLLOD equations of motion were developed[56], which lead to the following expression for the viscosity of a sheared fluid in a nonequilibrium stationary state (s.s.) in terms of its (s.s.) Lyapunov exponents[54,57]:

$$\eta(\gamma) = -\frac{k_B T}{V \gamma^2} \sum_{i=1}^{6N} \lambda_i^{ss}(\gamma) \tag{14}$$

Here the fluid consists of N atomic particles, γ is the shear rate in the fluid and λ_i^{ss} is the i-th of the $6N$ Lyapunov exponents of the fluid in the s.s. A symmetry of the Lyapunov spectrum, in the form of a pairing rule between corresponding pairs of Lyapunov exponents[57], reduces the r.h.s. of (14) to the expression:

$$\eta(\gamma) = -\frac{3nk_B T}{\gamma^2} [\lambda_{max}^{ss}(\gamma) + \lambda_{min}^{ss}(\gamma)] \tag{15}$$

where λ_{max}^{ss} and λ_{max}^{ss} are the (positive) maximum and (negative) minimum Lyapunov exponents of the fluid in the s.s. Eq.(15) holds not only in the linear regime for the Newtonian viscosity ($\gamma \rightarrow 0^+$), but also for the rheological (i.e., γ-dependent) viscosity. It allows a numerical determination of $\eta(\gamma)$ via its largest and its smallest Lyapunov exponents alone and agrees well (up to 5%) with a direct numerical determination of $\eta(\gamma)$ from the phenomenological (linear) law [57]. The origin of the relation (14) is that the phase space contraction of the dissipative fluid system is given both by the irreversible entropy production $\eta(\gamma)\gamma^2/k_B T$ and by the (negative) sum of all its Lyapunov exponents[54].

An expression similar to (15), but different in detail, has been obtained by Gaspard and Nicolis[58] for the linear diffusion coefficient of a point particle moving through a regular triangular array of fixed hard disks and has been generalized very recently to general transport coefficients[59]. The precise connection between these expressions for the transport coefficients is not clear at the moment.

V. Analogy Atomic Liquids and Concentrated Colloidal Suspensions

10. Cage-diffusion

A physical analogy between simple atomic liquids and complex concentrated (monodisperse) colloidal suspensions has been noticed in that cage diffusion plays a dominant role in both systems. The analogy can be expressed as a scaling analogy[60], which can qualitatively be understood in that the motion of a Newtonian and a Brownian particle in a dense fluid mainly differ in their respective diffusion time scales[1,61]. The analogy is further based on the existence of only one conserved quantity (the number density) in both dense systems for $k \approx k^*$, (where the colloidal static structure factor has a sharp maximum) and can be made quantitative by using an equivalent dense hard sphere fluid for both. The cage diffusion coefficients for the two fluid systems are then related for $k \approx k^*$ by[1,60,62a]:

$$\frac{D_c^{hs}(k)}{D_B} = \frac{D_c^{coll}(k)}{D_0} \tag{16}$$

where D_0 is the low concentration Stokes-Einstein diffusion coefficient of the colloidal suspension. Eq.(16) can be verified via neutron scattering data of atomic liquids (l.h.s. of (16)) and light scattering data of concentrated colloidal suspension (r.h.s. of (16))(cf.fig.6), since then $\sigma_a/\lambda_\ell \approx \sigma_c/\lambda_n$ where the subscripts a and c refer to atomic and colloid, respectively[62b]. This confirms very well the similarity of the cage diffusion processes in the two fluid systems.

11. Transport Coefficients

One of the major questions of nonequilibrium statistical mechanics is the theoretical computation of the transport coefficients of dense fluids. For that, it is necessary to know what the basic mechanisms are of mass, momentum and energy transfer in a dense fluid. The translational and collisional transfer as incorporated in the Boltzmann and Enskog equations, respectively, are the classically known transfer mechanisms. The vortex and especially the cage diffusion discussed above are two new mechanisms[60,62]. It appears that together they might comprise all relevant transport mechanisms in atomic liquids as well as in concentrated colloidal suspensions. So far, only two transport coefficients have been studied: D and η. While the vortex diffusion process is only important for the self-diffusion coefficient D[60], the cage diffusion process has to be taken into account in both D and η. The contributions to D and η from cage-diffusion are determined by the transfer of a particle or of the particle momentum, respectively, from one cage to the next. This involves not only the cage diffusion coefficient of the (central) particle in a cage but also that of the central

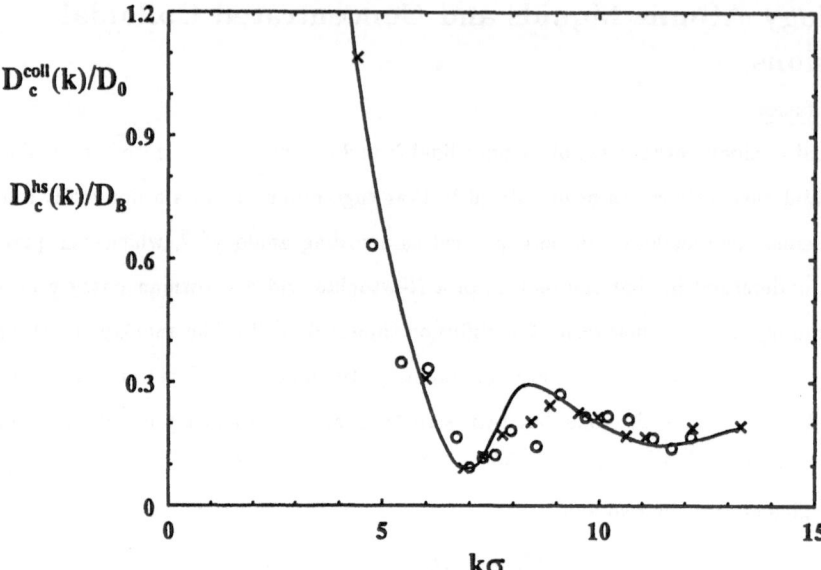

Fig. 6. Reduced cage-diffusion coefficients $D_c(k)$ as a function of the reduced wave number $k\sigma$, where σ is the equivalent hard sphere diameter. D_B and D_0 are the Boltzmann and the Stokes-Einstein diffusion coefficients, respectively. The x refer to the half-widths $\omega_H(k)$ of the neutron spectra of liquid Rb at 315 K with $\sigma = 4.44\text{Å}$ at a reduced density of $n\sigma^3 = 0.92$ [ref.1, fig.4]. The o are deduced from the decay rate of $S_{eq}^R(k,t)$ measured in the light spectra of polystyrene spheres in water with $\sigma = 6000\text{Å}$ at $n\sigma^3 = 0.92$ [ref.1,fig.4]. The solid line corresponds to the theoretical expression (12) for $D_c(k)$ for a dense hard sphere fluid.

particle in a neighboring cage, which is a wall particle of the first cage[60b]. Consequently two cage diffusion coefficients will appear, representing a mode-mode coupling contribution to the transport coefficients[60].

Thus one finds for the viscosity of a dense hard sphere fluid[60]:

$$\eta^{hs} = \eta_E + \int_{8t_0}^{\infty} dt \int_0^{\infty} dk \, V(k) e^{-2D_c^{hs}(k)k^2 t} \tag{17}$$

while for a concentrated colloidal suspension one obtains[60]:

$$\eta^{coll} = \eta_0 + \int_0^{\infty} dt \int_0^{\infty} dk \, V(k) e^{-2D_c^{coll}(k)k^2 t} \tag{18}$$

where

$$V(k) = \frac{k_B T}{60\pi^2} [\frac{k^2}{S(k)} \frac{dS(k)}{dk}]^2 \tag{19}$$

where η_0 is the viscosity of the solvent. The time integral in eq.(17) is not very sensitive to the precise choice of its lower limit and values between $6t_0$ and $10t_0$ can be used, larger

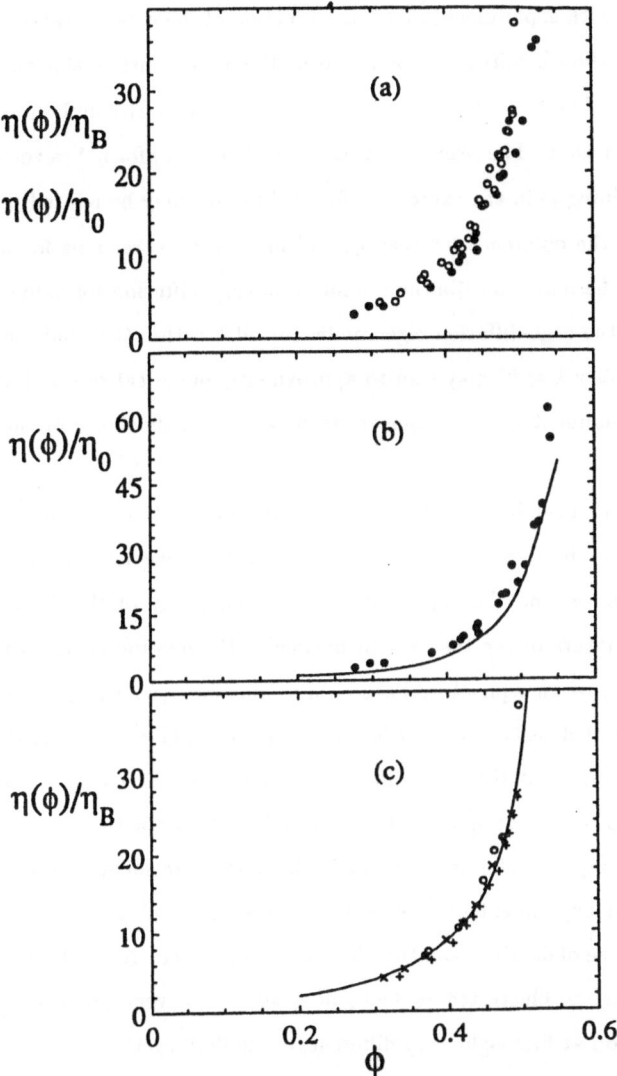

Fig. 7 Reduced viscosity as a function of volume fraction[60a,c,65,66]; (a) experiments: liquids (Ar, CH$_4$, Hard Spheres: open circles) and colloids (silica spheres in cyclohexane: closed circles) are virtually the same; (b) Theory and experiment: colloids (closed circles) and eq.(18) (solid line); (c) id.: liquids (Argon +, Methane x, Hard Spheres (o) and eq.(17) (solid line).

than the time scale $\approx 5t_0$ on which the contributions to η_E occur. The factor 2 in the exponents is due to the above mentioned mode-coupling and the amplitude $V(k)$ represents the deformability of the two neighboring cages in response to the external shear stress.

Eqs.(17) and (18) are compared with experiment in fig.7 as a function of the volume fraction $\phi = \pi n \sigma^3 / 6$. In the same figure the viscosities of Argon and Methane are plotted, showing that there is a far-going analogy between dense hard sphere fluids, simple atomic liquids and concentrated colloidal suspensions. The reason that, contrary to eq.(17) for η^{hs}, a zero lower limit appears in the time integral in the expression for η^{coll} in (18) may be related to the presence of the solvent and the difference between η_0 and η_E. Both expressions for η contain integrals over k, where the main contributions come from cage diffusion for values of $k \approx k^*$. However, the use of the cage-diffusion eigenmodes for all k rather than only in their appropriate domain of validity $k \approx k^*$ may lead to approximate numerical results for η, although they agree with experiment by what appears to be a reasonable choice of the lower limit of the time integral.

In this connection, one should bear in mind that a complete quantitative agreement between theory and experiment or between the experiments themselves over large ranges of the thermodynamic parameters cannot be expected[63]. The complexity of the dense fluid systems, the approximate nature of the theory – in particular the non-uniqueness of and the sensitivity to the choice of an equivalent hard sphere diameter (which appears as σ^3 in ϕ) – the limited range of density over which a hard sphere fluid can represent a real fluid at a given temperature and the experimental uncertainties, all conspire to limit the possibility of making a conclusive quantitative comparison between theory and experiment. In fact, not even the noble gas fluids correspond. Complete agreement between a hard sphere based theory and experiment or even between atomic liquid and colloid experiments over an extended range of densities can therefore not be expected. Nevertheless the good quantitative agreement, as illustrated in fig.6, does suggest a very analogous physical behavior of two classes of, at first sight very different, dense fluid systems.

Recently a beginning has been made of a theory of the visco-elastic (i.e. frequency dependent)[64,66] and rheological (i.e. shear rate dependent) behavior of neutral concentrated colloidal suspensions[65,66] based on the cage diffusion in an equivalent hard sphere fluid using eq.(12).

VI. Conclusion

I believe that in the last twenty-five years significant progress has been made in our understanding of dense - simple as well as some complex - fluids. Many interesting new phenomena have been or still have to be observed. Curiously, it have mainly been the microscopic or mesoscopic properties, as observed in neutron or light scattering that have been much better understood. How to obtain a similar full understanding of the macroscopic properties, such as the transport coefficients, is not entirely clear yet although it seems that relevant physical processes that determine these coefficients have been identified. Hopefully the next twenty-five years will bring a solution to this both theoretically and practically important problem.

Acknowledgement I am indebted to Professors J. R. Dorfman and M. H. Ernst as well as to Drs. R. Schmitz and A. Campa for helpful remarks and to Dr. I. M. de Schepper for his generous assistance. Financial support of the Department of Energy under Grant number DE-FG-02-88-ER13847 is also gratefully acknowledged.

References

1. (a) E. G. D. Cohen, "Fifty Years of Kinetic Theory", Physica A **194**, 229-257 (1994); (b) J. R. Dorfman and H. van Beijeren, "The Kinetic Theory of Gases", in: *Statistical Mechanics*, Part B, B. J. Berne, ed., Plenum Press, New York (1977) Ch.3, 65-179.

2. S. Chapman and T. G. Cowling, "The Mathematical Theory of Non-uniform Gases", 3rd ed. Cambridge Mathematical Library Series (Cambridge University Press, Cambridge, 1990).

3. See, e.g., J. R. Dorfman, "Kinetic and Hydrodynamic Theory of Time Correlation Functions" in: *Fundamental Problems in Statistical Mechanics* III", E. G. D. Cohen, ed. (North-Holland, Amsterdam, 1975) p.277-330; E. G. D. Cohen, "The Kinetic Theory of Dense Gases", in: ibid.II, p.228-275.

4. R. Zwanzig, "Time-Correlation Functions and Transport Coefficients in Statistical Mechanics", Ann. Rev. Phys. Chem.**16**, 67-102 (1965).

5. L. Onsager, "Reciprocal Relations in Irreversible Processes. I", Phys. Rev.**37**, 405-426 (1931); id., II, Phys. Rev.**38**, 2265-2279 (1931); S. R. de Groot and P. Mazur, "Non-equilibrium Thermodynamics", (North-Holland, Amsterdam (1962) p.100.

6. A current fluctuation is the difference between the actual current and the average current. Since the average currents we consider here all vanish, the current fluctuation equals the actual current.

7. The decay of $< j_L(t) >_{[j_L(o)]}$ can be more complicated than the exponential decay given here (cf. the next section).

8. R. Kubo, M. Yohota and S. Nakajima, "Statistical-Mechanical Theory of Irreversible Processes. II. Response to Thermal Disturbance", Phys. Soc. Jap. **12**, 1203-1211 (1957).

9. M. S. Green, "Markoff Random Processes and the Statistical Mechanics of Time-

Dependent Phenomena. II. Irreversible Processes in Fluids", J. Chem. Phys. 22, 398-413 (1954).

10. R. Kubo, "Statistical-Mechanical Theory of Irreversible Processes. I. General Theory and Simple Applications to Magnetic and Conduction Problems", J. Phys. Soc. Jap.12, 570-586 (1957).

11. M. H. Ernst, J. R. Dorfman and E. G. D. Cohen, "Transport Coefficients in Dense Gases I. The Dilute and Moderately Dense Gas", Physica31, 493-521 (1965).

12. N. G. van Kampen, "The Case Against Linear Response Theory", Physica Norvegica 5, 279-284 (1971).

13. (a) B.J. Alder and T.E. Wainwright, "Velocity Autocorrelations for Hard Spheres", Phys. Rev. Lett.18, 988-990 (1967); (b) id., "Enhancement of Diffusion by Vortex-like Motion of Classical Hard Particles", J. Phys. Soc. Jap. Suppl. 26, 267-269 (1969); (c) id., "Decay of Velocity Autocorrelation Function", Phys. Rev. A1, 18-21 (1970).

14. Y. Pomeau and P. Résibois, "Time Dependent Correlation Functions and Mode-Mode Coupling Theories", Phys. Rept. 19, 64-139 (1975).

15. J. R. Dorfman and E. G. D. Cohen, "Velocity Correlation Functions in Two and Three Dimensions", Phys. Rev. Lett.25, 1257-1260 (1970); id., "Velocity Correlation Functions in Two and Three Dimensions: Low Density", Phys. Rev. A6, 776-790 (1972); id., "Velocity Correlation Functions in Two and Three Dimensions: Higher Density", Phys. Rev. A12, 292-316 (1975); E. G. D. Cohen, "Kinetic Theory of Non-equilibrium Fluids", Physica A118, 17-42 (1983).

16. M. H. Ernst, E. H. Hauge and J. M. J. van Leeuwen, "Asymptotic Time Behavior of Correlation Functions", Phys. Rev. Lett. 25, 1254-1256 (1970); id., "Asymptotic Time Behavior of Correlation Functions. I. Kinetic Terms", Phys. Rev. A4, 2055-2065 (1971); id., "Asymptotic Time Behavior of Correlation Functions. II. Kinetic and Potential Terms", J. Stat. Phys.15, 7-22 (1976); id., "Asymptotic Time Behavior of Correlation Functions. III. "Local Equilibrium and Mode-Coupling Theory", J. Stat. Phys.15, 23-58 (1976).

17. Y. Pomeau, "A Divergence Free Kinetic Equation for a Dense Boltzmann Gas", Phys. Lett. A26, 336 (1968); id., "A New Kinetic Theory for a Dense Classical Gas", id., 27, 601-602 (1968).

18. L. P. Kadanoff and J. Swift, "Transport Coefficients Near the Liquid Gas Critical Point", Phys. Rev. 166, 89-101 (1968).

19. See ref. 1a, fig.1 and p.235.

20. I. Procaccia, D. Ronis and I. Oppenheim, "Light Scattering from Non-Equilibrium Stationary State: The Implication of Broken Time-Reversal Symmetry", Phys. Rev. Lett. 42, 287-291 (1979); I. Procaccia, D. Ronis, M. A. Collins, J Ross and I. Oppenheim, "Statistical Mechanics of Stationary States I. Formal Theory", Phys. Rev. A19, 1290-1306 (1979); D. Ronis, I. Procaccia and I. Oppenheim, "Statistical Mechanics of Stationary States II. Applications to Low Density Systems", Phys. Rev. A19, 1307-1323 (1979); id., "Statistical Mechanics of Stationary States III. Fluctuations in Dense Fluids with Applications to Light Scattering" , Phys. Rev. A19, 1324-1339 (1979); I. Procaccia, D. Ronis and I. Oppenheim, "Statistical Mechanics of Stationary States IV. Far From Equilibrium Stationary States and the Regression of Fluctuations", Phys. Rev. A20, 2533-2546 (1979); D. Ronis, I Procaccia and J. Machta, "Statistical Mechanics of Stationary States VI. Hydrodynamical Fluctuating Theory Far From Equilibrium", Phys. Rev. A22, 714-724 (1980).

21. G. Satten and D. Ronis, "Modification of Non-Equilibrium Fluctuations by Interaction with Surfaces", Phys. Rev. A26, 940-949 (1982).

22. D. Ronis and I. Procaccia, "Nonlinear Resonant Coupling Between Shear and Heat Fluctuations in Fluids Far From Equilibrium", Phys. Rev. A 26, 1812-1815 (1982).

23. T. R. Kirkpatrick, E. G. D. Cohen and J. R. Dorfman, "Kinetic Theory of Light Scattering from a Fluid not in Equilibrium", Phys. Rev. Lett. 42, 862-865 (1979); id., "Hydrodynamic Theory of Light Scattering from a Fluid in a Nonequilibrium Steady State", Phys. Rev. Lett. 44, 472-475 (1980); id., "Fluctuations in a Nonequilibrium Steady State: Basic Equations", Phys. Rev. A26, 950-970 (1982); id., "Light Scattering by a Fluid in a Nonequilbrium Steady State I: Small Gradients", Phys. Rev. A26, 972-994 (1982); id., "Light Scattering by a Fluid in a Nonequilbrium Steady State II: Large Gradients", Phys. Rev. A26, 995-1014 (1982); R. Schmitz and E. G. D. Cohen, "Fluctuations in a Fluid Under a Stationary Heat Flux I. General Theory", J. Stat. Phys. 39, 285-316 (1985).

24. A.-M. S. Tremblay, M. Arai and E. Siggia, "Fluctuations About Hydrodynamic Nonequilibrium Steady States", Phys. Lett. A76, 57-60 (1980); Phys. Rev. A23, 1451-1480 (1981).

25. G. van der Zwan, D. Bedeaux and P. Mazur "Light Scattering From a Fluid with a Stationary Temperature Gradient", Physica A107, 491-508 (1981).

26. (a) D. Beyssens, Y. Garrabos and G. Zalczer, "Experimental Evidence for Brillouin Aysmmetry Induced by a Temperature Gradient", Phys. Rev. Lett. 45, 403-406 (1980); (b) H. Kiefte, M. J. Clouter and R. Penney, "Experimental Confirmation of Nonequilibrium Steady-State Theory: Brillouin Scattering in a Temperature Gradient", Phys. Rev. B30, 4017-4020 (1984); (c) A.-M. S. Tremblay, "Theories of Fluctuations in Nonequilibrium

Systems", in: *Recent Developments in Nonequilbrium Thermodynamics*, J. Casas-Vazquez, F. Jou and G. Lebon, eds., *Lecture Notes in Physics 199*, Springer, New York (1984) p. 267-315.

27. B. M. Law, R. W. Gammon and J.-V. Sengers, "Light Scattering Observations of Long-Range Correlations in a Non-Equilibrium Liquid", Phys. Rev. Lett. 60, 1554-1557 (1988); B. M. Law, P. N. Segrè, R. W. Gammon and J. V. Sengers, "Light-scattering Measurements of Entropy and Viscous Fluctuations in a Liquid Far From Thermal Equilibrium", Phys. Rev. A 45, 816-824 (1990); P. N. Segrè, R. W. Gammon, J. V. Sengers and B. M. Law, "Rayleigh Scattering in a Liquid Far From Thermal Equilibrium", Phys. Rev. A45, 714-724 (1992); W. B. Li, P. N. Segrè, R. W. Gammon and J. V. Sengers, "Small-angle Rayleigh Scattering from Nonequilibrium Fluctuations in Liquids and Liquid Mixtures", Physica A204, 399-436 (1994).

28. J. R. Dorfman, T. R. Kirkpatrick and J. V. Sengers, "Generic Long-range Correlations in Molecular Fluids", Ann. Rev. Phys. Chem. 45, (1994).

29. R. Schmitz and E. G. D. Cohen, "Fluctuations in a Fluid Under a Stationary Heat Flux III. Brillouin Lines", J. Stat. Phys. 46, 319-348 (1987); id., "Brillouin Scattering From Fluids Subject to Large Thermal Gradients", Phys. Rev. A35, 2602-2610 (1987).

30. This is analogous to in quantum mechanics, where for a(n) (allowed) radiative transition of an electron between two energy eigenstates of an atom, i.e., for a coupling to the radiation field, to occur the product of the pair of eigenfunctions corresponding to this pair of states must have a multipole (dipole) character.

31. R. Schmitz and E. G. D. Cohen, "Fluctuations in a Fluid Under a Stationary Heat Flux II. Slow Part of the Correlation Matrix", J. Stat. Phys. 40, 431-482 (1985).

32. R. N. Segrè, R. Schmitz and J. V. Sengers, "Fluctuations in Inhomogeneous and Non-Equilibrium Fluids Under the Influence of Gravity", Physica A195, 31-52 (1993), curve (d) of fig.4.

33. I. M. de Schepper and E. G. D. Cohen, "Collective Modes in Fluids and Neutron Scattering", Phys. Rev. A22, 287-289 (1980); id., "Very-Short-Wavelength Collective Modes in Fluids", J. Stat. Phys. 27, 223-281 (1982).

34. E. G. D. Cohen and I. M. de Schepper, "Effective Eigenmode Description of Dynamical Processes in Dense Classical Fluids and Fluid Mixtures", Il Nuovo Cimento 12, 521-542 (1990).

35. See, e.g., S. W. Lovesey, *Theory of Neutron Scattering From Condensed Matter*, (Claren-

don Press, Oxford, 1984) Volume I; J. P. Hansen and I. R. McDonald, *Theory of Simple Liquids*, (Academic Press, London, 1990).

36. See, e.g., B. J. Berne and R. Pecora, *Dynamic Light Scattering*, (Wiley, New York, 1976) ch.X; R. D. Mountain, "Spectral Distribution of Scattered Light in a Simple Fluid", Rev. Mod. Phys.**38**, 205-214 (1966).

37. I. M. de Schepper, E. G. D. Cohen, C. Bruin, J. C. van Rijs, W. Montfrooij and L. A. de Graaf, "Hydrodynamic Time Correlation Functions for a Lennard-Jones Fluid", Phys. Rev. A**38**, 271-287 (1988).

38. U. Bafile, P. Verkerk, F. Barocchi, L. A. de Graaf, Y. B. Suck and H. Mutka, "Onset of Departure From Linearized Hydrodynamic Behavior in Argon Gas Studied With Neutron Brillouin Scattering", Phys. Rev. Lett. **65**, 2394-2397 (1990).

39. I. M. de Schepper, P. Verkerk, A. A. van Well and L. A. de Graaf, "Short-Wavelength Sound Mode in Liquid Argon", Phys. Rev. Lett. **50**, 974-977 (1983).

40. B. Kamgar-Parsi, E. G. D. Cohen and I. M. de Schepper, "Dynamical Processes in Hard-sphere Fluids", Phys. Rev. A**35**, 4781-4795 (1987).

41. A. Campa and E. G. D. Cohen, "Observable Fast Kinetic Eigenmode in Binary Noble-Gas Mixtures", Phys. Rev. Lett. 61, 853-856 (1988); id., "Kinetic-Sound Propagation in Dilute Gas Mixtures", Phys. Rev. A**39**, 4909-4911 (1989); id., "Fast Sound in Binary Fluid Mixtures", Phys. Rev. A **41**, 5451-5463 (1990); id., "Fast and Slow Sound in Binary Fluid Mixtures", Physica A **174**, 214-222 (1991).

42. W. T. Montfrooij, P. Westerhuijs, V. O. de Haar and I. M. de Schepper, "Fast Sound in a Helium-Neon Mixture Determined by Neutron Scattering", Phys. Rev. Lett.**63**, 544-550 (1989); W. T. Montfrooij, "From Visco-Elasticity Towards Thermal Relaxation", Thesis, Technical University Delft (1990) ch.VI; P. Westerhuijs, "Microscopic Dynamics in Dense Helium Mixtures", Thesis, Technical University Delft (1991).

43. G. H. Wegdam, A. Bot, R. P. C. Schram and H. M. de Schaink, "Observation of Fast Sound in Disparate-Mass Gas Mixtures by Light Scattering", Phys. Rev. Lett. **63**, 2697-2700 (1989); M. J. Clouter, H. Luo, H. Kiefte and J. A. Zollweg, "Light Scattering in Gas Mixtures: Evidence of Fast and Slow Sound Modes", Phys. Rev. A**41**, 2239-2242 (1990); G. H. Wegdam and H. M Schaink, "Light Scattering Study of Helium-Xenon Gas Mixtures: Slow Sound ", Phys. Rev. A**41**, 3419-3420 (1990); R. P. C. Schram, G. H. Wegdam and A. Bot, "Rayleigh Brillouin Light Scattering Study of Both Fast and Slow Sound in Binary Gas Mixtures", Phys. Rev. A**44**, 8063-8071 (1991); R. P. C. Schram and G. H. Wegdam, "Fast and Slow Sound in the Two-temperature Model", Physica A **203**, 33-52 (1994).

48

44. U. Frisch, B. Hasslacher and Y. Pomeau, "Lattice Gas Automata for the Navier-Stokes Equation", Phys. Rev. Lett.56, 1505-1508 (1986); U. Frisch, D. d'Humières, B. Hasslacher, P. Lallemand, Y. Pomeau and J. Rivet, "Lattice Gas Hydrodynamics in Two and Three Dimensions", Complex Systems 1, 649-707 (1987).

45. S. Wolfram, "Cellular Automata Fluids 1: Basic Theory", J. Stat. Phys. 45, 471-526 (1986).

46. S. A. Orszag and V. Yakhot, "Reynolds Number Scaling of Cellular-Automata Hydrodynamics", Phys. Rev. Lett.56, 1691-1693 (1986); V. Yakhot, B. J. Bayly, and S. A. Orszag, "Analogy Between Hyperscale Transport and Cellular Automata Fluid Dynamics" in: *Lattice Gas Methods for Partial Differential Equations*, G. D. Doolen, ed., (Addison-Wesley, New York, 1990) p.283-288.

47. D. H. Rothman and S. Zaleski, "Lattice-gas Models of Phase Separation: Interfaces, Phase Transitions and Multiphase Flow", Rev. Mod. Phys. (1994).

48. A. Lawniczak, D. Dab, R. Kapral and J. P. Boon, "Reactive Lattice Gas Automata", Physica D57, 132-158 (1991); R. Kapral, A. Lawniczak, and P. Masiar, "Reactive Dynamics in a Multispecies Lattice Gas Automata", J. Chem. Phys. 2762-2776 (1992); D. Dab, J. P. Boon and J.-X. Li, "Lattice Gas Automata for Coupled Reaction-Diffusion Equations", Phys. Rev. Lett. 66, 2535-2538 (1991); R. Kapral, A. Lawniczak and P. Masiar, "Oscillations and Waves in a Reactive Lattice Gas Automaton", Phys. Rev. Lett. 66, 2539-2542 (1991).

49. P. M. Binder, "Lattice Models of the Lorentz Gas: Physical and Dynamical Properties", Complex Systems 1, 559-574 (1987).

50. M. H. Ernst and G. A. van Velzen, "Lattice Lorentz Gas", J. Phys. A 22, 4611-4632 (1989); G. A van Velzen, "Lorentz Lattice Gases", Thesis, University of Utrecht, Utrecht, The Netherlands (1990).

51. Th. W. Ruijgrok and E. G. D. Cohen, "Deterministic Lattice Gas Models", Phys. Lett. A133, 415-418 (1988); E. G. D. Cohen, "New Types of Diffusion in Lattice Gas Cellular Automata" in: *Microscopic Simulations of Complex Hydrodynamic Phenomena*, M. Maréschal and B. L. Holian, eds., Plenum Press, New York (1992) p.137-152; E. G. D. Cohen and F. Wang, "Diffusion and Propagation in Lorentz Lattice Gases", in: *The Fields Institute Series*, Am. Math. Soc. (1994).

52. A number of previously published results (cf.ref.51) have to be revised, because computer experiments of much longer duration than before have shown a different behavior than previously reported: A. L. Owczarek and T. Prellberg, "Universality of Polymer Collapse in

Two Dimensions and Super-Diffusive Behavior in a Lorentz Lattice Gas", preprint (1994); F. Wang and E. G. D. Cohen, to be published.

53. L. A. Bunimovich and S. E. Troubetzkoy, "Recurrence Properties of Lorentz Lattice Gas Cellular Automata" J. Stat. Phys.67, 289-302 (1992); id., "Non Gaussian Behavior in Lorentz Lattice Gas Cellular Automata", in: *Proceedings of Dynamics of Complex and Irregular Systems*, Ph. Blanchard, ed. (World Scientific, Singapore, 1994); id., "Topological Dynamics of Flipping Lorentz Lattice Gas Models", J. Stat. Phys. 72, 297-308 (1993); id., "Rotators, Periodicity and Absence of Diffusion in Cyclic Cellular Automata", J. Stat. Phys. (1994).

54. H. A. Posch and W. G. Hoover, Lyapunov Instability of Dense Lennard-Jones Fluids", Phys. Rev. A38, 473-482 (1988); id., "Equilibrium and Nonequilibrium Lyapunov Spectra for Dense Fluids and Solids", Phys. Rev. A 39, 2175-2188 (1989); H. Posch and W. Hoover, "Nonequilibrium Molecular Dynamics of a Classical Fluid" in: *Molecular Liquids: New Perspectives in Physics and Chemistry*, J. Texeire-Dias, ed., Kluwer Academic Publishers (1992) p. 527-547.

55. W. G. Hoover and W. T. Ashurst, "Non-Equilibrium Molecular Dynamics", Theor. Chem. Adv. and Persp. 1, 1 (1975); W. G. Hoover, "Molecular Dynamics", *Lecture Notes in Physics 258*, (Springer, New York, 1986); id., "Non-Equilibrium Molecular Dynamics: the First 25 Years", Physica A194, 450-461 (1993); G. Ciccotti and G. Jacucci, "Direct Computation of Dynamical Response by Molecular Dynamics: The Mobility of a Charged Lennard-Jones Particle", Phys. Rev. Lett.35, 789-792 (1975).

56. D. J. Evans and G. P. Morriss, "Non-Newtonian Molecular Dynamics", Comput. Phys. Rep. 1, 300-343 (1984); id., *Statistical Mechanics of Nonequilibrium Liquids*, (Academic Press, New York, 1990).

57. D. J. Evans, E. G. D. Cohen and G. P. Morriss, "The Viscosity of a Simple Fluid From its Maximal Lyapunov Exponents", Phys. Rev. A42, 5990-5997 (1991).

58. P. Gaspard and G. Nicolis, "Transport Properties, Lyapunov Exponents and Entropy Per Unit Time", Phys. Rev. Lett.65, 1693-1696 (1990).

59. J. R. Dorfman and P. Gaspard, "Chaotic Scattering Theory of Transport and Reaction-Rate Coefficients", Phys. Rev. E (1995).

60. (a) E. G. D. Cohen and I. M. de Schepper, "Note on Transport Processes in Dense Colloidal Suspensions", J. Stat. Phys. 63 241-248 (1991); 65, 419 (1991); (b) id., "The Colloidal Many Body Problem: Colloidal Suspensions as Hard Sphere Fluids", in: *Recent Progress in Many Body Theories*, Vol.3, T. L. Ainsworth, C. Campbell, B. Clements and

E. Krotcheck, eds., (Plenum Press, New York, 1992) p.387-395; (c) id., "Transport Properties of Concentrated Colloidal Suspensions", in: *Slow Dynamics in Condensed Matter*, K. Kawasaki, M. Tokuyama and T. Kawakatsu, eds., Amer. Inst. Phys., New York (1992) p.359-369.

61. H. Löwen, J.-P. Hansen and J. N. Roux, "Brownian Dynamics and Kinetic Glass Transition in Colloidal Suspensions", Phys. Rev. A$\underline{44}$, 1169-1181 (1991).

62. (a) I. M. de Schepper, E. G. D. Cohen, H. N. W. Lekkerkerker and P. N. Pusey, "Long Time Diffusion Coefficient in Charged Colloidal Solutions", J. Phys. Cond. Matt. $\underline{1}$, 6503-6506 (1989); (b) P. N. Pusey, H. N. W. Lekkerkerker, E. G. D. Cohen and I. M. de Schepper, "Analogies Between the Dynamics of Concentrated Charged Colloidal Suspensions and Dense Atomic Liquids", Physica A$\underline{164}$, 12-27 (1990).

63. See ref. 1, p.251.

64. 1. M. de Schepper, H. E. Smorenburg and E. G. D. Cohen, "Viscoelasticity in Dense Hard Sphere Colloids", Phys. Rev. Lett. $\underline{70}$, 2178-2181 (1993).

65. I. M. de Schepper and E. G. D. Cohen, "Rheological Behavior of Concentrated Colloidal Suspensions and Dense Atomic Fluids", Phys. Lett. A$\underline{150}$, 308-310 (1990).

66. I. M. de Schepper and E. G. D. Cohen, "Viscoelastic and Rheological Behavior of Concentrated Colloidal Suspensions", Intern. J. of Thermophys. (1995).

Models for Dissipation in Quantum Mechanics

N.G. van Kampen

Inst. Theoretical Physics. University of Utrecht, The Netherlands.

The equations for dissipation and noise in a quantum mechanical system are obtained by including the interaction with a bath. The Markov character is usually obtained by means of an assumption, hopefully called "approximation". To investigate the validity of this assumption we here study two simple but exactly solvable model systems. It is shown that the interaction must be weak; the bath must contain a smooth distribution of frequencies; and the temperature must be so high that quantum effects do not get in the way. Then there is an approximatelky exponential dissipation described by the usual equation. There is not, however, a Langevin equation of the usual type.

1 The problem

In quantum mechanics an isolated system S is governed by a Hamilton operator H_S, and has a constant energy. In order to describe dissipation one has to add a second system acting as a heat bath B, with Hamilton operator H_B and an interaction H_I between both. The combined system has a total Hamiltonian

$$H_I = H_S + H_B + H_I \tag{1}$$

and can be described by a total density matrix $\rho_T(t)$, whose evolution is determined by

$$\dot{\rho}_T = -i[H_T, \rho_T] \tag{2}$$

One may write this shortly

$$\dot{\rho}_T = \mathcal{L}_T \rho_T, \tag{3}$$

where $\mathcal{L}_T$ is a superoperator, i.e., a linear map of an operator into another operator. This equation determines a mapping from $\rho_T(0)$ into $\rho_T(t)$, which constitutes a semigroup (even a group) with infinitesimal generator $\mathcal{L}_T$.

However we are interested in S alone; its density matrix is obtained by taking the trace with respect to the bath,

$$\rho_S(t) = \operatorname*{Tr}_B \rho_T(t). \tag{4}$$

There is no reason why this reduced density matrix should again be a semigroup obeying a differential equation. Yet in practice one always works with such an equation, calling it the Markov approximation. Our question is: How good is this approximation?

The usual derivation of this equation for ρ_S runs as follows [1]-[3]. Take an initial $\rho_T(0)$ that is a direct product of some initial $\rho_S(0)$ and the equilibrium density matrix for the bath:

$$\rho_T(0) = \rho_S(0) \otimes \rho_B^e. \tag{5}$$

This implies that no interaction H_I was operating before $t = 0$. From (2), (5), (4)

$$\rho_S(t) = \underset{B}{\mathrm{Tr}}\ e^{-itH_T}\rho_S(0)\rho_B^e e^{itH_T}, \tag{6}$$

which constitutes, for fixed t, a linear mapping of any initial $\rho_S(0)$ into the corresponding $\rho_S(t)$. The exponential embody the evolution of the combined system. They are not known explicitly, but for small time Δt they can be expanded to second order in the coupling. The linear map (6) then becomes

$$\rho_S(\Delta t) = \rho_S(0) + \Delta t \mathcal{L}_S \rho_S(0) + \mathcal{O}(\Delta t)^2. \tag{7}$$

So far everthing is above board, but now comes the crucial step: from (7) it is concluded that $\rho_S(t)$ obeys

$$\dot{\rho}_S = \mathcal{L}_S \rho_S(t) \tag{8}$$

The difficulty is manifest. The initial condition (5) is special in that it involves no correlations between system and bath, and the bath is in equilibrium. Under this condition (6) has been obtained, and therefore (7) can be asserted only to hold at $t = 0$. One may use (7) for computing $\rho_S(\Delta t)$, but not again for the next Δt. The net Δt does *not* start with an initial state of the form (5) since during the first Δt the interaction has created correlations between S and B. It is therefore necessary to *assume* that somehow these correlations do not affect the evolution of ρ_S and may be ignored; and to *assume* that the bath may be taken again in equilibrium, so that one may start the second interval Δt with the initial value

$$\rho_S(\Delta t) \otimes \rho_B^e \tag{9}$$

This randomness assumption has to be repeated after each Δt in order to assert (8).

The calculation outlined here is basically the same so that of Dirac and Pauli [4] (often referred to as 'the golden rule') but it is here applied to the density matrix rather than just its diagonal elements. There are other derivations, which give the same result but also have to rely, explicitly [5] or tacitly [6], on the repeated radomness assumption. Our puropose is to examine this assumption by working out some model systems for which (6) can be solved explicitly.

2 First model: free particle

As a solvable bath B take a set of independent harmonic oscillators labeled $n = 1, 2, 3, \cdots$, frequencies k_n, creators and annihilators $a_n^\dagger, a_n$. Let the system S be a free particle in one dimension, momentum P, and take a bilinear interaction [7]

$$H_T = \frac{P^2}{2M} + \sum_n k_n a_n^\dagger a_n + P \sum_n v_n(a_n + a_n^\dagger), \tag{10}$$

with coupling constant v_n. The bath equilibrium is

$$\rho_B^e = z^{-1} e^{-\beta \sum k_n a_n^\dagger a_n}, \tag{11}$$

$$z^{-1} = \prod_n \left(1 - e^{-\beta k_n}\right). \tag{12}$$

P commutes with H_T; hence the matrix elements of ρ_S in the P- representation decouple,

$$(P'|\rho_S(t)|P'') = (P'|\rho_S(0)|P'') \times \tag{13}$$
$$\times \mathrm{Tr} e^{-itH_T(P')} \rho_B^e e^{itH_T(p'')}.$$

This is worked out in the Appendix with the result

$$(P'|\rho_S(t)|P'') = (P'|\rho_S(0)|P'') \exp\left[-\frac{it}{2M}(P'^2 - p''^2)\right] \times \tag{14}$$
$$\times e^{-iF(t)(P'^2 - P''^2)} e^{-G(t)(P' - P'')^2},$$

where

$$F(t) = \sum_n \frac{v_n^2}{k_n}\left(t - \frac{\sin k_n t}{k_n}\right), \tag{15}$$

$$G(t) = \sum_n \frac{v_n^2}{k_n^2}(1 - \cos k_n t) \coth \frac{\beta k_n}{2}. \tag{16}$$

From (1) one finds on inspection the operator equation

$$\dot{\rho}_s(t) = -i\left[\left\{\frac{1}{2M} - \dot{F}(t)\right\} P^2, \rho_S(t)\right] - \tag{17}$$
$$+ \dot{G}(t)\left\{2P\rho_S(t)P - P^2\rho_S(t) - \rho_S(t) - \rho_S(t)p^2\right\}.$$

This is of the form (8) if and only if $\dot{F}$ and $\dot{G}$ are independent of t, and our task is to find the conditions for this to be true, at least approximately.

Clearly it is necessary that the frequencies k_n cover the semi-axis $0 < k < \infty$ densely and that the v_n vary smoothly. Define the strength function $g(k)$ by

$$g(k)\Delta k = \sum_{k < k_n < k + \Delta k} v_n^2. \tag{18}$$

The sums (15) and (16) may then be written as integrals:

$$F(t) = \int_0^\infty \frac{g(k)}{k} \left(t - \frac{\sin kt}{k} \right) dk \tag{19}$$

$$G(t) = \int_0^\infty \frac{g(k)}{k^2} (1 - \cos kt) \coth \frac{\beta k}{2} dk \tag{20}$$

Now suppose there exists a number k_0 such that for $k < k_0$

$$g(k) \approx \gamma^k \qquad (\gamma \text{ constant}, k < k_0) \tag{21}$$

Moreover one must have $\beta k_0 < 1$. Then one may argue [7] that F and G are proportional to t provided that

$$k_0 t < 1 \tag{22}$$

Thus there is a transient time $\vartheta \sim k_0^{-1}$, during which the effect of the special form of the initial value (5) survives. According to (21) the transient depends on the construction of the bath, but cannot be reduced to zero because the integrals (19), (20) diverge when $g(k) = \gamma k$ for all k. In addition there is the universal limitation $\vartheta > \hbar/kT$: the high frequencies in the bath are frozen by quantum mechanics. The influence of the bath *cannot* be represented by white noise for lack of high frequencies. This implies that Nyquist's formula for the spectral density of fluctuations [8, 9] looses its meaning at frequencies largen than $kT/\hbar$; the quantum correction factor with which it is often embellished is spurious.

Conclusion. The semigroup property is a valid approximation on a time scale on which ϑ may be neglected. The evolution can be represented as a Markov series with time steps $\Delta t > \vartheta$. In order that this difference equation can be replaced with a differential equation it is necessary that the change during each time step is small, which requires a weak interaction: $g\vartheta < 1$. These are the conditions for the existence of an approximate equation (8).

3 Second Model: Harmonic Oscillator

The preceding model is special in two ways. First H_I commutes with H_S, which simplifies the calculations. Secondly, H_S itself has a continous spectrum without discrete levels. We therefore now consider a second model, in which S is a harmonic oscillator with frequency Ω, which interacts with the same B as before. This model has been studied by numerous authors [10], but they were mainly concerned with obtaining the differential equation for the dissipation rather than with the limitations thereof.

Our model is defined by

$$H_T = \Omega a_0^\dagger a_0 + \sum_n k_n a_n^\dagger a_n + \sum_n v_n (a_n a_0^\dagger + a_n^\dagger a_0). \tag{23}$$

The coupling is again bilinear and is chosen of the form of a rotating wave interaction to simplify the algebra. The equations of motion of the Heisenberg operators are linear and may be solved, with the result [11].

$$a_0(t) = U(t)a_0(0) + \sum_m v_m(t)a_m(0) \tag{24}$$

$$a_n(t) = W_n(t)a_0(0) + \sum_m S_{nm}(t)a_m(0). \tag{25}$$

The time dependent coefficients may be written as complex integrals by means of the analytic function

$$G(z) = z - \Omega - \sum_n \frac{v_n^2}{z - k_n}, \tag{26}$$

whose zeros are the eigenvalues of the equations of motion. Then

$$U(t) = \frac{1}{2\pi i} \int_C \frac{e^{-izt}}{G(z)} dz, \qquad v_m(t) = \frac{1}{2\pi i} \int_C \frac{e^{-izt}}{(z - k_m)G(z)} dz, \tag{27}$$

where the contour C surrounds the positive axis. We do not need the coefficients in (25).

In principle this is the solution, but we ought to translate it into the density matrix. Unfortunately that requires quite some algebra and will not therefore be done here. We here merely look at the average energy $\Omega\langle a_0^\dagger a_0 \rangle = \Omega\langle N_0(t)\rangle$ of the spectrum, given that initialy the state had precisely M quanta:

$$\rho_T(0) = |M\rangle\langle M| \otimes \rho_B^e \tag{28}$$

One finds from (24)

$$\langle N_0(t)\rangle = |U(t)|^2 M + \sum_m |v_m(t)|^2 \left(e^{\beta k_m} - 1\right)^{-1}. \tag{29}$$

It remains to study the time dependence of $U(t)$ and $v_m(t)$.

In order to obtain the usual results we remember the lesson of section 2 and evaluate $U(t)$ and $v_m(t)$ for large t. First distort C into two parallel lines above and below the real axis. The lower line does not contribute when $t > 0$. The upper line is shifted down onto the real axis, so that use can be made of the identity

$$\sum_n \frac{v_n^2}{z - k_n} = \int_0^\infty \frac{g(k)dk}{x + i\varepsilon - k} = \fint_0^\infty \frac{g(k)dk}{x - k} - \pi i g(x). \tag{30}$$

Thus one has

$$U(t) = \frac{-1}{2\pi i} \int_{-\infty}^\infty \frac{e^{-ixt}dx}{x - \Omega - \fint_0^\infty \frac{g(k)dk}{x-k} + \pi i g(x)}, \tag{31}$$

which is still exact.

For small g it is clear that there is a resonance near Ω, which can be described by a pole below the real axis, roughly at

$$x = \Omega + \int_0^\infty \frac{g(k)dk}{\Omega - k} - \pi i g(\Omega)$$
$$= \Omega' + i\Gamma, \tag{32}$$

where Ω' is a Lamb shifted frequency and $\Gamma = \pi g(\Omega)$, the width of the resonance. If this pole provides the dominant contribution to (27) one has

$$U(t) = e^{-i\Omega't - \Gamma t} \tag{33}$$

$$v_m(t) = v_m \frac{e^{-\Omega't - \Gamma t} - e^{ik_m t}}{\Omega' - i\Gamma - k_m}. \tag{34}$$

The condition for this to hold is that g is small; and practically constant within the range (of order $1/t$) that contributes to the integral (31). This implies

$$g \ll \Omega, \qquad \Omega t \gg 1. \tag{35}$$

Under this condition we may use (33) and (34) as approximations. Substitution into (29) yields

$$\langle N_0(t) \rangle = M e^{-2\Gamma t} + \int_0^\infty \frac{g(k)dk}{e^{\beta k} - 1} \frac{e^{2\Gamma t} + 1 - 2e^{-\Gamma t}\cos(\Omega' - k)t}{(\Omega' - k)^2 + \Gamma^2} \tag{36}$$
$$= M e^{-2\Gamma t} + \frac{g(\Omega)}{e^{\beta \Omega'} - 1} \left\{ \frac{\pi}{\Gamma} \left(e^{2\Gamma t} + 1 \right) - \frac{2\pi}{\Gamma} e^{-2\Gamma t} \right\}$$
$$= M e^{-2\Gamma t} + \frac{1}{e^{\beta \Omega'} - 1} \left(1 - e^{2\Gamma t} \right).$$

For this calculation we assumed not only (35) but, in addition

$$\Gamma\beta \ll 1, \qquad \hbar\Gamma \ll kT. \tag{37}$$

Of course the line width has to be small compared to $1/\beta$ in order that the thermal equilibrium factor $(e^{\beta\Omega'} - 1)^{-1}$ makes sense. Equation (36) is the standard result: the energy of the system S dissipates exponentially until it is in equilibirum with the temperature of the heat bath. The damping coefficient 2Γ is the same as the transition probability that figures in the golden rule.
Conclusion. From (36) one sees

$$\partial_t \langle N_0(t) \rangle = -2\Gamma\{\langle N_0(t) \rangle - \langle N_0 \rangle^e\}, \tag{38}$$

which is the equation usually found with the aid of a repeated randomnees assumption. Here we have obtained it from honest dynamics, and it transpired that it is valid only when a transient time ϑ has elapsed since the initial condition (5). One therefore has again a discrete Markov series, which can be described by a differential equation only if $g\vartheta \ll 1$. In that case ϑ is short enough for the equation (38) to faithfully describe the damping. On the other hand, (35) tells us that $\vartheta \gg \Omega^{-1}$, so that there is *no* continuous time Markov process that reproduces the oscillations of (33). Hence *there exists no Langevin equations for the amplitude*, of the type

$$\partial_t a_0 = (-i\Omega' - \Gamma)a_0 + f(t) \tag{39}$$

with (nearly) whit noise $f(t)$, not even at high temperatures. This disappointing fact has been encountered before [12].

Appendix

In order to work out (13) first write (19) in the form

$$H_T = \frac{P^2}{2M} + \sum_n H_n(P) - P^2 \sum_n \frac{v_n^2}{k_n}, \tag{40}$$

$$H_n(P) = k_n(a_n^\dagger + u_n)(a_n + u_n), \tag{41}$$

where $u_n = Pv_n/k_n$. Substitution in (13) yields with the aid of (11)

$$(P'|\rho_S(t)|P'') = (P'|\rho_S(0)|P'') \exp\left[\frac{-it}{2M}(P'^2 - P''^2)\right]$$

$$\times \exp\left[it(P'^2 - P''^2)\sum_n \frac{v_n^2}{k_n}\right]$$

$$\times \prod_n (1 - e^{-\beta k_n})\mathrm{Tr}e^{-itH_n(P')}e^{-\beta k_n a_n^\dagger a_n}e^{itH_n(P'')}. \tag{42}$$

To compute this trace we temporarily omit the subscript n. The operator $U = e^{u(a-a^*)}$ has the property $UaU^{-1} = a + u$, so that

$$H(P') = U'ka^\dagger aU'^{-1}, \quad H(P'') = U''ka^\dagger aU''^{-1}, \tag{43}$$

where the primes refer to the parameters P', P'' hidden in U. Hence the trace in (42) becomes

$$\mathrm{Tr}U'e^{-itka^\dagger}e^{-u'(a-a^\dagger}e^{-\beta ka^\dagger}e^{u''(a-a^\dagger)}e^{itka^\dagger a}U''^{-1}. \tag{44}$$

Moreover one has the identities

$$e^{-itka^\dagger}ae^{itka^\dagger} = ae^{ikt}, e^{-itka^\dagger}a^\dagger e^{itka^\dagger a} = a^\dagger e^{-ikt}, \tag{45}$$

so that the trace may be written

$$\mathrm{Tr}U'e^{-u'(ae^{ikt}-a^\dagger e^{-ikt})}e^{-\beta ka^\dagger a}e^{-u''(ae^{ikt}-a^\dagger e^{-ikt})}U''^{-1}. \tag{46}$$

With the aid of Baker-Hausdorff the first two factors can be combined into a single exponential:

$$e^{u'\{a(1-e^{ikt})-a^\dagger(1-e^{ikt})\}}e^{-iu'^2\sin kt} \tag{47}$$

and we get for the trace

$$e^{-i(u'^2-u''^2)\sin kt}\mathrm{Tr}\left[e^{-\beta ka^\dagger a}\right.$$

$$\left.\times e^{u''\{a(e^{ikt}-1)-a^\dagger(e^{-ikt}-1)\}}e^{u'\{a(1-e^{ikt})-a^\dagger(1-e^{-ikt})\}}\right] \qquad (48)$$

In the first factor obtained here restore the subscript n and take the product over all n; it then combines with the second line of (42) to give

$$\exp\left[i(P'^2-P''^2)\sum_n \frac{v_n^2}{k_n}\left(t-\frac{\sin k_n t}{k_n}\right)\right] = e^{i(P'^2-P''^2=F(t)}. \qquad (49)$$

It remains to compute the trace in (48). The two exponentials combine into

$$e^{wa-w^*a^\dagger}, \qquad w = (u'-u'')(1-e^{ikt}). \qquad (50)$$

Subsequently

$$\begin{aligned}
\mathrm{Tr}\, e^{-\beta ka^\dagger a}e^{wa-w^*a^\dagger} &= e^{1/2ww^*}\mathrm{Tr}e^{-\beta ka^\dagger a}e^{wa}e^{-w^*a^\dagger}\\
&= e^{1/2ww^*}\sum_{N=0}^\infty e^{-\beta kN}\langle N|e^{wa}e^{-w^*a}|N\rangle\\
&= e^{1/2ww^*}\sum_{N=0}^\infty e^{-\beta kN}\sum_{m=0}^\infty \frac{(-ww^*)^m}{m!^2}\frac{(N+m)!}{N!}\\
&= e^{1/2ww^*}\sum_{N=0}^\infty e^{-\beta kN}(-1)^N\sum_{m=0}^\infty \frac{(-ww^*)^m}{m!}\begin{pmatrix}-m-1\\N\end{pmatrix}\\
&= e^{1/2ww^*}\sum_{N=0}^\infty \frac{(-ww^*)^m}{m!}\left(1-e^{-\beta k}\right)^{-m-1}\\
&= \left(1-e^{-\beta k}\right)^{-1}-\exp\left[ww^*\left(\tfrac{1}{2}-\frac{1}{1-e^{-\beta k}}\right)\right]\\
&= \left(1-e^{-\beta k}\right)^{-1}-\exp\left[-(u'-u'')^2\left(1-\cos kt\right)\coth\tfrac{\beta k}{2}\right].
\end{aligned}$$

Substitution in the third line of (42) produces the last factor on (14).

Note. There is *no* renormalization of the mass M, as stated erroneously in [7].

References

1. W.H. Louisell, "Quantum Statistical Properties of Radiation", Wiley, New York 1973, ch 6.
2. M. Sargent, M. Scully, W.E. Lamb Jr., "Laser Physics", Addison- Wesley, Reading, Mass. 1974, ch. 16.
3. F. Haake, in "Quantum Statistics and Solid-State Physics", Springer Tracts n. 66; Berlin 1973.
4. P.A.M. Dirac, Proc. Roy. Soc., **114**, 243 (1927); W. Pauli, in "Probleme der modernen Physik", Sommerfeld Festcschrift, Hirzel, Leipzig 1928.
5. P. und T. Ehrenfest, in "Encyklopädie der mathematischen Wissenschaften", **4**, 32 (Teubner, Leipzig 1912); N.G. van Kampen, Physica **20**, 603 (1954); R. Zwanzig, in "Lectures in Theoretical Physics, vol. III", Boulder, Colorado 1960 (Interscience, New York 1961).
6. L. van Hove, Physica, **21**, 517 (1955);**23**, 441 (1957); R. Kubo, J. Phys. Soc. Jap. **12**, 570 (1957); and others, e.g. references [3-9].

7. N.G. van Kampen, to be published in J. Stat. Phys.

8. H. Nyquist, Phys Rev., **32**, 10 (1928).

9. H.B. Callen and T.A. Welton; Phys. Rev., **83**, 34 (1951).

10. P. Hemmer, Thesis Trodheim (1959); P. Mazur and E. Braun, Physica, **30**, 1973 (1964); P. Ullersma, Physica **32** 27, 56, 74, 90 (1966).

11. N.G. van Kampen, "Stochastic Processes in Physics and Chemistry" (2nd. ed.; North-Holland, Masterdam 1992).

12. Ref.[2], p.316; F. Haake and R. Reibold, Phys. Rev., **A32**, 2462 (1985); N.G. van Kampen, Physica **A147**, 165 (1987); C.W. Gardiner, "Quantum Noise", Springer, Berlin 1991.

Brownian Motion Theory: Properties of Nonlinear Langevin Equations

P. Mazur

Instituut Lorentz. University of Leiden.
P.O. Box 9506, 2300RA Leiden, The Netherlands.

The instrumental role is discussed of causality and microscopic reversibility in determining the stochastic properties of the Langevin force in linear and nonlinear Langevin equations.

1 Introduction: Theory and theories of Brownian motion

A quantitative theory of Brownian motion was first given by Einstein [1] considering this phenomenon as a diffusive process. The foremost objective of this theory was to establish an expression for the mean square displacement of a Brownian particle. The formula found by Einstein and essentially also by Smoluchowski [2] (who obtained an expression of the same form but with a different numerical coefficient) made possible an experimental verification of the hypothesis underlying the theory of Brownian motion. This hypothesis, clearly formulated (as stated by Langevin [3]) by Gouy[4], sees in the continuous Brownian motion an echo of the thermal molecular agitation. The experimental verification was conclusively performed in the years 1908 and 1909 by J. Perrin who could then write [5] that "it becomes rather difficult to deny the objective reality of molecules (il deviant assez difficile de nier la réalite objective des molécules)". But already in 1908 Langevin [3] had shown, "that it is easy to give, by a completely different method, an infinitely simpler demonstration" [of Einstein's formula]. This derivation, in the short note in Comptes Rendus of less than three pages, is based, as almost all modern theory on Brownian motion, on the stochastic equation of motion for the Brownian particle now called the Langevin equation. This equation for the momentum $p(t)$ of the particle has the form

$$\frac{dp(t)}{dt} = -\beta p(t) + f(t), \tag{1}$$

where β is the friction coefficient. Langevin notes that the frictional force $-\beta p$ is in fact only an average and, due to collisions with molecules of the fluid in which the Brownian particle is suspended, oscillates around that value. For this reason he complements this hydrodynamic force with a 'complementary' random force (now commonly called Langevin force). The only property of the complementary force of interest to Langevin, for the purpose of his derivation, is that its mean value vanishes:

$$\langle f(t)\rangle_{p_0} = 0, \qquad t > 0 \tag{2}$$

The bracket $< \cdots >_{p_0}$ denotes an average over a subensemble of Brownian particles suspended in a liquid in equilibrium, all having the same initial momentum $p(t = 0) = p_0$.

Property (2) specifies only the first moment of $f(t)$. It was enough however for Langevin to derive in a simple way from eq. (1), by multiplying both members of this equation with the Brownian particle's coordinate $x(t)$, Einstein's expression for the mean square displacement which is an expression related to the second moment of $x(t)$.

Equation (1) with the specification of only the first moment of $f(t)$ should be considered as practically void. Indeed with (2) alone one cannot determine fully the stochastic process $p(t)$. For that purpose one customarily uses the specification of $f(t)$ postulated by Ornstein and Uhlenbeck, who implicitly assumes, in addition to property (2), that $f(t)$ is independent of the state $p(t)$. They assume moreover that the auto correlation function of $f(t)$ is given by

$$\langle f(t)f(t + \tau)\rangle_{p_0} = 2q\delta(\tau), \tag{3}$$

where q in a constant, and finally that

$$f(t) \qquad \text{is a gaussian process,} \tag{4}$$

whose first two moments are given by (2) and (3) respectively as well kwown the constant q in eq. (3) is determined [7] form the stationarity of the process $p(t)$ and is related to the friction constant β by the simple fluctuation-dissipation theorem

$$q = \beta < p^2 > \tag{5}$$

Here the bracket $< \cdots >$ denotes an average over an equilibrium ensemble of Brownian particles suspended in a liquid heat bath.

The brackets $< \cdots >_{p_0}$ and $< \cdots >$ are related as follows

$$< \cdots >_{p_0} = \frac{< \delta(p(0) - p_0) \cdots >}{< \delta(p(0) - p_0) >} = \frac{< \delta(p(0) - p_0) \cdots >}{P_0(p_0)} \tag{6}$$

where the equilibrium distribution $P_0(p)$ is given by the Maxwell distribution

$$P_0(p) = \langle \delta(p(0) - p)\rangle = (2\pi m k_B T)^{-1/2} \exp(p^2/2m k_B T) \tag{7}$$

Here m is the mass of the particle, k_B Boltzmann's constant and T the temperature.

The properties of the random force $f(t)$, independent of p, are frequently also specified as follows: $f(t)$ is a stationary gaussian process with mean value

$$\langle f(t)\rangle = 0 \tag{8}$$

and autocorrelation function

$$\langle f(t)f(t+\tau)\rangle = 2q\delta(\tau), \tag{9}$$

q being again determined to be given by the values (5). Note that conditions (2) and (3) automatically imply (8) and (9). Clearly the converse is, however, not necessarily true.

But if a causality condition were to hold in the sense that $p(t)$ or functions thereof at earlier times t cannot be correlated to the noise $f(t)$ (or products thereof) at later times, one would have in particular, using the definition (6),

$$\langle f(t)\rangle_{p_0} = \langle f(t)\rangle, \quad t > 0 \tag{10}$$

and

$$\langle f(t)f(t+\tau)\rangle_{p_0} = \langle f(t)f(t+\tau)\rangle, t > 0 \quad \tau > -t \tag{11}$$

so that (8) and (9) do imply (2) and (3) and conversely.

Since for the linear Langevin equation condition (8) must be trivially satisfied (the left hand side of eq. (1) vanishes in equilibrium as does the equilibrium average value of p), condition (2) represent a trivial and physically obvious form of the causality requirement. Such a requirement in its more general form stated above gives expression to the fact that it is the random force which generates the fluctuations of the momentum p.

In three previous papers [8],[9],[10] to be denoted hereafter as I, II and III respectively, Dick Bedeaux and I have shown that causality, stationarity and also microscopic reversibility are instrumental to determine the stochastic properties of the Langevin force, even when the systematic force in the Langevin equation is not linear. In Section 2 of this review, presented on the occasion of the 25th anniversary of the Sitges Conferences on non equilibrium statistical physics, we discuss the fact (c.f. I) that stationarity and causality alone completely determine the second moment of the random force in a (nonlinear) Langevin equation for a stationary random process $\alpha(t)$

$$\frac{d\alpha}{dt} = -B(\alpha) + f(t), \tag{12}$$

where $B(\alpha)$ has, without loss of generality, been defined in such a way that

$$\langle B(\alpha)\rangle = 0 \tag{13}$$

In Sec. 3 (cf. II) it is then shown that if one demands that $f(t)$ be independent of $\alpha(t)$ (a requirement henceforth to be indicated by a subindex zero), that if one demands that

$$f(t) = f_0(t), \tag{14}$$

it follows from causality that the noise is white (see also Sec. 2), with

$$\langle f_0(t)f_0(t+\tau)\rangle = 2\langle \alpha B(\alpha)\rangle\delta(\tau) \tag{15}$$

and that it is gaussian and white if and only if the function $B(\alpha)$ is related to the stationary (equilibrium) distribution function $P_0(\alpha)$ by

$$B(\alpha) = -Ld \ln P_0(\alpha)/d\alpha \tag{16}$$

where the constant L is given by

$$L = \langle \alpha B(\alpha) \rangle. \tag{17}$$

If furthermore microscopic reversibility holds, it is shown that eq.(16) must necessarily be satisfied if one also demands property (14).

But what if $B(\alpha)$ is not given by eq.(16)? And $f(t)$ can therefore be independent of $\alpha(t)$?. This case is discussed in Sec. 4 (c.f. III) assuming the random force to be of a multiplicative nature and to be of the form

$$f(t) = C(\alpha(t - \epsilon))f_0(t) \tag{18}$$

One can then prove that f_0 is again a gaussian white process and that the function $C(\alpha)$ is the solution of a differential equation which involves $P_0(\alpha)$. This equation can easily be solved when the function $P_0(\alpha)$ is gaussian.

2 The random force autocorrelation function

Consider now the Langevin equation in its original linear form (1) as applied to the problem of Brownian motion and let us only assume that the random force, Langevin's complementary force, has the property (2).

Let us then define the momentum autocorrelation function $R(\tau)$

$$R(\tau) \equiv \langle p(t)p(t + \tau) \rangle \tag{19}$$

Due to stationarity, invariance for translation in time, this function depends only on τ and has the property (replace t by $t - \tau$)

$$R(\tau) = R(-\tau) \tag{20}$$

Stationarity also implies that $[dg(t)/dt = \dot{g}(t)]$

$$\langle \dot{p}(t)\dot{p}(t + \tau) \rangle = -\ddot{R}(\tau) \tag{21}$$

$$\langle \dot{p}(t)p(t + \tau) \rangle = -\dot{R}(\tau) \tag{22}$$

Using the Langevin equation (1) as well as the last two relations, one may express the time correlation function (second moment) of the random force in terms of the correlation function $R(\tau)$ and its second time derivative

$$\langle f(t)f(t + \tau) \rangle = -\ddot{R}(\tau) + \beta^2 R(\tau) \tag{23}$$

Multiply on the other hand the Langevin equation (1) for $t = \tau$ by $p(0)$ and average. If we now apply condition (2), which, as we discussed in the introduction, can be viewed as a simple and obvious form of a causality requirement, one obtains the following relation

$$\dot{R}(\tau) = -\beta R(\tau), \quad \tau > 0 \tag{24}$$

With the stationarity property (20) this relation becomes for positive as well as negative τ

$$\dot{R}(\tau) = -S(\tau)\beta R(\tau) \tag{25}$$

where $S(\tau)$ is defined as

$$S(\tau) = \begin{cases} -1 \text{ for } \tau < 0 \\ +1 \text{ for } \tau > 0 \end{cases} \tag{26}$$

Differentiating eq.(25) with respect to τ one obtains

$$\ddot{R}(\tau) = \beta^2 R(\tau) - 2\beta < p^2 > \delta(\tau) \tag{27}$$

since $R(0) = < p^2 >$.

If one then substitutes this equation into eq.(23) one finds that the noise $f(t)$ is white

$$\langle f(t)f(t+\tau)\rangle = 2\beta < p^2 > \delta(\tau), \tag{28}$$

and this result for the second moment of the random force $f(t)$ follows therefore from the obvious condition (and causality requirement) (2), which is a condition for the conditional first moment of $f(t)$ and from stationarity alone.

If instead of eq.(1) one has to deal with a non linear Langevin equation of the form (12) one may show in the same simple way, as above, that condition (2) (causality) and stationarity then lead to

$$\langle f(t)f(t+\tau)\rangle = 2\langle \alpha B(\alpha)\rangle\delta(\tau) \tag{29}$$

For the proof of this result, which represents at the same time a non linear fluctuation dissipation theorem, we refer to paper I, where it is also shown that the result (29) can easily be generalized to a many variable case.

3 When and why is the random force a gaussian process?

We saw in the previous section that stationarity and causality completely determine the second moment of the random force in a Langevin equation of the form (12). Do these conditions, in particular the requirement of causality in the sense stated in Sec. 2, also determine higher moments of $f(t)$, or properties thereof?

To address this question we shall now explicitly demand that $f(t)$ does not depend on the state $\alpha(t)$ and is therefore of the form (14).

Causality in the sense that $p(t)$ or products of functions thereof at earlier times cannot be correlated to products of the noise at later times, then leads for the cumulants of the noise to the following property,

$$\langle f_0(t_1) f_0(t_2) \cdots f_0(t_n) \rangle_c = \gamma_n \delta(t_1 - t_n) \delta(t_2 - t_n) \cdots \delta(t_{n-1} - t_n) \qquad (30)$$

where cumulants, denoted as $< \cdots >_c$ are related to moments in the usual way [7].

Property (30) for $n > 2$ is a generalization of eq.(29) which is that property for $n = 2$. For the proof of eq.(30) we refer to paper II,. where it is also shown that the coefficients γ_n of the δ- correlated cumulants can be determined from stationarity, using also causality, and obey the following equation

$$n \langle \alpha^{n-1} B(\alpha) \rangle = \sum_{m=2}^{n} \frac{n!}{m!(n-m)!} \langle \alpha^{n-m} \rangle \gamma_m \qquad (31)$$

For $n = 2$, one recovers here the coefficient of the δ-function in eq.(29) for the second moment of $f(t)$. The general form of γ_n in terms of equilibrium averages can be obtained from eq.(31) by inversion and is

$$\gamma_n = n \langle \alpha^{n-1} B(\alpha) \rangle_c \qquad (32)$$

Suppose now that $B(\alpha)$ is of the form

$$B(\alpha) = -L \frac{d \ln P_0(\alpha)}{d\alpha}, \qquad (33)$$

with L, a constant independent of α. We then find by substitution of eq.(33) into eq. (31) and partial integration, that

$$\gamma_n = 2L\delta_{n_2} \qquad (34)$$

It follows therefore from causality that all cumulants of $f(t)$ except the second vanish, and that $f_0(t)$ is a gaussian process if $B(\alpha)$ is related to the equilibrium distribution by eq.(33).

On the other hand, if $f_0(t)$ is a gaussian process, and eq.(34) holds then $B(\alpha)$ must be, according to eq.(31), such that

$$\langle \alpha^{n-1} B(\alpha) \rangle = L \left\langle \frac{d}{d\alpha} \alpha^{n-1} \right\rangle \qquad (35)$$

Since

$$\langle g(\alpha) \rangle = \int d\alpha P_0(\alpha) g(\alpha), \qquad (36)$$

equation (35) can only be satisfied for all n if $B(\alpha)$ is given by eq.(33).

In conclusion it follows therefore from eq.(31), and ultimately from causality and stationarity, that $f_0(t)$ is a gaussian process if and only if $B(\alpha)$ is of the form (33).

According to the principles of non-equilibrium thermodynamics [11] the quantity $X(\alpha)$

$$X(\alpha) = \frac{\partial S(\alpha)}{\partial \alpha} = \frac{\partial \ln P_0(\alpha)}{\partial \alpha}, \qquad (37)$$

is the thermodynamic force, conjugate to the variable α with $S(\alpha)$ the entropy of the system as a function of α.

The result obtained above from causality then implies that the Brownian motion-like Langevin equation

$$\frac{\partial \alpha}{\partial t} = LX(\alpha) + f_0(t), \tag{38}$$

with L a constant Onsager coefficient and $f_0(t)$ a white gaussian process, consistently describes the fluctuations accompanying the laws of "linear thermodynamics" of irreversible processes. These laws, which are linear in the thermodynamic forces $X(\alpha)$ need not be linear in α, if $X(\alpha)$ is a non-linear function of α, or equivalently $P_0(\alpha)$ not a gaussian. But writing eq. (37) for $\alpha = p$, the momentum of a Brownian particle and with $P_0(p)$ the Maxwell distribution function, eq.(38) reduces to the linear Langevin equation with gaussian white noise.

It would now seem, again as a consequence of the theorem derived, namely that $f_0(t)$ is gaussian if and only if eq.(33) holds, that for an equation of the form (38), but with an α-dependent coefficient $L(\alpha)$, fluctuations may still be "added" through an α-independent random force $f_0(t)$, as long as this force is no longer a gaussian process. A further restriction however occurs for variables α physically of interest for which the property of microscopic reversibility is valid. This restriction makes such a procedure inconsistent.

To show that such is the case we derive the master equation obeyed by the conditional distribution function $P(\alpha_0|\alpha, t)$

$$P(\alpha_0|\alpha, t) = \langle \delta(\alpha(t) - \alpha) \rangle_{\alpha_0}, \tag{39}$$

where $\alpha(t)$ obeys the Langevin equation (12) with $f(t) = f_0(t)$.

This master equation has the form (c.f. II)

$$\frac{\partial P(\alpha_0|\alpha, t)}{\partial t} = \int d\alpha' \alpha' \left[T(\alpha|\alpha')P(\alpha_0|\alpha', t) - T(\alpha'|\alpha)P(\alpha_0|\alpha, t) \right], \tag{40}$$

where the transition probability $T(\alpha|\alpha')$ is given by

$$T(\alpha|\alpha') = \left(\frac{\partial}{\partial \alpha} B(\alpha) + \sum_{n=2}^{\infty} \frac{1}{n!}(-1)^n \gamma_n \frac{\partial^n}{\partial \alpha^n} \right) \delta(\alpha - \alpha'), \tag{41}$$

which is completely determined in terms of $B(\alpha)$ and $P_0(\alpha)$ (c.f. eq.(32)).

Now as a consequence of microscopic time reversal invariance $T(\alpha|\alpha')$ satisfies the property of microscopic reversibility[11]

$$T(\alpha'|\alpha)P_0(\alpha) = T(\tau\alpha|\tau\alpha')P_0(\alpha'), \tag{42}$$

if α is an even variable with $\tau = 1$, or if α is odd with $\tau = -1$.

Using eqs.(41) and (42) it then follows that

$$\gamma_n = 0 \quad \text{for} \quad n \geq 3, \tag{43}$$

so that according to the theorem found above $B(\alpha)$ must necessarily be of the form (33) if one requires $f(t)$ in equation (12) to be independent of α, and causality to hold.

Consequently in a Brownian motion-like Langevin equation of the form

$$\frac{d\alpha}{dt} = L(\alpha)\frac{d}{d\alpha} \ln P_0(\alpha) + f(t) \tag{44}$$

the random force must be of the form

$$f(t) = f(\alpha(t), t) \tag{45}$$

This will have to be the case e.g. if one wishes to discuss Brownian motion of a particle with a momentum dependent friction coefficient. In the next section we discuss the stochastic properties of the random force in such a case.

4 The genuinely non-linear case

Consider then a stationary process $\alpha(t)$, which is either even or odd under time reversal and obeys the stochastic differential equation (12), where $B(\alpha)$ which satisfies (13) is now an arbitrary non-linear function not necessarily of the form (33). As stated in Sec. 3, the random force can then no longer be taken independent of the state $\alpha(t)$.

We shall assume $f(t)$ to be of multiplicative character

$$f(t) = C(\alpha(t - \epsilon))f_0(t) \tag{46}$$

with $C(\alpha)$ a factor multiplied by a random function $f_0(t)$ which is independent of $\alpha(t)$, is causal and has zero mean value. Due to causality the average of $f(t)$, (46) vanishes as it should since we demand $B(\alpha)$ to satisfy condition (13). Note also that the form (46) of $f(t)$ corresponds to the prescription given by Itô[12] for the interpretation of the product $C(\alpha(t))f_0(t)$.

Explicitly the Langevin equation considered here has the form

$$\frac{d\alpha}{dt} = -B(\alpha) + C(\alpha(t - \epsilon))f_0(t) \tag{47}$$

We refer to a Langevin equation of this form as genuinely non linear because it contains a random function $f_0(t)$ which is multiplied by a function $C(\alpha)$.

Can one determine this function and also the stochastic properties of $f_0(t)$?. As a consequence of causality the cumulants of $f_0(t)$ have as before the property (30) (c.f. II), but the coefficients γ_n now satisfy the following equation

$$n\left\langle \alpha^{n-1} B(\alpha) \right\rangle = \sum_{m=2}^{\infty} \frac{n!}{n!(n-m)!} \left\langle \alpha^{n-m} C^m(\alpha) \right\rangle \gamma_m \tag{48}$$

This equation can be found using stationarity, causality and the Langevin equation (47) instead of (12) with (14).

If $B(\alpha)$ is of the form

$$B(\alpha) = -LC^2(\alpha)\frac{d\ln\{P_0(\alpha)C^2(\alpha)\}}{d\alpha} \qquad (49)$$

With L constant, we find by substitution into eq.(48) that

$$\gamma_n = 2L\delta_{n_2} \qquad (50)$$

and that $f_0(t)$ is a gaussian white process.

Conversely it follows (see the analogous derivation for the case $C = 1$, in Sec. 2) that $B(\alpha)$ must be of the form (49) if the process $f_0(t)$ is gaussian, so that eq.(49) holds.

But, as for the case $C = 1$, (c.f. Sec. 2) microscopic reversibility and causality lead again to the fact that $f_0(t)$ is a gaussian white process (e.q.(50)) so that equation (49) must hold. This equation, however, is a differential equation for $C(\alpha)$ which may be solved if the function $B(\alpha)$ and the equilibrium distribution functions are known. An alternative derivation of this result is given in paper III.

To summarize:

For a stationary random process $\alpha(t)$, either even or odd under time reversal and obeying the Langevin equation (47), it follows, using causality and microscopic reversibility, that $f_0(t)$ is a gaussian white process with

$$\langle f_0(t)\rangle = 0 \qquad \langle f_0(t)f_0(t+\tau)\rangle = 2L\delta(t-t') \qquad (51)$$

and that $C(\alpha)$ is a solution of the following differential equation,

$$B(\alpha) = -LC^2(\alpha)\frac{d\ln P_0(\alpha)}{d\alpha} - L\frac{dC^2(\alpha)}{d\alpha} \qquad (52)$$

A few remarks are in order:

1. If one takes $C(\alpha) = 1$ in eq.(47), or in other words if one insists that the noise $f_0(t)$ be purely additive then we recover from eq.(52) the result found in Sec. 3 namely that $B(\alpha)$ must be of the form (33).
2. If $P_0(\alpha)$ is gaussian, the differential equation can easily be solved.
   Take e.g. the case of a Brownian particle, $\alpha \equiv p$, with a momentum dependent friction coefficient $\beta(p) = \beta(-p)$. Then the function $B(p)$ is given by

$$B(p) = \beta(p)p, \qquad (53)$$

while the equilibrium distribution $P_0(t)$ is the Maxwell distribution function (7). With (53) and (7), eq.(52) for $\alpha = p$ has the solution

$$LC^2(p) = mk_BT \sum_{n=0}^{\infty} d^n\beta(p)/d[p^2/2mk_BT]^n \qquad (54)$$

The quantities L and C^2 are thus determined up to a constant multiplicative factor. This indeterminacy is removed if one chooses:

$$L = mk_BT\beta(0) \tag{55}$$

so that one has $C = 1$ if the friction coefficient is independent of p.

3. Suppose finally that $B(\alpha)$ is linear

$$B(\alpha) = \gamma\alpha \tag{56}$$

but $P_0(t)$ nongaussian. This is the case e.g. in decay processes. Then according to eq.(52) $C(\alpha)$ is not a constant independent of α. On the other hand, as a consequence of the causality requirement one obtains, taking the conditional average of both members of eq.(47) with (56),

$$\frac{d\langle\alpha\rangle_{\alpha_0}}{dt} = -\gamma\langle\alpha\rangle_{\alpha_0}, \tag{57}$$

This is a strictly linear relation corresponding to the phenomenological law of the process studied. One has therefore the seemingly paradoxical situation that for the genuinely non-linear case, genuinely non-linear because the stochastic differential equation has a multiplicative noise term, the equation obeyed by the average of the process considered may nevertheless be linear. We have seen that, at the same time, if $C(\alpha) = 1$, that is when the noise is purely additive and one finds oneself in the so-called linear, or quasilinear regime, the stochastic differential equation is indeed linear in the thermodynamic force conjugate to α but may be highly non-linear in α if the equilibrium distribution function $P_0(\alpha)$, is not gaussian. These fews remarks, with which we conclude this discussion of the stochastic properties of the noise in (nonlinear) Langevin equations, may undercore the relative value of the qualifications linear and nonlinear if not supplemented by more specific characterizations.

References

1. A. Einstein, Ann. Phys., **17**, 549 (1905); **19**, 371 (1906).
2. M. v. Smoluchowski, Ann. Phys., **21**, 766 (1906).
3. P. Langevin; Compt. Rend., **146**, 530 (1908).
4. T. Gouy; Compt. Rend., **109**, 102 (1889).
5. J. Perrin; J. Ann. Chim. Phys., **18**, 5 (1909).
6. G.E. Uhlenbeck and L.S. Ornstein; Phys. Rev., **34**, 823 (1930).
7. See e.g. N.G. v. Kampen, "Stochastic Processes in Physics and Chemistry", North Holland, Amsterdam, 1992.
8. P. Mazur; Phys. Rev., **A15**, 8957 (1992). (I)
9. P. Mazur and D. Bedeaux; Physica **A173**, 155 (1991). (II) See also erratum Physica **A188**, 693 (1992).
10. P. Mazur and D. Bedeaux; Langevin **8**, 2947 (1992). (III)
11. See e.g. S.R. de Groot and P. Mazur, "Non Equilibrium Thermodynamics", Dover, New York, 1983.
12. K. Itô; Proc. Imp. Acad. (Tokyo) **20**, 519 (1944); Mem. Am. Math. Soc. **4**, 51 (1951).

Non-Equilibrium, Phase Transition, and Related Problem

Kyozi Kawasaki

Department of Physics. Kyushu University 33. Fukuoka 812, Japan.

1 Introduction

On this occasion of commemorating the 25th year since the foundation of the
Sitges Conference, I would like to start my lecture by briefly reviewing developments of non-equilibrium statistical mechanics emphasizing the more recent
history. For this purpose I found it convenient to divide the entire period into
the following three periods which undoubtedly reflect my own personal view and
a universal acceptance is not claimed:

1. the ancient period which ends at 1965
2. the recent period, 1965-1980
3. the modern period, 1980-

As we have learned from text books the original motivation of statistical mechanics was to understand the macroscopic behavior of a large system in terms
of that of atoms and molecules which constitute the system. Since macroscopic
description requires only a small limited number of degrees of freedom whereas
a system possesses a great number of microscopic degrees of freedom, this in-
evitably involves elimination of a large number of microscopic degrees of freedom.
A text book example is the well established derivation of thermodynamics via
equilibrium statistical mechanics where we finally end up with describing the
system in terms of only a few independent thermodynamic variables. The non-
equilibrium case is similar though much less clear cut. In both cases the existence
of macroscopic descriptions, i.e., thermodynamics and macroscopic deterministic equations of motion like hydrodynamics, depends crucially on the enormous
separations of spatial as well as temporal macroscopic and microscopic scales,
which is possible only for systems with huge numbers of microscopic degrees of
freedom. Here we are helped by the law of large numbers which permits us to
neglect fluctuating deviations from macroscopic behavior [1].

The wide separation of scales can be most simply explained by considering
the motion of a large Brownian particle immersed in fluid. As in well-known,
this is described by the following Langevin equation for the velocity $v(t)$ of the
particle:

$$M\frac{d}{dt}v(t) = -\zeta v(t) + f(t), \tag{1}$$

where M is the particle mass, ζ is the friction constant and $f(t)$ is the random force exerted by the surrounding fluid molecules which has the following properties:

$$< f(t) > = 0,$$
$$< f(t)f(t') > = 2k_B T\zeta\delta(t - t'), \tag{2}$$

where the angular bracket is the equilibrium average and $k_B T$ is the Boltzmann constant times the absolute temperature. The macroscopic law simply follows from (1) by dropping $f(t)$ where we write $v^*(t)$ instead of $v(t)$,

$$M\frac{d}{dt}v^*(t) = -\zeta v^*(t). \tag{3}$$

Then we readily find

$$v^*(t) = v^*(0)e^{-\gamma t} \tag{4}$$

$$< [v(t) - v^*(t)]^2 > = \frac{k_B T}{M}(1 - e^{-2\gamma t}) \tag{5}$$

with

$$\gamma \equiv \frac{\zeta}{M}. \tag{6}$$

Assuming the Stokes law $\zeta = 6\pi\eta R$ with η the shear viscosity and R the radius of the Brownian particle, we find $\gamma \propto M^{-2/3}$

We are now ready to discuss validity of the macroscopic law. First we note that the Langevin equation (1) itself is the limiting equation in which the microscopic time scale τ_m that characterizes the rapid molecular motion of the surrounding fluid as revealed through $f(t)$ is infinitesimally small as composed with the macroscopic time scale $\tau_M = \gamma^{-1}$. Otherwise, the delta function on the rhs of (2) will be modified, for example, by a smooth function like

$$\delta(t - t') \rightarrow \frac{1}{2\tau_m}e^{-|t-t'|/\tau_m}, \tag{7}$$

and the friction term in (1) is also modified. Next we note from (5) that the fluctuating deviation from the macroscopic velocity $v^*(t)$ is of the order of the thermal velocity of the Brownian particle $v_T = (k_B t/M)^{1/2}$. Thus the macroscopic law (3) is found to be valid by making the Brownian particle sufficiently large where

$$\frac{\tau_m}{\tau_M} \propto M^{-2/3}, \qquad \frac{v_T}{|v^*(t)|} \propto M^{-1/2}. \tag{8}$$

Finally, we find by integrating (2), we find the important expression:

$$\zeta = \frac{1}{k_B T} \int_0^\infty < f(t)f(0) > dt, \tag{9}$$

which is an example of transport coefficient expressed as an integral of equilibrium time correlation function of random force.

The expression like (9) was extensively discussed starting around the mid fifties and culminated in the very elegant and comprehensive linear response theory of transport phenomena where expression of the type (9) have been found for all the transport coefficients. For instance, the shear viscosity η and the thermal conductivity λ of one-component classical fluid are

$$\eta = \frac{1}{k_B T V} \int \int dr dr' \int_0^\infty dt < j^{xy}(\mathbf{r}, t) j^{xy}(\mathbf{r}', 0) >, \tag{10}$$

$$\lambda = \frac{1}{k_B T^2 V} \int \int dr dr' \int_0^\infty dt < j_T^x(\mathbf{r}, t) j_T^x(\mathbf{r}', 0) >, \tag{11}$$

where V is the system volume and $j^{xy}(\mathbf{r}, t)$ and $j_T^x(\mathbf{r}, t)$ are the microscopic expressions for the xy component of the local stress tensor and the x component of the local heat fluxes, respectively. The expressions like (9) - (11) are important in the sense that they give formally exact connections between transport coefficients which enter macroscopic description and thermal fluctuations which deal with microscopic dynamics. These developments are concisely reviewed by Zwanzig in 1965 [2], the year I have chosen to be the final year of the ancient period.

It is important at this point to note that despite its general and formally exact character of the linear response theory, this does not mean that we have succeeded in deriving macroscopic laws microscopically. For this the integral expressions in (9) - (11) must converge to finite values. This in term requires that thermal fluctuations entering in the integrands must have sufficiently short spatio-temporal correlations, the quality required for true random forces. Since these time correlation functions cannot be analytically evaluated in general, the random properties of thermal fluctuations when viewed on the macroscopic scale were explicitly or implicitly assumed. In other words, clear cut separation of macroscopic and microscopic scales were assumed at the outset.

2 Recent Period

The recent period that started around 1965 begins with our realization that the basic assumption just mentioned may not be so obvious. As a matter of fact, Landau and Lifshitz treats in their text book on Fluid Mechanics (1953) [3] the motion of a sphere moving in an incompressible viscous fluid of the density ρ and obtain instead of (3) the following result:

$$\left[M + (\frac{2\pi}{3}\rho R^3)\right] \frac{d}{dt}v^*(t) = -\zeta v^*(t) - 6\pi R^2(\rho\nu/\pi)^{1/2} \int_{-\infty}^{t} \frac{ds}{(t-s)^{1/2}} \frac{d}{ds}v^*(s),$$

$$(12)$$

if we interpret this as the deterministic part of the corresponding generalized Langevin equation where $\nu = \eta/\rho$. This implies that the thermal noise term that should be added to the *rhs* of (12) in the generalized Langevin equation is no larger delta-correlated but has a component which decays slowly in time (long time tail). This can be expressed by making the friction constant ζ frequency dependent:

$$\zeta \rightarrow \zeta(\omega) = \zeta + 3\pi R^2(2\nu|\omega|)^{1/2}(i - \frac{\omega}{|\omega|}). \tag{13}$$

In this case this long-time tail arises from macroscopic fluid flow surrounding the moving sphere (large Brownian particle), which was ignored in the original Langevin equation (1), and hence more or less was expected. Therefore, one may argue that such anomalies should be absent in the transport phenomena of fluid and the correlation functions entering (10) and (11) will be short-ranged in time since there appear to be no such obvious fluid flow effects which might affect this expected short range character.

The first indication that the transport phenomena of fluids are not so simple came from realization in the mid sixties that the density expansion of transport coefficients of gas does not exist since collision processes involving only finite numbers of particles can give rise to divergent contributions to transport coefficients [4]. This forced us to take into account even in gases collisions involving indefinitely large numbers of particles from the beginning. It was also realized that the long time contributions to transport coefficients originating from such cooperative processes lead to various anomalies in transport coefficients such as divergences in lower spatial dimensionalities or non-analytic frequency dependences. The implication of this development is that the clear cut separation of spatio-temporal scales which formed the basis of microscopic derivations of macroscopic laws is now in doubt. In fact, the microscopic basis for such separation is the orthogonality of supposedly random flux entering expressions like (10) and (11) denoted typically as j and macroscopic variables typically denoted as A [5]:

$$< jA >= 0. \tag{14}$$

Therefore the flux j does not contain components slowly varying as A. However, this does not of course exclude a possibility that j contains components slowly varying as some nonlinear functions of A[6]. Such components of j produce large corrections to transport coefficients often called the renormalization of transport coefficients.

As a simple illustration we return to the Brownian particle discussed at the beginning, but here we shall not be concerned with the flow of surrounding fluid but with the effects of an added spatially periodic force field $F(x)$ with the zero mean value where x denotes the positions of the center of the Brownian particle. For simplicity we consider the over-damped case of very large friction such that the inertial term can be neglected. The Langevin equation is then, writing $\dot{x}(t)$ for $v(t)$ and ζ_0 for ζ,

$$\zeta_0 \, \dot{x}(t) = F(x(t)) + E + f(t), \tag{15}$$

with (2), where a constant external field E was added. In the absence of the periodic potential, we find by averaging (15) that the average particle velocity v is simply given by $v = m_0 E$ with the following unrenormalized mobility m_0 :

$$m_0 \equiv \zeta_0^{-1}. \tag{16}$$

When we switch on the periodic force, the Brownian particle performs random jumps from one potential minimum to the neighboring ones instead of unhindered random walk. However on length scales much longer than the period of $F(x)$ and on time scales such longer than the average residence time in a particular minimum, the particle velocity averaged over many jumps still is a constant proportional to E with the renormalized mobility m reduced from m_0, (16) because of the increased friction arising from necessity for the particle to overcome periodic potential barriers. In order to obtain the renormalized mobility, we follow Risken's treatment [7] and consider the Fokker-Planck equation for the distribution function $P(x,t)$,

$$\frac{\partial}{\partial t} P(x,t) = \zeta_0^{-1} \frac{\partial}{\partial x} \left[k_B T \frac{\partial}{\partial x} - F(x) - E \right] P(x,t). \tag{17}$$

This equation permits a stationary solutions $P_s(x)$ with a constant probability current J given by

$$J = \zeta_0^{-1} \left[F(x) + E - k_B T \frac{\partial}{\partial x} \right] P_s(x). \tag{18}$$

The physically meaningful solutions $P_s(x)$ which is finite at $x \to \pm\infty$ is shown to be periodic in x and is given by

$$P_s(x) = \frac{\zeta_0 J}{L_B t} e^{-U(x)/k_B T} \left\{ \frac{L_+}{1 - e^{-El/k_B T}} - \int_0^x dx' e^{U(x')/k_B T} \right\}, \tag{19}$$

with

$$L_\pm \equiv \int_0^l dx e^{\pm U(x)/k_B T} \tag{20}$$

and

$$U(x) = - \int^x F(x') dx - Ex, \tag{21}$$

l being the period of $F(x)$. The mean particle velocity v is then obtained from

$$v = Jl \int_0^l dx P_s(x) = \frac{k_B Tl}{\zeta_0} \left\{ (1 - e^{-El/k_B T})^{-1} L_+ L_- \right.$$
$$\left. - \int_0^l dx \int_0^x dx' \exp\left[(U(x') - U(x))/k_B T\right]^{-1} \right]. \tag{22}$$

For the mobility we only need the term on the *rhs* linear in E, the coefficient of which is the renormalized mobility m. Therefore we find for $\zeta = m^{-1}$,

$$\zeta = \frac{\zeta_0 L_+^0 L_-^0}{l^2}, \tag{23}$$

where $L_\pm^0$ is $L_\pm$ with $E = 0$. By the use of Schwarz's inequality we find $\zeta/\zeta_0 > 1$, the equality applying only for vanishing $F(x)$. In particular, for a sinusoidal force,

$$F(x) = \frac{2\pi}{l} U_0 \sin \frac{2\pi x}{l} \tag{24}$$

we obtain

$$L_\pm = l I_0 \left(\frac{U_0}{k_B T} \right), \tag{25}$$

I_0 being the modified Bessel function. Thus

$$\zeta = \zeta_0 I_0 \left(\frac{U_0}{k_B T} \right)^2. \tag{26}$$

For $|U_0| \gg k_B T$, ζ behaves as

$$\zeta \simeq \frac{\zeta_0}{2\pi} \left| \frac{k_B T}{U_0} \right| \exp \left(\frac{2U_0}{k_B T} \right). \tag{27}$$

This result shows the characteristic temperature dependence of an activation processes of the energy barrier $2U_0$.

We have discussed this case in some detail as a simple example where renormalization effects arising from nonlinear reversible force (mode coupling) can be studied exactly.

The most spectacular example of non-equilibrium phenomena where this kind of renormalization effects appears is found in the dynamical critical phenomena where the cooperativity is further enhanced by long-range spatial correlation of the order parameter fluctuations [8]. As a typical case we take up the thermal conductivity near the liquid-gas critical point of a single- component fluid, where we start from the expression (11). Away from criticality, the local heat flux $j_T^x(\mathbf{r}, t)$ is dominated by microscopic processes where kinetic energies of individual molecules and local interaction energies are transported over microscopic distances whose spatio-temporal scales are well separated from any macroscopic scales. Near criticality, however, the picture is radically different. First, large scale density fluctuations develop whose length scale ξ diverges as $|T - T_c|^{-\nu}$ if the average density is fixed at its critical value where T_c is the critical temperature and $\nu \approx 0.63$ the critical exponent. This length easily reaches thousands of Angstroms and can interface with macroscopic processes. Here $j_x(\mathbf{r}, t)$ is dominated by transport of such large fluctuating regions by instantaneous local fluctuating velocity field $v_x(\mathbf{r}, t)$, the point first realized by Fixman in other context [9]. Thus the dominant part of $j_x(\mathbf{r}, t)$ is

$$j_x(\mathbf{r}, t) = \rho T \delta s(\mathbf{r}, t) v_x(\mathbf{r}, t) \tag{28}$$

where ρ is the average velocity and $\delta s(\mathbf{r}, t)$ is the fluctuating part of the specific entropy. Here we note that the density fluctuation can be decomposed into δs and the pressure fluctuation and the latter propagates away rapidly as a sound wave and hence can be neglected. It turns out that δs changes very downy as compared to v_x due to critical slowing down and hence the time dependence of $\delta s(\mathbf{r}, t)$ can be neglected. Assuming v_x to obey the Navier-Stokes equation for incompressible fluid we find

$$\int_0^\infty v_x(\mathbf{r}, t) dt = \rho \int \mathbf{T}_{xx}(\mathbf{r} - \mathbf{r}') v_x(\mathbf{r}', 0) d\mathbf{r}', \tag{29}$$

with $\mathbf{T}(\mathbf{r})$ the Oseen tensor. Thus λ, (11), is expressed totally in terms of static correlations of δs and v_x which are assumed to be statistically independent as well the shear viscosity being assumed to be finite. Using the Ornstein-Zernike form for the static pair correlation of δs and the equipartition for that of v_x we finally find that λ diverges as ξ at the critically. In addition, λ shows marked nonlocality for the distance of the order of ξ signifying breakdown of Fourier's law of heat conduction[6].

Major advances in critical dynamics exemplified above as well as in equilibrium critical phenomena have already taken place by early seventies. The field has now matured into an exact science where research frontier is pushed to ever higher precision and minor details such as effects of gravity. The reason for this is that once the principle is understood, the universality of the phenomena permits us to make theoretical predictions as precise as we wish which is only limited by technical difficulties.

The most important consequence of the success in critical phenomena is that we are liberated from the central dogma of statistical mechanics which had forced us to seek a microscopic starting point. We have now learned that often we need not and in fact should not start from a microscopic model if we do not want to lose sight of the essential core of the problem. The great technical advances [10] that converted critical phenomena into an exact science have made this change of paradigm both convincing and respectable. This realization, I believe, provided the setting that ushered in the modern period.

3 Modern Period

The keyword that characterizes the modern period is "diversification". We can say that with regard to the types of problems being investigated, the methodologies employed to study them and the objectives of investigations. Here the original objective of statistical mechanics, that is, the microscopic understanding of macroscopic phenomena, can be found only in some corners. One can mention chaos and turbulence, growth kinetics like DLA and dendrite growth, phase ordering dynamics, complex fluids including foams and colloids, granular systems, fracture, tribology, complex systems including glasses, neural networks, biological problems, etc. As for the methodologies employed, in addition to theory and experiment, the ever increasing role played by computers is evident. Nowadays it is getting harder and harder to find common goals for those engaged in different sub-fields. For instance, the goals of mathematical physicists quite often appear to be closer to those of mathematicians than to those of physicists. The aims of materials science often overlap with these of engineers. Some physicists deal with problem outside physics, even outside natural sciences.

One can think of both positive and negative aspects about the trends that characterize the modern period. Among the positive aspects, we can count tremendous expansion of research frontiers of statistical mechanics and contacts with other disciplines and cultural activities such as arts. One can mention as negative aspects inadequate understanding of problems at the deepest level which is comparable, for instance, to that achieved in critical phenomena and division of physicists into separate communities with little mutual communications. I have no further comments or criticisms to make with respect to these trends of the modern period, since such comments and criticisms are usually bound to be wrong and can even be harmful. Instead, I will briefly describe our recent work of foam rheology which belongs to the modern period, which was done in collaboration with T. Okuzono [11]. Our interest in foam rheology arose from our long standing concern in phase ordering dynamics. Namely, coarsening of foam is considered as an example of phase ordering dynamics with infinite numbers of degenerate ordered phases.

4 Dynamics of two dimensional foams

Foam is an example of complex fluids (or more appropriately, structured fluids) where a building block is a single cell bounded by liquid films and filled by gas. When the volume fraction of gas is close to unity, the foam will exhibit a certain rigidity for small external deformations as long as effects of topology changes of the liquid film network can be neglected. Our model for studying mechanical properties of foams is very simple, especially in two-dimensions. Thus we consider a two dimensional liquid film network where each cell is filled with inviscid and incompressible gas. We assume the quality of the foam is so good that no degradation of the foam through diffusion of gas across liquid films can occur. We are interested in very slow processes where any inertia effects of gas or liquid can be neglected. Then dynamics is basically governed by the free energy of change and the dissipation rate. The free energy is stored in the form of surface tension σ of liquid and gas and the dissipation arises from viscous dissipation due to flow of liquid near every intersection called Plateau border of three liquid films. Here we introduce our basic simplifying assumption that curvature of all the liquid film can be neglected. Although this is not exact, we have verified fairly good quality of this approximation for foam coarsening [12]. Then the free energy is

$$E = \sum_{<ij>} 2\sigma |\mathbf{r}_{ij}| \qquad (30)$$

with $\mathbf{r}_{ij} \equiv \mathbf{r}_i - \mathbf{r}_j$, $\mathbf{r}_i$ being the position vector of the intersection (also referred to as vertex) i and summation being over all the straight liquids films (edges). The factor 2 in front takes into account two sides of liquid films.

The energy dissipation occurring in the part of the edge $< ij >$ near the vertex i was obtained elsewhere [13] and is given by

$$Q_{ij}^P = (3\eta)^{2/3}\sigma^{1/3}|\mathbf{v}_i^{(j)}, \hat{\mathbf{r}}_{ij}|^{5/3} \left[I_e\theta(\mathbf{v}_i^{(j)}, \hat{\mathbf{r}}_{ij}) + I_c\theta(-\mathbf{v}_i^{(j)}, \hat{\mathbf{r}}_{ij}) \right], \qquad (31)$$

where η is the shear viscosity of liquid in films, $\hat{\mathbf{r}}_{ij} \equiv \mathbf{r}_{ij}/|\mathbf{r}_{ij}|$, $\theta(x)$ is the step function equal to 1 (or 0) for $x > 0$ (or $x < 0$), $I_e = 1.2215$ and $I_c = 1.1866$ and $\mathbf{v}_i^{(j)} = \mathbf{v}_i - \mathbf{u}_{ij}$. Here $\mathbf{v}_i$ is the velocity of the vertex i and $\mathbf{u}_{ij}$ is the velocity of the edge $< ij >$ far from its two vertices.

The static force is readily obtained as

$$\mathbf{f}_i^S = -\frac{\partial E}{\partial \mathbf{r}_i} = -2\sigma\hat{\mathbf{r}}_{ij}. \qquad (32)$$

The friction force $\mathbf{f}_i^P$ acting on the vertex i is the sum of the three forces $\mathbf{f}_{ij}^P$ coming from three edges connected to the vertex where

$$\mathbf{f}_{ij}^P \cdot \mathbf{v}_i^{(j)} = Q_{ij}^P. \qquad (33)$$

That is,

$$\mathbf{f}_{ij}^P = (3\eta)^{2/3}\sigma^{1/3}|\mathbf{v}_i^{(j)}, \hat{\mathbf{r}}_{ij}|^{2/3} \left[I_e\theta(\mathbf{v}_i^{(j)}, \hat{\mathbf{r}}_{ij}) - I_c\theta(-\mathbf{v}_i^{(j)}, \hat{\mathbf{r}}_{ij}) \right] \hat{\mathbf{r}}_{ij}. \qquad (34)$$

By symmetry considerations and Galilean invariance we find $\mathbf{u}_{ij} = (\mathbf{v}_i + \mathbf{v}_j)/2$. One additional force on the vertex i arises from the constraint that every cell should have a constant area required by the absence of gas diffusion across every film. The vertex equation of motion is then expressed as a force balance equation as follows:

$$\mathbf{f}_j^P = \mathbf{f}_i^S + \sum_\alpha \lambda_\alpha \frac{\partial}{\partial \mathbf{r}_i} A_\alpha \qquad (35)$$

$$\frac{d}{dt} A_\alpha = 0, \qquad (36)$$

where A_α is the area of the cell α and λ_α the Lagrange multiplier.

The set of equations (35) and (36) was numerically simulated under the boundary condition of the Lees-Edwards type with a constant rate of shear deformation. In order to prepare initial states of the system for simulation, we first choose a system with great numbers of cells ($\simeq 10^4$ cells) and let it coarsen by permitting gas diffusion across liquid film until the regime of scale invariant coarsening is reached. We then switch the dynamics to (35) and (36) but without shear deformation and equilibrate. We are left with equilibrated systems containing roughly one thousand cells (exact numbers slightly differ from one initial state to another). These are the initial states for our simulation. We have chosen a small shear rate of $10^{-4}\sigma/\eta a^{1/2}$ with a the average area of a cell. Some snapshots showing locations and velocities of cell centers are displayed in Fig. 1

A striking feature is an initially quiescent flow pattern suddenly develops into a violent flow pattern and then calms low subsequently. Physical origin of this behavior is not difficult to find. As the system evolves under a steadily increasing shear deformation, the energy is slowly fed into the system. After sufficient accumulation of the energy a topology change somewhere in the system triggers a large scale release of energy like an avalanche, which one may call "foam quake" since there is a close analogy with earthquakes[14].

Appearance of such intermittent turbulent flow is intimately connected with Galilean invariance of our dynamical model. Whenever there is a topological

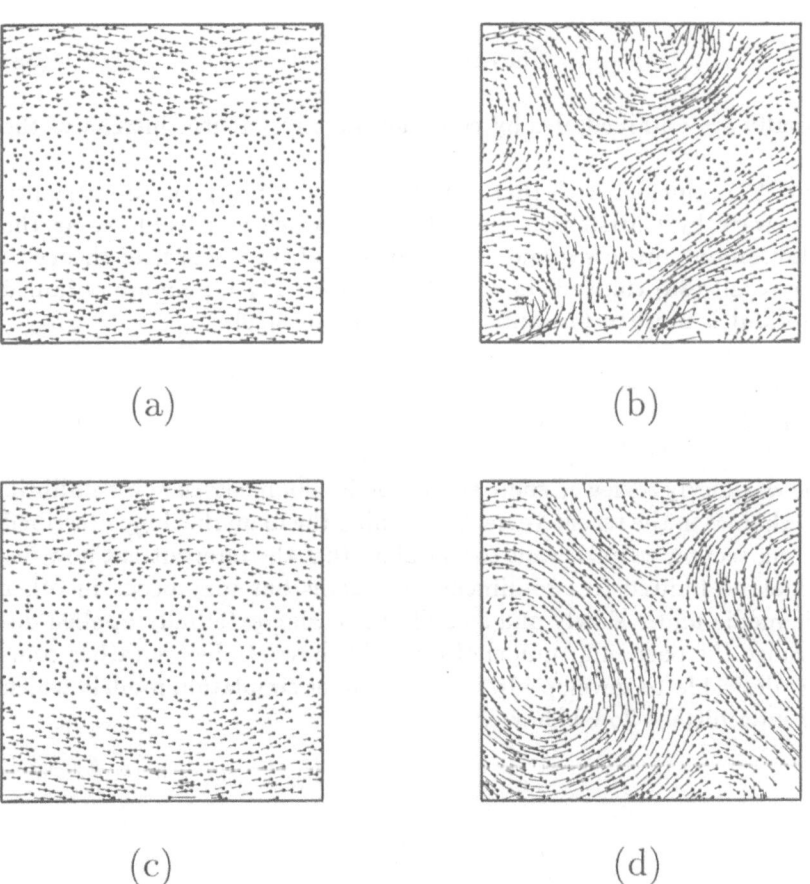

(a)   (b)

(c)   (d)

change leading to rapid vertex motion of those involved in the process accompanied by energy release, slight unbalance of forces due to random structure is enough to induce large scale organized motion which is easier to excite than disorganized motion due to Galilean invariance. Indeed, in our earlier simulation without Galilean invariance, such large scale flow was not observed.[15]

In order to make this analogy more quantitative, we consider the energy functions $E(t)$. This function shows steady increase while the energy is fed by steadily increasing external strain until a topology change triggers an avalanche leading to rapid drop of $E(t)$. $E(t)$ again starts to increase at the end of the avalanche. Thus we can take the drop of $E(t)$ as a measure of the avalanche size which we denote as s. The probability distribution $P(s)$ is approximated by the following power law:

$$P(s) \simeq s^{-3/2}. \tag{37}$$

On the other hand, the power spectrum of $E(t)$ itself is also shown to be represented by the power law as

$$S(\omega) \simeq \omega^{-2}. \tag{38}$$

The same results are obtained also by choosing the stress tensor instead of the energy.

The result (37) coincides with that of the model of self- organized-criticality (SOC) put forward by Christensen and Olami[16] which is solvable using the analogy with branding processes and also with that of the mean field theory of SOC[17]. The result (38) contradicts with the so-called $1/f$ type noise obtained earlier [17, 18] but is more in line with the later findings[19].

5 Discussion

In this article we quickly reviewed some aspects of developments of non-equilibrium statistical mechanics in the recent years. Concerning the foam rheology described in the preceding section we just want to speculate that the intermittent flow behavior may not be limited to two-dimensional foams but may occur in other elasto-plastic media under slowly and steadily increasing externally applied deformations. Since the particular model of foams we have studied is rather complicated, it is desirable to develop alternative simpler simulation models which show similar intermittent flow behavior.

References

1. J.L. Lebowitz, E. Presutti and H. Spohn, J. Stat. Phys. **51**, 841 (1988).
2. R. Zwanzig, Ann. Rev. Phys. Chem., **16**, 67 (1965).
3. L.D. Landau and E.M. Lifshitz, "Fluids Mechanics", Pergamon Press, Oxford (1959), section 24.
4. J.R. Dorfman and H. van Beijeren: in "Statistical Mechanics, Part B", ed. B.J. Berne, Plenum Press, New York (1977).
5. H. Mori, Prog. Theor. Phys (Kyoto), **33**, 423 (1965).
6. K. Kawasaki, Ann. Phys.(NY) **61**, 1 (1970).
7. H. Risken, "The Fokker-Planck Equation", Springer-Verlag, Heidelberg, (1984).
8. K. Kawasaki and J.D. Gunton: in "Progress in Liquid Physics", ed. C.A. Croxten, Wiley, New York (1978). P.C. Hohenberg and B.I. Haperin,. Rev. Mod. Phys. **49**, 435 (1977).
9. M. Fixman, J. Chem. Phys. **36**, 310 (1962). See also L.P. Kadanoff and J. Swift, Phys. Rev., **166**, 89 (1968).
10. M. Lévy, J.C. Le Guillou and J. Zinn-Justin, eds. "Phase Transitions". Plenum Press, New York (1982). M. Suzuki, in: "Evolutionary Trends in Physical Sciences", eds. M. Suzuki and R. Kubo, Springer Verlag, Heidelberg (1991).
11. T. Okuzono and K. Kawasaki in: "Trends in Statistical Physics" ed., J. Menon, Council of Scientific Integration, Trivandrum, India (1994), and submitted to Phys. Rev. E.
    K. Kawasaki and T. Okuzono, in: "Proceedings of International Conference on Dynamical Systems and Chaos-Tokyo, 1994", ed. K. Shiraiwa (to be published).
12. K. Kawasaki, T. Nagai, T. Okuzono and K. Fuchizaki, in "Modeling of Coarsening and Grain Growth", eds. S.P. Marsh and C.S. Pande, The Minerals, Metals & Materials Society, (1993).
13. L.W. Schwartz and H.M. Princen, J. Colloid. Interface Sci. **118**, 201 (1987).
14. J.M. Carlson and J.S. Langer, Phys. Rev., **A40**, 6470 (1989).
15. T. Okuzono, K. Kawasaki and T. Nagai, J. Rheol. **37**, 751 (1993).
16. K. Christensen and Z. Olami, Phys. Rev., **E48**, 3361 (1993)
17. C. Tang and P. Bak, J. Stat. Phys. **51**, 797 (1988).
18. P. Bak, C. Tang and K. Wiesenfeld, Phys. Rev. Lett. **59**, 381 (1987).
19. K. Christiansen, H.C. Fogedly and H.J. Jensen, J. Stat. Phys., **63**, 653 (1991)

SIMULATION OF THE CONSISTENT BOLTZMANN EQUATION FOR HARD SPHERES AND ITS EXTENSION TO HIGHER DENSITIES

Francis J. Alexander, Alejandro L. Garcia⋆ and Berni J. Alder

Institute for Scientific Computing Research L-416
Lawrence Livermore National Laboratory
Livermore, California 94550

The direct simulation Monte Carlo method is modified with a post-collision displacement in order to obtain the hard sphere equation of state. This leads to consistent thermodynamic and transport properties in the low density regime. At higher densities, when the enhanced collision rate according to kinetic theory is introduced, the exact hard sphere equation of state is recovered, and the transport coefficients are comparable to those of the Enskog theory. The computational advantages of this scheme over hard sphere molecular dynamics are that it is significantly faster at low and moderate densities and that it is readily parallelizable.

1 Introduction

The direct simulation Monte Carlo (DSMC) method is a particle-based, numerical scheme for solving the nonlinear Boltzmann equation [1, 2, 3]. Rather than exactly calculating successive hard sphere (HS) collisions, as in molecular dynamics (MD) [4], DSMC generates collisions stochastically with scattering rates and post-collision velocity distributions determined from the kinetic theory of a dilute gas. DSMC encounters the usual inconsistency of the Boltzmann equation, namely, it yields the transport properties for a dilute gas of hard spheres of diameter σ, yet results in an *ideal gas* equation of state (implying $\sigma = 0$) [5]. In this paper, a modification to DSMC is introduced which removes this inconsistency and, in fact, recovers the exact HS equation of state at *all* densities.

The DSMC method solves the Boltzmann equation by using a representative random sample drawn from the actual velocity distribution. In the simulation, the state of the system is given by the positions and velocities of particles, $\{r_i, v_i\}$. The system evolves in two steps, advection (or free streaming) and collision. In free streaming, particles are propagated for a time Δt as if they did not interact. In other words, their positions are updated to $r_i + v_i \Delta t$. Any particles that reach a boundary are reflected according to the boundary condition (e.g., specularly or diffusely).

⋆ Permanent Address: Department of Physics, San Jose State University, San Jose, CA 95192-0106.

After the advection step, the particles are sorted into cells to evaluate the collisions in the gas. Particles within a cell are randomly selected as collision partners according to the collision probabilities derived from dilute hard sphere kinetic theory. Conservation of momentum and energy provide four of the six equations needed to determine the post-collision velocities. The remaining two conditions are selected stochastically with the assumption that the direction of the post-collision relative velocity is uniformly distributed on the unit sphere. The spatial "coarse-graining" of particles into cells allows two particles to collide by simply being located within the same cell. Since only the magnitude of the relative velocity between particles is used in determining their collision probability, even particles that are moving away from each other may collide.

The DSMC scheme is only accurate when the time step is a fraction of the mean collision time and the cell volume is a fraction of a cubic mean free path. Because each particle in the simulation represents an effective number of molecules in the physical system, macroscopic systems may be accurately modeled by using as few as $10^4 - 10^5$ particles, with at least 20 particles per cubic mean free path [6]. A more detailed description of the standard DSMC method may be found in References [1] and [2].

The DSMC method was developed for use in rarefied gas dynamics to compute flows at high Knudsen number (ratio of mean free path to characteristic length) [7]. The algorithm has been thoroughly tested over the past 20 years and found to be in excellent agreement with both experimental data [8, 9] and molecular dynamics computations [10, 11]. Recently, it was proved that DSMC is equivalent to a Monte Carlo solution of an equation "close" to the Boltzmann equation [3]. The DSMC method has also been useful in the study of nonequilibrium fluctuations [12], chemically reacting systems [13, 14] and nanoscale hydrodynamics [15].

2 Non-ideal Gas DSMC

To obtain a consistent equation of state, DSMC must be modified in the collision step to include the extra separation, $\mathbf{d}$ ($|\mathbf{d}| = \sigma$), that the particles would have experienced if they had collided as hard spheres. Consider for simplicity a one-dimensional system with two hard rods of length σ initially traveling toward each other. They collide when their centers are a distance σ apart. After the collision, the distance between centers will be larger than the separation between similarly colliding point particles by a distance 2σ [16]. For hard spheres in three dimensions this effect generalizes to a displacement, $\mathbf{d}$:

$$\mathbf{d} = \frac{(\mathbf{v'}_1 - \mathbf{v'}_2) - (\mathbf{v}_1 - \mathbf{v}_2)}{|(\mathbf{v'}_1 - \mathbf{v'}_2) - (\mathbf{v}_1 - \mathbf{v}_2)|}\sigma = \frac{\mathbf{v'}_r - \mathbf{v}_r}{|\mathbf{v'}_r - \mathbf{v}_r|}\sigma, \tag{1}$$

where the incoming velocities of the colliding particles 1 and 2 are $\mathbf{v}_1$ and $\mathbf{v}_2$, and the post-collisional velocities are $\mathbf{v'}_1$ and $\mathbf{v'}_2$ respectively; $\mathbf{v}_r$ is the relative velocity. Thus, particle 1 is displaced by the vector distance $\mathbf{d}$ and particle 2 by $-\mathbf{d}$. See Figure 1 for an example. In the low density limit the displacement yields the correct second virial coefficient. The average projection of the velocity change onto the line connecting centers of colliding particles after displacement, $\langle \mathbf{r}_{ij} \cdot \Delta\mathbf{v}_i \rangle$, the virial, resulting from this procedure is that of hard spheres at all densities.

3 Dense Gas DSMC

If, in addition to the displacement, $\mathbf{d}$, the Boltzmann collision rate is scaled by the so-called Y-factor, the enhanced probability of a collision due to the volume occupied by the spheres, a model in the spirit of Enskog results [17]. This density dependent Y factor can be obtained from the HS equation state as determined by Monte Carlo and MD simulations and expressed in the Padé form [18]

$$Y(n) = \frac{1 + 0.05556782 b_2 n + 0.01394451 b_2^2 n^2 - 0.0013396 b_2^3 n^3}{1 - 0.56943218 b_2 n + 0.08289011 b_2^2 n^2}, \qquad (2)$$

where $b_2 = (2/3)\pi\sigma^3$ is the HS second virial coefficient. Collisions within a cell are generated with a rate $\Lambda(n^*) = Y(n^*)\Lambda_{00}(n^*)$, where $n^* = n\sigma^3$ is the reduced particle number density and Λ_{00} is the Boltzmann collision rate:

$$\Lambda_{00}(n) = 2N_c n\sigma^2 \sqrt{\pi k_B T/m}. \qquad (3)$$

In this expression k_B is the Boltzmann constant, and T is the temperature, m is the particle mass and N_c is the number of particles in a given cell. In the Enskog approximation, the mean free path for a dense gas is $\lambda = 1/(\sqrt{2}\pi n Y(n)\sigma^2)$ [17].

4 Computer Simulations

A series of computer simulations tested this model with the units determined by setting $m = 1$, $\sigma = 1$, and $k_B T = 1$. The equilibrium pressure as a function of density can be determined from the virial and also by measuring the normal momentum transfer across a plane. Both procedures yield the HS equation of state within 1% for all densities (see Fig. 2) when the time step is less than 0.03 mean collision times. From the hydrodynamic expression for the direct scattering function, $S(k, \omega)$, [19], the sound speed obtained from the location of the Brillouin peak is in agreement with HS MD at low densities. At the higher densities, the Rayleigh and Brillouin peaks are not well separated, and accurate measurements of the sound speed cannot be made in this way. Furthermore, the radial distribution (pair correlation) function is that of a perfect gas so that the compressibility, as determined from the density fluctuations in a volume V, $\chi_T = \langle \delta n^2 \rangle V/k_B T n^2$ is that of a perfect gas and does not agree with $\chi_T = (\partial \log n/\partial p)_T$ as obtained directly from the equation of state.

The self-diffusion coefficient, D, is measured using the Einstein relation,

$$D = \frac{1}{6t} \langle \frac{1}{N} \sum_i^N (\mathbf{r}_i(t) - \mathbf{r}_i(0))^2 \rangle, \qquad (4)$$

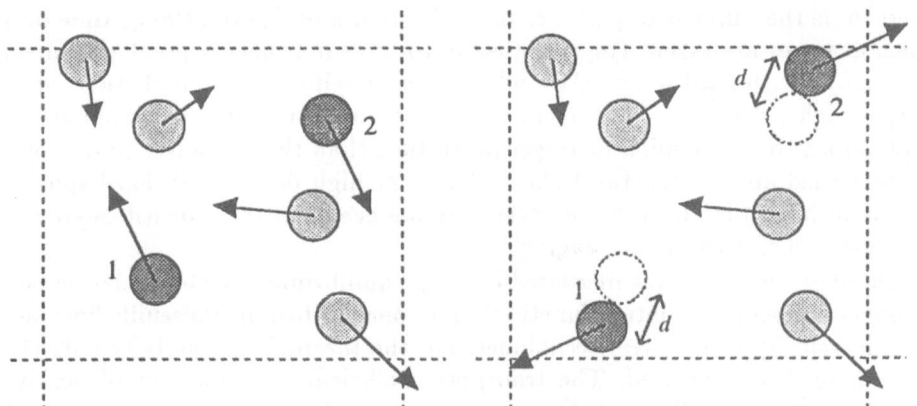

– Figure 1. Schematic illustration of the displacement occurring after a collision.

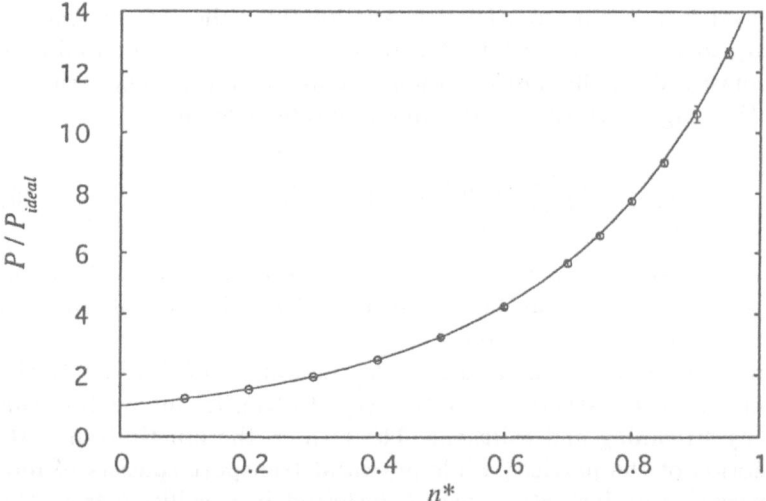

– Figure 2. Pressure (normalized by ideal gas pressure) as a function of number density for a time step of $\Delta t = 0.04\lambda/\langle v \rangle$ and the cell width is λ. The solid line is HS MD.

where N is the number of particles in the system and t is the (long) time over which averages are taken. For densities up to $n^* \approx 0.3$ there is good agreement (within 5%) with hard sphere MD and the Enskog self-diffusion prediction. However, at higher densities, the agreement fails because the post-collisional particle displacement $\mathbf{d}$ is of similar or larger magnitude than the mean free path. Also, as mentioned above, structural effects found at high densities in hard spheres are absent in this model and the backscattering events at these densities are not reproduced (i.e., there is no "caging").

The shear viscosity was measured by both equilibrium (Einstein relation and transverse current correlation function) and nonequilibrium (Poiseuille flow and relaxing velocity sine waves) techniques. For the thermal conductivity only the Einstein relation was used. The transport coefficients as functions of density are shown in Figures 3 and 4. For the shear viscosity, there is good agreement with both Enskog theory and HS MD at lower densities. At higher densities the measured shear viscosity shows better agreement with HS MD than does Enskog theory.

Poiseuille flows for various densities were generated in a channel by applying a constant external force on the particles parallel to the walls. At the walls, a thermal boundary condition was used; that is, particles colliding with a wall were emitted with a biased Maxwellian distribution at temperature T. The resulting velocity profile (See Figure 5) was fit assuming a parabolic form,

$$U(x) = (\frac{nF}{2\eta})((L/2)^2 - x^2) + U_{slip}, \tag{5}$$

where U_{slip} is the slip velocity at the walls, F is the force applied to the fluid, and L is the channel width. As can be seen in Fig 3, the viscosity obtained in this way agrees with alternative methods.

The Einstein relation allows one to assess the separate contributions to the transport coefficients. In HS MD there are two ways to transfer momentum and energy, namely by streaming and collisions. The former, the kinetic transport, is due to the motion of the particle, while potential transport consists of momentum and energy being instantaneously transferred in a collision from the center of one sphere to the center of its collision partner. The shear viscosity and thermal conductivity may then be decomposed into three distinct parts: the kinetic, potential, and cross contributions [20]. These separate terms can also be determined from the Enskog theory of hard spheres [17].

In the model presented in this paper the kinetic contribution to the fluxes is the same as that for uncorrelated hard spheres as given by the Enskog theory. The collisional transport, however, has two parts: exchange between colliding particles (which are in the same cell) and post-collision displacement. The viscosity, for example, then has the form

$$\eta = \frac{m^2}{2Vk_BTt}\langle[\int_0^t \sum_i^N v_{xi}(s)v_{yi}(s)ds + \sum_{coll.pairs} (v'_{xi} - v_{xi})y_{ij} \tag{6}$$

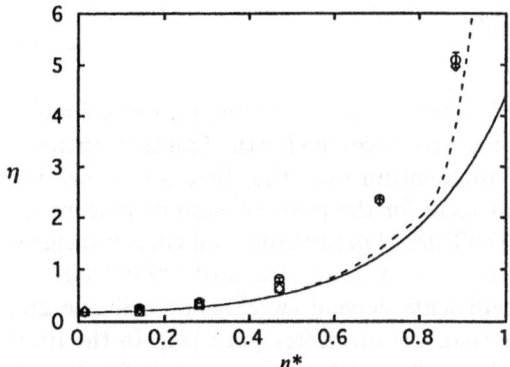

– Figure 3. Viscosity versus number density as measured using the Einstein relation (circles), transverse current correlation function (diamonds), velocity sine wave decay (triangles) and Poiseuille flow (squares); the solid line is Enskog theory and the dashed line is HS MD.

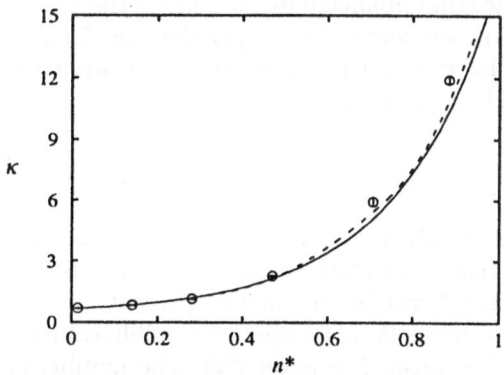

– Figure 4. Thermal conductivity versus number density as measured using the Einstein relation (circles); the solid line is Enskog theory and the dashed line is HS MD.

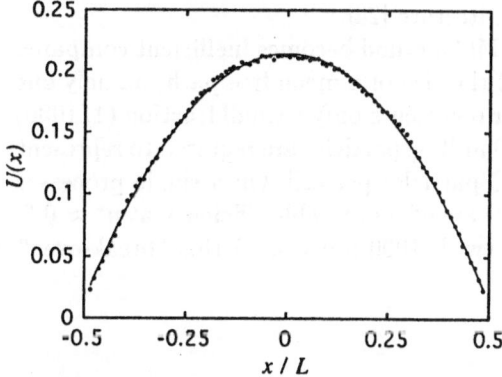

– Figure 5. Velocity versus position in a channel of length L for $n^* = 0.1414$. The solid line is the quadratic fit of the data. The cell width is 0.38λ, and the time step is $\Delta t = 0.038\lambda/\langle v \rangle$.

$$+ \sum_{coll.pairs} (v'_{xi}d_y - v'_{xj}d_y)]^2 \rangle, \qquad (7)$$

where y_{ij} is the y-component of the distance between colliding particles i and j and d_y is the y-component of d. The first term accounts for the kinetic transport; the second term for the transfer of momentum over the distance separating colliding particles i and j, and the last term for the post-collision displacement. The second term in (5) corresponds to collisional momentum and energy transfer on a length scale on the order of a cell size. In both standard DSMC and its dense gas extension, the transport coefficients depend (weakly) on cell size Δy, yet this effect is small when Δy is less than the mean free path [21]. In the limit of cell size tending to zero, this "grid error" vanishes (since $y_{ij} \rightarrow 0$). For all cases shown in Figure 3, the grid error was within the error bars of the measured transport coefficients.

Good agreement with Enskog theory is found for the kinetic and cross terms of the shear viscosity and thermal conductivity at all densities [22]. The potential term in the shear viscosity is about twice that predicted by the Enskog theory; for thermal conductivity the potential term was about 25% larger than the Enskog predicted value. A kinetic theory explanation for these differences between the Enskog model and the present model is in progress.

5 Efficiency

The model presented here runs with nearly the same efficiency as standard DSMC at low densities. The calculation of displacements and the use of the Y factor only increase the computational cost by one or two percent. At low densities, HS MD is inefficient because of the large number of possible collision partners within a neighborhood of a few mean free paths [23]. The number of operations per collision per particle with hard sphere dynamics grows as n^{-2} at low densities, while it is independent of density for DSMC. In comparison with a scalar hard spheres molecular dynamics code, the dense gas DSMC scheme runs two orders of magnitude faster for $n^* = 0.01414$. This advantage can be further enhanced by running on a parallel architecture [24].

At high densities, the dense gas DSMC method becomes inefficient compared with HS MD. The reason is that a cell the size of a mean free path, namely one which is roughly 1/10 of a HS diameter represents only a small fraction (1/1000) of a single hard sphere particle. Thus 20 million particles are required to represent 1000 HS particles, assuming 20 DSMC particles per cell. On a single processor computer, HS MD and dense gas DSMC are of comparable efficiency at $n^* \approx 0.3$, while on a massively parallel machine (with 1000 processors) this "break-even" density increases to $n^* \approx 0.7$.

6 Conclusions

In this paper a modification of the DSMC algorithm which extends the method to dense gases is described. Computer simulations of this method yielded the equilibrium thermodynamic and nonequilibrium transport properties. In general, for all properties good agreement was found with HS MD at densities less than $n^* = 0.3$. Further exploration of the effects of time step, spatial grid, effective number, and overall system size is necessary for more quantitative comparisons.

Direct simulation Monte Carlo has been a popular method for the simulation of hydrodynamic flows of high Knudsen number where conventional Navier-Stokes solvers are inaccurate. Since most DSMC applications have been in rarefied flows, the method's restriction to ideal gases has not been viewed as a major drawback. This dense gas version of DSMC will extend the method's utility to a variety of new problems, which involve not only very low density gases, but moderate density as well. These include the study of cold boundary layers in high altitude flows and strong shocks [25].

7 Acknowledgments

We thank G. L. Eyink, A. J. C. Ladd, M. Malek Mansour and M. Mareschal for a number of very helpful discussions. This work was carried out under the auspices of the Department of Energy at Lawrence Livermore National Laboratory under Contract $\#W - 7405 - ENG - 48$.

References

1. G. A. Bird, *Molecular Gas Dynamics and the Direct Simulation of Gas Flows* (Clarendon, Oxford, 1994).
2. A. L. Garcia, *Numerical Methods for Physics*, Chapter 10 (Prentice Hall, Englewood Cliffs NJ, 1994).
3. W. Wagner, *J. Stat. Phys.* **66**, 1011 (1992).
4. M.P. Allen and D.J. Tildesley, *Computer Simulation of Liquids*, (Clarendon, Oxford, 1987).
5. Modified collision rates are commonly used in DSMC to reproduce the transport properties of non-hard sphere gases *e.g.*, the Maxwell molecule and Bird's variable hard sphere model [1]. However, these modifications retain the ideal gas equation of state. See also F. Baras, M. Malek Mansour and A.L. Garcia, *Phys. Rev. E* **49** 3512 (1994).
6. M. A. Fallavollita, D. Baganoff, and J. D. McDonald *J. Comp. Phys.* **109** 30 (1993).
7. E.P. Muntz, *Ann. Rev. Fluid Mech.* **21** 387 (1989).
8. D. A. Erwin, G. C. Pham-Van-Diep and E. P. Muntz, *Phys. Fluids A* **3** 697 (1991).
9. D. C. Wadsworth, *Phys. Fluids A* **5**, 1831 (1993).
10. D. L. Morris, L. Hannon and A. L. Garcia, *Phys. Rev. A* **46**, 5279 (1992).
11. E. Salomons and M. Mareschal, *Phys. Rev. Lett.* **69**, 269 (1992).
12. M. Malek Mansour, A. L. Garcia, G. C. Lie and E. Clementi, *Phys. Rev. Lett.* **58**, 874 (1987).
13. S. M. Dunn and J. B. Anderson *J. Chem. Phys.*, **99** 6607 (1993).
14. M. M. Mansour and F. Baras *Physica A*, **188** 253 (1992).
15. F. J. Alexander, A. L. Garcia and B. J. Alder *Phys. Fluids* (to appear) (1994).

16. Moving to contact, *each* point particle travels an extra distance $\sigma/2$ (as compared with hard rods). Moving apart after the collision, each point particle must also travel an additional distance $\sigma/2$.

17. P. Resibois and M. De Leener, *Classical Kinetic Theory of Fluids*, (John Wiley and Sons, New York, 1977).

18. J. J. Erpenbeck and W. W. Wood, *J. Stat. Phys.* **35** 321 (1984); J. J. Erpenbeck and W. W. Wood, *J. Stat. Phys.* **40** 787 (1985).

19. B. J. Berne and R. Pecora *Dynamic Light Scattering*, Krieger Publ., Malabar, Fla. (1976).

20. B. J. Alder, D. M. Gass, and T. E. Wainwright, *J. Chem. Phys.* **53** 3813 (1970).

21. The relative magnitude of the grid error goes as $(\Delta y/\lambda)^2$; this undesirable grid effect may be minimized by using a cell/subcell hierarchy [1].

22. The cross and potential contributions exclude the grid error (second term in (6)) and only count the post collisional displacement (third term in (6)).

23. M. Reed and K. Flurchick, *Comp. Phys. Comm.* **81** 56 (1994).

24. M. A. Fallavollita, J. D. McDonald and D. Baganoff, *Comp. Sci. Eng.* **3** 283 (1992).

25. A. Frezzotti and C. Sgarra, *J. Stat. Phys* **73** 193 (1993).

Over Two Decades of Pattern Formation, a Personal Perspective

Guenter Ahlers

Department of Physics and Center for Nonlinear Studies, University of California, Santa Barbara, CA 93106

Abstract: Patterns are ubiquitous in the world that surrounds us. They can form *via* bifurcations, for instance from the spatially uniform state, as a control parameter is varied. Their nature generally is determined by *nonlinear* terms in the relevant equations of motion, and thus their elucidation is a non-trivial goal in nonlinear physics. In the early 1970's, there was a revival of interest in the condensed-matter physics community in chaos and pattern formation in nonlinear dissipative systems. Experimentalists and theorists brought the tools of their field to bear on these challenging problems. Mostly in terms of his own experiences, the author of this paper reviews some of the issues that have been addressed, some of the techniques that have been applied, and some of the progress that has been made by experimentalists during the two-and-a-half decades since then, and the relationship which these results have to our present-day understanding of nonlinear systems.

1 Introduction

When a system is removed far from equilibrium by subjecting it to a stress, it will often undergo a transition from a spatially uniform state to a state with spatial variation. We refer to this variation as a "pattern". Pattern formation is generally associated with nonlinear effects. These are of great fundamental interest to physicists because they often lead to qualitatively new phenomena which do not exist in linear systems and which are not yet fully understood.

Another fascinating aspect of pattern-formation phenomena is that many of them have a universal character. Even though the system under investigation may for instance be of physical, chemical, or biological nature, many common features are encountered. Pattern formation has been studied in such diverse fields as fluid mechanics, optics, chemistry, biology, solid-state physics, gas

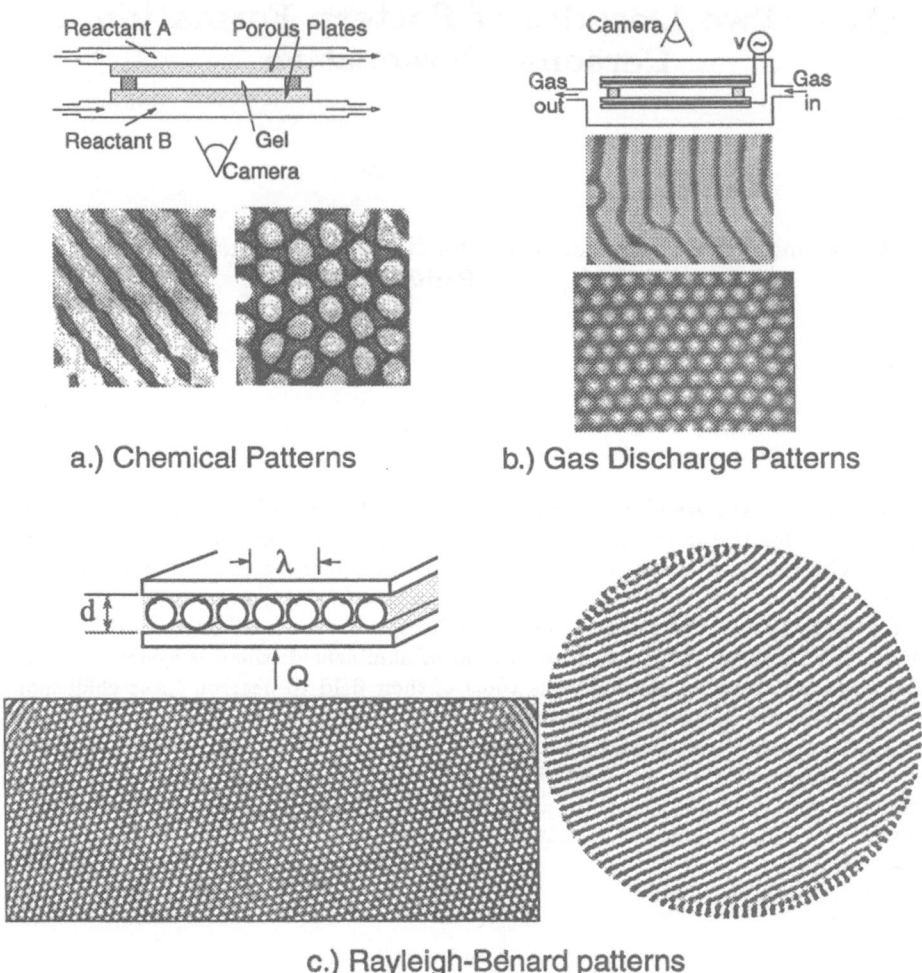

a.) Chemical Patterns   b.) Gas Discharge Patterns

c.) Rayleigh-Bénard patterns

Fig. 1. Three systems in which rolls (also called stripes) or hexagons occur over different parameter ranges. a): Chemical patterns obtained by Ouyang and Swinney. [1] b): Patterns in a gas discharge system obtained by Breazeal, Flynn, and Gwinn. [2] c): Patterns in Rayleigh-Bénard convection obtained by Bodenschatz et al. [3] (hexagons) and by Hu et al. [4] (rolls).

discharges, and others. Figure 1 illustrates some of these. In Fig. 1a chemical patterns obtained by Ouyang and Swinney [1] are illustrated. Two different mixtures of reactants are circulated through two channels across the outside of porous plates. Between these plates, a layer of a gel permits mixing by diffusion of the chemicals, which leads to a spatially dependent chemical reaction. A spatial variation develops in the concentration fields. Among other patterns, hexagons and "stripes" or "rolls" are obtained, depending on the parameter ranges. In Fig. 1b we see patterns, obtained by Breazeal, Flynn, and Gwinn, [2] of visible light emitted from a gas subjected to a large alternating electric field. Here also hexagons and stripes are observed over a range of parameter values. Finally, in Fig. 1c, we see the result of Rayleigh-Bénard convection. A fluid is contained between two parallel horizontal plates separated by a distance d and heated from below by a heat current Q. When the fluid properties do not vary significantly over the applied temperature difference, convection rolls are obtained. [4] These are similar to the stripes in Figs. 1a and 1b. When there is a significant variation of the fluid properties, hexagons occur. [3, 5] The communality (or universality) between these diverse systems is of course provided by the common features of the mathematical equations which describe these systems. A scientist interested in the fundamental aspects of these ubiquitous effects will choose a particular system which is well suited to detailed, quantitative study. The universal aspects of patterns provoke hope of a deeper understanding in very complicated systems (where the equations of motion might not even be known) from the study of simpler ones. For instance one might expect to learn something about the formation of biological patterns which are difficult to study under highly controlled conditions from work on simpler physical or chemical systems where boundary conditions and control parameters can be specified precisely in an experiment.

Aside from the intellectually fascinating aspects of pattern formation, there are pratical reasons for being interested in this field. Our physical surroundings are dominated by patterns. We see ripples on a pond, regular or irregular cloud formations, the remarkable shapes of snowflakes, and the patterns in animal skins or coats. The surprising aspect is the great regularity which is often encountered. A deeper understanding of these natural phenomena will enrich our lives and deepen our appreciation of our environment. There are also more practical benefits for our society which will arise from a more fundamental understanding of patterns. Their better understanding enables us to prevent undesirable or promote advantageous patterns in industrial processes. Examples include the displacement of oil by water in a porous oil reservoir, in which finger-like patterns formed by the water can reduce the yield. Others can be found in crystal-growth processes, where patterns in the melt or the solid phase may be undesirable. Patterns occur in the wake of airplane wings, in chemical reaction fronts, flames, leaching of minerals from ore beds, fronts between infected and unifected portions of a population, and in many other technologically, economically, biologically, or sociologically important situations. In the long run their fundamental understanding clearly will have far-reaching consequences for our society. An extensive review of the physics of pattern formation in physical, chemical, and biological systems was published recently by Cross and Hohenberg. [6]

I have been very grateful for my serendipitous introduction to the field of nonlinear nonequilibrium systems, which eventually led to my interest in the experimental study of patterns. In 1970, I was a staff member at Bell Laboratories. My interests at that time were primarily in the field of critical phenomena, with an emphasis on the superfluid transition in liquid ^{4}He. Late in that year, my department head Paul Fleury had a visitor from Belgium by the name of Jean Pierre Boon. Jean Pierre told Paul, me, and some of our colleagues about

Rayleigh-Bénard convection (RBC). [7] It seems difficult to imagine from our present vantage point; but to my knowledge none of us had ever heard about this phenomenon as an interesting physical system even though of course all of us were familiar with convection from everyday experiences. Today RBC has evolved into one of the primary model systems for the experimental study of pattern formation. Another serendipitous aspect of Jean Pierre's visit was that I had an apparatus in my laboratory at that time which I used for the study of the thermal conductivity of liquid ^{4}He near the superfluid transition at 2.172 K. [8,9] At that temperature, helium has a negative expansion coefficient, and thus I was heating the sample from below so as to *avoid* convection. It was a simple matter to increase the temperature by a fraction of a degree and then to study the interesting effect of convection upon the heat transport in a temperature range where the expansion coefficient is positive. Since my apparatus was cold and operational at the time of Jean Pierre's visit, I was able to obtain heat-transport data which were a great deal more precise than previous results in the literature [10] within a day or two. Some of these are shown in Fig. 3 below. They revealed a bifurcation at a well defined value ΔT_c of the temperature difference ΔT. There was a little rounding of the bifurcation which initially provoked interesting speculations about fluctuation effects; but this rounding is now well understood in terms of small imperfections in the sample geometry. The value of ΔT_c agreed well with the theoretical prediction.

In the remainder of this paper I would like to present some of the events in the study of pattern formation which I witnessed since the visit by Jean Pierre in 1970. It is my hope that I can convey to the reader the excitement of the field by describing some my own experiences. Under no circumstances should this account be seen as a review of the field. Rather, it is a presentation of some of the important events from a highly personal point of view, as I experienced them, and it neglects many major contributions to the field. Having said this, I hope that I do not even have to apologize to those of my friends whose work I am omitting.

2 The 1970's and Before

Although there is a long history of the study of bifurcations and pattern formation in fluid mechanics by the applied mathematics and engineering community, physicists for the most part had not really appreciated the interesting aspects of this field prior to about 1970. When physicists finally learned about this fascinating area of study, I believe they soon began to play an important role. The experimentalists did not feel constrained by the practical needs of the engineer and felt free to concentrate on problems which were just complicated enough to be challenging but still simple enough to be amenable to theoretical analysis and quantitative experimental study. Thus, their hallmark became precise experiments on *relatively* simple systems which lent themselves to a detailed comparison with theory. Before long this led to a comparison between experiments and theoretical or numerical results at the 0.1% level. In earlier days, researchers often felt that they were on the right track when they achieved agreement at the level of perhaps 10%.

So far as I know, there was relatively little quantitative experimental work by physicists on patterns in the 1960's (a notable exception to this is the seminal work by Donnelly and collaborators on Taylor-vortex flow which started [11–13] a decade earlier). However, I want to mention two important *theoretical* results of that decade which had a great impact on subsequent experimental activity. The first is the work of Lorenz, [14] through which it became widely appreciated that systems describable by coupled nonlinear ordinary differential equations can exhibit non-periodic time dependence. Investigations of such systems fall into the field of Dynamical Systems, which by now has reached a certain level of maturity. [15] Far less is understood about irregular variations in systems which are extended also in space (*i.e.* in patterns), and which thus must be described by *partial* differential equations. In this latter case, there can be non-periodic variation both in space and in time, and this phenomenon frequently is referred to as spatio-temporal chaos (STC). I mention the work of Lorenz and the field of dynamical systems because there has been some as yet unsubstantiated hope that STC can somehow be understood by using what we know about dynamical systems as a starting point. The elucidation of STC is still very much at the forefront of reseach today, and will no doubt remain so for some time to come. From my point of view, another important theoretical result of the 1960's was contained in the paper of Schlüter, Lortz, and Busse, [16] which showed that a stable pattern above the onset of convection in a Boussinesq system consists of parallel straight rolls of arbitrary orientation when boundary effects can be neglected. A good approximation to this state can indeed be found in experiments, and is illustrated in Fig. 1c. A second paper which to me was equally important was the one by Busse [17] in 1967, which among other issues presented a systematic perturbation calculation for non-Boussinesq convection [18, 19] in which hexagons become stable at the bifurcation from the conduction state. Such a hexagonal pattern is also shown in Fig. 1c. It was found to be stable over a range of ϵ which is consistent with the calculation. [3]

The experimental developments of the 1970's had, I think, a profound impact in the long run on the study of spatio-temporal complexity. Some of us brought the experimental tools of condensed-matter physics to bear on the problems of fluid mechanics. In particular, the precision measurement-techniques which had been developed for the study of critical phenomena were applied to the study of bifurcations, chaos, and STC. I already described how I was drawn into this field; other experimental physicists who entered it included Pierre Bergé and Monique Dubois, Jerry Gollub, Albert Libchaber, and Harry Swinney. Perhaps another serendipitous occurrence is that all of these people chose problems from fluid mechanics for their initial projects. It turned out that fluid systems lend themselves extremely well to the quantitative investigations which the experimentalists wanted to undertake. In well chosen fluid systems, extremely precise geometries can be specified, boundary conditions can be well controlled, and control parameters can be held extremely constant or varied in very small steps. Later on of course it became desirable to look at other systems; but fluid mechanics offered many advantages initially.

Although the conventional tools of solid-state physics, such as high-resolution thermometry, lock-in amplifiers, light scattering, and others played an important role, I believe that the most important experimental development of the 1970's was the advent of the computer in the laboratory. It enabled us to carry out projects which we never would have considered in earlier decades, even though by today's standards these early efforts at automation and data collection may seem primitive. Already early in 1971, we used a home-made data acquisition system to collect time series of a scalar quantity (the temperature of the bottom plate of a convection cell) which contained several thousand values, and used fast-Fourier-transform methods to obtain their power spectra. [20, 21] A little later we progressed to an LSI-11 based system which was interfaced to a PDP-11/45. [22, 23] Similar developments may well have taken place earlier in other laboratories; but for us they revolutionized the kind of projects that could be tackled. Thus, they not only provided us a new tool, but they also gave us a completely new perspective on what types of experiments to do. On some kind of a logarithmic scale, this was as great a step forward as another which I witnessed as a graduate student a decade earlier, when we started to use DC amplifiers and chart recorders, instead of recording individual voltages in our laboratory notebook after obtaining them by looking through a 10-meter telescope at a ballistic galvanometer. Subsequent changes in computer technology have been perhaps as great or greater; but somehow we began to take these for granted because the data-handling capability of computers was growing continuously. By now this has evolved to the point where we routinely program experiments to run unattended for weeks when long time scales are involved, and collect hundreds of Mbytes of data in such a run.

Let me now get to the physics that was being done. Some of the important scientific results of the 1970's were, I believe, the quantitative measurements by Bergé and Dubois of various components of the velocity field for RBC in pseudo-one-dimensional systems, $i.e.$ in relatively narrow and long cells where the convection rolls line up parallel to the short (y) axis with their wave vector parallel to the x-axis. [24, 25] Examples of their results are shown in Fig. 2. They confirm that the velocity amplitude varies with the mean-field exponent 0.5 as $\epsilon \equiv (\Delta T - \Delta T_c)/\Delta T_c$ is increased. Similarly, these authors measured the characteristic length ξ over which the velocity field grows from zero amplitude at a solid boundary to its bulk value in the interior, and the temporal rate σ at which the system responds to changes in the control parameter. They found the mean-field exponents -0.5 and 1 respectively. This work, as well as some measurements in our laboratory, [26, 27] culminated [28] in a quantitative comparison of the RB system with predictions based on a Ginzburg-Landau (GL) equation [29], and firmly established the analogy between bifurcations in pattern-forming systems on the one hand and critical phenomena in the mean-field limit on the other. Not only the exponents (which of course follow from Landau's general assumption of analyticity [29] and are not system specific), but also the coefficients which had been calculated specifically for RBC from the Navier-Stokes equation [27, 28], agreed. Similar work was being done with Taylor-vortex flow [30] by Gollub and Freilich, [31 − 33] by Rehberg, [34] and by Pfister and Rehberg. [35] Taylor-vortex flow (TVF) is also a one-dimensional pattern-forming system, but it differs from RBC in narrow, long cells in that the velocity amplitude at the ends does not vanish, but rather grows to a value of order one. This renders the bifurcation imperfect, and was discussed in terms of a GL equation by Graham and Domaradzki. [36] Rehberg extended the comparison with a GL equation to the secondary Hopf bifurcation to azimuthally travelling waves in TVF. The GL equation has played an enormously important role in the study of patterns during subsequent years, and I believe it is the work which I have described which put its applicability to physical systems on a firm experimental foundation.

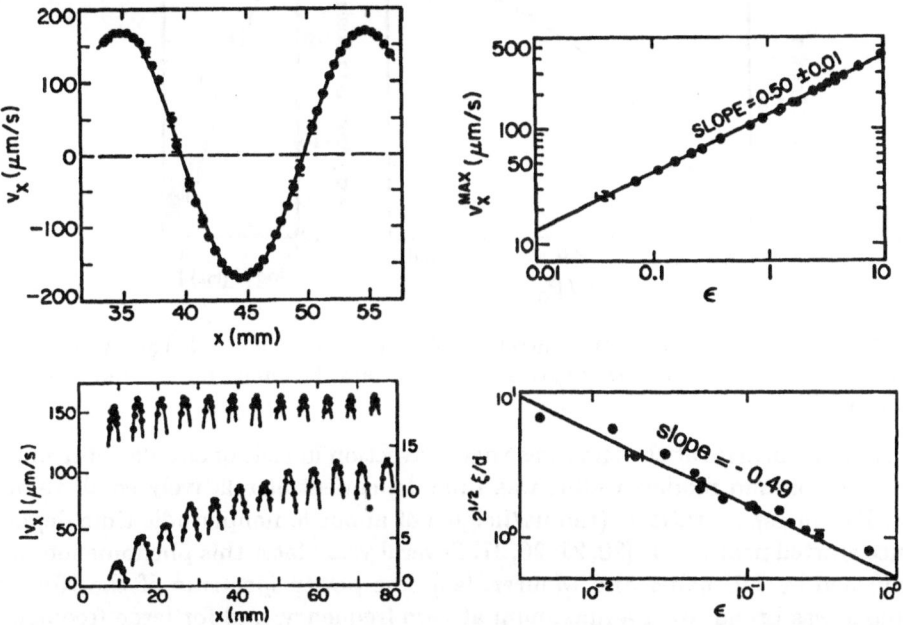

Fig. 2. Laser-Doppler velocimetry measurements of the velocity in a Rayleigh-Bénard cell by Dubois and Bergé. [24, 25] Left top: horizontal velocity component v_x near the center of the cell as a function of position x (x is in the direction of the roll wavevector). Right top: amplitude of the velocity variation near the center of the cell as a function of $\epsilon = \Delta T / \Delta T_c - 1$. Left bottom: horizontal velocity component v_x as a function of position x near a sidewall. Right bottom: the healing length ξ over which the pattern grows from zero amplitude at the wall to its bulk amplitude.

At Bell Laboratories, I pursued my studies of convection at cryogenic temperatures. Before long, Robert Behringer and then Robert Walden joined me in this effort. [20, 21, 23, 26, 27, 37 − 43] During the very first measurements using liquid helium, the versatility of cryogenic fluids immediately became apparent.

In some of the work we used gaseous [4]He because an easily applied change in the pressure could alter ΔT_c, and thus the extent to which the Boussinesq approximation was satisfied, by a large amount. [21, 26, 40] The opportunity to change the fluid properties, and in particular the Prandtl number, by approaching the critical point, was also exploited. [38, 39] As an illustration of some of the early work, results for the Nusselt number [26] are shown in Fig. 3. By changing the sample pressure, precision measurements for Rayleigh numbers up to $150R_c$ were made. Related cryogenic work with an emphasis on even larger Rayleigh numbers was done simultaneously by Threlfall [44], but we were not aware of it until it appeared in the literature.

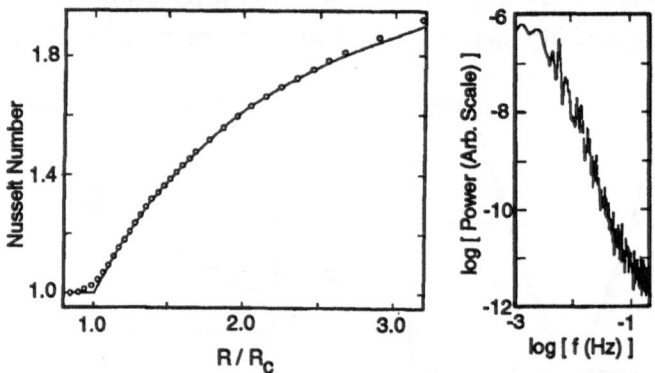

Fig. 3. Left: Time average of the Nusselt number in a cell with $\Gamma = 5.3$ as a function of $R/R_c = 1 + \epsilon$. Right: The power spectrum of the Nusselt number for $\epsilon = 1.4$. Adapted from Ref. 26.

A great surprise at the time was that convection in cells of circular horizontal cross section and modest radius was time dependent at relatively small values of ϵ. For an aspect ratio Γ (radius/height) of about 5, nonperiodic time dependence started near $\epsilon = 1$. [20, 21, 26, 37] Several years later this phenomenon was confirmed by Libchaber and Maurer. [45] The power spectrum of the Nusselt number was broad, with a maximum at zero frequency, and for large frequency it fell off as f^{-4}. This is shown in the right portion of Fig. 3, for $\epsilon = 1.4$. Since here we had found an experimental system with chaotic (*i.e.* broad band) time dependence, there was a great temptation initially to try to find a connection between the observations and the chaotic behavior of the Lorenz model. This model had been investigated in detail by McLaughlin and Martin. [46, 47] Later we realized that the apparently algebraic falloff was surprising because simple models of chaos in deterministic systems with relatively few degrees of freedom, such as the Lorenz model, had a spectrum with an exponential falloff. [48, 49] As already noted in 1972, [20] it seems likely that the onset of time dependence was associated with wavenumber adjustments as a function of ϵ which caused the system to cross an instability boundary, from our present vantage point most likely the skewed-varicose (SV) instability. [50] The apparently algebraic falloff of the spectrum presumably is then attributable to the presence of a large but finite number of interacting modes in the spatially extended system which turns out to lead to effectively algebraic decay at *intermediate* f; but so far as I know a quantitative explanation of this phenomenon is still lacking. At very high frequencies the spectrum should then still decay exponentially; but there the power may well be so small as to be unmeasurable. Also still unexplained seems to be the fact that the system remains in the chaotic skewed-varicose-unstable regime, instead of reducing its wavenumber so as to enter once more the regime of stable rolls (this latter phenomenon occurs in the one-dimensional case of a narrow rectangular cell where the SV instability leads to the expulsion of a roll pair and a consequent reduction of the wavenumber).

When ϵ was increased beyond 3.7 with the cells of $\Gamma \simeq 5$, the spectrum of the Nusselt number was modified by a shoulder which developed in the frequency range of algebraic decay. [39] This was also observed in the experiments of Libchaber and Maurer. [45] The power due to this excess contribution grew continuously from zero as ϵ was increased, and thus was consistent with a forward Hopf bifurcation (albeit with a frequency which was broadened by the underlying chaotic state). The shoulder was at the proper frequency to be attributable to the oscillatory instability (OI) of Clever and Busse. [51] This was so even though of course the calculation was based on the assumption of a bifurcation from the time-independent parallel-roll state.

Much additional work at low temperatures on time-dependent flows in cells of modest Γ was done after these early observations, and has been reviewed in detail by Behringer. [52] Here I mention only a surprising result obtained in a very large cell ($\Gamma = 57$). Non-periodic time dependence was found even when the threshold had been exceeded by only 10% or so. [38, 39] At the time this seemed quite contrary to the extensive stability analysis of Busse and coworkers. [16, 51] It is really only now gradually being understood [4, 53, 54] in terms of sidewall effects which cause roll curvature, focus instabilities, and local compression of rolls in the cell interior beyond the skewed-varicous instability. Of course it was a great disadvantage of the cryogenic samples that the patterns could not be observed. Today we can do at room temperature with flow visualization much of what was done in the 70's with liquid helium, [3, 4, 53 − 56] and the cryogenic work would be justified only for those few experiments which cannot be done readily at ambient temperature. But two decades ago, the high resolution at low temperatures did open up some new vistas.

Another problem which caught my early interest was the effect of time dependent heating on RBC. The experiments were done with a constant temperature at the top of the cell, and with a step, a temporal ramp, or time-periodic modulation, of the heat current. I was fortunate to be able to interest my colleague Pierre Hohenberg in this problem. He and Jack Swift had worked on the effect of external noise on the Rayleigh-Bénard instability, [57] and their collaboration had produced the now well known Swift-Hohenberg equation. Direct observation of the fluctuations below the bifurcation (which are the response of the system to the external noise) was regarded by most physicists to be nearly impossible because the predicted amplitudes [57 − 62] were so small. Interestingly, at the present time, *i.e.* 20 years later, these amplitudes have just been measured quantitatively in a Rayleigh-Bénard system. [63] I think Pierre's interest in the ramping experiments was related to the fact that the flow which evolves as ϵ is swept slowly through zero should evolve from these fluctuations if deterministic imperfections were small enough. In that case the time evolution of the flow could be used to estimate the size of the fluctuations which prevailed when ϵ was negative. It was difficult to devise a proper model for the experiment because the pattern which evolved was not known. Nonetheless, with the help of Mike Cross and Sam Safran we carried out a rather detailed analysis. [60] I think in the end we became convinced that deterministic imperfections dominated the early experiments, and it seems to me that the main virtue of the early experimental effort and its analysis was to motivate later, more refined measurements in more carefully constructed cells. [64 − 66] Another important consequence was, however, Pierre's longterm close interest in my results on nonequilibrium systems, which has been of immense value to all of my subsequent work on patterns. Our direct collaboration on temporally modulated flows continued into the next decade, when Manfred Lücke joined our effort while he and Pierre were at the Institute for Theoretical Physics in Santa Barbara. [67 − 69]

A noteworthy event which I should mention is that early in the 1970's, Albert Libchaber came to visit us at Bell Laboratories, and apparently became convinced that convection in liquid or gaseous helium had much to offer to the experimentalist. But unlike us, he had the wisdom to understand that without flow visualization the global studies of heat transport, particularly in spatially extended systems, would bring only limited results. Thus, he and Maurer developed *local* temperature probes. [45] However, soon they concentrated on small systems which could accomodate only about one pair of convection rolls. Naively, I had always thought that spatially extended systems would be much more interesting. The work by Libchaber and Maurer on small systems led to the now famous discovery of period-doubling bifurcations, and together with the theoretical work by Mitch Feigenbaum [70] it really opened up the field of dynamical systems. [71] Exciting as it is, this work does not involve patterns, and I will not discuss it further here. [15]

Finally, I want to say a word or two about conferences which were important to me during the 70's. In May of 1973, there was a meeting organized by Jim Gunton and Mel Green at Temple University. It dealt primarily with critical phenomena, and emphasized the new ideas based on the mode-coupling and renormalization-group methods. For this meeting, I had come from Jülich, Germany where I was spending the calendar year on a sabbatical leave from Bell Labs. This gave me a chance to show my experimental results for the $\Gamma \simeq 5$ cell to Paul Martin, who had previously heard about them through Pierre Hohenberg. I think that Paul's genuine interest in my results played an important role in convincing me that the study of STC should be taken seriously. Evidence of my prior rather casual attitude toward it can be found in the fact that, except for a couple of talks at the January 1972 APS meeting in San Francisco,

[20, 21] I had not really published my results. Paul's interest, and his work with McLaughlin, [46] finally convinced me that I had to write up my work as soon as I returned to Bell Labs early in 1974. [37]

The most important meeting for the field as a whole was, I believe, the NATO ARW organized by Tormod Riste in Geilo, Norway in April of 1975. It was one in a series of workshops on critical phenomena, and was entitled *Fluctuations, Instabilities, and Phase Transitions*. But Tormod had the vision to see how appropriate it was at that time to devote it primarily to the emerging field of instabilities and patterns. In addition to a large number of theoretical papers, much of the early experimental work by Bergé, Goldberg, Gollub, Whitehead, myself, and others was presented, and this was the first time that some of us met each other. In a sense, one could regard this ARW as the beginning of the study of patterns as a coherent subfield of condensed-matter physics. The increasing appreciation for the importance of this field is also reflected in the fact that a major fraction of a Solvay conference [72] was devoted to it in 1979.

3 The 1980's

Looking back now, it is apparent that the 1980's brought both qualitative and quantitative advances to the field of pattern formation. Advances in computer technology, and in particular in affordable storage capacity, permitted more and more advanced experiments to be performed. Instead of taking time series of one or a few scalars, we now could actually take two-dimensional images, or time series of one-dimensional images. "Contour plots", which showed the time evolution of a one-dimensional array, became popular for the documentation of the dynamics of one-dimensional patterns. One of the early visualizations of two-dimensional convection patterns was done by Gollub and Steinman, [73] and was based on laser-doppler velocimetry; but that method turned out to be too tedious and time consuming when the primary interest was the pattern and not the quantitative amplitudes of the flow. Instead, the shadowgraph technique was developed into a very sensitive tool capable of visualizing even extremely feeble flows very near onset. Of course the method had a long history. After having been employed in the 1950's by Silveston, [10, 74, 75] it was developed further for the investigation of well developed patterns by Busse and Whitehead. [76] Its application in a more sensitive form to the study of patterns very close to onset is, however, more recent. [77, 78] By now, we can under favorable circumstances resolve temperature fluctuations a good bit smaller than a part per million of the critical temperature difference. The nature of the problems under experimental investigation also became more advanced. As I will describe below, more advanced bifurcation phenomena and one-dimensional patterns were one area of interest; but the problem of pattern formation in two dimensions was also beginning to be addressed.

An important development during this decade was the addition of many new experimental physicists to the field. At the risk of omitting some and in random order, I mention that Robert Behringer at Duke, Mike Gorman at Houston, David Andereck at Ohio State, Bob Walden, Cliff Surko, and Paul Kolodner at Bell Laboratories, Victor Steinberg in Israel, Ingo Rehberg in Germany, Sergio Ciliberto in Italy (now France), Carlos Perez-Garcia in Spain, Vincent Croquette, Alain Pocheau, Eduardo Wesfreid, and others in France, and Shoichi Kai in Japan all set up laboratories for the study of patterns in RBC, in electro-convection in nematic liquid crystals, in TVF, or in other pattern-forming systems.

At the beginning of 1980, I moved from Bell Laboratories to the University of California at Santa Barbara. After devoting a year or so to setting up my low-temperature laboratory, I returned to the study of non-equilibrium systems. In collaboration with my colleague Dave Cannell, we started a program of investigations of non-equilibrium fluid-mechanical systems at or near room temperature. Dave had tremendous experience from his work on critical phenomena in the control of temperature, design of complex apparatus, and precision measurements especially by optical means. For instance, it turned out to be possible to control the temperature of a water bath within about 10^{-4}°C, [79] which is 0.3 ppm of the absolute temperature and in that sense within a factor of three or so of what at that time could be achieved at low temperatures. Initially, we focused on RBC, using water as the fluid. During this time Victor Steinberg was a postdoctoral researcher in our laboratory and played an important role in the development of our instrumentation. Soon we added Taylor-vortex flow (TVF) because it offered exceptional opportunities for the quantitative study of one-dimensional patterns. One of its great virtues is that it is truly periodic in the y-direction (*i.e.* in the azimuthal direction). This is an advantage over RBC when it comes to comparing with simple mathematical models which often assume periodicity in that direction. TVF had of course been used very extensively by many others before, including Koschmieder, Cole, Snyder, Gollub, Swinney, Donnelly, and their numerous coworkers. [80] But much of that work had focused on somewhat higher Reynolds numbers where secondary bifurcations and chaotic regimes exist, whereas we concentrated more on a quantitative study near the primary bifurcation of the Taylor-vortex state itself. Dave had designed and our machine shop had built an excellent laser-doppler velocimetry system (with the limited funding which we were able to attract we could not afford a commercial one), which served us well in this work.

One of the hallmarks of this decade was the study of advanced bifurcation phenomena, and their application to spatially extended systems. One could reasonably ask whether all the phenomena allowed by the normal forms of bifurcation theory really occurred in physical systems. The answer that became apparent was that certainly very many of them did. Of interest to mathematicians at this time was the area of multi-parameter bifurcations. In 1985, an entire conference in Arcata, CA was devoted to this topic. [81] When there are two control parameters which can be adjusted, it is sometimes possible to choose the value of one so that the bifurcation provoked by a change in the other is of a special nature. For instance, it may be possible to adjust a parameter so that the coefficient g_3 of the cubic term in the Landau equation relevant to a particular system is zero or negative. In that case, the next-higher order (quintic) term has to be retained in the amplitude expansion in order to obtain a stable solution. For the marginal case $g_3 = 0$ the amplitude of the flow is expected to grow as the fourth root of ϵ instead of as the second. Based on the work of Benjamin and Mullin [82], we saw that this phenomenon should occur in the flow between concentric cylinders when the aspect ratio L (the system length measured in units of the gap between the cylinders) has a particular small value. For L less than about two, there are only two vortices. Since these are provoked by the cylinder ends, they are not really Taylor, but rather Ekman vortices, and the system might be referred to as Ekman-vortex flow. As the inner-cylinder speed ω is increased beyond an L-dependent value ω_1, the system undergoes a bifurcation from a state of two symmetric vortices of equal size to a state of broken symmetry where one vortex is larger than the other. One can define an order parameter Ψ which measures the asymmetry between the two vortices. As L is varied from larger to smaller values through a special value L_t, the bifurcation changes from supercritical, where Ψ grows continuously from zero as $\epsilon = \omega/\omega_1 - 1$ is increased through zero, to subcritical where Ψ jumps from zero to a finite value near $\epsilon = 0$. The marginal case $L = L_t$ is often called tricritical, in analogy to the equivalent

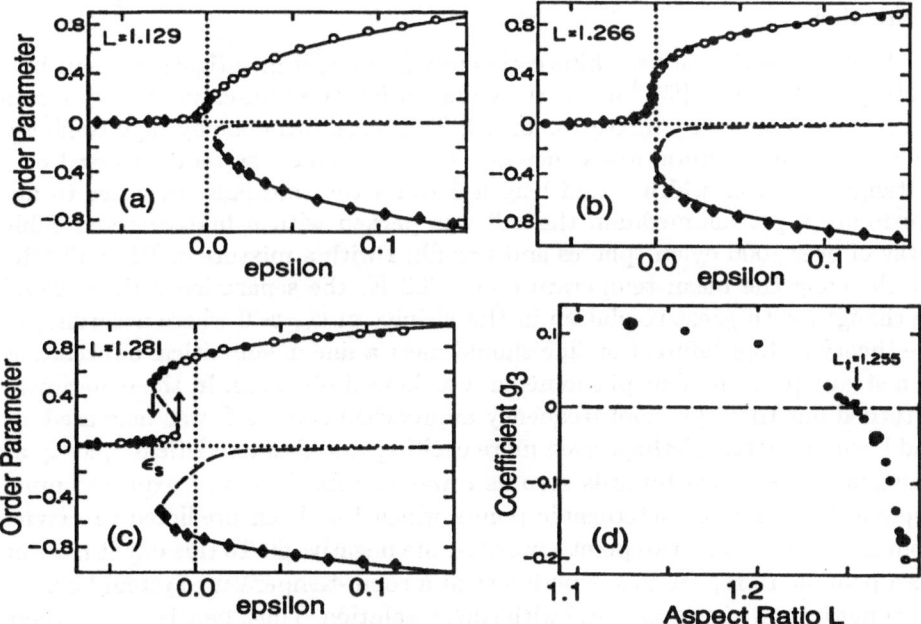

Fig. 4. (a) to (c): The order parameter Ψ as a function of ϵ for three aspect ratios L. (d): The coefficient g_3 of the cubic term of the Landau equation as a function of L. Adapted from Ref. 83.

phenomenon in equilibrium critical phanomena. With Anneli Aitta, [83] a visitor from Finland in our group, we investigated this system quantitatively by laser-doppler velocimetry measurements of the axial component v_z of the fluid velocity as a function of axial position. Figures 4a to 4c show the experimental results for steady-state values of the order parameter $\Psi \equiv \int_0^L v_z dz / \int_0^L |v_z| dz$, together with fits to the Landau equation $d\Psi/dt = h + \epsilon\Psi - g_3\Psi^3 - g_5\Psi^5 = 0$. Here the "field" h was included because the bifurcation in the physical system was not quite ideal, but rather "imperfect". We actually viewed the imperfection positively, since it gave us the additional opportunity to study the effect of an imperfection on bifurcations of different types. An interesting consequence of having $h \neq 0$ is the unfolding of the bifurcation. In the experimental results, the two branches became disconnected, as they should according to the model. As can be seen, the fits to the Landau equation are excellent. They provide the coefficients of the equation as a function of L. The coefficient g_3 of the cubic term is shown in Fig. 4d. The tricritical value of L, where $g_3 = 0$, turned out to be 1.255. At about the time of our experiments, the bifurcation diagram for this system was calculated by numerical integration of the Navier-Stokes equation. [84, 85] This work yielded [84] $L_t = 1.259$, in quantitative agreement with the experiment. A bit later, Anneli extended the analysis of her data to the transients of Ψ, i.e. to the *dynamics* of these bifurcations. [86] This provided beautiful illustrations of the difference in the temporal evolution of Ψ at fixed ϵ which is associated with the different shapes of the potentials for the subcritical and the supercritical bifurcation.

Another multiparameter-bifurcation problem under investigation in our laboratory at that time [87, 88] was the codimension-two bifurcation which occurs in convection of binary mixtures. [89, 90] This work was done by Ingo Rehberg, who was a postdoctoral researcher in our group. Ingo's system consisted of a rectangular cell of width 4 and length 8 times the thickness. In order to approximate a porous medium, the cell was packed with a bodycentered cubic array of over 1000 nylon spheres and was filled with a mixture of ^{3}He and ^{4}He. By changing the mean temperature near 2.2 K, the separation ratio Ψ could be changed with great resolution in the vicinity of $\Psi_{ct} \simeq 0$ where according to the theory a Hopf-bifurcation line should meet a line of subcritical bifurcations to a steady pattern. This phenomenon was indeed observed. In the experiment it turned out that the Hopf frequency approached zero as Ψ was increased, as had been predicted. Perhaps even more exciting, the characteristic frequency at constant Ψ decreased towards zero as ϵ was increased, as was expected upon approaching a line of heteroclinic points which had been predicted to extend from the codimension-two point towards more negative Ψ. To this day it has not been possible to repeat this experiment in a room-temperature system because ψ cannot be varied so easily and with such resolution. Thus, here is a case where the extra effort and cost of the low-temperature experiment was justified in spite of the disadvantage of the lack of flow visualization.

We extended the cryogenic-mixture work also to a bulk fluid (i.e. without the nylon spheres) in a larger cell of width 6.5 and length 26. [91] This work led to the unexpected conclusion that the bifurcation to steady convection apparently has a tricritical point at **positive** Ψ, a result which is difficult to understand and as yet not explained. This work was continued by Tim Sullivan, [92, 93] a new postdoctoral associate in our group, after Ingo joined the institute of Professor Busse in Bayreuth, Germany. Tim's results show that the Hopf frequency at the codimension-two point of the bulk mixture is at least an order of magnitude greater than had been calculated. [94 – 96] It is clear that the codimension-two point in the bulk mixtures is not well understood at this time, and well worthy of further investigations. One may hope that in the future room-temperature techniques can be refined sufficiently to shed some light on the problem.

Although the work on multi-parameter bifurcations which I have described was very exciting, it is perhaps fair to say that another area of activity of ours, involving stability ranges and wavenumber adjustments of spatially extended systems, had a greater impact on the field of patterns in non-equilibrium systems. We became involved in this on the one hand because of our interests in studying the TVF state. But on the other, a great impetus was provided by a program at the Institute for Theoretical Physics (ITP) at Santa Barbara in 1982 which was run by Pierre Hohenberg and Jim Langer. This program brought many of the theorists active in pattern formation to town, and in addition some who were just thinking of getting into this field. It was well known at this time that there was a continuous *band* of stable states above a typical primary bifurcation from a spatially uniform state, limited in the case of TVF by the Eckhaus instability. [97] A particularly interesting issue for us which evolved from the ITP Program was the question of whether a spatial ramp in the control parameter from below to above critical would allow the selection of any one of the states within the stable band, or whether a particular state would be preferred over the others. Theoretically, this issue was discussed, and illustrated in the context of nonlinear reaction-diffusion equations, in a paper by Lorenz Kramer, Eshel Ben-Jacob, Helmut Brand, and Mike Cross, [98] all of whom were at the ITP. As the result of many stimulating lunchtime discussions at the Arbor on our Campus, we started an experimental program to address this issue. During the next several years, an intense and very fruitful interaction between the theorists and experimentalists developed which, I think, brought much reward to all involved. The experiments soon revealed a unique wavenumber when the outer cylinder of the TVF apparatus had a gentle ramp in its diameter. [99] Here I might say that it is a great tribute to our machine shop and to the skills of Rudy Stuber (the head of our shop) that the required precise geometries [100] could be produced. After the initial excitement, the issue of stability ranges and selection processes became a serious program for us. Marco Dominguez-Lerma, a graduate student in our group, designed and built a superb TVF apparatus with a near-perfect geometry. He made highly quantitative measurements of the Eckhaus-instability line in the ϵ-wavenumber plane. Obviously, this was necessary before any selected states could be appreciated. Figure 5 shows his results. [100, 101] Although the Eckhaus boundary had been investigated before in a number of systems, [102 – 104] really only the essentially simultaneous work by Lowe and Gollub [104] on electroconvection in a nematic liquid crystal can be regarded as quantitive. Marco's measurements were in excellent agreement with calculations based on the NS equations and a Galerkin method by Riecke and Paap [105] which are shown by the solid line in Fig. 5. Figure 6 shows again the Eckhaus boundary, and in addition as plusses the data obtained by Marco [99, 100] for the state selected by a ramp in the outer-cylinder radius.

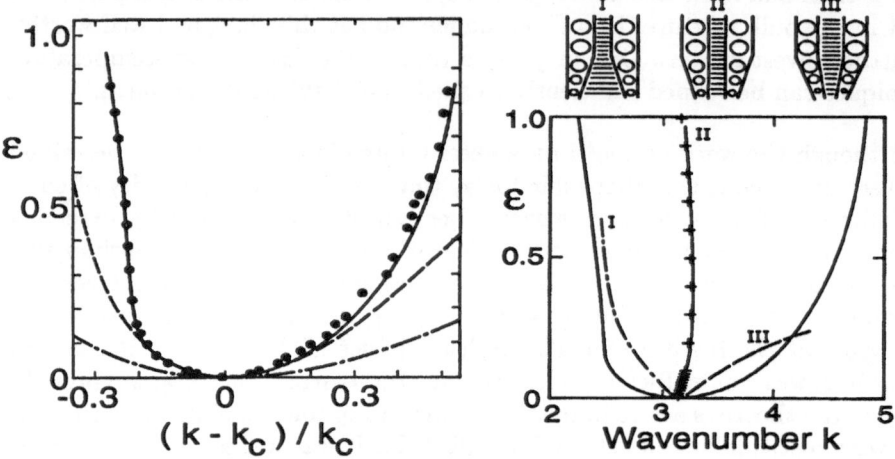

Fig. 5. The Eckhaus instability boundary for TVF. The solid points are from Ref. 100. The solid line is the calculation by Riecke and Paap. [105].

Fig. 6. The Eckhaus instability boundary for TVF as in Fig. 5, and the predictions for the selected wavenumbers for the three different ramps in the cylinder radii which are shown schematically at the top of the figure. The plusses are the experimental results from Ref. 100.

Although the investigation of wavenumber selection by ramps was at least in part motivated initially by a hope of finding something equivalent to an extremum principle for these nonequilibrium systems, it soon became clear from the theoretical work [98, 106, 107] that the unique wavenumber had a different origin. Somewhere along the ramp, one has $\epsilon = 0$, and at that point only the critical wavenumber can occur. From there on into the interior of the system (where $\epsilon > 0$) the axial variation of the phase of the pattern, and thus the change in its gradient which is the wavenumber change, is fixed by the equations of motion. This spatial phase variation could be calculated conveniently using a phase-equation approach. [107]

The above realization also soon led to the prediction that the selected state depends on precisely **how** ϵ is varied from below to above critical. Thus, for instance, in the TVF case one could have a ramp in the outer or inner cylinder, with ramp angles α_o and α_i respectively. Riecke and Paap [107] predicted that the selected state depends upon $r = \alpha_o/\alpha_i$, and that by changing this ratio, any wavenumber inside the Eckhaus-stable band could be selected. The three cases $r = 0$, $r = \infty$, and $r = 2$ are illustrated schematically in the top portion of Fig. 6, and the predicted wavenumbers are given in Fig. 6 as cases I, II, and III respectively. Case III is particularly intriguing. Here the prediction is that the selected wavenumber should be unstable for $\epsilon \gtrsim 0.2$. This case, as well as case I, was investigated experimentally by Li Ning, a student in our group. [108] Some of Ning's results in the stable range for case III are shown in Fig. 7. They agree well with the prediction. For case III and $\epsilon > 0.2$, the selection of the unstable state was predicted to lead to a repeated occurrence of the loss of a roll pair *via* the Eckhaus mechanism in the system interior. It was expected that these transitions would lead to a travelling wave of vortices coming up the ramp so as to provide the supply of vortices needed to sustain the vortex-pair losses. This is precisely what Ning found in his experiments. Figure 8 shows the frequency of the vortex-pair loss, together with a calculation of this frequency based on the phase equation by Riecke and Paap (the two curves are for two slightly different geometries and represent the uncertainty in the dimensions of the apparatus). We see that even the creation of a *dynamic* state *via* the selection of an *unstable* state could be calculated quantitatively using phase equations. Thus, even though the search for an extremum principle was in vain, the results of the joint theoretical and experimental effort may well be regarded as a significant advance in our fundamental understanding of patterns.

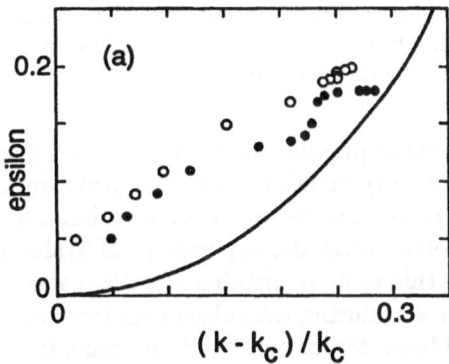

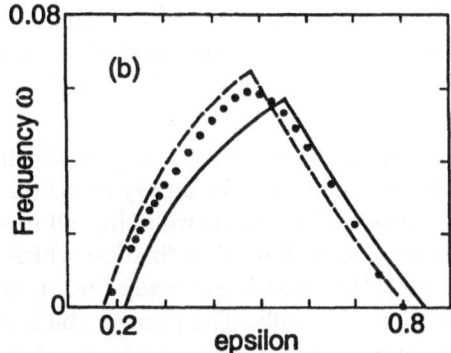

Fig. 7. Experimental results in the stable range $\epsilon < 0.2$ for the wavenumbers selected by ramps in the inner and outer cylinders with $\alpha_o/\alpha_i = .2$ (case III). Adapted from Ref. 108.

Fig. 8. The frequency of the travelling wave in the ramped section and of the vortex-pair loss in the interior. The lines are the predictions for two slightly different geometries. Adapted from Ref. 108.

An interesting related experiment during this time was one of the first in the laboratory of Ingo Rehberg in Bayreuth, where a Rayleigh-Bénard system with different ramps on the two ends was investigated. [109] The two ramps selected different wavenumbers, and thus created a wavenumber gradient in the system interior. From the phase-diffusion equation this is expected to lead to a travelling wave of convection rolls, which was indeed observed in the experiment. The travelling wave is shown in the right portion of Fig. 9, which is a contour plot of the pattern amplitude as revealed by a shadowgraph method as a function of position and time. For comparison, the left portion of Fig. 9 gives a similar contour plot for Ning's experiment on Case III and for $\epsilon = 0.30$. Here the ramp is at negative values of the axial position z, and contains the travelling wave. Vortex-pair losses occur in the uniform supercritical section at the position indicated by the arrow.

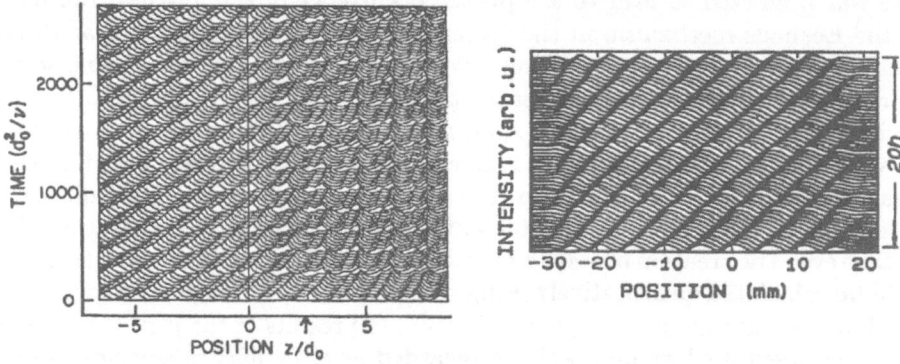

Fig. 9. Left: A contour plot of the pattern observed in Ref. 108 for Ramp III and $\epsilon = 0.3$. The ramp is to the left of the vertical line. Right: A contour plot for travelling RBC rolls in the experiment of Rehberg *et al.* (Ref. 109) where competing ramps at the two ends cause a wavenumber gradient in the system interior.

One of the fascinating topics which became popular in this decade was the study of convection in binary mixtures. I already mentioned the cryogenic work near the codimension-two point. Of much broader scope was work at room temperature with flow visualization which evolved after the experiment of Walden *et al.* [110] made it generally known that this system sustains *travelling* waves of convection rolls. The papers which appeared during the subsequent few years are simply too numerous to be mentioned here. Major contributions came from Kolodner and Surko at Bell Laboratories. Victor Steinberg, after moving from Santa Barbara to Rehovot, Israel early in 1984, built convection apparatus very similar to ours and started binary-mixture work. The field was particularly exciting because of the very close interaction between experimentalists and theorists. The theoretical work of Cross, Knobloch, Lücke, Brand, Deissler, and others on some occasions was stimulated by experiment, and on others motivated new measurements. At Santa Barbara, we initially studied ethanol-water mixtures in narrow rectangular containers with Richard Heinrichs, a post-doctoral associate in our group. Richard made the remarkable discovery that the travelling waves which evolve from the conduction state can focus into a stable localized pulse. [111, 112] The pulse has a stationary or very slowly moving envelope, and travelling convection rolls form at one end and leave it by decaying to zero amplitude at the other. The same discovery was made at about the same time and quite independently in Israel by the Steinberg group. [113] Naturally there were some skeptics who suggested that the very existence of these pulses might be attributable to the short sidewall towards which the waves were travelling. Thus we as well as others [114] immediately wanted to do experiments in annular containers which lack the short sidewall of the rectangular cell. Just about then we went through some extremely frustrating times because the NSF Engineering Directorate decided to stop funding the kind of work which we physicists were doing. Our group, Gollub, Swinney, and Donnelly were all eliminated from their program in spite of excellent reviews of our proposals. To me at least this appeared as one of the low points in the history of funding of fundamental research in the United States. After considerable delay, we were able to get support from another agency with a more enlightened policy. But of course it took time to get going again because new personnel had to acquire the skills to do these sophisticated experiments. In the end we were able to show that the pulses existed even in a periodic system. [115] The left portion of Fig. 10 gives a shadowgraph image of a three-pulse state in this geometry which was obtained by Joe Niemela in our group. The experimental discovery of pulses let to considerable theoretical activity. Soon after they were found in the experiments, pulses were observed numerically by Thual and Fauve [116, 117] as a solution of a Ginzburg-Landau equation with a destabilizing cubic and a stabilizing quintic term. Such an equation might be regarded as a reasonable approximation (albeit not a systematically derived envelope equation) for binary-mixture convection. Much further interesting numerical work based on model equations was carried out by Brand and Deissler. [118, 119] Lücke and collaborators studied pulses by the numerical integration of the Navier-Stokes equations for binary-mixture convection. [120] It has been suggested that pulses are related to the solitons of the nonlinear Schrödinger equation, modified by the dissipative contributions to the Ginzburg-Landau equation or the physical system. [121] A detailed theoretical discussion of fronts and pulses was given by van Saarloos and Hohenberg. [122] These localized structures certainly are one of the more interesting pattern-formation phenomena which I have encountered.

Although chronologically it really belongs into the next chapter, I mention here that we extended the study of pulses in binary mixtures to a two-dimensional system. This is work by Kristina Lerman, a student in our group. [56, 123] One of her two-dimensional pulses is shown in the right portion of Fig. 10. They formed spontaneously from the conduction state when the threshold for convection was exceeded by a fraction of a percent. This seemed particularly interesting because it is known that the two-dimensional nonlinear Schrödinger equation does not have solutions which are localized in two dimensions. Here it is interesting to note, however, that long-lived or stable pulses were found numerically by Deissler and Brand [124] in two-dimensional model equations, although these equations are quite different from ones which might represent the experiment on RBC in binary mixtures. Kristina's pulses turned out not to be stable in the long term. After existing for many hours (the vertical diffusion time is only a minute or so), they split into two parts *via* a transverse instability at their tail end where the travelling rolls formed. Often one half survived and reformed into a pulse while the other decayed; but eventually a more complicated structure of disordered travelling convection rolls developed. So far as I know, the formation of two-dimensional pulses is not understood theoretically at this time.

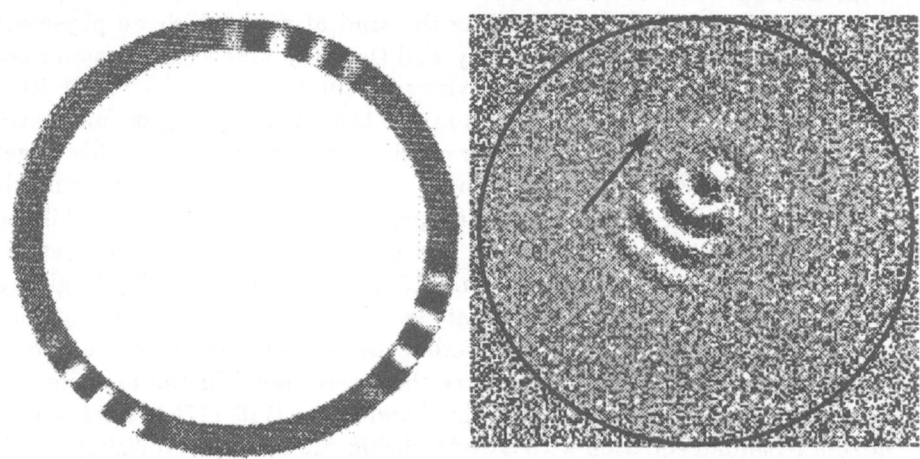

Fig. 10. Left: three pulses of travelling convection rolls in an annular cell (Ref. 115). Right: a two-dimensional pulse (Ref. 123) in a circular cell of large radius (the solid circle gives the cell boundary and the arrow indicates the direction of travel of the rolls).

I have descibed a lot of progress in our understanding of *one*-dimensional patterns. However, much was learned also with RBC in samples which were extended in two dimensions. Particularly noteworthy are the important contributions by Vincent Croquette and coworkers, [53, 125] who studied RBC in compressed Argon with a Prandtl number $\sigma \simeq 0.7$. This is a range not accessible in classical liquids, but similar to σ in the early liquid helium experiments. However, in Vincent's experiments it was possible to visualize the flow. He studied concentric rolls stabilized by sidewall forcing, as well as straight-roll patterns, in cylindrical cells with radius-to-height ratios Γ=7.66 and 20. For the "straight" rolls in a small-Γ cell, there were two focus singularities bracketing the rolls. The rolls tended to bend toward the singularities as ϵ was increased, resulting in compression of rolls along the line connecting the singularities. When the compression became too large, the rolls in the center of the cell, where the compression was most severe, underwent a skewed-varicose-like instability. For $\Gamma = 7.66$, this first occurred at $\epsilon = 0.13$. In the large-Γ case, the onset of time dependence was at the lower $\epsilon = 0.085$. One compressed roll pair got pinched off and nucleated two defects. These two defects climbed along the roll axis to within 1 or 2 wavelength of the sidewall and then glided along the wall to disappear into the focus singularities. When the defects disappeared, the singularities nucleated a new roll pair, resulting in a pattern containing the same number of rolls as before. This process then repeated itself. It seems likely that similar phenomena were responsible for many of the various time dependences observed in the early liquid helium experiments. [38, 39, 52] The onset of this instability has been demonstrated to be strongly dependent on large-scale flows which are associated with roll curvature. [126] When the flow was suppressed, the tendency for the rolls to end more perpendicular to the sidewall at higher ϵ was also suppressed, and the straight-roll state was observed to be stable to ϵ as large as about 0.6.

Finally I want to mention just briefly the very interesting work on pattern formation in electro-convection (EC) in nematic liquid crystals (NLC) which experienced an upsurge during this decade. The system consists of a thin layer of a NLC confined between transparent plates which are covered with transparent electrodes. The director is aligned uniformly in a specified direction, usually parallel to the confining plates. This is accomplished by appropriate treatment of the surfaces. For parallel alignment, a voltage applied to the electrodes will induce hydrodynamic flow if the dielectric constant has a negative anisotropy and if there are ionic impurities present in the sample. In contrast to the RBC case, which is invariant under rotation in the horizontal plane, this system has a preferred direction and for that reason exhibits quite different pattern-formation phenomena. Important work in this area was done by Kai, Rehberg, Ribotta, Steinberg, and their collaborators. [127, 128] They studied the orientation of the convection rolls relative to the director, stationary and travelling patterns, defect-chaos states, and many other interesting effects. This system has the potential for quantitative contact between experiments and weakly nonlinear theories, *i.e.* Ginzburg-Landau equations. [128, 129] The equations of motion are much more complicated than the NS equations of RBC, and for that reason this system provides a more severe test of modern theories of pattern formation. However, there are also more factors such as the conductivity of the sample which can have a decisive influence on the phenomena which occur. Often the experimental foundation for a quantitative comparison with the theory has not really been provided. Usually the experimentalists did not adequately characterize their samples and they did not obtain convincing evidence about the nature of the primary bifurcation which occurred in their samples. Thus, although much has been learned already, there is tremendous potential here for the quantitative study of more advanced pattern-formation phenomena.

4 The 1990's

The present decade naturally has brought further technological advances to the experimental study of pattern formation. Most of these were again in the data-processing area. Even with a modest budget, we can now have faster computers (486's) in the laboratory than ever before, and disk storage has become even more affordable. It is now common to take time series of *two-dimensional* images and to play them back as movies, whereas a decade ago only time series of one-dimensional images were readily accessible to us. This turns out to be a very important capability. Particularly in RBC, many processes are very slow, and within a human attention span a pattern may seem steady. Playing back a movie at a rate much greater than real time often reveals interesting dynamics which might have been missed otherwise. Also the digital data processing has advanced significantly. For instance, it is now well within the reach of the average experimentalist to compute three-dimensional Fourier transforms of two-dimensional image time-sequences on his own work station. So he can extract $S(\mathbf{k}, \omega)$ from the data. A decade ago we did not have this capability.

One of the new directions in our laboratory was the study of convection in gases under pressure. As I described above, this had been pioneered by Vincent Croquette a few years earlier. [53, 125] John deBruyn started to develop and his successors Eberhard Bodenschatz, Stephen Morris, Yu-Chou Hu, and Mingming Wu perfected a rather complicated apparatus for the study of gas convection at pressures up to 100 bar or so. [130] It differed from Croquette's apparatus in that the water bath outside the sample cell was also pressurized, thus providing hydrostatic conditions for the entire cell. With this design there was no pressure differential across the cell top and bottom, and we were able to achieve large-diameter cells (up to 9 cm) with no distortion of the confining top and bottom surfaces. Our interest at first was in finding a binary mixture of gases with a negative separation ratio. This would have had the advantage of a Lewis number of order one rather than 10^{-2}, and for instance would have made codimension-two phenomena much more accessible to experiment. The transport properties of gas mixtures are not very well known, and so it was difficult to find an appropriate mixture. While we were searching, we became discouraged in our effort by a theoretical development. It turns out that in gas mixtures the Dufour effect [131] becomes important. Calculations by Hort *et al.* [132] showed that it would be unlikely that travelling waves would exist near a codimension-two point. Nonetheless, there would be interesting phenomena to be explored; but for the present at least we let ourselves be distracted by a variety of other fascinating problems involving RBC in pure gases.

One of the aspects of convection in gases is that a change in the pressure will produce a dramatic change in ΔT_c. Alternatively, the same ΔT_c can be achieved with a range of sample thicknesses d by varying the pressure. This flexibility meant that thin samples with radius to height ratios of order 100 and with extremely short time scales (vertical diffusion times of order one second) could be obtained. So it was possible to study two-dimensional stationary processes in samples of unprecedented size, hopefully approaching the theoretical ideal of a laterally unbounded system. Varying ΔT_c provided the opportunity to vary the extent of departures from the Boussinesq approximation.

Fig. 11. Left and Middle: One- and two-armed giant spirals at $\epsilon = 0.12$ (Ref. 55) and 0.15 (Ref. 3) respectively. Right: Spiral-defect chaos for $\epsilon \simeq 0.7$ (Ref. 133).

Eberhard Bodenschatz developed and used our apparatus to study the formation and stability range of hexagons due to non-Boussinesq effects. [3] It turned out that the hexagonal lattice of convection cells which is stable above onset is perfect so far as we can tell. Any defects which may form initially when ΔT_c is exceeded migrate to the sample perimeter and disappear. A portion of one of Eberhard's images was shown already in Fig. 1 at the beginning of this paper. The entire image contains about 5000 convection cells, forming an essentially perfect lattice. So far as we know, this is the largest defect-free dissipative structure that has been produced in the laboratory. Although still small on the scale of crystal lattices in solid-state physics, it suggests that "Broken Symmetry" [134] may exist after all in nonequilibrium dissipative systems. The hexagons were found to be stable over a range of ΔT which was consistent with the calculations by Busse which I mentioned near the very beginning of this paper. [17] When ΔT was increased beyond the stability limit of the hexagons, Eberhard found that, after transients had decayed, a single giant spiral filled his sample cell. The spiral could have a number of arms, and after many turns away from the center each arm terminated in a dislocation. The left and middle portions of Fig. 11 illustrate such structures with one and two arms. The giant spirals rotated slowly, with a characteristic frequency of order 10^{-3} when time is measured in terms of the vertical diffusion time. Interestingly, the outer terminations of the arms and the tips rotated synchronously, and thus the entire structure was stable. Little is known about these interesting patterns, and serious theoretical and much more detailed experimental work on them surely would be warranted. For instance, one would like to know whether the radial location of the dislocation has a unique value for given conditions, and whether the rotation is caused by a wavenumber gradient in the radial direction or by large-scale flows induced primarily near the tip. It seems to me that these problems fall within that category I mentioned earlier of being complicated enought to be interesting and simple enough to be soluble. There are many similarly interesting issues associated with these large-scale structures which should be pursued. One of them is the temporal behavior of a defect deliberately introduced into the otherwise perfect hexagonal lattice. This defect is known to be associated with two of the three sets of rolls which make up the hexagons. Since these are not orthogonal to each other, the defect can neither climb nor glide. We know that in time it moves out of the system, and its motion must be a combination of climbing and gliding. Will it undergo some sort of irregular chaotic diffusive motion?

Stephen Morris was Eberhard's successor as a postdoctoral associate. During a period of overlap they worked together and made a remarkable discovery. [133] When a large Boussinesq RBC system of radius ratio 78 and with a Prandtl number near one is taken about 26 % above onset, the pattern turns into a state consisting of many small spirals and other defects which interact with each other and form a state which we call spiral-defect chaos (SDC). A typical snapshot of SDC is shown in the right portion of Fig. 11. Stephen made extensive measurements of its statistical properties. Its mean wavenumber as a function of $\epsilon = \Delta T/\Delta T_c - 1$ falls right into the middle of the wavenumber band over which, according to the theory, straight rolls are stable. [135] In the end we concluded that there are two attractor basins for the solutions to the equations of motion of the system. One is the straight-roll attractor of Busse and Clever. The other, found in our experiments, is a previously unknown chaotic attractor corresponding to SDC. Much of the experiment has recently been reproduced numerically, both in model equations [136, 137] and by integration of the Navier-Stokes equations in the Boussinesq approximation. [138] I think the discovery of SDC fundamentally changes our view about RBC in systems with $\sigma \simeq 1$. Previously, we had known that there was slow time dependence in this system; [38, 39, 53] but this was thought to be provoked by locally (in the cell interior) exceeding the skewed-varicose instability because of roll "pinching" due to the curvature of the rolls which in turn was caused by the lateral walls of the cell, [4, 53, 54] and by the emission of rolls from sidewall foci [54]. Now it appears that SDC has nothing to do with the walls and instead corresponds to a stable attractor with a large attractor basin even for the a system with periodic boundary conditions, [138] and presumably for the infinite system.

In a collaboration with Bob Ecke at Los Alamos, we meanwhile had duplicated our gas-convection apparatus and started a second set of experiments. Our student Yu-Chou Hu took up residence in Los Alamos, and carried out this work. Yu-Chou's primary goal was to study convection in the presence of rotation, which leads to the Küppers-Lortz instability [139 − 141] and which was interesting to us because it gives spatio-temporal chaos at onset *via* a forward bifurcation. This work is not yet completed, although preliminary results [56] have already provoked interesting theoretical work by Tu and Cross. [142] In order to understand the rotating state, we felt it necessary to study also the system without rotation, [4, 54] although we expected it to be quite well understood. [53] Yu-Chou developed pattern-analysis methods which could be used to determine the onset of SDC objectively from the pattern statistics. [143] He found that in the smaller system with a radius ratio $\Gamma = 40$ SDC occurred for $\epsilon \gtrsim 0.6$, whereas in the system with $\Gamma = 78$ it had been observed [133] all the way down to $\epsilon \simeq 0.26$. This is consistent with the fact that Croquette, in a system with the smaller $\Gamma = 20$, did not report any SDC even though he worked up to quite large ϵ-values. The question naturally arises of how low the onset of SDC would be in the laterally infinite system. Clearly, the determination of the aspect-ratio dependence of the SDC onset is an interesting problem for the future. One would also like to understand why the onset is so sensitive to Γ when Γ is already so large. Presumably the answer to this latter question will be that sidewalls suppress mean flows, and that mean flows are essential to the sustenance of SDC. Interestingly, Yu-Chou found that the onset of SDC occurred at somewhat smaller ϵ when the system was rotated around a vertical axis. In addition, rotation broke the chiral symmetry of the spirals in the sense that the numbers of right-handed and left-handed spirals in the SDC state were no longer equal.

I believe that one of the experimental accomplishments of this decade is that we are finally learning about the influence of thermal noise on macroscopic hydrodynamic systems. From a theoretical viewpoint, this has been an active issue in the 1970's, [57 − 59] but as I mentioned above, the effects were generally considered too small to be observable in real experiments because the energy $k_B T$ is many orders of magnitude smaller than typical energies involved in macroscopic fluid motion. Nonetheless, by judicial choices of experimental systems it has now been possible to observe and measure directly the consequences of fluctuating fluid-velocity fields near bifurcations which are the response of the system to thermal noise. The first measurements were made in electro-convection in a nematic liquid crystal. [144] In that system the conditions are particularly favorable because the thickness of the fluid layer, and thus the typical fluctuating volume, is small. In addition, the relevant macroscopic dissipative energy is determined by the Frank elastic constants, [145] which are also relatively small. [146] In the liquid crystal, there is a preferred direction corresponding to the average alignment of the molecules, and this is reflected in the fluctuating convection rolls. Thus, the wavevectors of these rolls are centered about two distinct positions in the two-dimensional Fourier space. More recently, quantitative measurements of the fluctuation amplitudes in RBC have been made, using compressed CO_2 as the fluid. [63] In this case, the system is isotropic in the plane, and therefore wavevectors of all possible orientations but with a modulus close to the critical wavenumber are equally represented in the fluctuations. It turns out that the fluctuation amplitudes vary as $1/|\epsilon|^{0.25}$, precisely as predicted on the basis of a stochastic GL or Swift-Hohenberg equation with white noise. The experimentally measured amplitude of this power law agrees within 20 % or so with predictions [61, 62] based on the formulation of this problem by Landau and Lifshitz [131].

An important characteristic of the present decade is that physicists are beginning to tackle more complicated, and thus in some sense more practically relevant, systems. As I described above, the previous two decades had been used to establish a firm foundation for this. One of the interesting directions being pursued is open flows. The particular case of TVF with an imposed axial flow has been a most fruitful example for quantitative laboratory study. [147 − 153] Here too noise plays a central role. In the absence of noise there is a range of ϵ and throughflow Reynolds-numbers over which, although disturbances have a positive growthrate, the throughflow velocity is large enough to sweep away an isolated disturbance faster than it can grow. In this "convectively" unstable regime, there would be no spatial structures if they were not continuously provoked by external noise. It turns out that even microscopic noise can create "noise-sustained structures" [154]. A quantitative explanation of the observed phenomena was possible in terms of a stochastic complex Ginzburg-Landau equation. [148, 149] One may hope that this work will provide guidance in the understanding of more complicated open-flow systems.

Another thrust towards understanding more complex pattern-formation phenomena which is beeing pursued at present is the study of RBC in nematic liquid crystals. We became interested in this problem when I learned about it during a sabbatical leave in 1990 at the University of Bayreuth, where Professors Kramer and Pesch had started a program of theoretical analysis of this system. The equations of motion are significantly more complicated than the Navier-Stokes equations for isotropic fluids. The usual viscosity and conductivity are replaced by five independent viscosities and two conductivities, and the equations for momentum and energy balance must be coupled to an equation for the director field which contains three elastic constants. In spite of these complexities, it has been possible to carry out quantitative stability analyses of these equations, and under some conditions predictions in the weakly nonlinear regime have been made. [128, 155, 156] The system is very rich since the director (average alignment direction of the molecules) can be horizontal (known as parallel alignment) or vertical (known as homeotropic alignment). In the homeotropic case convection can occur when heating from below or above. In addition, a magnetic field in the same direction as the director can be used as a second control parameter and has a profound effect on the expected phenomena. Several of the predictions have been confirmed by recent experiments. [157] For instance, the measured critical Rayleigh numbers R_c as a function of the magnetic field strength H agree quantitatively with the predictions for both the parallel and the homeotropic case. In the parallel case, there is excellent agreement between experiment and calculations for the orientation of the convection rolls relative to the field and the director. Many opportunities exist in this systemn for additional interesting experiments, and some of these are under way.

5 Outlook

At this point of a review it is customary to make predictions of major break-throughs which will serve to solve the primary outstanding problems. I have no reason to believe that these are likely to occur. The account which I have given of the last two decades or so has been one of gradual progress. In the 1970's we considered the simplest possible problems, namely how time and length scales evolve as the threshold of a forward bifurcation is crossed. [28] Some issues of temporal complexity were raised in spatially extended systems, [37,38] but really no progress towards their illucidation was made. In the 1980's we built upon what had been learned, and gained more insight into complicated bifurcation phenomena [83,87] and into the simplest stability limits of spatially extended systems [100]. During the end of the last and the beginning of the current decade we have learned about non-trivial spatially-extended structures such as pulses and fronts, [111,113,115] we have made great progress in understanding textured patterns, [4,53,54,125] and we have made an attempt to penetrate the secrets of spatio-temporal complexity [133]. I think that this gradual process will continue. As in the past, some will rush ahead and tackle problems well beyond current comprehension; but others will advance more cautiously and study the next-more complicated problem on the horizon in detail. I hope that this work will include further experiments on open systems which can be well controlled in the laboratory, [149] because these are important not only from the scientific viewpoint, but also in relation to practical issues involving industrial processes and engineering applications. I also hope that our understanding of patterns in more complex fluids will increase. A step in that direction is the recent work on RB- and electro-convection in nematics. I hope that this will be continued, and extended to such systems as ferrofluids and visco-elastic fluids. One of the Holy Grails is of course, as it has been in bygone years, the elucidation of spatio-temportal chaos. Are there universality classes which may serve to categorize these complex phenomena? Can these effects be captured in the solutions of simple model equations such as complex Ginzburg-Landau and Swift-Hohenberg equations? Will the much more advanced understanding of dynamical systems be helpful here? To what extent do stochastic effects which come from outside the usual hydrodynamic equations play a role? These are some of the issues which I expect to be addressed during the next decade, albeit one step at a time lest we should get lost at dead ends. In more concrete terms and from the experimentalist's viewpoint, I am hopeful that the phenomena observed in hydrodynamic instabilities in nematic liquid crystals [127,128,158] and perhaps some other anisotropic systems, which *a priori* appear much more complicated than RBC in isotropic fluids, will in a sense provide a simplification by virtue of the preferred orientation. Breaking the rotational symmetry of RBC in isotropic fluids, although it leads to mathematically more complicated equations, may well lead to simpler physical phenomena. As an example I mention here the STC discovered recently by Mike Dennin in our group and illustrated in Fig. 12. [158] In electro-convection in a particular liquid crystal (I52 over the right temperature range and appropriately doped at the right level) one

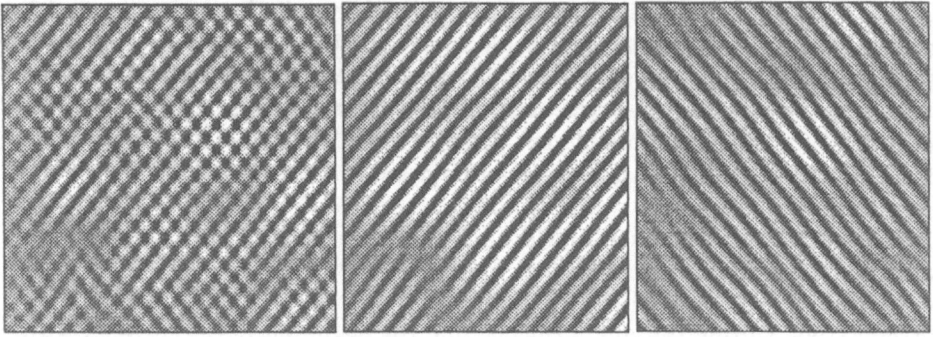

Fig. 12. Spatio-temporal chaos in electro-convection (Ref. 158). The image on the left is Fourier-decomposed into its two components in the middle and on the right.

obtains a forward bifurcation to a state of STC. The state consists of travelling convection rolls which have an oblique orientation to the director. In that case, the two modes with angles Θ and $-\Theta$ are equally likely to occur. Under some conditions, both modes occur simultaneously, on the one hand stabilizing each other and on the other influencing each other's amplitudes in a complex manner. A state of amplitude-chaos arises where the mode amplitudes will, as time passes, go through zero at some spatial locations. At these points, defects can form in the roll structure. Since this state arises *via* a forward bifurcation, one may hope to capture these rich phenomena in a *relatively* simple model, which may consist of a weakly nonlinear theory involving two coupled complex GL equations, one for each of the two roll orientations. It is my hope that such relatively simple examples of STC will serve to lay a foundation for the gradual classification and understanding of these complex phenomena. If significant progress is made in this direction during the next decade, I believe that we as a community should be very satisfied with our accomplishments. I regard as a very hopeful sign for progress in that direction the encouraging number of young people who have entered the field during the last few years.

REFERENCES

1. Q. Ouyang and H.L. Swinney, *Chaos* 1, 411 (1991).
2. W. Breazeal, K.M. Flynn, and E.G Gwinn, to be published.
3. E. Bodenschatz, J. de Bruyn, G. Ahlers, and D.S. Cannell, *Phys. Rev. Lett.* 67, 3078 (1991).
4. Y. Hu, R. Ecke, and G. Ahlers, *Phys. Rev. E* 48, 4399 (1993).
5. S. Ciliberto, E. Pampaloni, and C. Perez-Garcia, *Phys. Rev. Lett.* 61, 1198 (1988).
6. M.C. Cross and P.C. Hohenberg, *Rev. Mod. Phys.* 65, 851 (1993).
7. A large literature pertaining to this field has evolved. Particularly useful as introductions to early work are the reviews by E.L. Koschmieder, *Adv. Chem. Phys.* 26, 177 (1974); and in *Order and Fluctuations in Equilibrium and Nonequilibrium Statistical Mechanics*, XVIIth International Solvay Conference, ed. by G. Nicolis, G. Dewel, and J.W. Turner (Wiley, NY, 1981), p. 168; and by F. Busse, in *Hydrodynamic Instabilities and the Transition to Turbulence*, edited by H.L. Swinney and J.P. Gollub (Springer, Berlin, 1981), p. 97; and in *Rep. Prog. Phys.* 41, 1929 (1978).
8. G. Ahlers, *Phys. Rev. Lett.* 21, 1159 (1968).
9. G. Ahlers, in *Proceedings of the Twelfth International Conference on Low Temperature Physics*, edited by E. Kanda (Academic, Tokyo, 1971), p. 21.
10. P.L Silveston, *Forsch. Ing.-Wes.* 24, 29 (1958).
11. R.J. Donnelly, *Phys. Rev. Lett.* 3, 507 (1959).
12. R.J. Donnelly and N. Simon, . Fluid Mech. 7, 401 (1960).
13. R.J. Donnelly and D. Fultz, *Proc. Nat. Acad. Sci., Wash.* 46, 1150 (1960).
14. E.N. Lorenz, *J. Atmos. Sci.* 20, 130 (1963).
15. An excellent overview of dynamical systems and temporal chaos is provided by *Order within Chaos*, P. Bergé, Y. Pomeau, and C. Vidal (Wiley, NY, 1986).
16. A. Schlüter, D. Lortz, and F. Busse, *J. Fluid Mech.* 23, 129 (1965).
17. F. Busse, *J. Fluid Mech.* 30, 625 (1967).
18. Earlier theoretical investigations of non-Boussinesq convection were carried out by several theorists, starting to my knowledge with Palm [19]; but Busse's calculation was the first which systematically included the temperature dependence of all relevant fluid properties.
19. E. Palm, *J. Fluid Mech.* 8, 183, (1960).
20. G. Ahlers and J.E. Graebner, *Bull. Am. Phys. Soc.* 17, 61 (1972).
21. G. Ahlers, *Bull. Am. Phys. Soc.* 17, 59 (1972).
22. B.D. Wonsiewicz, A.R. Storm, and J.D. Sieber, *Bell Sys. Tech. J.* 57, 2209 (1978).
23. R.P. Behringer and G. Ahlers, *J. Fluid Mech.* 125, 219 (1982).
24. P. Bergé, in *Fluctuations, Instabilities, and Phase Transitions*, edited by T. Riste, (Plenum, NY, 1975), p. 323.
25. M. Dubois and P. Bergé, *J. Fluid Mech.* 85, 641 (1978).

26. G. Ahlers, in *Fluctuations, Instabilities, and Phase Transitions*, edited by T. Riste (Plenum, NY, 1975), p. 181.

27. R.P. Behringer and G. Ahlers, *Phys. Lett.* **62A**, 329 (1977).

28. J. Wesfreid, Y. Pomeau, M. Dubois, C. Normand, and P. Berge, *J. Physique (Paris)* **39**, 725 (1978).

29. L.D. Landau, *Phys. Z. Sowjet* **11**, 26 (1937); **11**, 545 (1937); V.L. Ginzburg and L.D. Landau, *Zh. Eksp. Teor. Fiz.* **20**, 1064 (1950). See also *Collected Papers of L.D. Landau*, edited by D. Ter Haar (Gordon and Breach, New York, 1965) for English translations.

30. A sizable literature now exists dealing with this system. A comprehensive review has been given by R.C. DiPrima and H.L. Swinney, in *Hydrodynamic Instabilities and Transitions to Turbulence*, edited by H.L. Swinney and J.P. Gollub (Springer, Berlin, 1981). Important early papers in this field are numerous; but particularly noteworthy are D. Coles, *J. Fluid Mech.* **21**, 385 (1965); and H.A. Snyder, *J. Fluid Mech.* **35**, 273 (1969); and J.E. Burkhalter and E.L. Koschmieder, *Phys. Fluids* **17**, 1929 (1974).

31. J.P. Gollub and M.H. Freilich, in *Fluctuations, Instabilities, and Phase Transitions*, edited by T. Riste (Plenum, NY, 1974).

32. J.P. Gollub and M.H. Freilich, *Phys. Rev. Lett.* **33**, 1465 (1974).

33. J.P. Gollub and M.H. Freilich, *Phys. Fluids* **19**, 618 (1976).

34. I. Rehberg, Diplom Thesis, Universität Kiel, Kiel, W. Germany, 1980 (unpublished).

35. G. Pfister and I. Rehberg, *Phys. Lett.* **83A**, 19 (1981).

36. R. Graham and J.A. Domaradzki, *Phys. Rev. A* **26**, 1572 (1982).

37. G. Ahlers, *Phys. Rev. Lett.* **33**, 1185 (1974).

38. G. Ahlers and R. P. Behringer, *Prog. Theor. Phys. Suppl.* **64**, 186 (1978).

39. G. Ahlers and R. P. Behringer, *Phys. Rev. Lett.* **40**, 712 (1978).

40. G. Ahlers, *J. Fluid Mech.* **98**, 137 (1980).

41. G. Ahlers, in *Systems Far From Equilibrium*, edited by L. Garrido and J. Garcia (Springer, NY, 1980).

42. G. Ahlers and R.W. Walden, *Phys. Rev. Lett.* **44**, 445 (1980).

43. R.W. Walden and G. Ahlers, *J. Fluid Mech.* **109**, 89 (1981).

44. D.C. Threlfall, *J. Fluid Mech.* **67**, 17 (1975).

45. A. Libchaber and J. Maurer, *J. Phys. Lett.* **39**, L-369 (1978).

46. J.B. McLaughlin and P. Martin, *Phys. Rev. Lett.* **33**, 1189 (1974).

47. J.B. McLaughlin and P. Martin, *Phys. Rev. A* **12**, 186 (1975).

48. H.S. Greenside, G. Ahlers, P.C. Hohenberg, and R.W. Walden, *Physica* **5D**, 322 (1982).

49. U. Frisch and R. Morf, *Phys. Rev. A* **23**, 2673 (1981).

50. F.H. Busse and R.M. Clever, *J. Fluid Mech.* **91**, 319 (1979).

51. R.M. Clever and F.H. Busse, *J. Fluid Mech.* **65**, 625 (1974).

52. R. P. Behringer, *Rev. Mod. Phys.* **57**, 657 (1985).

53. V. Croquette, *Contemp. Phys.* **30**, 153 (1989).

54. Y.C. Hu, R. Ecke, and G. Ahlers, *Phys. Rev. Lett.* **72**, 2191 (1994).

55. E. Bodenschatz, S. Morris, J. de Bruyn, D.S. Cannell, and G. Ahlers, in *Proceedings of the KIT International Workshop on the Physics of Pattern Formation in Complex Dissipative Systems*, edited by S. Kai (World Scientific, Singapore, 1992), p. 227.

56. E. Bodenschatz, D. S. Cannell, J. R. de Bruyn, R. Ecke, Y-C. Hu, K. Lerman, and G. Ahlers, *Physica D* **61**, 77 (1992).

57. J.B. Swift and P.C. Hohenberg, *Phys. Rev. A* **15**, 319 (1977).

58. V.M. Zaitsev and M.I. Shliomis, *Zh. Eksp. Teor. Fiz.* **59**, 1583 (1970) [English translation: *Sov. Phys. JETP* **32**, 866 (1971)].

59. R. Graham, *Phys. Rev. A* **10**, 1762 (1974).

60. G. Ahlers, M.C. Cross, P.C. Hohenberg, and S. Safran, *J. Fluid Mech.* **110**, 297 (1981).

61. H. van Beijeren and E.G.D. Cohen, *J. Stat. Phys.* **53**, 77 (1988).

62. P.C. Hohenberg and J.B. Swift, *Phys. Rev. A* **46**, 4773 (1992).

63. M. Wu, G. Ahlers, and D.S. Cannell, to be published.

64. C.W. Meyer, G. Ahlers, and D.S. Cannell, *Phys. Rev. Lett.* **59**, 1577 (1987).

65. C.W. Meyer, D.S. Cannell, G. Ahlers, J.B. Swift, and P.C. Hohenberg, *Phys. Rev. Lett.* **61**, 947 (1988).

66. C. W. Meyer, G. Ahlers, and D. S. Cannell, *Phys. Rev. A* **44**, 2514 (1991).

67. G. Ahlers, P.C. Hohenberg, and M. Lücke, *Phys. Rev. Lett.* **53**, 48 (1984).

68. G. Ahlers, P.C. Hohenberg, and M. Lücke, *Phys. Rev. A* **32**, 3493 (1985).

69. G. Ahlers, P.C. Hohenberg, and M. Lücke, *Phys. Rev. A* **32**, 3519 (1985).

70. M. Feigenbaum, *J. Stat. Phys.* **19**, 25 (1978).

71. A. Libchaber and J. Maurer, *J. Phys. Colloq. C3* **41**, C3-51 (1978).

72. For the proceedings of this conference, see *Order and Fluctuations in Equilibrium and Nonequilibrium Statistical Mechanics*, XVIIth International Solvay Conference, edited by G. Nicolis, G. Dewel, and J. W. Turner (Wiley, NY, 1981).

73. J. P. Gollub and J.F. Steinman, *Phys. Rev. Lett.* **47**, 505 (1981).

74. For a description of early work by Silveston [10], see Sect. II.18 of Ref. [75].

75. S. Chandrasekhar, *Hydrodynamic and Hydromagnetic Stability*, (Oxford University Press, London, 1961).

76. See, for Instance, F.H. Busse and J.A. Whitehead, *J. Fluid Mech.* **66**, 67 (1974).

77. See, for instance, V. Steinberg, G. Ahlers, and D.S. Cannell, *Phys. Script.* **32**, 534 (1985).

78. S. Rasenat, G. Hartung, B. L. Winkler, and I. Rehberg, *Experiments in Fluids* **7**, 412 (1989).

79. H. R. Haller, C. Destor, and D. S. Cannell, *Rev. Sci. Instrum.* **54**, 973 (1983).

80. For a detailed review, see R.C. DiPrima and H.L. Swinney, in *Hydrodynamic Instabilities and Transitions to Turbulence*, edited by H.L. Swinney and J.P. Gollub (Springer, Berlin, 1981).

81. M. Golubitsky and J. Guckenheimer, editors, *Multiparameter Bifurcation Theory*, *Contemp. Math.* **56** (Am. Math. Soc., Providence, RI 1986).
82. T.B. Benjamin and T. Mullin, *Proc. Roy. Soc. Lond. A* **377**, 221 (1981).
83. A. Aitta, G. Ahlers, and D.S. Cannell, *Phys. Rev. Lett.* **54**, 673 (1985).
84. K.A. Cliffe, *J. Fluid Mech.* **135**, 219 (1983).
85. M. Lücke, M. Mihelcic, K. Wingerath, and G. Pfister, *J. Fluid Mech.* **140**, 343 (1984).
86. A. Aitta, *Phys. Rev. Lett.* **62**, 2116 (1989).
87. I. Rehberg and G. Ahlers, *Phys. Rev. Lett.* **55**, 500 (1985).
88. I. Rehberg and G. Ahlers, *Contemporary Mathematics* **56**, 277 (1986).
89. H. Brand and V. Steinberg, *Phys. Lett.* **93A**, 333 (1983).
90. H.R. Brand, P.C. Hohenberg, and V. Steinberg, *Phys. Rev. A* **27**, 591 (1983).
91. G. Ahlers and I. Rehberg, *Phys. Rev. Lett.* **56**, 1373 (1986).
92. T.S. Sullivan and G. Ahlers, *Phys. Rev. Lett.* **61**, 78 (1988).
93. T.S. Sullivan and G. Ahlers, *Phys. Rev. A* **38**, 3143 (1988).
94. S.J. Linz and M. Lücke, *Phys. Rev. A* **36**, 3505 (1987).
95. E. Knobloch and D.R. Moore, *Phys. Rev. A* **37**, 860 (1988).
96. M.C. Cross and K. Kim, *Phys. Rev. A* **37**, 3909 (1988).
97. W. Eckhaus, *Studies in Nonlinear Stability Theory* (Springer, NY, 1965).
98. L. Kramer, E. Ben-Jacob, H. Brand, and M.C. Cross, *Phys. Rev. Lett.* **49**, 1891 (1982).
99. D.S. Cannell, M.A. Dominguez-Lerma, and G. Ahlers, *Phys. Rev. Lett.* **50**, 1365 (1983).
100. M.A. Dominguez-Lerma, D.S. Cannell, and G. Ahlers, *Phys. Rev. A* **34**, 4956 (1986).
101. G. Ahlers, D.S. Cannell, M.A. Dominguez-Lerma, and R. Heinrichs, *Physica* **23D**, 202 (1986).
102. M. Boucif, J.E. Wesfreid, and E. Guyon, *J. Phys. Lett.* **45**, 413 (1984).
103. H.A. Snyder, *J. Fluid Mech.* **35**, 273 (1969).
104. M. Lowe and J.P. Gollub, *Phys. Rev. Lett.* **55**, 2575 (1985).
105. H. Riecke and H.G. Paap, *Phys. Rev. A* **33**, 547 (1986).
106. L. Kramer and H. Riecke, *Z. Phys.* **B59**, 245 (1985).
107. H. Riecke and H.G. Paap, *Phys. Rev. Lett.* **59**, 2570 (1987).
108. L. Ning, G. Ahlers, and D.S. Cannell, *Phys. Rev. Lett.* **64**, 1235 (1990).
109. I. Rehberg, E. Bodenschatz, B. Winkler, and F.H. Busse, *Phys. Rev. Lett.* **59**, 282 (1987).
110. R. W. Walden, P. Kolodner, A. Passner, and C. M. Surko, *Phys. Rev. Lett.* **55**, 496, (1985).
111. R. Heinrichs, G. Ahlers, and D.S. Cannell, *Phys. Rev. A* **35**, 2761 (1987).
112. G. Ahlers, D.S. Cannell, and R. Heinrichs, *Nucl. Phys. B (Proc. Suppl.)* **2**, 77 (1987).
113. E. Moses, J. Fineberg, and V. Steinberg, *Phys. Rev. A* **35**, 2757 (1987).
114. P. Kolodner, D. Bensimon, and C.M. Surko, *Phys. Rev. Lett.* **60**, 1723 (1988).

115. J.J. Niemela, G. Ahlers, and D.S. Cannell, *Phys. Rev. Lett.* **64**, 1365 (1990); and unpublished.

116. O. Thual and S. Fauve, *J. Phys. (France)* **49**, 1829 (1988).

117. S. Fauve and O. Thual, *Phys. Rev. Lett.* **64**, 282 (1990).

118. H.R. Brand and R.J. Deissler, *Phys. Rev. Lett.* **63**, 2801 (1989).

119. H.R. Brand and R.J. Deissler, *Phys. Rev. A* **41**, 5478 (1990).

120. W. Barten, M. Lücke, and M. Kamps, *Phys. Rev. Lett.* **66**, 2621 (1991).

121. V. Hakim, P. Jakobsen, and Y. Pomeau, *Europhys. Lett.* **11**, 19 (1990).

122. W. van Saarloos and P.C. Hohenberg, *Physica D* **56**, 303 (1992).

123. K. Lerman, E. Bodenschatz, D.S. Cannell, and G. Ahlers, *Phys. Rev. Lett.* **70**, 3572 (1993).

124. R.J. Deissler and H. Brand, *Phys. Rev. A* **44**, 3411 (1991).

125. V. Croquette, *Contemp. Phys.* **30**, 113 (1989).

126. F. Daviaud and A. Pocheau, *Europhys. Lett.* **9**, 7 (1989).

127. For a recent review of pattern formation in electro-convection of nematic liquid crystals, see I. Rehberg, B.L. Winkler, M. de la Torre-Juarez, S. Rasenat, and W. Schöpf, *Festkörperprobleme-Advances in Solid State Physics* **29**, 35 (1989).

128. L. Kramer and W. Pesch, *Annu. Rev. Fluid Mech.* **27**, in print.

129. E. Bodenschatz, W. Zimmermann, and L. Kramer, *J. Phys. France* **49**, 1875 (1988).

130. J.R. deBruyn, E. Bodenschatz, S. Morris, S. Trainoff, Y.-C. Hu, D.S. Cannell, and G. Ahlers, in preparation.

131. L.D. Landau and E.M. Lifshitz, *Fluid Mechanics* (Addison-Wesley, Reading MA, 1959).

132. W. Hort, S.J. Linz, and M. Lücke, *Phys. Rev. A* **45**, 3737 (1992).

133. S.W. Morris, E. Bodenschatz, D.S. Cannell, and G. Ahlers, *Bull. Am. Phys. Soc.* **37**, 1734 (1992); *Phys. Rev. Lett.* **71**, 2026 (1993).

134. P.W. Anderson, in *Order and Fluctuations in Equilibrium and Nonequilibrium Statistical Mechanics*, XVIIth International Solvay Conference, edited by G. Nicolis, G. Dewel, and J. W. Turner (Wiley, NY, 1981)

135. F.H. Busse and R.M. Clever, in *Recent Developments in Theoretical and Experimental Fluid Mechanics-Compressible and Uncompressible Flows*, edited by U. Mueller, K.G. Roesner, and B. Schmidt (Springer, 1979).

136. M. Bestehorn, M. Fantz, R. Friedrich, and H. Haken, *Phys. Lett. A* **174**, 48 (1993)

137. H. Xi, J.D. Gunton, and J. Viñals, *Phys. Rev. Lett.* **71**, 2030 (1993).

138. W. Decker, W. Pesch, and A. Weber, *Phys. Rev. Lett.* **73**, 648 (1994).

139. G. Küppers and D. Lortz, *J. Fluid Mech.* **35**, 609 (1969).

140. R.M. Clever and F.H. Busse, *J. Fluid Mech.* **94**, 609 (1979).

141. K.E. Heikes and F.H. Busse, *Annals of the N.Y. Academy of Sciences* **357**, 28 (1980).

142. Y. Tu and M.C. Cross, *Phys. Rev. Lett.* **69**, 2515 (1992).

143. Y.C. Hu, R. Ecke, and G. Ahlers, to be published.

144. I. Rehberg, S. Rasenat, M. de la Torre-Juarez, W. Schöpf, F. Hörner, G. Ahlers, and H.R. Brand, *Phys. Rev. Lett.* **67**, 596 (1991); and unpublished.
145. P.G. de Gennes, *The Physics of Liquid Crystals* (Clarendon Press, Oxford, 1973).
146. R. Graham, in *Fluctuations, Instabilities, and Phase Transitions*, edited by T. Riste (Plenum, NY, 1975), p. 215.
147. K. Babcock, G. Ahlers, and D.S. Cannell, *Phys. Rev. Lett.* **67**, 3388 (1991).
148. K.L. Babcock, D.S. Cannell, and G. Ahlers, *Physica D* **61**, 40 (1992).
149. K. Babcock, G. Ahlers, and D.S. Cannell, *Phys. Rev. E*, submitted.
150. A. Tsameret and V. Steinberg, *Europhys. Lett.* **14**, 331 (1991).
151. A. Tsameret and V. Steinberg, *Phys. Rev. Lett.* **67**, 3392 (1991).
152. A. Tsameret and V. Steinberg, *Phys. Rev. E* **49**, 1291 (1994).
153. A. Tsameret, G. Goldner, and V. Steinberg, *Phys. Rev. E* **49**, 1309 (1994).
154. R.J. Deissler, *Physica* **25D**, 233 (1987).
155. Q. Feng, W. Pesch, and L. Kramer, *Phys. Rev. A* **45**, 7242 (1992).
156. Q. Feng, W. Decker, W. Pesch, and L. Kramer, *J. Phys. France* **II 2**, 1303 (1992).
157. L.I. Berge, G. Ahlers, and D.S. Cannell, *Phys. Rev. E* **48**, R3236 (1994).
158. M. Dennin, D.S. Cannell, and G. Ahlers, *Molec. Cryst. Liq. Cryst.*, in print.

Fluctuations in the steady state

Robert Graham

Fachbereich Physik, Universitaet GH Essen
45117 Essen, Germany

Abstract

Fluctuations in the steady state of macroscopic systems can be described
by non-equilibrium potentials, whose theory has been developed considerably
during the last two decades. In the present lecture we review some recent developments connected with the appearance of plateaus in the non-equilibrium
potential in regions of configuration space containing saddles of the deterministic
dynamics.

1 Introduction

Thermodynamic equilibria are always beset with fluctuations, which occur in-
evitably, simply because the equilibrium is not static, i.e. a state of absolute
rest, but dynamic, i.e. a state of perpetual microscopic motion. These fluctuations are small and may not be noticeable for macroscopic variables; they are
enhanced near bifurcations or phase transitions. Fluctuations in thermodynamic
equilibrium states all share one symmetry property, namely detailed balance or,
in systems with some variables and/or parameters transforming odd under time
reversal, the appropriately generalized form of detailed balance.

For systems in stationary non-equilibrium states (steady states) the above
mentioned properties of fluctuations in thermodynamic equilibrium still hold except for the detailed balance property. The consequences of this exception are
important: the phenomenology of fluctuations becomes much richer and fluctuations may have more drastic influences; see e.g. the collection of reviews in
[1]. Let us consider some consequences of the absence of detailed balance. Any
steady state, like a state of thermodynamic equilibrium, can be characterized
by an invariant, i.e. time-independent, probability measure, and we call W the
corresponding probability density, if it exists. The probability current density
g in the steady state is sourceless, $\nabla \cdot g = 0$, where ∇ is the gradient in the
configuration space of the macroscopic variables $\mathbf{x}$. If detailed balance holds,
then either $\mathbf{g} = 0$, or, if there are variables which transform odd under time
reversal, then $\mathbf{g}$ may be non-zero, but changes sign under time-reversal. In other

words, only reversible probability currents may flow in thermodynamic equilibrium. In steady states, on the other hand, the probability current is not reversed under time reversal and describes irreversible cyclic fluctuation processes. The rate vector of these cyclic fluctuation processes is defined as $\mathbf{r} = \mathbf{g}/W$. This rate vector has sources and sinks given by

$$-\nabla \cdot \mathbf{r} = \mathbf{r} \cdot \nabla \ln W. \tag{1}$$

The right hand side may be interpreted as the local production rate of $\ln W$ along the cyclic fluctuation process, which gives the strength of the sinks of $\mathbf{r}$. The appearance of these cyclic non-reversible rates $\mathbf{r}$ is the most characteristic feature of stationary non-equilibrium states. In some cases they may completely dominate the properties of a system. E.g. in the weak-noise limit the rate $\mathbf{r}$ does not vanish, in general, and is responsible for the possibility of an appearance, in that limit, of attractors more complicated than fixed points (in which $\mathbf{r}$ would have to vanish), like limit cycles or even strange attractors. Together with the possibility of the occurence of such attractors, there comes a much richer phenomenology of bifurcations and concomitantly enhanced fluctuations with new types of critical behavior and scaling properties. Mean field-type considerations and the weak-noise limit, in which the noise intensity, say ϵ, is taken to zero asymptotically [2, 3, 4], play an important role in non-equilibrium systems, where correlation lengths are usually large, even far from bifurcation points, so that the local averaging implied by mean field descriptions leading to small noise is justified. In fact, to my knowledge, non-classical critical exponents (which would, of course, signal a break-down of the mean-field) have never been observed in non-equilibrium systems. Instead, the interesting behavior in such systems arises from the appearance of the cyclic non-reversible rates $\mathbf{r}$. This gives rise to the possibility of instabilities and bifurcations with new and different scaling properties without counterpart in thermodynamic equilibrium states. To discuss the weak-noise limit we write

$$W(\mathbf{x}) \sim \exp\left(-\frac{\phi(\mathbf{x})}{\epsilon}\right) \tag{2}$$

whereby the nonequilibrium potential $\phi(\mathbf{x})$ is introduced. Furthermore, we define

$$\mathbf{r}_0(\mathbf{x}) = \lim_{\epsilon \to 0} \mathbf{r}(\mathbf{x}). \tag{3}$$

For $\epsilon \to 0$ the relaxation dynamics to the deterministic attractors is described by some dynamical equations of the form

$$\dot{\mathbf{x}}\big|_{\mathrm{rel}} = \mathbf{K}(\mathbf{x}) = \mathbf{r}_0(\mathbf{x}) + \mathbf{d}(\mathbf{x}) \tag{4}$$

where the right-hand side defines $\mathbf{d}(\mathbf{x})$. While $\mathbf{r}_0(\mathbf{x})$ describes cyclic transitions and does not lead to relaxation, the part $\mathbf{d}(\mathbf{x})$ is responsible for the approach to the deterministic attractors. Fluctuations away from the deterministic attractors occur very rarely, for $\epsilon \to 0$, and if they occur then with overwhelming probability along paths which satisfy $\dot{\mathbf{x}}\big|_{\mathrm{fluct}} = \mathbf{r}_0(\mathbf{x}) - \mathbf{d}(\mathbf{x})$. Here, only the part $-\mathbf{d}(\mathbf{x})$ allows the system to get away from the deterministic attractor. Thus, if $\mathbf{r}_0(\mathbf{x}) \neq 0$, then the fluctuation paths are not simply the relaxation paths in reverse. The cyclic rate $\mathbf{r}_0(\mathbf{x})$ is just the mean of the fluctuation rate and the relaxation rate

$$\mathbf{r}_0(\mathbf{x}) = \frac{1}{2}\left(\dot{\mathbf{x}}\big|_{\text{fluct}} + \dot{\mathbf{x}}\big|_{\text{rel}}\right) \tag{5}$$

The non-equilibrium potential $\phi(\mathbf{x})$ defined by (2) is the most convenient object for characterizing a steady state in the weak-noise limit.

For Markoffian models such as stochastic flows of the form (in the sense of Ito)

$$\dot{\mathbf{x}} = \mathbf{K}(\mathbf{x}) + \epsilon \sum_k \mathbf{g}_k(\mathbf{x})\xi_k(t)$$

$$\langle \xi_k(t)\rangle = 0$$
$$\langle \xi_k(t)\xi_l(t')\rangle = \delta_{kl}\delta(t - t'), \text{ Gaussian}, \tag{6}$$

ϕ has the property

$$\frac{d\phi(\mathbf{x})}{dt}\bigg|_{\text{rel}} \leq 0. \tag{7}$$

(A similar property $\phi(\mathbf{x}') - \phi(\mathbf{x}) \leq 0$ holds for discrete stochastic maps $\mathbf{x} \to \mathbf{x}'$.) Furthermore, in such models the production rate of ϕ for the cyclic processes in the steady state vanishes for $\epsilon \to 0$

$$\mathbf{r}_0 \cdot \nabla\phi = \epsilon\nabla \cdot \mathbf{r}_0 \to 0 \tag{8}$$

$\mathbf{r}_0$ is related with $\nabla\phi$ via

$$\mathbf{r}_0(\mathbf{x}) = \mathbf{K}(\mathbf{x}) + \frac{1}{2}\mathbf{Q} \cdot \nabla\phi. \tag{9}$$

Here the components of the tensor $\mathbf{Q} = \sum_k \mathbf{g}_k \otimes \mathbf{g}_k$ in configuration space form a set of generalized transport coefficients. ϕ is defined by eqs. (8), (9) from

$$\mathbf{K} \cdot \nabla\phi + \frac{1}{2}\nabla\phi \cdot \mathbf{Q} \cdot \nabla\phi = 0 \tag{10}$$

as a single valued function only on the Lagrangean manifold $\mathbf{p} = \nabla\phi(\mathbf{x})$ in the phase-space $(\mathbf{p}, \mathbf{x})$. The projection down on the configuration space, in general, leads to a multi-valued function, of which the envelope from below gives the highest probability in (2) and must therefore be selected. In this form discontinuities in $\nabla\phi$ arise, which also entail discontinuities in $\mathbf{r}$. All these properties of ϕ are rather well known and have been summarized here very briefly, in order to present some backround. More detail can be found in the quoted literature.

In the bulk of the present lecture I wish to focus on some novel aspects of the problem of the escape out of a basin of attraction due to fluctuations. The escape time, for $\epsilon \to 0$, satisfies an Arrhenius-type formula

$$\tau \simeq \tau_0 \exp\left(\frac{\Delta\phi}{\epsilon}\right) \tag{11}$$

where τ_0 is a deterministic time and $\Delta\phi$ is the difference of the smallest value of the non-equilibrium potential on the basin boundary and its value on the attractor. Quite often $\Delta\phi$ is determined by a potential plateau, i.e. a whole region where $\phi = $ const, which develops in the region of configuration space where the exit from the domain of attraction is most likely to occur. Such potential plateaus have been noted and discussed in stochastic maps and stochastic flows. In the section 2 we shall first discuss the appearance of a potential plateau in simple examples of stochastic flows and then mention more examples occuring in maps. Then I want to turn to the discussion of a phenomenon intimately related to potential plateaus, which was considered in [5] and has been called 'noise-induced attractor explosion' there. To this end section 3 first develops the concept of a 'noisy attractor'. Section 4 introduces 'attractor explosions'. Finally, section 5 describes some universal scaling of potential plateaus and the related escape times near bifurcations. The discussion of section 3-4 is based on the work of ref. [5].

2 Potential plateaus

Potential plateaus develop rather easily and have been found in the steady states of both continuous stochastic systems and in stochastic maps. A very simple example is provided by the overdamped stochastic pendulum under a constant external torque described by the stochastic differential equation

$$\gamma\dot{x} = F - U_0\sin(x - x_0) + \xi(t) \tag{12}$$

with Gaussian white noise of intensity $\langle\xi(t)\xi(t')\rangle = 2kT\gamma\delta(t-t')$. The coordinate x is defined in $-\pi < x \leq \pi$, i.e. it is a phase angle. We introduced x_0 to put the unstable equilibrium of the deterministic part of (12) in $x_s = \pi$, i.e. $F - U_0\sin x_0 = 0$. The steady-state probability density (1), and hence the non-equilibrium potential, is easily calculated for this system [6]. Using the parameter $\epsilon = \frac{kT}{U_0}$ to characterize the noise strength, the non-equilibrium potential takes the form

$$\phi = \begin{cases} -\frac{F}{U_0}x & -\cos(x - x_0) + \phi_0 & \bar{x}_1 \leq x \leq \pi \\ \phi_0 - \frac{F}{U_0}\pi & +\cos x_0 & -\pi < x \leq \bar{x}_1 \end{cases} . \tag{13}$$

Here ϕ_0 is chosen in such a way that $\phi = 0$ in the minimum of the potential in the interval $(-\pi, \pi]$. Such a minimum exists if $\frac{F}{U_0}$ does not surpass a critical size. The potential has a plateau in the whole domain $-\pi < x \leq \bar{x}_1$ where $\bar{x}_1$ is the only solution $\bar{x}_1 \neq -\pi, \pi$ of

$$-\frac{F}{U_0}\bar{x}_1 - \cos(\bar{x}_1 - x_0) = -\frac{F}{U_0}\pi + \cos x_0 \tag{14}$$

in the interval $-\pi < \bar{x}_1 < \pi$. In the first domain $\bar{x}_1 \leq x \leq \pi$ the circular transition rate r vanishes. At $x_s = \pi$, which is to be identified with $\bar{x}_s = -\pi$, it starts to develop a non-vanishing value

$$r(x) = \frac{1}{\gamma}\left(F - U_0\sin(x - x_0)\right) \tag{15}$$

in a continuous way and jumps discontinuosly to zero at $x = \bar{x}_1$, where according to eq. (13) $\partial\phi/\partial x$ also has a discontinuity. Apart from allowing a domain with $r(x) \neq 0$, the potential plateau in the present simple example does not have drastic consequences.

The following more interesting example has been discussed recently by Maier and Stein [7]. To discuss it let us first consider a case with detailed balance, which might therefore arise in an equilibrium system. We consider the noisy gradient flow

$$\dot{\mathbf{x}} = -\frac{1}{2}\nabla\phi + \sqrt{\epsilon}\,\boldsymbol{\xi} \tag{16}$$

with normalized Gaussian white noise $\boldsymbol{\xi}$ and ϕ of the form

$$\phi = -\left(x^2 + \frac{1}{2}y^2\right) + \frac{1}{2}\left(x^4 + \frac{1}{2}y^4\right). \tag{17}$$

This system has $\mathbf{r} = 0$. The deterministic dynamics is completely symmetric under $x \to -x$ and $y \to -y$. It has attractors in $(\pm 1, 0)$, saddles in $(0, \pm 1)$, and a repeller in $(0, 0)$. Since $\mathbf{r} = 0$ it follows that the most probable fluctuation paths satisfy $\dot{\mathbf{x}}|_n = +\frac{1}{2}\nabla\phi$. Therefore, starting in the attractor $(1, 0)$ say, they climb the potential to the saddles at $(0, \pm 1)$ in order to leave the basin of attraction there. A seemingly similar problem, but without detailed balance, is

$$\dot{\mathbf{x}} = \begin{pmatrix} x - x^3 \\ y - y^3 - 2x^2y \end{pmatrix} + \sqrt{\epsilon}\,\boldsymbol{\xi}. \tag{18}$$

The deterministic dynamics has the same symmetries, and the same attractors (at $(\pm 1, 0)$), saddles (at$(0, \pm 1)$), and repeller (at $(0, 0)$). The non-equilibrium potential for this problem can be calculated, and has the form

$$\phi = \begin{cases} -x^2 + \frac{x^4}{2} + x^2y^2 - y^2 + \frac{y^4}{2} + 1 & y^2 \geq 1 \\ -x^2 + \frac{x^4}{2} + x^2y^2 + \frac{1}{2} & y^2 \leq 1 \end{cases}. \tag{19}$$

The cyclic rate in the two regions is then given by

$$\mathbf{r} = \begin{cases} \begin{pmatrix} xy^2 \\ -x^2y \end{pmatrix} & y^2 \geq 1 \\ \begin{pmatrix} xy^2 \\ y - y^3 - x^2y \end{pmatrix} & y^2 \leq 1 \end{cases}. \tag{20}$$

The first thing to remark is that $\phi = \frac{1}{2}$ on the line-segment $x = 0$, $-1 \leq y \leq +1$, i.e. there is a potential plateau. We also note that this 'strict' plateau is embedded in a larger region around the y-axis for $-1 \leq y \leq 1$, where the potential hardly changes. The next thing to observe is that the escape paths, e.g. from $(1, 0)$, are drastically modified by $\mathbf{r} \neq 0$. As discussed in detail in [7], they don't climb via the saddles, where they would be diverted by $\mathbf{r} \neq 0$, but instead pass along the x-axis, where it happens that $\mathbf{r} = 0$, and pass very closely by the repeller $(0, 0)$ which is embedded in the potential plateau. The crossing of the y-axis into the neighboring domain of attraction of $(-1, 0)$ occurs with about equal probability everywhere in the domain $-1 \leq y \leq 1$ due to the presence of the plateau. Quantitatively this probability is then determined by the y-dependence of the prefactor in (11), which in fact turns out to be independent of y.

Potential plateaus are, in fact, quite common in non-equilibrium systems. Let us only mention two more examples from systems described by discrete stochastic maps [8, 9, 10, 11, 12]

$$\mathbf{X}_{n+1}^{(\epsilon)} = \mathbf{F}\left(\mathbf{X}_n^{(\epsilon)}\right) + \boldsymbol{\xi}_n^{(\epsilon)} \tag{21}$$

with

$$\langle \boldsymbol{\xi}_n^{(\epsilon)} \otimes \boldsymbol{\xi}_{n'}^{(\epsilon)} \rangle = \epsilon \, \underline{1} \, \delta_{nn'} \tag{22}$$

and

$$W^{(\epsilon)} \sim \exp\left(-\frac{\phi(\mathbf{x})}{\epsilon}\right) . \tag{23}$$

E.g. in the period-3 window of the logistic map (and all one-dimensional maps in its universality class) there is a fractal repeller which is embedded in a potential plateau [11]. Another example are the very intricate basin boundaries of coexisting attractors in two-dimensional maps. For complex analytic maps these are the Julia sets. As shown for a particular example (Newton's root-finding algorithm for the solution of $z^3 - 1 = 0$) in [11, 12], the Julia set is embedded in a potential plateau, which itself is embedded in a larger region of configuration space in which the potential hardly changes. Further examples will be mentioned in the following sections, where we turn to the discussion of one striking consequence of potential plateaus, the sudden enlargement, or 'explosion', of 'noisy attractors' with the increase of the noise strength beyond a critical value.

3 Noisy attractors

The following three sections will present a discussion of some recent work done in collaboration with Andreas Hamm and Tamas Tèl [5] in which potential plateaus feature prominently. This work is related to earlier work by Sommerer and coworkers [13, 14, 15] and by Beale [16].

The work reported in [5] makes use of a notion of 'noisy attractor', which is defined as the set

$$A_{\epsilon,x} = \left\{\mathbf{x} : W^{(\epsilon)}(\mathbf{x}) \geq \chi\right\} . \tag{24}$$

Here χ is a small number, which defines the lowest value of $W^{(\epsilon)}(\mathbf{x}) \geq 0$ which can be reliably resolved in a given experiment. For weak noise the noisy attractor can be approximated by the set

$$\bar{A}_{\epsilon,x} = \left\{\mathbf{x} : \phi(\mathbf{x}) \leq \epsilon \ln \chi^{-1}\right\} \tag{25}$$

which is defined in terms of the non-equilibrium potential alone. This approximation is valid if $|\phi - \epsilon \ln W^{(\epsilon)}| << \epsilon \ln \chi^{-1}$, i.e. for ϵ and χ both sufficiently small. One can argue that in any physical system in the steady state what one observes is the noisy attractor (24) (approximated by (25)), rather than the attractor of the idealized deterministic dynamical system. Of course, in may cases the noise, even though always present, is so small as to constitute an irrelevant complication, and the most constructive point of view then is to neglect it. Here we are interested in the opposite case.

The interesting feature of $A_{\epsilon,\chi}$ and $\bar{A}_{\epsilon,\chi}$ is that it changes with the noise-strength. For sufficiently small noise the noisy attractor is localized around the *global* minimum of ϕ. In general, this will not comprise all deterministic attractors of the system, which correspond to local minima of ϕ only, but only the most stable one under the noisy perturbations of the given system. With increasing noise the size of the noisy attractor grows locally, and in addition new disconnected components may appear, corresponding to local minima of ϕ, which now contribute in (25). Finally, a sudden increase of the size of the noisy attractor may occur once the noise strength surpasses a critical value. This happens if the noise strength becomes sufficiently large to excite the system over a potential barrier defined by a plateau. The whole region containing the plateau is then suddenly added to the noisy attractor. For examples of this phenomenon I refer to [5].

4 Potential plateaus and escape times near bifurcations

For an 'attractor explosion' of the indicated type to occur it is necessary that the barrier defined by the plateau be sufficiently low to be reached even for small noise. This condition is naturally satisfied near bifurcations where the barrier-height goes to zero. The first studies were made by Arecchi et al [17] and by Sommerer et al [13] for crisis points [18]. In [5] this and the case of tangent bifurcations have been considered and we shall discuss the latter. Let $F_\lambda(\mathbf{x})$ be a map with a tangent bifurcation at $\lambda = 0$ such that the unstable periodic orbit for $\lambda > 0$, is embedded in a Cantor saddle. In this case there will be a potential plateau in which the Cantor saddle and the unstable periodic orbit are both embedded. Near a tangent bifurcation the dynamics, restricted to its center manifold, may be described the 1-dimensional map

$$(\lambda, y) \rightarrow (\lambda, y + \lambda - y^2) \tag{26}$$

where the parametrization of the center manifold by (λ, y) has to be suitably chosen in order to achieve this normal form. For $\lambda > 0$ and small there is a stable and an unstable fixed point of this map whose distance Δy satisfies $\Delta y \sim \sqrt{\lambda}$. If the deterministic map is perturbed by (weak) Gaussian white noise ξ with a distribution

$$\psi(\xi) \sim \exp\left(-\frac{\xi^2}{2\epsilon}\right) \tag{27}$$

the non-equilibrium potential rises quadratically from the minimum at the stable fixed point. Therefore the height of the plateau, which coincides with the value of the potential at the unstable fixed point of (26), is given by

$$\Delta\phi_\lambda(\Delta y) \sim C(\lambda)|\Delta y|^2. \tag{28}$$

The coefficient can be estimated [5] by methods given in [11] with the result

$$C(\lambda) \sim \sqrt{\lambda} \qquad (29)$$

which gives the result that the height of the potential plateau scales like

$$\Delta\phi_\lambda \sim \lambda^{3/2} \qquad (30)$$

for all bifurcations of this kind. This leads to the prediction that the critical noise-strength for the predicted 'noisy attractor explosion' scale also as

$$\epsilon_c \sim \lambda^{3/2}. \qquad (31)$$

Furthermore, the escape time from the deterministic attractor should scale as

$$\tau^{(\epsilon)} = \tau_0(\lambda) \exp\left(\frac{\Delta\phi_\lambda}{\epsilon}\right) \qquad (32)$$

where $\tau_0(\lambda)$ is a deterministic time associated with the bifurcation which scales as [19]

$$\tau_0(\lambda) \sim \frac{1}{\sqrt{\lambda}}. \qquad (33)$$

Hence, one predicts the existence of a universal scaling function

$$\sqrt{\lambda}\tau^{\epsilon\lambda^{3/2}} \sim \exp\left(\frac{\text{const}}{\epsilon}\right) \qquad (34)$$

which is well born out by numerical experiments [5].

It is interesting, that in the derivation of this result the assumption of Gaussian white noise had to be used explicitly. Indeed, in[5] the derivation was repeated for non-Gaussian noise with a distribution

$$\psi(\xi) \sim \exp\left(-\frac{|\xi|^r}{r\epsilon^{r/2}}\right) \qquad (35)$$

which contains the Gaussian case for $r = 2$, but also covers, for instance, the case of a uniform box-distribution for $r \to \infty$. The non-equilibrium potential then is defined by

$$W^{(\epsilon)}(x) \sim \exp\left(-\frac{\phi(x)}{\epsilon^{r/2}}\right). \qquad (36)$$

The scaling of the height of the potential plateau is now obtained as

$$\Delta\phi_\lambda(\Delta y) \sim \lambda^{\frac{r-1}{2}}|\Delta y|^r. \qquad (37)$$

For the critical noise strength of the attractor-explosion one finds accordingly

$$\epsilon_c \sim (\Delta\phi_\lambda)^{\frac{2}{r}} \sim \lambda^{2-\frac{1}{r}} \qquad (38)$$

and the escape time becomes

$$\tau^{(\epsilon)} \sim \frac{1}{\sqrt{\lambda}} \exp\left(\frac{\Delta\phi_\lambda}{\epsilon^{r/2}}\right). \qquad (39)$$

The scaling relation (34) therefore changes and becomes

$$\sqrt{\lambda} \tau^{\epsilon \lambda^{2-1/r}} \sim \exp\left(\frac{\text{const}}{\epsilon^{r/2}}\right). \qquad (40)$$

This result seems to be in conflict with an earlier result in the literature [19] in which a scaling law of the form (34) was found to hold independent of the noise distribution with a universal (but unspecified) scaling function. This seeming contradiction was clarified in [5], by observing that the limits $\lambda \to 0$ and $\epsilon \to 0$ do not commute. If $\lambda \to 0$ *before* $\epsilon \to 0$, then the random perturbations in all time steps add up and Gaussian behavior is obtained by the central limit theorem irrespective of the form of $\psi(\xi)$, and the result (34) becomes universal as claimed in [19]. However, if $\lambda \to 0$ *after* $\epsilon \to 0$, as is assumed when adopting the description via a non-equilibrium potential, then the central limit theorem is not applicable and non-Gaussian scaling behavior is obtained. Numerical simulations reported in [5] bear out this conclusion.

5 Conclusion

In the preceeding paragraphs we have discussed fluctuations in steady states under conditions where these fluctuations are small and the ideal deterministic behavior of the limiting noiseless system still exerts a strong influence in spite of the presence of noice. In such cases the non-equilibrium potential ϕ and the circular transition rate r_0 are meaningful and useful objects. A lot has been learned about this description in the last two decades and the literature on this subject continues to grow. I have concentrated here on one aspect only – the frequent appearance of potential plateaus. They frequently appear in regions where the deterministic dynamics has saddles or repellers and therefore they may exert a dominating influence on escape times and escape paths. Furthermore, if low-lying potential plateaus appear, like in systems with strange saddles approaching a bifurcation point, or like in systems close to a crisis, then a slight increase of noise, or a further approach to the bifurcation point, may lead to a sudden enlargement of the observed 'noisy attractor'. This is accompanied by a universal scaling of the plateau height, the critical noise strength, the critical size of the bifurcation parameter, and the escape time of the attractor. Often this escape leads to a wandering of the system in the region of the potential plateau (if a strange repeller or saddle is embedded in that region) which is observed as noise-induced intermittency [5].

I may end with expressing my wish that in the next twenty years the progress with these methods may even surpass that achieved in the past two decades.

Acknowledgement

I wish to acknowledge my long-standing and fruitful collaboration in this field with Andreas Hamm and Tamas Tél whose results and insights contributed enormously to my own understanding of the subjects of this lecture. This work was supported by the Deutsche Forschungsgemeinschaft through the Sonderforschungsbereich 'Unordnung und große Fluktuationen'.

References

1. F. Moss and P.V.E. McClintock eds.: Noise in Nonlinear Dynamical Systems. (Cambridge University Press, Cambridge 1989) Vol. 1-3
2. M.I. Freidlin and A.D. Wentzell: Random Perturbation of Dynamical Systems. (Springer Verlag, Berlin 1984)
3. R. Graham: Noise in Nonlinear Dynamical Systems. (Cambridge University Press, Cambridge 1989) Vol. 1, p. 225
4. P. Talkner and P. Hänggi: Noise in Nonlinear Dynamical Systems. (Cambridge University Press, Cambridge 1989) Vol. 2, p. 87
5. Hamm, A., Tél, T., and Graham, R.: Phys. Rev. **A185** (1994) 313-320
6. Graham, R., and Tél, T.: Phys. Rev. **A33** (1986) 1322-1337
7. Maier, R.S., and Stein, D.L. Phys. Rev. **E48** (1993) 931
8. Reimann, P., And Talkner, P. Helv. Phys. Acta **63** (1990) 845
9. Reimann, P., and Talkner, P. Phys. Rev. **A44** (1991) 6348
10. Graham, R., and Hamm, A.: in 'From Phase Transitions to Chaos', G. Györgyi et al eds., (World Scientific, Singapore 1992) p. 449
11. Hamm, A., Graham, R.: J. Stat. Phys **66** (1992) 689-725
12. Hamm, A.: Thesis (Universität GH Essen, 1993; in German)
13. Sommerer, J.C., Ott, E., and Grebogi, C.: Phys. Rev. **A43** (1991) 1754
14. Sommerer, J.C., Ditto, W.L., Grebogi, C., Ott, E., Spano, M.L.: Phys. Rev. Lett **66** (1991) 1947
15. Sommerer, J.C.: Phys. Lett. **A176** (1993) 85
16. Beale, P.D.: Phys. Rev. **A40** (1989) 198
17. Arecchi, F.T., Badii, R., and Politi, A.: Phys. Lett. **A103** (1984) 3
18. Grebogi, C., Ott, E., and Yorke, J.A.: Physica D **7** (1983) 181
19. Pomeau, Y., Manneville, P.: Commun Math. Phys **77** (1980) 189; Manneville, P., Pomeau, Y.: Physica **D1** (1980) 219
20. Eckmann, J.-P., Thomas, L., and Wittwer, P.: J. Phys. **A14** (1981) 3153

Slow non-equilibrium dynamics

Giorgio Parisi

Dipartimento di Fisica, Università *La Sapienza*
INFN Sezione di Roma I
Piazzale Aldo Moro, Roma 00187

1 Introduction

Many systems (among them glasses, rubber, spin glasses...) show at low temperature a very slow approach to equilibrium. A very interesting phenomenon is the following: if the temperature is decreased fast enough, the system goes into an amorphous state that has practically infinite mean life; moreover under very slow cooling some of these systems go into an ordered crystal phase. The crystal phase does not exist in general, it exists only if the parameters of the interactions among the atoms are *intentionally* chosen in such way this phase is energetically favoured. Independently from the possible existence of the crystal phase, the dynamics in the glassy phase becomes extremely slow at low temperature.

These effects cannot be understood purely using only the tools of standard equilibrium statistical mechanics, because they are non-equilibrium phenomena. In this note I will present a coherent picture of the dynamics of these systems. The results have been obtained in these recent years mainly in the framework of spin glasses theory [1, 2] and I will show how they can be extended to other systems like glasses. Some of the statements are well proven, while other are still conjectural.

In the second section of this note I will give a general qualitative description of the dynamics based on the hypothesis that systems evolve in time jumping from one to an other quasi-equilibrium state. In the third section I will review some very interesting results obtained by a direct computation of the dynamical evolution of the system using more powerful methods; I will also compare these results with those obtained in the previous section. Finally in the fourth section I will address to the old problem of comparing the behaviour of a system with random Hamiltonian (like spin glasses), with that of system that have a fixed Hamiltonian (like glasses). This comparison will be done in a model system: one finds that the properties of the system with fixed Hamiltonian are very similar to that of the systems with random Hamiltonian. The only difference is the possible existence of a crystal phase for specific choices of the parameters of the fixed Hamiltonian; the cristal phase does not exists for the random Hamiltonian. Some brief conclusions are presented at the end.

2 Local Equilibrium States

In this note we will consider Ising spin model in which there are N variables σ_i, which take the values ± 1 [1] There are many possible kinds of Hamiltonian that we can write down. In the simplest case the interaction involves only two spins,

$$H_J(\sigma) = -\sum_{i,k} J_{i,k}\sigma_i\sigma_k, \tag{1}$$

and all the pairs i, k are equivalent.

If the J are Gaussian variables randomly distributed, with variance $1/N$, we obtain the SK model, otherwise one has a different model. We can also consider the p-spin model where the interaction involves p spin:

$$H_J(\sigma) = -\sum_{i,k,l} J_{i,k,l}\sigma_i\sigma_k\sigma_l. \tag{2}$$

At low temperature all these models have a corrugated free energy landscape, with many local minima, separated by high barriers. In this situation general arguments imply a very slow dynamics because the system may be trapped in a valley and it takes quite a long time to escape from it.

The jumping from a valley to an another valley is a controlled by the height of the barriers and the time need (neglecting prefactors) is

$$\tau = \exp(\beta \Delta F), \tag{3}$$

where ΔF is the minimum barrier in free energy that the system has to cross in going from one valley to an other valley [2].

The best characterisation of a valley [3] is a region of the phase space in which the system spends a long time. In this case it reasonable to define the local magnetizations in the valley α as

$$m_i^\alpha = \langle \sigma_i \rangle_\alpha, \tag{4}$$

where the average is taken inside the valley α.

In long range models these magnetizations satisfies the mean field equations, which (neglecting the Onsager reaction field [4]) in the $p = 2$ case are

$$m_i = th(\beta \sum_k J_{i,k} m_k) \tag{5}$$

---

[1] Similar consideration can also be done for interfaces or manifolds in a random medium [3].

[2] This is not the only method for having a slow dynamics. For alternative possibilities see [4, 5].

[3] Sometimes one uses the teminology *local equilibrium state* or *quasi-equilibrium state*.

[4] If we add the Onsager reaction field we obtain the TAP equations

Alternatively we can define a free energy as function of the m_i:

$$F[m] = -\sum_{i,k} J_{i,k} m_i m_k - T \sum_i s(m_i), \qquad (6)$$

where the local entropy is simply given by

$$s(m) = -\frac{1+m}{2} \ln(\frac{1+m}{2}) - \frac{1-m}{2} \ln(\frac{1-m}{2}). \qquad (7)$$

The solutions of the mean field equations are the local minima of the free energy $F[m]$.

The free energy landscape can be characterised by the structure of the set of the solutions (which we will label by Greek indices). Generally speaking the most important parameters are the free energy of the solution (f^α, the local magnetizations m_i^α in a given solution, the overlap among two solutions α and γ ($q_{\alpha,\gamma} = \frac{1}{N} \sum_{i=1,N} m_i^\alpha m_i^\gamma$) and the self overlap ($q_{EA} = q_{\alpha,\alpha}$), which in most of the models is independent from the solution.

At equilibrium it is reasonable to assume that different valleys may be populated, and the probability that the system is in one of this valley is given by

$$w_\alpha \propto w(f_\alpha) \equiv \exp(-\beta f_\alpha). \qquad (8)$$

It is evident that

$$\sum_\alpha w_\alpha = 1. \qquad (9)$$

In many disordered systems the number of valleys as function of the free energy ($\mathcal{N}(f)$) increases near the minimum value f_0 as

$$\mathcal{N}(f) \sim \exp(y(f - f_0)). \qquad (10)$$

At a given temperature there are two possibilities

- (a) If $y < \beta$ the integral

$$\int df \mathcal{N}(f) w(f) \qquad (11)$$

is dominated by f near f_0. In this case only few valleys dominates the sum in equation(9), although an infinite number of them give a non zero contribution. In this situation we say that the replica symmetry is broken.
- (b) If $y > \beta$ the integral

$$\int df \mathcal{N}(f) w(f) \qquad (12)$$

is dominated by $f - f_0$ of order of the size of the system. In this case the number of valleys which dominates the sum in equation (9) is exponentially large and each of the valley has a weight which is exponentially small. In this case the replica symmetry is not broken.

Let us consider the case where the valleys are separated by very high barriers (e.g. diverging with N) in the region where the temperature T is smaller than T_G. Depending on the nature of the problem we may enter or in the region (a) or (b) when we decrease the temperature from above to below T_G.

If we enter in the region (a), as it happens in the usual SK model for spin glasses, a phase transition is present from the equilibrium point of view at T_G.

On the contrary, if we enter in the region (b), no phase transition is present from the equilibrium point of view at T_G, and a transition for the static will be present only at smaller temperature, T_R, where y becomes smaller that β and we pass from region (a) to region (b). In this last case we have two transition one for the static and the other for the dynamics. If we quench the system at temperature smaller than T_G coming from an high temperature region, the internal energy (for an infinite system) does not go to the equilibrium value and the system remains trapped in a metastable state [6, 7].

It is also possible that there is an other isolated solution to the mean field equation, with a free energy density f_C smaller than f_0 and there are no solutions in the region

$$f_C < f < f_0. \tag{13}$$

In many cases this isolated solution describes an highly ordered state, which we call the cristalline state. The existence of this states does not change the properties of the system in the region where $f > f_0$.

¿From this point of view in order to compute the approach to equilibrium one should evaluate the free energy barriers which separate one valley from an other valley. This computation is rather difficult, especially if we take care that the system is still slightly out of equilibrium. However in the next section we shall see that a direct computation of the non equilibrium properties can be done.

3 The non-equilibrium equations

It was found quite recently [6, 7] that the non equilibrium behaviour of the system can be described directly for the infinte system (i.e. after having taken the limit $N \to \infty$), by introducing the average correlation function and the response function defined as

$$C(t, t') = \lim_{N \to \infty} \frac{\sum_{i=1,N} \sigma_i(t)\sigma_i(t')}{N},$$

$$G(t, t') = \lim_{N \to \infty} \frac{\sum_{i=1,N} \frac{\delta \sigma_i(t')}{\delta h_i(t)}}{N}. \tag{14}$$

In the equilibrium regime time translation invariance implies that these functions depend only on the time difference. We consider here the case where the system at time zero starts from a random configuration.

Closed equation can be written for these two functions. They have the form

$$\frac{\partial C(t, t')}{\partial t'} = E_C[C, G],$$

$$\frac{\partial G(t, t')}{\partial t'} = E_G[C, G], \tag{15}$$

where $E_C[C, G]$ and $E_G[C, G]$ have an explicit form (non local in time) which depend on the problem. In some case one can expand $E_C[C, G]$ and $E_G[C, G]$ in powers of C and G. The solution of these equations can be computed numerically and one can obtain a great amount of information in this way.

From the analytic point of view one can study these equation in the adiabatic approximation, where one set to zero the time derivative. This approximation is justified in the large time region. The resulting equations are non trivial; they are

$$E_C[C, G] = E_G[C, G] = 0. \tag{16}$$

The solution of these equations can be simplified [3, 6, 7] by noting that they are *reparametrization invariant*, i.e. if C and G are a solution also the functions

$$C_h(t, t') = C(h(t), h(t')),$$

$$G_h(t, t') = G(h(t), h(t')) \frac{dh(t')}{dt'}, \tag{17}$$

are an other solution of the adiabatic equations, for an arbitrary choice of the function h.

Reparametrization invariance strongly simplifies the study of the adiabatic equations and many results can be obtained in this limit. In some case it can be proved that for large times the internal energy tends at the equilibrium value, while in other case finds that there is dynamical transition at a temperature T_G. At lower temperature one finds that the dynamical energy does *not* tend at large times at the equilibrium value and therefore metastable states are present.

The evaluation of the reparametrization invariant quantities morally corresponds to the evaluation of the properties of the solutions of the mean fields equations of the previous section, although it contains more information.

The more difficult part, which at the present moment we can do only numerically, consists in computing quantities that are not reparametrization invariant, as the time dependence of the energy. This computation morally corresponds to the evaluation of the barriers separating the solutions of the mean fields equations of the previous section and it is not a surprise that it turns out to be much more difficult. Technically one ends up with a well defined and difficult mathematical problem, very similar in spirit, but more complicated, of the non linear velocity selection problem, which has been widely studied in the past.

It is extremely satisfactory that the very difficult problem of computing analytically the non equilibrium dynamics in these systems is now under control and I am convinced that the mathematical difficulties may be surmounted, may be with some help from our more mathematically minded friends.

The results obtained from this dynamical studies have the advantage to be easily compared with those obtained experimentally in spin glasses, where the condition N very large with rexpect to t is certainly satisfied. A very interesting phenomenon which appears is aging, i.e. the dependence of the experimental results on the age of the system [8, 9, 10]. A detailed discussion of this point would make this note too long.

4 Real Glasses

In spin glasses the Hamiltonian is random as an effect of quenched random disorder. In real glasses the Hamiltonian is not random and the quenched disorder is dynamically generated at low temperature. We can therefore ask how much of the qualitative and quantitative results which have obtained in spin glasses may be transferred to glasses.

In order to understand this point we have started to study models in which the Hamiltonian does not contain quenched disorder and to compare these results with those coming of random Hamiltonian [11].

In general our strategy is the following. We want to study the properties of a given Hamiltonian H_G which is not random. We consider a class of Hamiltonians H_R, of which H_G is a particular case. We choose the class H_R in such a way that the statistical properties of H_G and that of a generic Hamiltonian in H_R are as similar as possible. In the best case we can obtain that the two corresponding free energies coincides in the high temperature expansion. In general the behaviour of the system can be better controlled in the high temperature phase.

After having constructed H_R in an appropriate way, we can suppose that the statistical properties of H_G and H_R are the same or, if they are different, we can construct a perturbative expansion which compute this difference. It is clear that this approach may be successfully in the high temperature region (more or less by construction) and it may also reproduce the behaviour in the glassy region, included the dynamic and static transitions. However it is cannot certainly reproduce the possible existence of a crystal phase.

I will present now a simple model in which this approach works very well at all the temperature and it misses only the crystal phase, which exists only for *intentionally* chosen Hamiltonians H_G [12]. The Hamiltonian is:

$$H = \sum_{i,k} J_{i,k} \sigma_i \sigma_k, \tag{18}$$

which is the same Hamiltonian as in eq. (1).

In the case of H_G we have

$$J_{i,k} = N^{-1/2} \sin(\frac{2\pi i k}{N}), \tag{19}$$

while in the case of H_R we have that J is a random orthogonal symmetric matrix, i.e. a symmetric matrix such that

$$\sum_k J_{i,k} J_{k,j} = \delta_{i,j}. \tag{20}$$

It is easy to check that the first Hamiltonian is a particular case of the first one.

A details computation (partially analytic and partially numeric) for the Hamiltonian H_G shows that in the low temperature phase one obtains two different limits when N goes to infinity, one for generic N and an other for N odd, such that $p = 2N + 1$ is prime. Only in this second case there is a crystalline phase at low temperature. The model for generic N undergoes a dynamical glassy transition and a replica symmetry breaking transition; at all temperature it behaves in the same way as the model described by H_R. In this way one finds that the strategy of computing the properties of a given system by approximating it by a random system works very well in the glassy region and it looses the crystal phase, which exists only for non generic values of N.

The proposed stategy for approximating a given Hamiltonian with a random Hamiltonian works very well in the this case. There are many other system which have similar properties [13, 14], but I will not discuss them.

5 Conclusions

We have seen that we begin to control the off equilibrium dynamics in many model systems. There are cases in which we have a glass transition with metastability. This feature is present only in the mean approximation, which is correct in the infinite range limit.

A crucial and open problem is to understand how this feature of the infinite range approximation appears in more realistic finite range models, in which metastable states cannot exists. In this case the predictions for the time dependence should more realistic and a comparison with the real experiments should be possible. The study of off equilibrium dynamics in these models has started only recently an many surprices and interesting results are expected in the future.

6 Acknowledgements

It is a pleasure for me to thank for many useful discussions and the very pleasant collaboration on these problems L. Cugliandolo, J. Kurchan, E. Marinari, F. Ritort and M. Virasoro.

References

1. M. Mézard, G. Parisi and M. A. Virasoro, *Spin Glass Theory and Beyond* (World Scientific, Singapore 1987).
2. G. Parisi, *Field Theory, Disorder and Simulations*, World Scientific, (Singapore 1992).
3. S. Franz and M. Mézard; *Off equilibrium glassy dynamics: a simple case*, LPTENS 93/39, *On mean-field glassy dynamics out of equilibrium*, cond-mat 9403004, LPTENS 94/05.

4. L. F. Cugliandolo, J. Kurchan and G. Parisi,*Off equilibrium dynamics and aging in unfrustrated systems*, cond-mat preprint (1994)
5. M. Virasoro, unpublished, quoted in [4]
6. L. F. Cugliandolo and J. Kurchan; Phys. Rev. Lett. **71**, 1 (1993) and references therein.
7. L. F. Cugliandolo and J. Kurchan; *'On the out of equilibrium relaxation of the Sherrington - Kirkpatrick model'*, cond-mat **9311016**. L. F. Cugliandolo and J. Kurchan; *' Weak-ergodicity breaking in mean-field spin-glass models'*, Talk given at Fifth International Workshop on Disordered Systems, Andalo, Trento, 1994; cond-mat **9403040**, to be published in Philosophical Magazine.
8. J.-P. Bouchaud; J. Phys. France **2** 1705, (1992).
9. J-P Bouchaud, E. Vincent and J. Hammann; *'Towards an experimental measure of the number of metastable states in spin-glasses?'*, cond-mat 9303023, Saclay S93/016.
10. L. F. Cugliandolo, J. Kurchan and F. Ritort; Phys. Rev. **B49**, 6331 (1994).
11. E. Marinari, G. Parisi and F. Ritort, *Replica Field Theory for Deterministic Models: Binary Sequences with Low Autocorrelation* Hep-th/9405148, summited to J. Phys. **A** (Math. Gen.).
12. E. Marinari, G. Parisi and F. Ritort, *Replica Field Theory for Deterministic Models (II):A non-random Spin Glass with glassy Behaviour* Hep-th/9406074, summited to J. Phys. **A** (Math. Gen.).
13. G. Migliorini, *Sequenze Binarie in Debole Autocorrelazione*, Tesi di Laurea, Università di Roma *Tor Vergata* (Roma, March 1994); and to be published.
14. G.Parisi,*D-dimensional Arrays of Josephson Junctions, Spin Glasses and q-deformed Harmonic Oscillators*, (1994), J. of Phys. A (to be published).

Theory of glass transition in spin glasses, orientational glasses and structural glasses

K. Binder

Institut für Physik, Johannes Gutenberg–Universität Mainz

D–55099 Mainz, Staudinger Weg 7, Germany

Abstract: Theoretical concepts about the glass transition are briefly reviewed, and the test of these ideas by Monte Carlo simulations of simple lattice models is described, with an emphasis on isotropic and anisotropic orientational glasses, and the bond fluctuation model of polymer melts.

It is suggested that orientational glasses do have an equilibrium phase transition at zero temperature (in d = 3 dimensions!) only, in contrast to the Ising spin glass which orders at nonzero temperature. A diverging glass correlation length is identified that is responsible for the anomalous slowing down. For the Potts glass, the divergence seems to be exponential, implying that the model is at its lower critical dimensionality.

Choosing a Hamiltonian that prefers a long bond of the bond fluctuation model of polymer chains, the resulting "geometric frustration" in systems of densely packed chains leads to a glassy freeze–in, surprisingly similar to experiment. Expectations of various theories are qualitatively verified. It is suggested that also in this model a diverging length can be identified.

1   Introduction: basic facts and concepts

Glasses and other amorphous materials are known for a long time and of wide–spread technical use [1, 2]. But the theoretical understanding of the glassy state of condensed matter and of the glass transition remains a big challenge [3, 4]. There are two basic observations: the short range order of atoms in the glass (as measured, e.g. by the radial pair distribution function) hardly differs from that of the corresponding supercooled fluid above the glass transition temperature T_g, although dynamical (and mechanical) properties are so different. This immediately yields the question, is there a static (structural) quantity, that distinguishes fluid and glass: an "order parameter" that is zero in the fluid, nonzero in the glass. If such a quantity exists, analogy with second order phase transitions would lead one to expect that the correlation length of this glass order parameter fluctuations grows as one approaches the transition from the disordered phase. Growth of this length scale could then be responsible for the dramatic growth of the relaxation time (as measured by the increase of the viscosity near the structural glass transition over many decades). Such a phase transition scenario (where glassy relaxation results from strong critical slowing down) is believed to hold for spin glasses [5]. For the structural glass transition, however, evidence for the existence of a growing correlation length mostly has been lacking [6, 7], although there have been conflicting views [8, 9].

The second fact which we emphasize here is the surprising similarity of phenomena in systems which differ very much in chemical respects: window glass, amorphous semiconductors, metallic glasses, organic glasses, polymers, ... [1]–[3]. Thus it is tempting to search for a "universal" theory explaining these phenomena. It is still controversial to what extent such an approach makes sense, however.

The (historically) first theory [10, 11] tried to explain the "Kauzmann paradox" [12]: if one extrapolated the measured excess entropy of the supercooled fluid relative to the crystal to lower temperature, it seems to become negative unless one postulates an underlying static transition at some temperature $T_o < T_g$. One feels the glass is more disordered than the crystal and thus this entropy difference should not be negative. This temperature T_o often nearly coincides with the temperature T_{VF} when the (extrapolated) viscosity would diverge, if one describes it with the "Vogel–Fulcher–law" [13],

$$ln\, \eta \propto (T - T_{VF})^{-1}, \; T_{VF} \approx T_o \, . \tag{1}$$

Gibbs and di Marzio [10, 11] use a lattice model for polymer chains to construct an approximate expression for the configurational entropy $S(T)$ that exhibits a temperature T_o where $S(T)$ vanishes. Linking the decrease of this entropy for $T > T_o \{S(T) \propto T - T_o\}$ to a growing size of "cooperatively rearranging regions", arguments to justify Eq. (1) are given [11]. Of course, this cannot be taken as a first–principles derivation of Eq. (1), which rather is an empirical relation just as another "law" that ubiquitously describes the decay of relaxation and correlation functions in the context of glassy systems, the "Kohlrausch law" [14]

$$q(t) \propto exp\left[-(t/\tau)^y\right] , \tau \propto \eta \; , y = "Kohlrauschexponent" < 1. \tag{2}$$

The theory of Gibbs and di Marzio [10, 11] cannot explain Eq. (2) but it predicts that for polymers T_o should vary inversely proportional with chain length N. Indeed one observes a corresponding variation of T_g experimentally,

$$T_g(N) = T_g(\infty)(1 - \text{const}/N) \, , \tag{3}$$

where T_g empirically is defined as the temperature where $\eta(T) = 10^{13}$ Poise [3]. Of course, this approach is "taylored" to polymers and hence cannot contribute to understand why the glass transition is so similar in many different systems.

In experiments one observes a kink at T_g in various properties. This is attributed to the fact that the system falls out of equilibrium when the relaxation time τ becomes comparable to the time constant of the measurement (defined e.g. by the cooling velocity). Thus T_g depends somewhat on the cooling rate Q, and cooling very slowly one can observe a lower $T_g(Q)$ – but it is unclear whether $T_g(Q)$ really settles down at a well defined temperature T_o. Also the properties of the "quenched fluid" produced in this way are somewhat cooling rate dependent and differ hence from an "ideal glass". These difficulties complicate comparison between experiment and all theories further.

An alternative theory of Cohen et a. [15] is based on the concept that in the supercooled fluid the "free volume" decreases with decreasing temperature. Each atom is in a cell whose volume V depends on the distances to its neighbors. If V exceeds some critical value, the cell contains "free volume" and is "liquid–like"; otherwise it is "solid–like". Only liquid–like cells contribute to the transport in the fluid. The percolating cluster of liquid–like cells gets thinner and thinner as $T \to T_g$ from above, thus explaining the dramatic increase of the viscosity. Just below T_g, there is no longer any percolation of liquid–like cells, and the system is frozen. This theory is rather successful but very phenomenological and it is hard to give its basic parameters a precise physical meaning.

The only theory that can be linked to microscopic theories about the dynamics of liquids is the mode coupling theory due to Goetze et al. [16]–[18]. Assuming that the only slow variables of interest are the fourier components of the density, an approximate nonlinear kinetic equation for their correlation function is derived. In the "idealized" version of the theory, the glass transition is described as a kinetic singularity at a critical temperature $T_c > T_g$, where the viscosity diverges,

$$\eta \propto (T - T_c)^{-\Delta} \quad , \tag{4}$$

and for $T < T_c$ a glass order parameter ("nonergodicity parameter") appears discontinuously. In the more refined version of the theory [18] this power–law divergence is rounded off, and (due to "hopping processes" which release the atoms from the "cages" formed by their environment) there is only a smooth crossover at T_c, with $\eta(T_c)$ being of the order of 10^3 Poise only. While this theory has encouraging success in the description of many observations in various systems in this temperature region $T \gtrsim T_c$, the description of the region near the real glass transition ($T \approx T_g$ where $\eta(T) \approx 10^{13}$ Poise!) remains an unsolved problem.

As stressed above, some experimental evidence exists for all the above theories. But often it is difficult to characterize the system precisely, and unknown parameters have to be adjusted – thus the usefulness of comparing theory and experiment is limited.

The present work emphasizes Monte Carlo simulations, since there one focuses on a simple well–defined model hamiltonian: all interaction parameters are known precisely, ambiguities of characterizing the systems are avoided, as well as inessential complications. System configurations are available in full microscopic detail, and one can make "measurements" of quantities inaccessible in real experiments [5, 19, 20].

These advantages are clearly demonstrated for orientational glasses [19] in Sec. 2, showing that there a static glass correlation length $\xi_G(T)$ exists and leads to anomalous relaxation via critical slowing down [21]

$$\tau \propto [\xi_G(T)]^z , \quad z = \text{"dynamic exponent"}. \tag{5}$$

It will be shown that different "universality classes" can be identified for orientational glasses, depending on the spatial dimensionality and the symmetry properties of the degrees of freedom, similar to standard phase transitions [21, 22].

In Sec. 3 a simple lattice model for the structural glass transition of supercooled polymer melts is presented. It is shown that the bond fluctuation model is a reasonable compromise between simulational efficiency and realistic chemical detail [20],[23]–[30]. A potential energy that favors one particular length of the effective bonds connecting the effective monomers creates a conflict between configurational entropy of dense packing (satisfactory for the hard–core repulsive part of the potential) and the energetic tendency of the effective bonds to stretch as the temperature is lowered. The glass transition resulting from this conflict is in surprisingly good qualitative agreement with many experimental observations [26]–[31]. The model also allows significant comparison with mode coupling theory [33, 34], but despite this agreement there is also evidence for a growing glass correlation length [34, 35]. However, at this point statements on the existence of static phase transitions and "universality classes" cannot yet be made, as discussed further in the conclusions (Sec. 4).

2 Orientational glasses

Both spin glasses [5] and orientational glasses [19, 36] are crystals, as far as the geometrical arrangement of atoms is concerned. The glassy state of these materials refers to the random orientation of magnetic dipole moments in spin glasses (Fig. 1) and of electric quadrupole moments in orientational glasses [37, 38]. As a result of this quenched disorder resulting from the random dilution in such magnets, sometimes the ferromagnetic interactions win, sometimes the antiferromagnetic ones: no simple spin alignment is then really favorable, and so the system freezes in a randomly aligned phase. Similarly, the random dilution disrupts the regular alignment of quadrupole moments in molecular crystals $(o - H_2, N_2$ etc.), it leads to a freezing–in of these quadrupole moments in random directions over a wide concentration range for the diluted materials $\{(oH_2)_x(pH_2)_{1-x}, (N_2)_x Ar_{1-x}$, etc. [36]$\}$.

It is controversial whether these systems really are prototypes for the glassy state of matter. But the relaxation phenomena observed exhibit great qualitative similarity with those of structural glasses. In any case, for spin and orientational glasses a classification of models and physical systems is possible, and the concept that glassy relaxation is a special kind of critical slowing down {Eq. (5)} can be verified. But the nature of these static glass correlations described by $\xi_G(T)$ is rather special, and often static phase transitions occur at zero temperature only

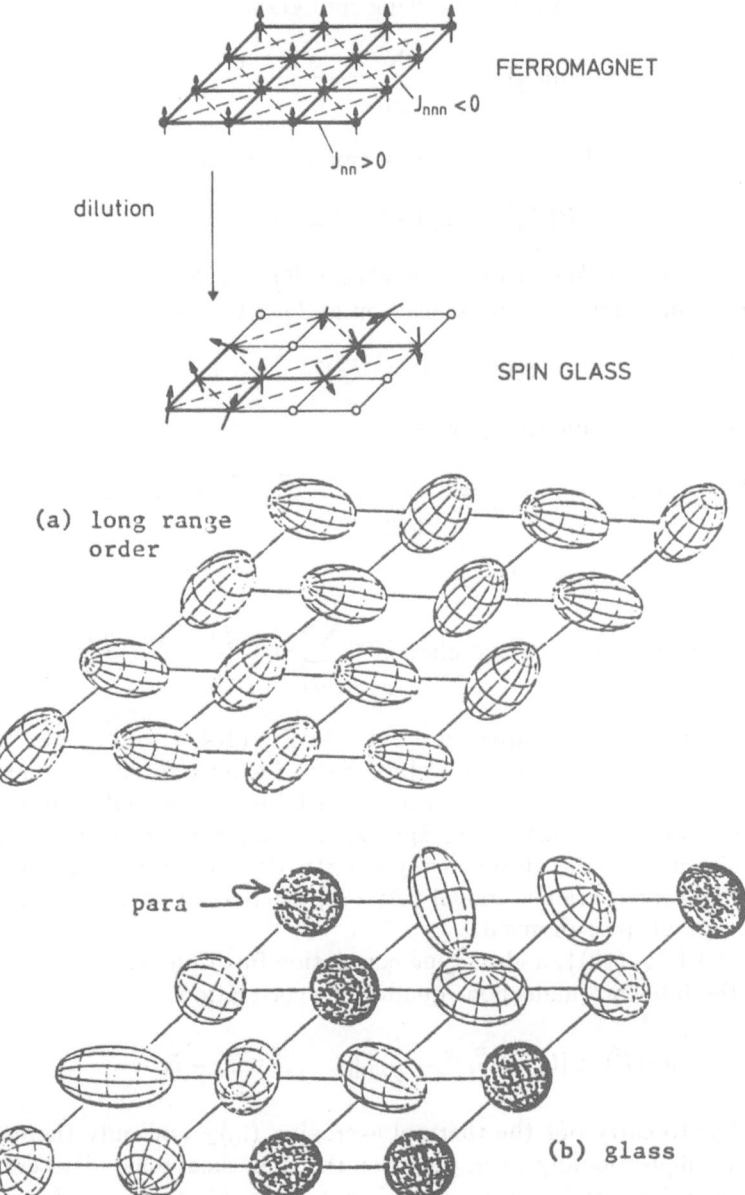

FERROMAGNET

$J_{nnn} < 0$

$J_{nn} > 0$

dilution

SPIN GLASS

(a) long range order

para →

(b) glass

Figure 1: A spin glass results from random dilution of a ferromagnet (magnetic atoms are shown by black dots) with non magnetic atoms (open circles), if there is competing exchange (e.g. in EuS the exchange is ferromagnetic between nearest neighbors ($J_{nn} > 0$) but antiferromagnetic between next nearest neighbors ($J_{nnn} < 0$) and a spin glass state occurs in $Eu_x Sr_{1-x} S$ for $x \lesssim 0.5$ [37] [upper part]

An orientational glass results from random dilution of a molecular crystal (e.g. ortho hydrogen, (a)) with spherical atoms (e.g. para hydrogen, (b)). From Sullivan [38].

– with the notable exception of the Ising spin glass,

$$H_{\text{Ising spin glass}} = -\sum_{\langle i,j \rangle} J_{ij} S_i S_j \ , S_i = \pm 1, \tag{6}$$

where either $J_{ij} = \pm J$ or drawn from a gaussian distribution

$$P(J_{ij}) \propto \exp[-J_{ij}^2/2(\Delta J)^2] \ . \tag{7}$$

While this model of anisotropic spin glasses [5] is believed to have in $d = 3$ dimensions a finite temperature transition at $T_f > 0$

$$\xi_G(T) \propto (T - T_f)^{-\nu} \ , \tag{8}$$

both models for isotropic spin glasses

$$H_{\text{Heisenberg spin glass}} = -\sum_{\langle i,j \rangle} J_{ij} \vec{S}_i \cdot \vec{S}_j, \ \vec{S}_i = \text{unit vector} \tag{9}$$

and isotropic orientational glasses

$$H_{\text{isotropic orientational glass}} = -\sum_{\langle i,j \rangle} J_{ij}[(\vec{S}_i \cdot \vec{S}_j)^2 - 1/3] \tag{10}$$

have transitions at zero temperature only. In Eq. (10), a unit vector $\vec{S}_i$ in the direction of the uniaxial molecule at lattice site i is introduced. The interaction of the squared scalar product form means a bilinear tensorial coupling of the quadrupole moments. Just as for spin glasses [5], it is convenient to replace the "site disorder" of the atomistically realistic description (Fig. 1) by a "bond disorder" of a coarse–grained model {Eq. (7)} where each lattice site is taken by a dipole or quadrupole moment.

In this model {Eq. (10)}, a glass type correlation function can be defined as the square of the bilinear quadrupole–quadrupole correlation,

$$g_G(\vec{R}) \equiv [\langle(\vec{S}_i \cdot \vec{S}_j)^2 - \frac{1}{3}\rangle_T^2]_{av} \ , \vec{R} \equiv \vec{r}_i - \vec{r}_j \ . \tag{11}$$

First one has to carry out the thermal averaging $\langle ... \rangle_T$ and only thereafter the square is configurationally averaged over the quenched bond disorder, $[...]_{av}$. Such a high–order correlation function is not accessible to any real experiment – "computer experiments" have a distinct advantage here, since Monte Carlo methods are the only methods by which problems such as Eqs. (10), (11) can be studied [19, 39, 40] (apart from the infinite range version for which a mean field theory can be developed [41]).

It is seen (Fig. 2) that this glass correlation function decays rapidly with increasing distance R, $g_G(\vec{R}) \propto \exp(-R/\xi_G(T))$, except at very low temperature: a static phase transition (where $\xi_G(T) \to \infty$) occurs at $T = 0$ only. At low temperatures the relaxation times in this model are nevertheless

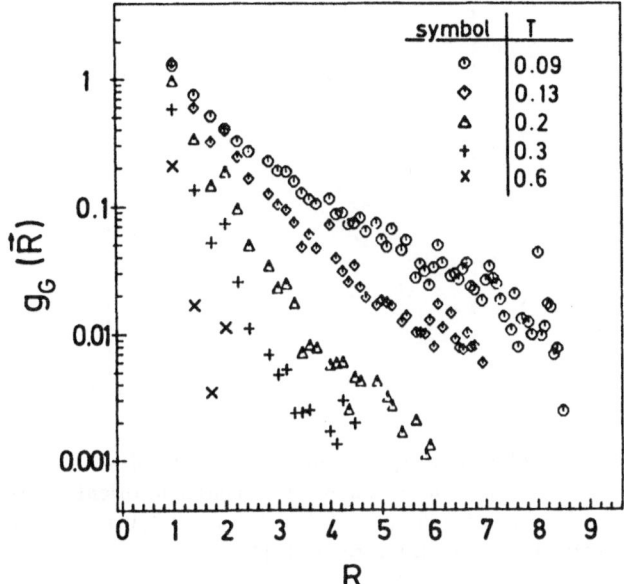

Figure 2: Semilog plot of the glass correlation function $g_G(\vec{R})$ vs. distance R in units of the lattice spacing, for Eq. (10) in $d = 3$. Five temperatures are indicated, choosing units $\Delta J/k_B \equiv 1$. Lattice sizes were 12^3 for the four higher temperatures and 18^3 for $T = 0.09$, respectively. From Hammes et al. [40].

very large. This dramatic slowing down shows up in the time autocorrelation function $q(t)$

$$q(t) = \frac{1}{L^3} \sum_i [\langle (\vec{S}_i(0) \cdot \vec{S}_i(t))^2 - \frac{1}{3} \rangle_T]_{av} \qquad (12)$$

of the quadrupole moments (Fig. 3). This function is seen to exhibit a nonexponential decay consistent with Eq. (2) over many decades in time: Plotting $ln\, q(t)$ versus time on a double–logarithmic plot one expects straight lines, the slope of which yields the "Kohlrausch exponent" y, Fig. 3. The decrease of this slope $y(T)$ with decreasing temperature means that the spectrum of relaxation times becomes broader. Indeed such a broadening is familiar both from experiments on spin glasses and from structural glasses.

The computer simulations allow to roughy estimate the temperature dependence of the glass correlation length $\xi_G(T)$ and associated glass susceptibility $\chi_G(T) = \sum_{\vec{R}} g_G(\vec{R})$. For the isotropic orientational glass power laws are found [39, 40],

$$\xi_G(T) \propto T^{-\nu_o}, \quad \chi_G(T) \propto T^{-\gamma_o}, \quad \tau \propto T^{-\nu_o z_o}, \quad T \to 0 \qquad (13)$$

where in $d = 2$ $\nu_o \approx 0.63$, $\gamma_o \approx 1,35$, $\nu_o z_o \approx 4.3$ and in $d = 3$ $\nu_o \approx 1.02$, $\gamma_o \approx 2.7$, $\nu_o z_o \approx 6.8$. The large values of the dynamic exponents z_o are characteristic for glassy systems – at low temperatures macroscopically large times result. Since the data (Figs. 2,3) are in a regime, where $\xi_G(T)$ is still rather small,

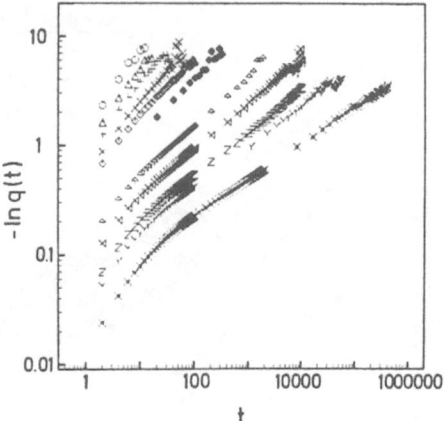

Figure 3: Log–log plot of $\{-lnq(t)\}$ vs. time t {in units of attempted Monte Carlo steps per site (MCS)}, for the three–dimensional isotropic orientational glass. Temperatures shown are $T = 1.5$, 1.0, 0.8, 0.6, 0.5, 0.4, 0.25, 0.2, 0.15, 0.12, and 0.09 (from left to right), respectively. From Hammes et al. [40].

however, it is not sure that the asymptotic region where Eq. (13) holds has been reached. The accuracy of the above exponent estimates thus is rather uncertain. It is very natural to consider also anisotropic orientational glasses: suppose the uniaxial molecules can align only in the x, y, or z–direction of a cubic crystal. This leads to the three–state Potts glass [42]–[49] if we assume that the interaction depends on the relative angle between the molecules only. The three states $(1, 2, 3)$ of this model correspond to the three orientations (x, y, z axes),

$$H_{\text{Potts glass}} = - \sum_{\langle i,j \rangle} J_{ij} \delta_{n_i n_j} , n_i \epsilon \{1,2,3\} . \qquad (14)$$

Since this can be viewed as a generalization of the Ising spin glass (two states: $S_i = \pm 1$) to a three state model, and one knows that the lower critical dimension $d_l < 3$ for the Ising spin glass [5], one expects a similar behavior for the Potts glass, too. Remember that $T_f > 0$ for $d > d_l$ but $T_f = 0$ for $T < d_l$, and a scaling theory of Mc Millan [50] predicts an exponential divergence at d_l

$$\log\xi_G(T) \propto T^{-\sigma} , \sigma = 2, \ d = d_l , T \to 0. \qquad (15)$$

Surprisingly, a finite size scaling analysis of corresponding Monte Carlo simulations [45] suggested that $d_l \approx 3$ for the Potts glass, i.e. the data are better consistent with Eq. (15) than with any power law {Eq. (13), if $d < d_l$, or Eq. (8), if $d > d_l$}, see Fig. 4.

Now it is important to verify this conclusion that $d_l = 3$ for the Potts glass by checking that a different behavior results for other dimensionalities – a task impossible for a real experiment, but readily feasible for simulation [49]! Indeed one finds that Eq. (13) holds for $d = 2$, and Eq. (8) holds for $d = 4$. However, the behavior in $d = 4$ (Fig. 5) where $T_f = 0.25$, $\nu = 0.8 \pm 0.3$ and $2 - \eta = 4.0 \pm 0.5$

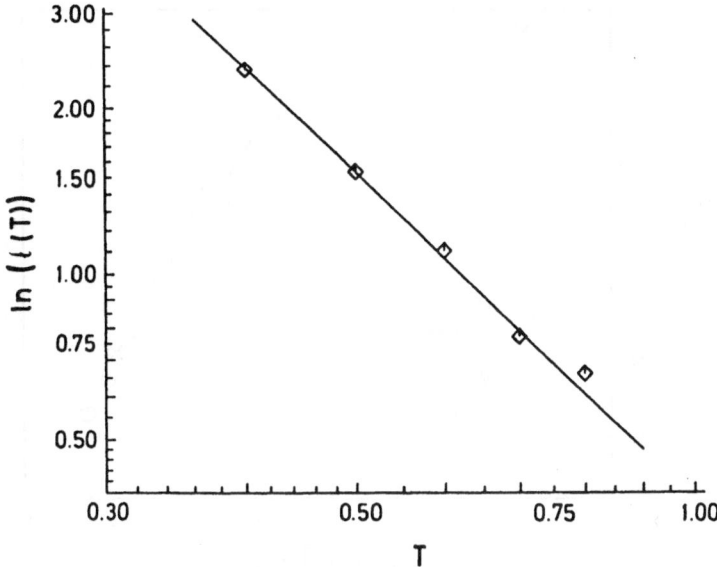

Figure 4: Log–log plot $\ln l(t)$ vs. T, for the three–state Potts glass in $d = 3$ dimensions. Here $l(T)$ is a characteristic length extracted from a finite size scaling analysis and is proportional to $\xi_G(T)$. Straight line indicates $\sigma = 2$ {Eq. (15)}. From Scheucher et al. [45].

[49] is rather puzzling, since the hyperscaling law $2 - \eta = d - 2\beta/\nu$ then implies $\beta = 0$, i.e. the transition would be first order. In fact, mean field theory [42, 43, 46] predicts for Potts glasses with p states for $p < p_c = 4$ an unusual first order transition, where a nonzero glass order parameter appears discontinuously (though there is no latent heat!).

Another unusual property [51, 52] of the Potts glass relates to its lack of spin inversion symmetry: in Ising and Heisenberg systems the free energy in the disordered phase is an even function of the field h conjugate to the magnetization m, i.e. in an expansion

$$m(h) = \chi_o h + \chi_2 h^2 + \chi_{nl} h^3 + \dots \tag{16}$$

only odd powers occur ($\chi_2 \equiv 0$). For symmetric distributions of exchange constants {such as Eq. (7)} we furthermore have $[\langle S_i S_j \rangle_T]_{av} = \delta_{ij}$, and hence $\chi_o = \sum [\langle S_i S_j \rangle_{av}/T = 1/T$ simply follows the Curie law [5], while the non–linear susceptibility χ_{nl} is simply related to the glass susceptibility [5], $\chi_{nl} T^3 = (2/3 - \chi_G)$. For Potts glasses, however, $\chi_2 \neq 0$, both χ_o and χ_2 have a nontrivial (critical [52]?) temperature dependence, and there is no simple relation between χ_{nl} and χ_G. Eq. (16) is of practical interest, since in spin glasses χ_{nl} is obtained from measuring the magnetic equation of state, and in this way experimental evidence for a divergent susceptibility associated with a static spin glass transition at $T_f > 0$ has been obtained [5]. For orientational glasses, the order parameter corresponding to m is the tensor describing the molecular orientations and can only be probed indirectly (via rotation–translation couplings) by measuring elastic

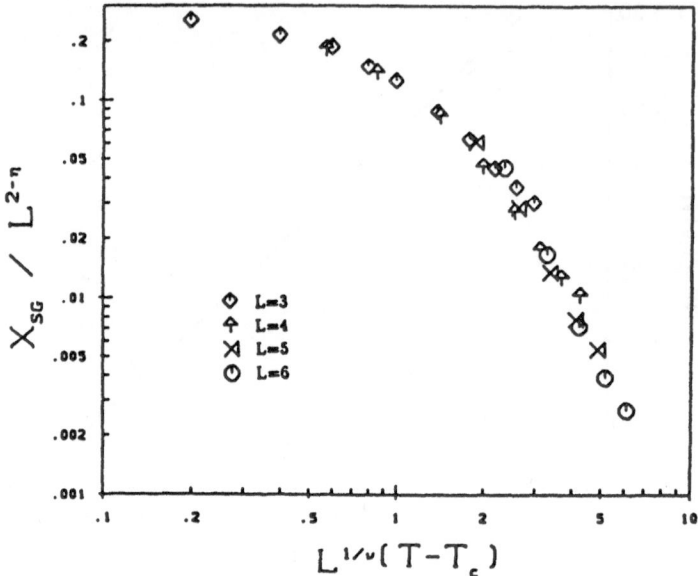

Figure 5: Finite size scaling plot of the glass susceptibility (defined here as $L^d[\langle q^2 \rangle_T]_{av}$ where q is the glass order parameter in the finite system, see [45, 49], using exponents $\nu = 0.8, 2 - \eta = 4$, and $T_c = 0.25$, for the 4-dimensional three-state Potts glass. Four different lattice linear dimensions are used, averaging over 400–2000 bond configurations. From Scheucher and Reger [49].

response [53]. A divergent nonlinear shear compliance in $(KBr)_{0.59}(KCN)_{0.41}$ [53] so far is the single piece of evidence that the phase transition scenario developed for spin glasses holds for orientational glasses, too. The situation indeed is complicated, since random dilution (Fig. 1) for quadrupole moments should not only lead to disorder of the "random bond" type {as assumed in Eqs. (10), (14)} but also [54] of the "random field" type [55]. Thus a lot remains to be done, both on the theoretical and the experimental side, before a full understanding of orientational glasses will be reached!

3 Towards the modeling of the glass transition in amorphous polymers

Again the problem is too complex to try a computer modeling with full atomistic detail [20]: thus we introduce a coarse–graining, grouping successive chemical monomers together into effective bonds, and put the resulting polymer chain on a lattice, for the sake of a very fast simulation [24]. In this bond fluctuation model, the length of the bonds can fluctuate, and dynamics is introduced by randomly displacing the effective monomers by one unit in the $\pm x, \pm y$ or $\pm z$ direction (Fig. 6). Excluded volume and entanglement restrictions are taken into account, if double occupancy of lattice sites by these effective monomers is forbidden. We now introduce temperature into this model via a potential for the bond length.

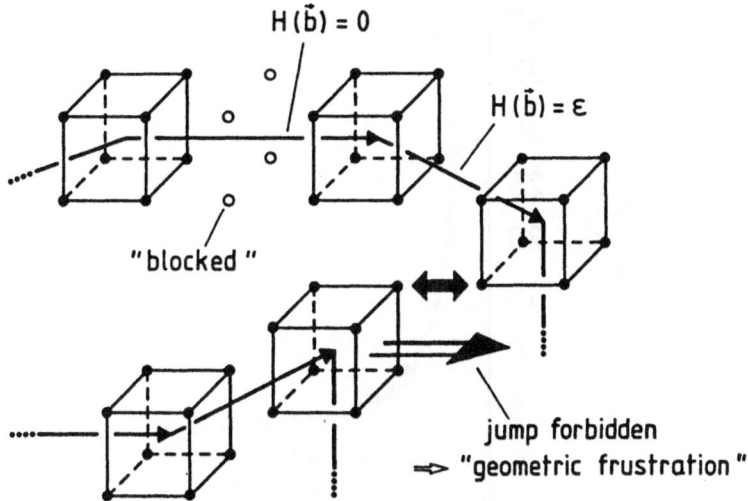

Figure 6: Sketch of a possible configuration of effective monomers in the three-dimensional bond fluctuation model. All bond vector $\vec{b}$ have the energy $H(\vec{b}) = \epsilon$ except the vectors $(\pm 3, 0, 0)$, $(0, \pm 3, 0)$ and $(0, 0, \pm 3)$ which have $H(\vec{b}) = 0$ and belong to the ground state. All lengths are measured here in units of the lattice spacing. These vectors block four lattice sites in between two effective monomers (marked by empty circles), which are no longer available for any other monomers, since two monomers must not overlap. Due to this excluded volume interaction, the jump in the direction of the large arrow is forbidden. In a dense system such constraints do not allow that all bonds reach their groundstate. There is a conflict between bond length energetics and the entropy of dense packing ("geometric frustration"). From Baschnagel et al. [28].

In $d = 2$, where the bond length can vary from $2 \leq b \leq \sqrt{13}$, a parabolic potential has been chosen, $U_{eff}(b) = U_o(b - \sqrt{10})^2$, with $U_o = (2 - \sqrt{10})^{-2}$. In $d = 3$, where the allowed bond lengths are $b = \{2, \sqrt{5}, \sqrt{6}, 3, \sqrt{10}\}$, we choose a two level hamiltonian (Fig. 6). A volume fraction $\Phi = 0.8\,(d = 2)$ or $\Phi = 0.533\,(d = 3)$ of occupied sites of the square (simple cubic) lattice is used.

The competition between bond length energetics and packing constraints is enough to find a glass transition in this model. The selfdiffusion constant decreases with decreasing temperature and essentially vanishes near $T_g \approx 0.2$ (for $d = 2$) [26]. However, this does not mean that all motions are frozen in: the acceptance rate for the jumps of the effective monomers is still nonzero for $T < T_g$, local motions are still possible. But on a global scale the configuration is frozen in a metastable minimum: the energy does not decrease further, and hence not all bonds can reach their ground state. The accessible free volume (which can be defined precisely for a lattice model) somewhat decreases in the fluid phase and also exhibits a kink at the glass transition [26]. And comparing snapshot pictures of the polymer melt at high temperatures with corresponding pictures of the glassy material, we indeed recognize more local order than in the fluid. This is no long range order of any simple kind, but a short range order of the bond

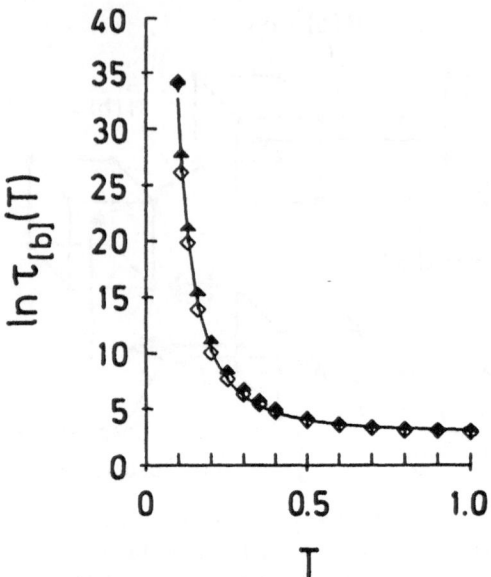

Figure 7: Logarithm of the relaxation time $\tau_{[b]}(T)$ of the bondvector autocorrelation function plotted vs temperature. Full and open symbols correspond to polymer chain lengths $N = 25$ and $N = 10$, respectively. The curve represents the equation $\ln \tau_{[b]}(T) = a + bT^{-2}$, with $a = 2.82$, $b = 0.30$. From Wittmann et al. [26].

vector orientations [26]. This means that indeed the configurational entropy has strongly decreased: thus some evidence is found for the concepts of all theories on the glass transition in this simulation: decrease of configurational entropy, as predicted by Gibbs and di Marzio [10]; decrease of free volume, as predicted by Cohen and Grest [15]; a cage effect, as predicted by the mode coupling theory [16]. But none of these theories accounts for all the findings of the simulation simultaneously.

Particularly interesting is the time auto–correlation function of the bond vector orientations $q(t)$. One finds that over intermediate times it can be fitted to the Kohlrausch law, Eq. (2). The Kohlrausch exponent has a kink at the effective glass transition temperature: but the relaxation time τ is not singular there, Fig. 7, it rather seems to diverge only as $T \to 0$, with a law [56] that is stronger than an Arrhenius law, $\ln \tau \propto T^{-2}$. Thus the freezing at $Tg \approx 0.2$ is a purely kinetic phenomenon, on the observable time scale (defined by a slow cooling where the temperature is lowered according to $1/T(t) = \Gamma_Q t/T_f$, with $T_f = 0.05$ and $\Gamma_Q = 4.10^{-7}$) the system falls out of equilibrium at $T_g \approx 0.2$, while an equilibrium transition (characterized by a divergence of the relaxation time) probably occurs at $T = 0$ only.

For $d = 3$ dimensions, however, the data are compatible with the Vogel–Fulcher law, Eq. (1). Fig. 8 shows data for the selfdiffusion constant. Note that the sim-

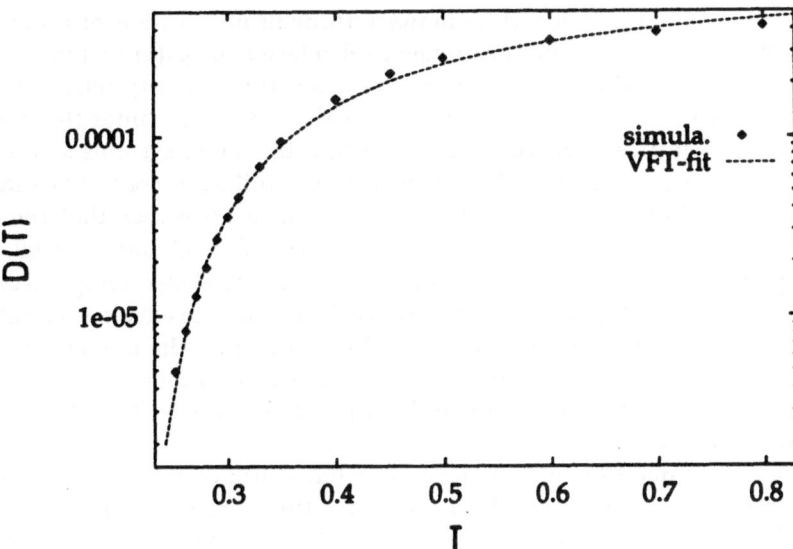

Figure 8: Plot of the selfdiffusion coefficient of the three–dimensional model versus reduced temperature (units chosen such that $\epsilon/k_B = 1$ in Fig. 6). Broken curve is a fit to $D = D_\infty \exp\{-A/(T - T_{VF})\}$ with $T_{VF} = 0.17$. From Baschnagel et al. [28].

plicity of the model (Fig. 6) allowed a very good statistics (in $d = 3$ we used [28] 288000 monomers) and checked carefully for thermal equilibrium by varying the cooling rate over a wide range. Defining a glass transition temperature $T_g(\Gamma_Q)$ from kinks in the temperature dependence of suitable observables, one finds a cooling rate dependence of $T_g(\Gamma_Q)$ that is slower than logarithmic [28], again compatible with corresponding experiments.

The static structure factor $S(\vec{q})$ of the model [30] also is perfectly reasonable, it is very small for wave vector $q \to 0$ (due to the small compressibility of the glass, which decreases with decreasing temperature only slightly). Then a diffuse maximum of the scattering intensity occurs, similar to the "amorphous halo" seen in the experiments. There is only a weak temperature dependence of this peak, and clearly the system does not crystallize. Many other quantities (including the radial pair distribution function, single chain structure factor, etc. [30]) have also been studied, and it thus has been shown that the model is not plagued too much by lattice artefacts, and that despite the shortness of the chains ($N = 10$) the model shows the main characteristics of disordered polymer melts.

We thus have established that the model shows a glass like freezing similar to experiment – but is there also a glass correlation length linked to the transition, as in the case of spin glasses and orientational glasses?

First evidence for such a length is indirect through observation of a finite size effect [32]. Fig. 9 shows the mean square displacement $g_3(t)$ of the center of mass in $d = 2$ divided by 4 t versus inverse time: this quantity settles down at the diffusion constant for $t \to \infty$. It is seen that for short times the data are independent of lattice size, while at large times the small systems settle down at a plateau earlier (and therefore yield a larger diffusion constant) than the larger systems. Note that $g_3(t > \tau)/4t = D$, and hence it is seen that the relaxation time τ for the smaller lattices is reduced (for $L = 40$ and $L = 60$, there is hardly any difference in Fig. 9, implying that at the shown temperature the asymptotic behavior for $L \to \infty$ is nearly reached). No size effects occur at high temperatures, and the size effect shown in Fig. 9 becomes the more pronounced the lower T. We interpret this finding as a finite–size rounding of Eq. (5): when $\xi_G(T)$ increases and becomes comparable to L, the law $\tau \propto [\xi_G(T)]^z$ starts to cross over to $\tau \propto L^z$.

Of course, a key question is to understand what static correlation function one has to study in order to derive this glass correlation length $\xi_G(T)$ for structural glasses directly. As a first attempt, we have taken the same type of quantity as for orientational glasses, Eq. (11) now $\vec{S}_i$ denoting a unit vector in the direction of an effective bond [35]. It is gratifying that one finds a strong increase of $\xi_G(T)$ at those temperatures where $D(T)$ strongly decreases (Fig. 8). However, the data obtained so far [35] do not allow to distinguish between a zero–temperature transition {Eq. (13)} or a finite temperature transition {Eq. (8)} yet. It turns out that the equilibration for $T \lesssim 0.25$ is very difficult – if one cools the system too fast $\xi_G(T)$ is reduced {consistent with Eq. (5), also the relaxation time τ is reduced by too fast cooling [33]}. This fact explains, why previous Molecular Dynamics work on other models [6, 7] – which corresponds to even much faster cooling than used in our Monte Carlo work – could not see any increasing length $\xi_G(T)$ at all.

Although our model thus supports the concept that the glass transition is related to a static phase transition in the supercooled fluid with a growing correlation length $\xi_G(T)$, it is nevertheless compatible with the mode coupling theory as well. A study of the incoherent intermediate scattering function $\Phi_q^s(t)$ (choosing q values at or near the peak of the amorphous halo) yields the characteristic decay with time in two steps for T near T_{VF} [33, 34]: the first step is attributed to "β–relaxation", the second step (i.e., the final decay) to "α–relaxation", and can well be fitted to corresponding formulas of the mode coupling theory [33, 34], with reasonable parameters. But at the same time, we also find evidence [31] for Eq. (3), the chain length dependence, predicted by the theory of Gibbs and Di Marzio [10]. Since the simulations at the same time give evidence in favor of quite different theories, one should not conclude about the validity and merits of these theories too hastily!

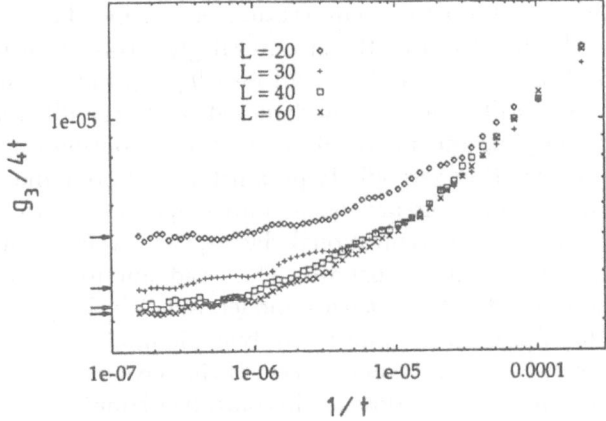

Figure 9: Log–log plot of the effective chain diffusion constant $D(t) \equiv g_3(t)/4t$ $\{g_3(t)$ is the meansquare displacement of the center of mass of a chain$\}$ versus time, for the two–dimensional model at $T = 0.22$. Four lattice sizes are shown as indicated. Arrows show the corresponding estimates of the diffusion constant $D(L, T)$. From Ray and Binder [32].

4   Concluding remarks

The theoretical understanding of glass transitions is still very incomplete. One does not really know whether the superficial similarity of the slowing down near the freezing in of spin glasses and orientational glasses and the various kinds of structural glasses reflects an underlying common principle or is more or less accidental. In spin and orientational glasses, first steps towards classification of possible phase transitions have been taken: both classes of systems have equilibrium transitions to a glass phase at nonzero temperature in mean field theory (applicable to the model with infinite interaction range). For the spin glass (vector character of the degrees of freedom) this is a second order transition, for the orientational glass (tensor character of the degrees of freedom) a first order transition also is possible (Potts glass with $q > q_c = 4$ states), though without latent heat. For short range models, only the Ising spin glass seems to have a finite temperature transition, while the Potts glass seems to have $d = 3$ as lower critical dimension. For isotropic spin and orientational glasses, $d = 3$ is below the lower critical dimension, and hence both the glass correlation length and the associated relaxation time diverge at $T \to 0$ only, although for practical purposes a finite temperature nonequilibrium glass transition may occur (at $T_g > 0$ the relaxation time becomes macroscopically large). The understanding of practical materials is hampered by the existence of interactions with power law decay (dipolar or Rudermann–Kittel interactions in spin glasses, elastic interactions among molecules in orientational glasses), so the applicability of the theoretical models sometimes is somewhat uncertain. For orientational glasses, the existence of random fields at least produces a rounding of possible transitions and further complicates comparison between theory and experiment.

Both recent simulations [32, 35] and experiment [57], where a size–dependent depression of the glass transition temperature in polymer films was observed, suggest that the slowing down at the structural glass transition may be linked to diverging length as well. This is an old idea [8, 11], but the nature of such glass–like correlation still is not well understood, and very different concepts – such as the mode coupling theory which implies the discontinuous appearance of a glass order parameter ("nonergodicity parameter") – have remarkable success. While Monte Carlo simulations have contributed significantly to clarify the behavior of models for spin and orientational glasses, the development of models for the structural glass still is under debate. The bond–fluctuation model of polymer melts is a good candidate for such a model, but it lacks both phonon–type excitations (in the frozen phase) and hydrodynamic modes (in the fluid phase) and thus it may miss some important aspects of the physics of glasses. Although this model, for the first time, allows well–controlled simulations where due to high statistical effort and very long runs the effects of incomplete equilibration due to too fast cooling can be sorted out, one cannot yet make any statements whether or not it shows an equilibrium glass transition at nonzero temperature. Thus there is still much need for further work.

Acknowledgements: The author is very grateful to J. Baschnagel, H.-O. Carmesin, P. Ray, M. Scheucher and H.-P. Wittmann for an intensive and fruitful collaboration on the research described here. Valuable contributions also are due to E. Andrejew, F. Haas, D. Hammes, K. Kremer, B. Lobe, J.D. Reger, K. Vollmayr, and A.P. Young. This research is partially supported by the Deutsche Forschungsgemeinschaft, Sonderforschungsbereich 262, and was only possible through generous grants of supercomputer time from the Höchstleistungsrechenzentrum Jülich (HLRZ) and the Regionales Hochschulrechenzentrum Kaiserslautern (RHRK).

References

[1] Zallen, R.: The Physics of Amorphous Solids (Wiley, New York 1983)

[2] Zarzycki, J. (ed): Materials Science and Technoloy, Vol. 9 (VCH Pub., Weinheim 1991)

[3] Jäckle, J.: Rep. Progr. Phys. **49**, 171 (1986)

[4] Hansen, J.P., Levesque, D. and Zinn–Justin S. (eds.): Liquids, Freezing and the Glass Transition (North–Holland, Amsterdam 1991)

[5] Binder, K. and Young, A.P.: Rev. Mod. Phys. **58**, 801 (1986)

[6] Dasgupta, C., Indrani, A.V., Ramaswamy, S. and Phani, M.K.: Europhys. Lett. **15**, 307 (1991)

[7] Ernst, R.M., Nagel, S.R. and Grest, G.S.: Phys. Rev. **B43**, 8070 (1991)

[8] Donth, E.: J. Non–Cryst. Sol. **53**, 325 (1982)

[9] Fischer, E.W., Donth, E. and Steffen W.: Phys. Rev. Lett. **68**, 2344 (1992)

[10] Gibbs, J.H. and Di Marzio, E.: J. Chem. Phys. **28**, 373, 807 (1958)

[11] Adam, G. and Gibbs, J.H.: J. Chem. Phys. **43**, 139 (1965)

[12] Kauzmann, W.: Chem. Rev. **43**, 219 (1948)

[13] Vogel H.: Phys. Z. **22**, (1921); Fulcher, G.S.: J. Am. Ceram. Soc. **8**, 339 (1925)

[14] Kohlrausch, R.: Ann. Phys. (Leipzig) **12**, 393 (1847)

[15] Grest, G.S. and Cohen, M.H. in: *Advances in Chemical Physics* (Prigogyne, I., and Rice, S.A., eds.) (Wiley, New York 1981)

[16] Götze, W. and Sjogren, L.: Rep. Progr. Phys. **55**, 241 (1992)

[17] Götze, W.: Z. Physik **B56**, 139 (1984); Leutheusser, E.: Phys. Rev. **A29**, 2765 (1984)

[18] Fuchs, M., Götze, W., Hildebrand, S. and Latz, A.: J. Phys.: Condens. Matter **4**, 7709 (1992)

[19] Binder, K. and Reger, J.D.: Adv. Phys. **41**, 547 (1992)

[20] Binder, K.: Macromol. Chem., Macromol. Symp. **50**, 1 (1991)

[21] Hohenberg, P.C. and Halperin, B.I.: Rev. Mod. Phys. **49**, 435 (1977)

[22] Fisher, M.E.: Rev. Mod. Phys. **46**, 597 (1974)

[23] Carmesin, I. and Kremer, K.: Macromol. **21**, 2819 (1188)

[24] Wittmann, H.P. and Kremer, K.: Computer Phys. Commun. **61**, 309 (1990)

[25] Deutsch, H.P., and Binder, K.: J. Chem. Phys. **94**, 2294 (1991)

[26] Wittmann, H.P., Kremer, K. and Binder, K.: J. Chem. Phys. **96**, 6291 (1992)

[27] Wittmann, H.P., Kremer, K. and Binder, K.: Makromol. Chem., Theory & Simul. **1**, 275 (1992)

[28] Baschnaagel, J., Binder, K. and Wittmann, H.P.: J. Phys. Condens. Matter **5**, 1597 (1993)

[29] Ray, P., Baschnagel, J. and Binder, K.: J. Phys.: Condens. Matter **5**, 5731 (1993)

[30] Baschnagel, J. and Binder, K.: Physica **A204**, 47 (1994)

[31] Lobe, B., Baschnagel, J. and Binder, K.: Macromol. (1994, in press)

[32] Ray, P. and Binder, K.: Europhys. Lett. (1994, in press)

[33] Baschnagel, J.: Phys. Rev. **B49**, 135 (1994)

[34] Baschnagel, J. and Fuchs, M.: preprint

[35] Andrejew, E., Baschnagel, J. and Binder, K.: in preparation

[36] Höchli, U.T., Knorr, K. and Loidl, A.: Adv. Phys. **39**, 405 (1990)

[37] Maletta, H. and Convert, P.: Phys. Rev. Lett. **42**, 108 (1979)

[38] Sullivan, N.S.: AIP Conf. Proc. **103**, 121 (1983)

[39] Carmesin, H.O. and Binder K.: Z. Physik **B58**, 375 (1987); Europhys. Lett. **4**, 269 (1987)

[40] Hammes, D., Carmesin, H.O. and Binder, K.: Z. Physik **B76**, 115 (1989)

[41] Goldbart, P. and Sherrington, D.: J. Phys. **C18**, 1923 (1985)

[42] Elderfield, D.J. and Sherrington, D.: J. Phys. **C16**, L497, L971, L1169 (1983); Elderfield, D.J.: J. Phys. **A17**, L517 (1984)

[43] Gross, D.J., Kanter, I. and Sompolinsky, H.: Phys. Rev. Lett. **55**, 305 (1985)

[44] Carmesin, H.O. and Binder, K.: J. Phys. **A21**, 4035 (1988)

[45] Scheucher, M., Reger, J.D., Binder, K. and Young, A.P.: Phys. Rev. **B42**, 6881 (1990)

[46] Cwilich, G.: J. Phys. **A23**, 5029 (1990); Cwilich, G. and Kirkpatrick, T.R.: J. Phys. **A22**, 4971 (1989)

[47] Scheucher, M. and Reger, J.D.: Phys. Rev. **B45**, 2499 (1922)

[48] Scheucher, M., Reger, J.D., Binder, K. and Young, A.P.: Europhys. Lett. **14**, 119 (1991); **20**, 343 (1992)

[49] Scheucher, M. and Reger, J.D.: Z. Physik **B91**, 383 (1993)

[50] McMillan, W.L.: J. Phys. **C17**, 3179 (1984)

[51] Vollmayr, K., Schreider, G., Reger, J.D. and Binder, K.: J. Noncryst. Solids (1994, in press)

[52] Haas, F., Vollmayr, K. and Binder, K.: in preparation

[53] Hessinger, J. and Knorr, K.: Ferroelectrics **29**, 127 (1992)

[54] Michel, K.H.: Phys. Rev. **B35**, 1405, 1414 (1987); Z. Physik **B68**, 259 (1987)

[55] Nattermann, T. and Villain, J.: Physe Transitions **11**, 5 (1988)

[56] Bässler, H.: Phys. Rev. Lett. **58**, 767 (1987)

[57] Keddie, J.L., Jones, R.A.L. and Cory, R.A.: preprint.

Evolution of Order Parameters in Disordered Spin Systems - a Closure Procedure

D. Sherrington and A.C.C Coolen

Dept. of Physics - Theoretical Physics, University of Oxford
1 Keble Road, Oxford OX1 3NP, U.K.

We discuss a theory to describe the dynamics of disordered spin systems in terms of deterministic flow equations for macroscopic parameters, and use it to study the Hopfield [1] neural network model near saturation, the Sherrington-Kirkpatrick spin-glass [2], and an exactly solvable toy model. Two assumptions, based on the removal of microscopic memory effects, allow us to close the macroscopic laws. The theory produces dynamical generalisations of AT- and zero-entropy lines and of Parisi's overlap distribution $P(q)$, and in equilibrium recovers the standard results from equilibrium statistical mechanics. For homogeneous initial conditions the theory is shown to capture the main characteristics of the flow in the order parameter plane, the impact of microscopic memory effects appears to be mainly an overall slowing down.

1 Derivation of Closed Macroscopic Flow Equations

1.1 Ising Spin Systems

The simplest type of model describes N Ising spins $\sigma_i \in \{-1, 1\}$ with the usual Hamiltonian

$$H(\boldsymbol{\sigma}) = -\sum_{i<j} \sigma_i J_{ij} \sigma_j - \theta \sum_i \sigma_i \tag{1}$$

The natural dynamics is a stochastic sequential alignment of the spins à la Glauber [3] to local fields $h_k(\boldsymbol{\sigma})$. The vector $\boldsymbol{\sigma} = (\sigma_1, \ldots, \sigma_N)$ denotes the *microscopic state* of the system. The stochastic process is described by a master equation for the microscopic probability distribution $p_t(\boldsymbol{\sigma})$:

$$\frac{d}{dt} p_t(\boldsymbol{\sigma}) = \sum_{k=1}^{N} [p_t(F_k \boldsymbol{\sigma}) w_k(F_k \boldsymbol{\sigma}) - p_t(\boldsymbol{\sigma}) w_k(\boldsymbol{\sigma})] \tag{2}$$

with the transition rates and the local fields:

$$w_k(\boldsymbol{\sigma}) = \frac{1}{2} [1 - \sigma_k \tanh[\beta h_k(\boldsymbol{\sigma})]] \qquad h_k(\boldsymbol{\sigma}) = \sum_{l \neq k} J_{kl} \sigma_l + \theta$$

and the spin-flip operator $F_k \Phi(\sigma) \equiv \Phi(\sigma_1, \ldots, -\sigma_k, \ldots, \sigma_N)$. The dynamics (2) leads to the Boltzmann equilibrium distribution $p_{eq}(\sigma) \sim \exp[-\beta H(\sigma)]$.

Our aim is to study the dynamics at a macroscopic level, by choosing an appropriate *finite* set of n macroscopic state variables $\Omega(\sigma) \equiv (\Omega_1(\sigma), \ldots, \Omega_n(\sigma))$, with corresponding macroscopic probability distribution

$$\mathcal{P}_t[\Omega] \equiv \sum_\sigma p_t(\sigma) \delta[\Omega - \Omega(\sigma)] \tag{3}$$

Upon substitution of the microscopic dynamic equation (2) one obtains the Kramers-Moyal expansion for the macroscopic distribution (see e.g. [4]):

$$\frac{d}{dt}\mathcal{P}_t[\Omega] = \sum_{\ell \geq 1} \frac{(-1)^\ell}{\ell!} \sum_{k_1=1}^{n} \cdots \sum_{k_\ell=1}^{n} \frac{\partial^\ell}{\partial \Omega_{k_1} \cdots \partial \Omega_{k_\ell}} \left\{ \mathcal{P}_t[\Omega] F^{(\ell)}_{k_1 \cdots k_\ell}[\Omega; t] \right\} \tag{4}$$

which is defined in terms of sub-shell averages $\langle f(\sigma) \rangle_{\Omega; t}$:

$$F^{(\ell)}_{k_1 \cdots k_\ell}[\Omega; t] \equiv \langle \sum_{j=1}^{N} w_j(\sigma) \Delta_{jk_1}(\sigma) \cdots \Delta_{jk_\ell}(\sigma) \rangle_{\Omega; t}$$

$$\langle f(\sigma) \rangle_{\Omega; t} \equiv \frac{\sum_\sigma p_t(\sigma) \delta[\Omega - \Omega(\sigma)] f(\sigma)}{\sum_\sigma p_t(\sigma) \delta[\Omega - \Omega(\sigma)]} \qquad \Delta_{jk}(\sigma) \equiv \Omega_k(F_j \sigma) - \Omega_k(\sigma)$$

If in the limit $N \to \infty$ the $\ell > 1$ terms in (4) vanish, we are on finite time-scales left with the Liouville equation

$$\frac{d}{dt}\mathcal{P}_t[\Omega] = -\sum_{k=1}^{n} \frac{\partial}{\partial \Omega_k} \left\{ \mathcal{P}_t[\Omega] F^{(1)}_k[\Omega; t] \right\}$$

the solutions of which are δ-distributions, describing *deterministic* flow:

$$\frac{d}{dt}\Omega(t) = \langle \sum_{j=1}^{N} w_j(\sigma) [\Omega(F_j \sigma) - \Omega(\sigma)] \rangle_{\Omega(t); t} \tag{5}$$

1.2 Disordered Ising Spin Systems

In disordered spin systems like spin-glasses and neural networks, the disorder resides in the long-range exchange interactions J_{ij}, which are drawn at random from a given distribution. We will restrict our discussion to systems in which these random interactions are of the form

$$J_{ij} = J_0/N + J z_{ij}/\sqrt{N} \qquad \langle z_{ij} \rangle_{\{z_{kl}\}} = 0, \quad \langle z_{ij}^2 \rangle_{\{z_{kl}\}} = 1 \qquad (i < j) \tag{6}$$

(averaging over the disorder is denoted by $\langle \cdots \rangle_{\{z_{kl}\}}$). We can now separate the disorder-dependent from the disorder-independent term in the Hamiltonian (1):

$$H(\sigma)/N = \overbrace{-\frac{1}{2}J_0 m^2(\sigma) - \theta m}^{\text{disorder-indep.}} \underbrace{-Jr(\sigma)}_{\text{disorder-dep.}} + \mathcal{O}(N^{-1}) \tag{7}$$

with

$$m(\boldsymbol{\sigma}) = \frac{1}{N}\sum_i \sigma_i \qquad r(\boldsymbol{\sigma}) = \frac{1}{N\sqrt{N}}\sum_{i<j} \sigma_i z_{ij}\sigma_j$$

The pair of intensive state variables (m, r) will define a *macroscopic state*. Moreover, we find $m(F_i\boldsymbol{\sigma}) - m(\boldsymbol{\sigma}) = \mathcal{O}(N^{-1})$ and $r(F_i\boldsymbol{\sigma}) - r(\boldsymbol{\sigma}) = \mathcal{O}(N^{-1})$, so for the Ising case the requirements for having deterministic macroscopic flow on finite time-scales are met. Working out (5) gives the corresponding $N \to \infty$ flow equations:

$$\frac{d}{dt}m = \int dz\ D_{m,r;t}[z]\tanh\beta\left(J_0 m + J z + \theta\right) - m \tag{8}$$

$$\frac{d}{dt}r = \int dz\ D_{m,r;t}[z]z\tanh\beta\left(J_0 m + J z + \theta\right) - 2r \tag{9}$$

with the subshell-averaged distribution of the disordered contributions $z_i(\boldsymbol{\sigma})$ to the local fields:

$$h_i(\boldsymbol{\sigma}) = J_0 m(\boldsymbol{\sigma}) + J z_i(\boldsymbol{\sigma}) + \theta + \mathcal{O}(\frac{1}{N}), \qquad z_i(\boldsymbol{\sigma}) = \frac{1}{\sqrt{N}}\sum_{j\neq i} z_{ij}\sigma_j$$

$$D_{m,r;t}[z] = \lim_{N\to\infty}\frac{\sum_{\boldsymbol{\sigma}} p_t(\boldsymbol{\sigma})\delta\left[m - m(\boldsymbol{\sigma})\right]\delta\left[r - r(\boldsymbol{\sigma})\right]\frac{1}{N}\sum_i \delta\left[z - z_i(\boldsymbol{\sigma})\right]}{\sum_{\boldsymbol{\sigma}} p_t(\boldsymbol{\sigma})\delta\left[m - m(\boldsymbol{\sigma})\right]\delta\left[r - r(\boldsymbol{\sigma})\right]} \tag{10}$$

1.3 Closure of the Macroscopic Laws

The flow equations (8,9) are exact, but, due to the explicit time-dependence in the distribution (10), as yet only closed in the limit $J \to 0$ where the disorder is removed. We now introduce two transparent assumptions to close the hierarchy of macroscopic equations:

1. We assume the evolution of the macroscopic state (m, r), and therefore the field distribution $D_{m,r;t}[z]$, to be *self-averaging* with respect to the microscopic realisation of the disorder (i.e. the interactions $\{J_{ij}\}$).
2. As far as evaluating the distribution $D_{m,r;t}$ is concerned, we assume *equipartitioning of probability* in the macroscopic (m, r) sub-shells of the ensemble.

As a result of assumption 1 we may average $D_{m,r;t}[z]$ (10) over the disorder. For the models and initial conditions we will consider in subsequent sections, numerical simulations clearly support this assumption. As the system size is increased, sample to sample fluctuations in the (m, r) trajectories do eventually disappear and well defined flow lines emerge. Assumption 2 states that the probability fluctuations *within* the macroscopic subshells can be neglected. It allows us to eliminate the explicit time dependence in $D_{m,r;t}[z]$ (10). This is clearly the most crucial and dangerous assumption, which amounts to precisely eliminating the microscopic memory effects. It is guaranteed to be true in equilibrium (here we have equipartitioning in the *energy* subshells, which is an even stronger statement) and at $t = 0$, upon choosing appropriate initial conditions. Away from $t = 0$ and $t = \infty$ the assumption can only be tested by comparing the predictions of the resulting theory for the distribution (10) and the actual (m, r) flow, with what is found in numerical simulations.

Combination of the two assumptions gives:

$$D_{m,r;t}[z] \rightarrow D_{m,r}[z] = \langle \frac{\sum_\sigma \delta\,[m - m(\sigma)]\,\delta\,[r - r(\sigma)]\,\frac{1}{N}\sum_i \delta\,[z - z_i(\sigma)]}{\sum_\sigma \delta\,[m - m(\sigma)]\,\delta\,[r - r(\sigma)]} \rangle_{\{z_{ij}\}}$$

$$(11)$$

We have now obtained a closed set of dynamical equations (8,9,11), which are by construction exact at $t = 0$, at $t = \infty$ and in the limit $J \rightarrow 0$ for all times.

1.4 Replica Theory

The distribution (11) can be calculated analytically with the replica method. We use the identity

$$\langle \Phi(\sigma) \rangle_W \equiv \frac{\sum_\sigma \Phi(\sigma) W(\sigma)}{\sum_\sigma W(\sigma)} = \frac{\sum_\sigma \Phi(\sigma) W(\sigma)}{\sum_\sigma W(\sigma)} \cdot \frac{[\sum_\sigma W(\sigma)]^{n-1}}{[\sum_\sigma W(\sigma)]^{n-1}}$$

$$= \lim_{n \to 0} \sum_{\{\sigma^\alpha\}} \Phi(\sigma^1) \prod_{\alpha=1}^n W(\sigma^\alpha)$$

which allows us to replace having to perform an average of a *fraction* involving the measure $W(\sigma) = \delta[m - m(\sigma)]\delta[r - r(\sigma)]$ over the disorder, by having to average *powers* of $W(\sigma)$ over the disorder. The replica calculation leads in the usual way to a saddle-point problem, with $n(n-1)$ integration variables $q_{\alpha\beta}$ $(\alpha \neq \beta)$ (details can be found in [6, 7]). The physical meaning of the saddle-point is, a la Parisi [8], given in terms of the disorder-averaged distribution of overlaps between microscopic states in two identical systems, constrained in an (m, r) subshell:

$$P_{mr}(q) =$$

$$\langle \frac{\sum_{\sigma,\sigma'} \delta\,[q - \frac{1}{N}\sum_k \sigma_k \sigma_k']\,\delta\,[m - m(\sigma)]\,\delta\,[r - r(\sigma)]\,\delta\,[m - m(\sigma')]\,\delta\,[r - r(\sigma')]}{\sum_{\sigma,\sigma'} \delta\,[m - m(\sigma)]\,\delta\,[r - r(\sigma)]\,\delta\,[m - m(\sigma')]\,\delta\,[r - r(\sigma')]} \rangle_{\{z_{ij}\}}$$

$$= \lim_{n \to 0} \frac{1}{n(n-1)} \sum_{\alpha \neq \beta} \delta\,[q - q_{\alpha\beta}]$$

For an ergodic system this distribution will have to be a δ-function. In this way we are led, according to the above expressions for $P_{mr}(q)$, to the so-called 'replica-symmetric' (RS) ansatz for the saddle-point:

$$q_{\alpha\beta} = q \quad (\alpha \neq \beta) \tag{12}$$

This ansatz is certainly correct for $r \to 0$. As r is raised, the so-called de Almeida-Thouless (AT) instability [9] occurs when the RS saddle-point (12) ceases to dominate the saddle-point integration, in favour of a continuously bifurcating saddle-point $q_{\alpha\beta} = q + \delta q_{\alpha\beta}$ without replica-symmetry. This corresponds to a second-order transition to a spin-glass like state, where ergodicity is broken in a highly non-trivial manner.

The replica method also allows us to calculate a 'dynamical entropy' $\tilde{S}$, defined as:

$$\tilde{S} \equiv \lim_{N \to \infty} \frac{1}{N} \log \sum_{\sigma} \delta \left[m - m(\sigma) \right] \delta \left[r - r(\sigma) \right] \tag{13}$$

We will find that at fixed-points of the flow (8,9,11) this quantity reduces to the usual thermodynamic entropy. Since our microscopic model is discrete, neither the thermodynamic entropy nor the dynamic entropy $\tilde{S}$ can be negative.

1.5 Specific Models

The Hopfield [1] neural network model, in which the spins σ_i represent neurons and the interactions J_{ij} represent synaptic efficacies, corresponds to choosing

$$J_{ij} = \frac{1}{N} \sum_{\mu=1}^{p} \xi_i^{\mu} \xi_j^{\mu}, \qquad \xi_i^{\mu} = \pm 1 \text{ (random)}, \quad \mu = 1, \ldots, p = \alpha N \tag{14}$$

The random vectors $\boldsymbol{\xi}^{\mu} = (\xi_1^{\mu}, \ldots, \xi_N^{\mu}) \in \{-1, 1\}^N$ represent patterns, stored in the system by virtue of the choice (14). The degree to which the microscopic system state σ resembles any of these patterns, is measured by the so-called 'overlaps' $m_{\mu}(\sigma)$:

$$m^{\mu}(\sigma) = \frac{1}{N} \sum_{k=1}^{N} \xi_k^{\mu} \sigma_k$$

As in the equilibrium studies by Amit et al. [5] we are interested in the case where only a finite number n of these (the 'condensed' patterns) are of order unity, the $p - n$ remaining overlaps (the 'uncondensed' patterns) being of order $N^{-\frac{1}{2}}$. For simplicity in this paper we choose only pattern 1 to be condensed. A simple gauge transformation

$$\sigma_i \to \sigma_i \xi_i^1 \qquad J_{ij} \to \xi_i^1 \xi_j^1 J_{ij} \qquad \frac{1}{N} \sum_k \xi_k^1 \sigma_k \to \frac{1}{N} \sum_k \sigma_k$$

now transforms the Hopfield model into the form (6) which we have introduced in the previous subsections (see the table below).

Our second example will the Sherrington-Kirkpatrick spin-glass model [2], which is simply obtained by choosing in (6) the quenched variables z_{ij} to be drawn independently from a Gaussian distribution.

In both models the disorder is of the type (6), they differ only in the statistics of the disorder-variables $\{z_{ij}\}$ and the choices made for the parameters J_0, J and θ:

Hopfield model :	Sherrington−Kirkpatrick model :
$z_{ij} = \frac{1}{\sqrt{p}} \sum_{\mu>1}^{p=\alpha N} \xi_i^1 \xi_i^{\mu} \xi_j^1 \xi_j^{\mu}$	z_{ij} indep. Gaussian variables
$J_0 = 1, \ J = \sqrt{\alpha}, \ \theta = 0$	

2 Results for the Hopfield Model

Full details of the derivations can be found in [6]. In order to arrive at the uniform presentation of the previous section and suppress repetition, we have in the present paper defined $D_{mr}[z]$ and $r(\sigma)$ slightly differently from [6]. To arrive at the notation of [6], we now put $z \to \frac{1}{\sqrt{\alpha}}z$ and $r \to \frac{1}{\sqrt{\alpha}}r + 1$.

2.1 Replica-Symmetric Flow Equations

Following the strategy outlined in the previous section and upon making the RS ansatz (12), we arrive for the Hopfield model [1] at the following expression for the distribution (11):

$$
D_{m,r}^{RS}[z] = \frac{e^{-\frac{1}{2}(\Delta+z)^2/\alpha r}}{2\sqrt{2\pi\alpha r}} \left\{ 1 - \int Dy \, \tanh\left[\lambda y \sqrt{\frac{\Delta}{\alpha\rho r}} + (\Delta+z)\frac{\lambda^2}{\alpha\rho r} + \mu \right] \right\}
$$

$$
+ \frac{e^{-\frac{1}{2}(\Delta-z)^2/\alpha r}}{2\sqrt{2\pi\alpha r}} \left\{ 1 - \int Dy \, \tanh\left[\lambda y \sqrt{\frac{\Delta}{\alpha\rho r}} + (\Delta-z)\frac{\lambda^2}{\alpha\rho r} - \mu \right] \right\} \quad (15)
$$

where we have introduced the Gaussian measure $Dx = (2\pi)^{-\frac{1}{2}}e^{-\frac{1}{2}x^2}dx$, and where the parameters Δ, λ, ρ and μ are the solutions of the saddle-point equations:

$$
\Delta = \alpha\rho r - \lambda^2/\rho \qquad r = \frac{1-\rho(1-q)^2}{[1-\rho(1-q)]^2} \qquad \lambda = \frac{\rho\sqrt{\alpha q}}{1-\rho(1-q)}
$$

$$
m = \int Dy \, \tanh(\lambda y + \mu) \qquad\qquad q = \int Dy \, \tanh^2(\lambda y + \mu)
$$

The distribution (15) is clearly in general not Gaussian, in contrast to what has often been assumed in previous studies (it is Gaussian only for $r = 1$).

If we insert (15) into (8,9), and perform some transformations of integration variables, we can write the corresponding RS flow equations in the form

$$
\frac{d}{dt}m = \iint Dx Dy \, M(m,r;x,y) - m \quad (16)
$$

$$
\frac{1}{2}\frac{d}{dt}r = \iint Dx Dy \, R(m,r;x,y) + 1 - r \quad (17)
$$

in which the kernels $M(m,r;x,y)$ and $R(m,r;x,y)$ depend on the macroscopic state (m,r) in a complicated way, through the various parameters that appear in the RS saddle-point equations.

The physical region in the (m,r) plane, outside of which no microscopic states σ exist, is bounded by the line where $q = 1$. This line turns out to be given by

$$
\sqrt{r} = 1 + \sqrt{\frac{2}{\alpha\pi}}e^{-[\text{erf}^{\text{inv}}(m)]^2} \quad (18)
$$

(the zero entropy line, $\tilde{S} = 0$, is found to be located slightly below (18)). The region of validity in the (m, r) plane of the RS theory is bounded by the AT instability [9], the location of which is marked by the continuous bifurcation of the so-called 'replicon' mode. The resulting condition for the AT instability is

$$\alpha = \rho^2 \left[\alpha + \Delta\right]^2 \int Dy \ \cosh^{-4}[\lambda y + \mu] \tag{19}$$

Finally one can show [6] that (i) the RS fixed-point equations $dm/dt = dr/dt = 0$ corresponding to (16,17) are equivalent to the RS equations derived from equilibrium statistical mechanics by Amit et al [5], (ii) at fixed-points, both retrieving $(m > 0)$ and non-retrieving $(m = 0)$, the AT instability condition (19) reduces to the corresponding equilibrium one derived in [5], and finally (iii) at fixed-points the RS dynamic entropy (13) reduces to the thermodynamic one. Apparently the asymptotic state of the process (16,17) is the one obtained from equilibrium statistical mechanics, as summarised in the phase diagram derived by Amit et al. [5].

2.2 Comparison with Numerical Simulations

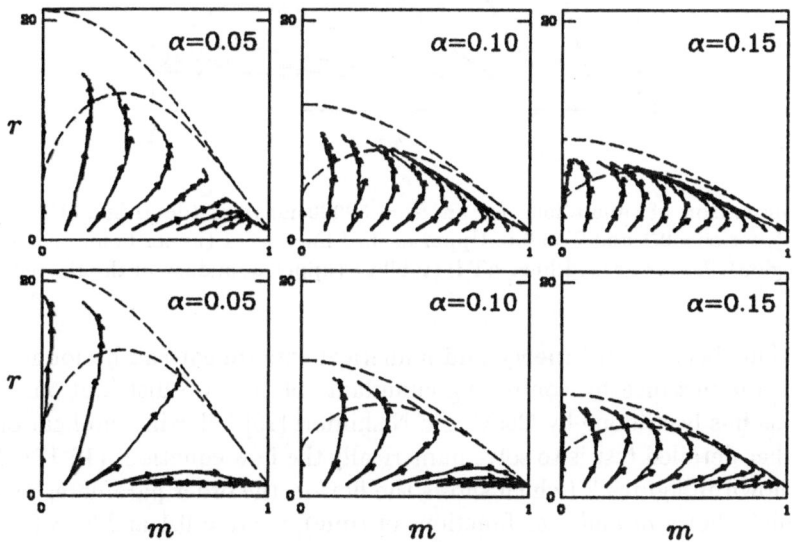

Fig. 1. Comparison of simulations for an $N = 16000$ network (solid lines) and RS theory (arrows). Top row of graphs: $T = 0.5$, bottom row: $T = 0.0$. Upper dashed line: $q = 1$ (upper boundary of the physical region). Lower dashed line: AT instability (upper boundary of the RS region).

We are now in a position to confront our theory with numerical experiments. Figure 1 shows the result of performing numerical simulations of an $N = 16000$

system for $0 \leq t \leq 10$ iterations/spin (solid lines), starting from initial configurations where the N spin states are generated independently according to $p_0(\sigma_i) = \frac{1}{2}[1+m_0]\delta_{\sigma_i,1}+\frac{1}{2}[1-m_0]\delta_{\sigma_i,-1}$. Arrows indicate the velocities predicted by the theory (16,17), calculated at intervals of 1 iteration/spin. It appears that, at least qualitatively, the equations (16,17) give a good description of the flow diagram, particularly in the region where replica symmetry holds (below the AT line).

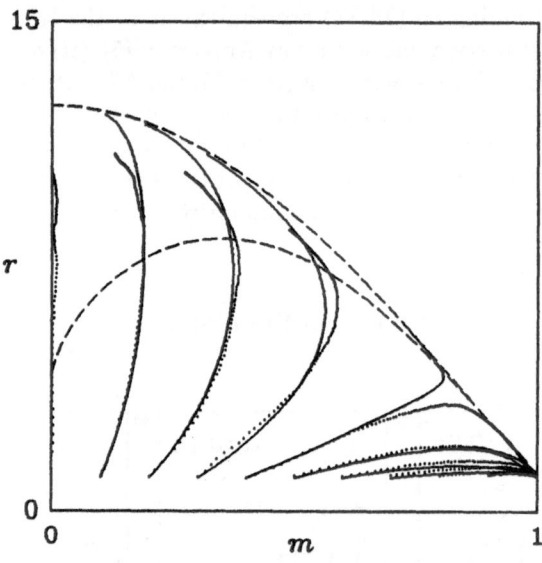

Fig. 2. Comparison of simulations for an $N = 32000$ network (dots) and RS theory (solid lines), for $\alpha = 0.1$ and $T = 0.0$. Upper dashed line: $q = 1$ (upper boundary of the physical region). Lower dashed line: AT instability (upper boundary of the RS region).

Deviations between RS theory and numerical experiment can be found and quantified, for instance by comparing cumulants of the two distributions (10) and (11), as has been done by Ozeki and Nishimori [10] following publication of [11]. Another detailed test is to solve numerically the flow equations (16,17). The result is shown in figures 2 (which shows the flow in the order parameter plane) and 3 (which shows m and r as functions of time), for $\alpha = 0.1$ and $T = 0$. The simulations are done for an $N = 32000$ system. We observe again a good, but nor perfect, agreement in terms of the flow in the (m, r) plane. Figure 3 shows perfect agreement in terms of the temporal dependence for those trajectories that lead towards a ferromagnetic state. However, for trajectories leading away from the ferromagnetic state the theory fails to reproduce an overall slowing down of the flow which, according to the simulations, seems to set in before the AT instability occurs (to which it therefore cannot be ascribed).

In fact figure 3 shows a remarkable resemblance to the results obtained for this model by Horner et al. [12], using a path-integral solution of the microscopic

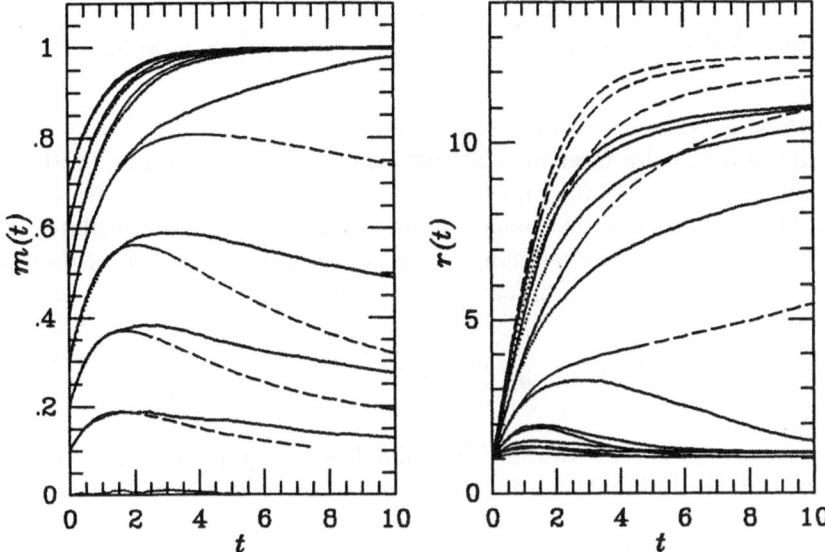

Fig. 3. Comparison of simulations for an $N = 32000$ network (dots) and RS theory (solid lines as long as RS is correct, dashed when RS is broken), now shown as functions of time, for $\alpha = 0.1$ and $T = 0.0$. Left graphs: $m(t)$, right graphs: $r(t)$.

master equation (2). Although their methods, formulated in terms of correlation- and response-functions, are quite orthogonal to the present one, the final results (the graphs of $m(t)$) seem almost identical. It is an interesting, but as yet open, question to find out quantitatively why and to what extent the two methods describe the same flow.

3 Results for the Sherrington-Kirkpatrick Model

Full details can be found in [7].

3.1 Replica Calculation of the Field Distribution

Calculating the distribution (11) with replica-theory leads in the limit $N \to \infty$, via a steepest descent integration, to the following expression:

$$D_{m,r}[z] = \int \frac{dx}{2\pi} e^{-\frac{1}{2}x^2 + ixz} \lim_{n \to 0} \frac{\langle e^{-ix\sum_\gamma \sigma_\gamma \rho_\gamma q_{1\gamma} + \sum_\gamma \mu_\gamma \sigma_\gamma + \frac{1}{2}\sum_{\gamma\delta} q_{\gamma\delta}\rho_\gamma\rho_\delta\sigma_\gamma\sigma_\delta}\rangle_\sigma}{\langle e^{\sum_\gamma \mu_\gamma \sigma_\gamma + \frac{1}{2}\sum_{\gamma\delta} q_{\gamma\delta}\rho_\gamma\rho_\delta\sigma_\gamma\sigma_\delta}\rangle_\sigma}$$

(20)

in which the parameters $\{q_{\alpha\beta}\}$, $\{\rho_\alpha\}$ and $\{\mu_\alpha\}$ are the solutions of the saddle-point equations

$$q_{\alpha\beta} = \frac{\langle \sigma_\alpha \sigma_\beta e^{\sum_\gamma \mu_\gamma \sigma_\gamma + \frac{1}{2}\sum_{\gamma\delta} q_{\gamma\delta}\rho_\gamma\rho_\delta\sigma_\gamma\sigma_\delta}\rangle_\sigma}{\langle e^{\sum_\gamma \mu_\gamma \sigma_\gamma + \frac{1}{2}\sum_{\gamma\delta} q_{\gamma\delta}\rho_\gamma\rho_\delta\sigma_\gamma\sigma_\delta}\rangle_\sigma}$$

$$m = \frac{\langle \sigma_\alpha e^{\sum_\gamma \mu_\gamma \sigma_\gamma + \frac{1}{2}\sum_{\gamma\delta} q_{\gamma\delta} \rho_\gamma \rho_\delta \sigma_\gamma \sigma_\delta} \rangle_\sigma}{\langle e^{\sum_\gamma \mu_\gamma \sigma_\gamma + \frac{1}{2}\sum_{\gamma\delta} q_{\gamma\delta} \rho_\gamma \rho_\delta \sigma_\gamma \sigma_\delta} \rangle_\sigma} \qquad \sum_\beta q_{\alpha\beta}^2 \rho_\beta = 2r$$

that maximise the exponent of the steepest descent integration. Fixed-points of the flow are obtained by inserting (20) into the dynamic equations (8,9) and subsequently requiring $dm/dt = dr/dt = 0$.

In equilibrium statistical mechanics (following [2] and [8]) one calculates the disorder-averaged free energy per spin $\overline{f}$, which also leads to a steepest descent integration. Here one has to minimise a function $F(\boldsymbol{m}, \boldsymbol{q})$:

$$F = \frac{J_0}{2n}\sum_\alpha m_\alpha^2 + \frac{\beta J^2}{4n}\sum_{\alpha\beta} q_{\alpha\beta}^2 - \frac{1}{\beta n}\log\langle e^{\beta \sum_\alpha \sigma_\alpha(J_0 m_\alpha + \theta) + \frac{1}{2}\beta^2 J^2 \sum_{\alpha\beta}\sigma_\alpha\sigma_\beta q_{\alpha\beta}} \rangle_\sigma$$

In Parisi's theory [8] the relevant saddle-point has the property $m_\alpha = m$ (the magnetisation). The link between the two approaches turns out to be [7]:

$$\mu_\alpha = \beta(J_0 m + \theta) \qquad \rho_\alpha = \beta J$$

from which the following can be derived [7]: (*i*) the two sets of saddle-point equations and the two objects to be minimised are equivalent, (*ii*) at fixed-points the RSB dynamic entropy reduces to the thermodynamic one, $\tilde{S}_{eq} = \beta^2 \partial_\beta \overline{f}$. We conclude that our dynamical theory in equilibrium reproduces the full RSB thermodynamic theory.

3.2 Replica-Symmetric Flow Equations

If for the SK spin-glass we make the replica-symmetric ansatz, we obtain expressions very similar to the ones we found for the Hopfield model:

$$D_{mr}^{RS}[z] = \frac{e^{-\frac{1}{2}[z+\rho(1-q)]^2}}{2\sqrt{2\pi}}\left\{1 + \int Dy \ \tanh\left[\rho y\sqrt{q(1-q)} - \rho q\,[z+\rho(1-q)] - \mu\right]\right\}$$

$$+ \frac{e^{-\frac{1}{2}[z-\rho(1-q)]^2}}{2\sqrt{2\pi}}\left\{1 + \int Dy \ \tanh\left[\rho y\sqrt{q(1-q)} + \rho q\,[z-\rho(1-q)] + \mu\right]\right\} \quad (21)$$

with

$$m = \int Dz \ \tanh[\rho\sqrt{q}z + \mu] \qquad q = \int Dz \ \tanh^2[\rho\sqrt{q}z + \mu] \qquad \rho\left[1 - q^2\right] = 2r$$

The RS flow equations, obtained by insertion of (21) into (8,9), can again be written in the form

$$\frac{d}{dt}m = \int\int Dx Dy \ M(m, r; x, y) - m \qquad (22)$$

$$\frac{d}{dt}r = \int\int Dx Dy \ R(m, r; x, y) - 2r \qquad (23)$$

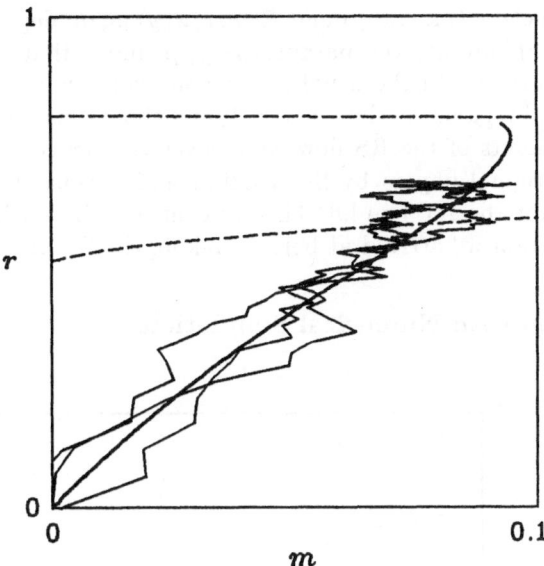

Fig. 4. Cooling in a field. Comparison of three independent simulations for an $N = 3200$ system (fluctuating lines) and RS theory (smooth line), for $J = 1$, $J_0 = 0$ and $T = 0.1$, with an external field $\theta = 0.1$. Upper dashed line: $q = 1$ (upper boundary of the physical region). Lower dashed line: AT line (upper boundary of the RS region).

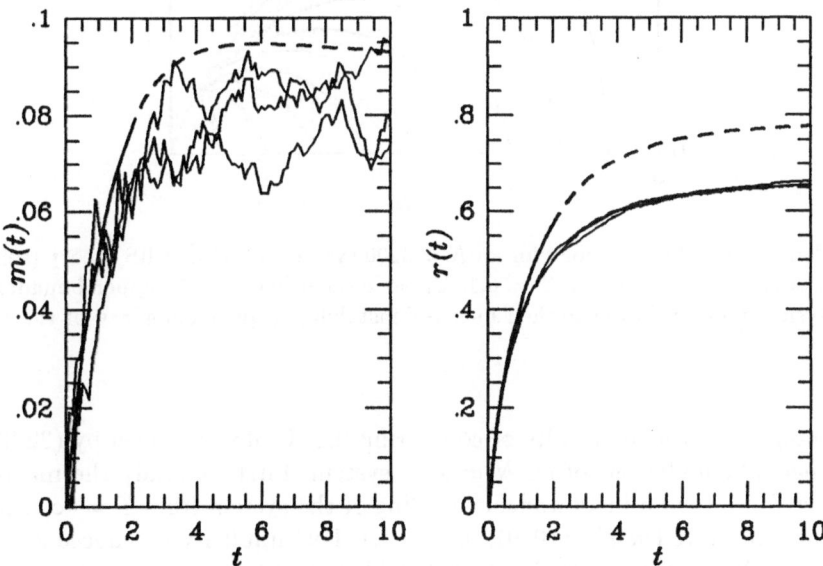

Fig. 5. Cooling in a field. Three independent simulations for an $N = 3200$ system (fluctuating lines) and RS theory (smooth solid lines as long as RS is correct, dashed when RS is broken, as distinguished by crossing the dynamic AT line) for $J = 1$, $J_0 = 0$ and $T = \theta = 0.1$, now shown as functions of time.

in which the kernels $M(m, r; x, y)$ and $R(m, r; x, y)$ again depend on the macroscopic state (m, r) through the parameters ρ, μ and q that appear in the RS saddle-point equations. In the usual manner we can now calculate the boundary of the physical region, i.e. the line $q = 1$, and the AT instability in the (m, r) plane. At fixed-points of the RS flow we recover the RS equilibrium results as derived in the original papers by Sherrington and Kirkpatrick [2, 13], as well as the equation for the AT line [9]. This, of course, is inevitable in view of the general identification we arrived at before making the RS ansatz.

3.3 Comparison with Numerical Simulations

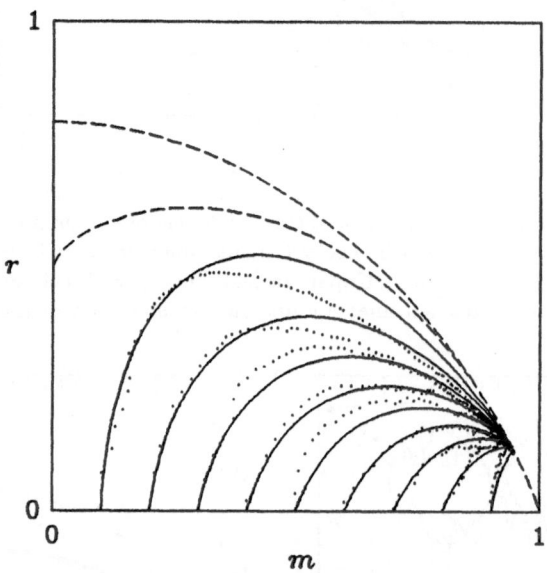

Fig. 6. Comparison of simulations for an $N = 3200$ system (dots) and RS theory (solid lines), $J_0 = 2$, $J = 1$, $\theta = 0$ and $T = 0.1$. Upper dashed line: $q = 1$ (upper boundary of the physical region). Lower dashed line: AT instability (upper boundary of the RS region).

We will now present some results of comparing the RS flow as given by (22,23) with numerical simulations of an $N = 3200$ system. First we study the macroscopic flow following an instantaneous cooling of the system from $T = \infty$, with a paramagnetic state $(m, r) = (0, 0)$, to $T = 0.1$. For simplicity we choose $J = 1$ and $J_0 = 0$. We apply a small external field $\theta = 0.1$, to induce a non-zero magnetisation. The results are shown in figures 4 and 5. As with the Hopfield model, the flow in the (m, r) plane is described quite well in the region where replica-symmetry is stable (below the AT line), but the theory fails to describe an overall slowing down that eventually sets in. The slowing down appears to occur before the AT line is reached in these simulations.

As a second example, in figure 6 we show the flow in the (m, r) plane for the case where there is a ferromagnetic bias $J_0 = 2$ in the interactions, for $T = 0.1$ (without external field). Again in terms of the macroscopic flow there is qualitative agreement between theory and experiment. Unfortunately, there are also non-negligable finite-size fluctuations, the infinite range of the interactions in the SK model at present preventing us from studying larger systems.

4 An Exactly Solvable Toy Model

Our RS theory appears to fit the simulations reasonably well, but with deviations above and near the AT lines. Above the AT lines this could be attributed to ignored replica symmetry breaking (RSB) effects, as indeed it might also below the AT lines if there is first-order RSB. As an extra test, however, we finally turn to an exactly solvable toy model [14] for which RSB is not required, to understand further the potential and the restrictions of our closure procedure.

4.1 The Model and its Solution

We consider a system of soft spins $\sigma_i \in \langle -\infty, \infty \rangle$ obeying the linear Langevin equation

$$\frac{d}{dt}\sigma_i = -\frac{\partial \tilde{H}}{\partial \sigma_i} + \eta_i = \sum_j J_{ij}\sigma_j - \mu\sigma_i + \eta_i \tag{24}$$

$$\tilde{H} = H + \frac{1}{2}\mu \sum_i \sigma_i^2 \qquad \langle \eta_i(t)\eta_j(t') \rangle = 2T\delta_{ij}\delta(t - t')$$

with H again given by (1), but with μ chosen large enough to prevent run-away modes. We concentrate on SK-distributed $\{J_{ij}\}$, as given by (6), with $J_0 = \theta = 0$. This implies $\mu \geq 2J$ since the eigenvalue spectrum of $\{J_{ij}\}$ obeys Wigner's semi-circular law, bounded by $\pm 2J$ in the thermodynamic limit. Transformation to the basis where $\{J_{ij}\}$ is diagonal leads to an immediate solution.

In analogy with the earlier formalism we choose as our macroscopic order parameters the two individual contributions to the Hamiltonian $\tilde{H}$, or equivalently (following [14]) the disorder-independent energy per spin $\frac{1}{2}Q$ and the total energy per spin E:

$$Q = \frac{1}{N}\sum_i \sigma_i^2 \qquad E = -\frac{1}{2N}\sum_{ij}\sigma_i J_{ij}\sigma_j + \frac{\mu}{2N}\sum_i \sigma_i^2$$

These turn out to evolve in time according to

$$Q(t) = \int d\lambda\, \rho(\lambda)\sigma_\lambda^2(0)e^{-2t(\mu-\lambda)} + T\int d\lambda \frac{\rho(\lambda)}{\mu - \lambda}\left[1 - e^{-2t(\mu-\lambda)}\right] \tag{25}$$

$$E(t) = \frac{1}{2}\int d\lambda\, \rho(\lambda)\sigma_\lambda^2(0)(\mu-\lambda)e^{-2t(\mu-\lambda)} + \frac{1}{2}T\left[1 - \int d\lambda\, \rho(\lambda)e^{-2t(\mu-\lambda)}\right] \tag{26}$$

Here $\rho(\lambda)$ denotes Wigner's semi-circular law for the eigenvalue spectrum of the matrix $\{J_{ij}\}$, and $\sigma_\lambda^2(0)$ is the contribution per degree of freedom to $Q(0)$ from eigenspace λ, i.e. $Q(0) = \frac{1}{N}\sum_i \sigma_i^2(0) = \int d\lambda\rho(\lambda)\sigma_\lambda^2(0)$.

4.2 The Closure Procedure

Our formalism proceeds as in the Glauber case, with the master equation (2) being replaced by the Fokker-Planck equation of the process (24), and considering the macroscopic distribution $P(Q, E)$. One finds deterministic evolution with

$$\frac{d}{dt}Q = -4\left[E - \frac{1}{2}T\right] \qquad \frac{d}{dt}E = \mu T - \langle h^2 \rangle_{Q,E;t} \qquad (27)$$

$$\langle h^2 \rangle_{Q,E;t} = \frac{\int d\sigma\, p_t(\sigma)\, \delta\left[Q - Q(\sigma)\right] \delta\left[E - E(\sigma)\right] \frac{1}{N} \sum_i \left[\sum_j J_{ij}\sigma_j - \mu\sigma_i\right]^2}{\int d\sigma\, p_t(\sigma)\, \delta\left[Q - Q(\sigma)\right] \delta\left[E - E(\sigma)\right]}$$

We close these equations by assuming self-averaging with respect to the realisation of the interactions, and, at least for evaluating $\langle h^2 \rangle_{Q,E;t}$, equipartitiong of probabilty in the (Q, E) subshells of the ensemble. The RS calculation involves the usual order parameter q, here defined in terms of integrals over spin configurations as opposed to summations. One finds three relevant saddle-points, each dominating the steepest-descent integral in one particular region of the (Q, E) plane:

region	q	dE/dt
$E < \frac{1}{2}Q(\mu - J)$	> 0	$-2\mu(E - \frac{1}{2}T) + (\mu - 2J)(\mu Q - 2E)$
$\frac{1}{2}Q(\mu - J) < E < \frac{1}{2}Q(\mu + J)$	0	$\mu T - QJ^2 - 4E^2/Q$
$E > \frac{1}{2}Q(\mu + J)$	> 0	$-2\mu(E - \frac{1}{2}T) + (\mu + 2J)(\mu Q - 2E)$

$$(28)$$

Replica-symmetry turns out to be stable in the $q = 0$ region, and marginally stable in the two $q > 0$ regions, so there is no need for replica symmetry breaking. The physical region (where microscopic states indeed exist) is $\frac{1}{2}Q(\mu - 2J) \leq E \leq \frac{1}{2}Q(\mu + 2J)$. The (correct) fixed-point is located in the $q = 0$ region.

4.3 Comparison with the Exact Results

Comparison of equations (27,28) with the exact result (25,26) in terms of the flow in the (Q, E) plane leads to figure 7 (in this example $\mu = 8$, $T = 3$ and $J/\mu \in \{3/8, 1/2\}$). The outer dashed lines indicate the boundaries of the physical region, the inner dashed lines are the boundaries of the $q = 0$ region. Away from the critical case $\mu = 2J$ the agreement is quite good. For $\mu = 2J$ (zero modes), however, the theory breaks down for inhomogeneous initial conditions in the low energy region of the (Q, E) plane.

For the present model, however, we can make more detailed quantitative comparisons, by analysing the equations (27,28) and (25,26) for small and large times. These results (with $x = 2J/\mu$), clearly confirm that the slowing down of the flow, observed for the Hopfield and SK models but not described by our replica-symmetric flow equations, can arise as a direct consequence of the elimination in our closure procedure of the microscopic memory effects, without the need to invoke finite-size effects or RS instabilities.

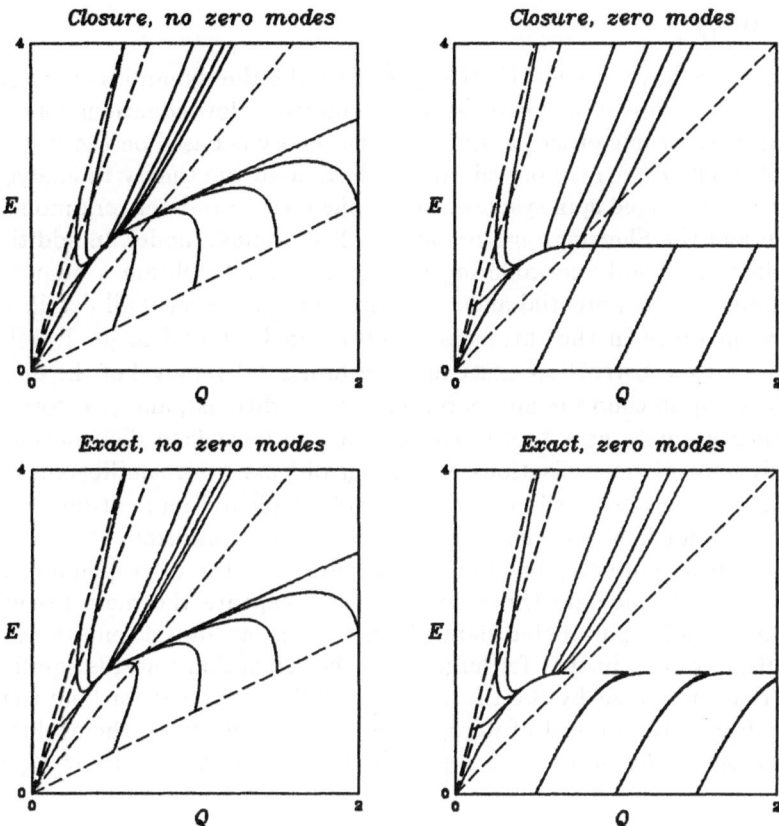

Fig. 7. Comparison of the exact (Q, E) flow and the flow according to our closure procedure, for $\mu = 8$, $T = 3$ and $J = 3\mu/8$ (left, no zero modes) versus $J = \mu/2$ (right, zero modes). Two sorts of initial conditions are shown: homogeneous, where $\sigma_\lambda^2(0) = Q_0$ for all λ, so $E_0 = \frac{1}{2}\mu Q_0$, and inhomogeneous, with only one $\sigma_\lambda^2(0) \neq 0$ for which we choose the extreme modes $\lambda = \pm 2J$, so $E_0 = \frac{1}{2}Q_0(\mu \pm 2J)$.

Short times :

hom. init. cond. :	$E_{\text{closure}}(t) - E_{\text{exact}}(t) = \mathcal{O}(t^3)$
inhom. init. cond. :	$E_{\text{closure}}(t) - E_{\text{exact}}(t) = \mathcal{O}(t^2)$

Large times :

$\mu > 2J$:	exact :	$e^{-t/\tau}$	$(2\mu\tau)^{-1} = 1 - x$
	closure :	$e^{-t/\tau}$	$(2\mu\tau)^{-1} = \frac{1}{2}(1-x) + \frac{1}{2}\sqrt{1-x^2}$
$\mu = 2J$:	exact :	$t^{-\alpha}$	$\alpha = 1/2$
	closure :	$t^{-\alpha}$	$\alpha = 1$

5 Summary

In this paper we have discussed a theory to describe the dynamics of disordered spin systems in terms of a closed set of deterministic flow equations for a *finite* number of macroscopic order parameters. The theory is based on the systematic removal of *microscopic* memory effects. We have used the theory to analyse two archetypical disordered spin systems: the Hopfield [1] neural network model near saturation and the Sherrington-Kirkpatrick [2] spin-glass model. In addition we have studied an exactly solvable toy model, in order to obtain a quantitative understanding of the potential and restrictions of our theory. Full details of the derivations involved in the three case studies can be found in [6, 7, 14]. Our equations are by construction exact in three limits: (*i*) removal of the disorder, (*ii*) for $t = 0$ (upon choosing appropriate initial conditions), and (*iii*) for $t = \infty$. Replica theory enters as a tool in calculating the local field distribution, and involves dynamical generalisations of familiar objects from equilibrium replica theory, like the overlap distribution $P(q)$ and of AT [9] and zero-entropy lines. At fixed-points of our flow equations we recover the full equilibrium replica theory, including replica symmetry breaking if it occurs, and the corresponding phase diagrams. For intermediate times our equations capture the main features of the flow in the order parameter plane (for homogeneous initial conditions). The theory fails, however, in that for large times the relaxation towards equilibrium is slower than predicted by the theory. Our results show that for homogeneous initial conditions the impact of microscopic memory effects on the evolution of the macroscopic order parameters appears to be mainly an overall slowing down.

Acknowledgement:
It is a pleasure to thank Silvio Franz for many interesting discussions on spin-glass dynamics and for permitting the reporting here of results from [14].

References

1. Hopfield J.J. (1982) *Proc. Natl. Acad. Sci. USA* **79** 2554
2. Sherrington D. and Kirkpatrick S. (1975) *Phys. Rev. Lett.* **35** 1792
3. Glauber R.J. (1963) *J. Math. Phys.* **4** 294
4. van Kampen N.G. (1981) *Stochastic Processes in Physics and Chemistry* (North-Holland, Amsterdam)
5. Amit D., Gutfreund H. and Sompolinsky H. (1987) *Ann. Phys.* **173** 30
6. Coolen A.C.C. and Sherrington D. (1994) *Phys. Rev. E* **49** 1921
7. Coolen A.C.C. and Sherrington D. (1994) *preprint Univ. of Oxford* OUTP-94-29S
8. Parisi G. (1983) *Phys. Rev. Lett.* **50** 1946
9. de Almeida J.R.L. and Thouless J. (1978) *J. Phys. A* **11** 983
10. Ozeki T. and Nishimori H. (1994) *preprint Tokyo Inst. of Technology*
11. Coolen A.C.C. and Sherrington D. (1993) *Phys. Rev. Lett.* **71** 3886
12. Horner H., Bormann D., Frick M., Kinzelbach H. and Schmidt A. (1989) *Z. Phys. B* **76** 381
13. Kirkpatrick S. and Sherrington D. (1978) *Phys. Rev. B* **17** 4384
14. Coolen A.C.C. and Franz S. (1994) *preprint Univ. of Oxford* OUTP-94-24S

ELASTIC INSTABILITIES IN CRYSTAL GROWTH

Jacques Villain*, Christophe Duport*, Philippe Nozières**

* CEA, Département de Recherche Fondamentale sur la Matière Condensée, SPSMS, MDN, CENG, 17 rue des Martyrs, F-38054 Grenoble Cédex 9
** Institut Laue-Langevin, Grenoble

ABSTRACT. This article is a short introduction to instabilities arising from elasticity in crystal growth. A simple description of the basic mechanisms is presented without mathematical details. Special attention is given to growth by molecular beam epitaxy (MBE) of a crystal limited by a singular surface or a surface with an orientation close to a singular one. It is found that instabilities which take place in the absence of growth do not always occur in the growth regime.

1. Introduction

Epitaxial layer growth is a common technique in semiconductor industry, and one generally tries to get layer by layer growth (Fig. 1 a). However, this is not always possible. In other words, a planar interface between a substrate and an adsorbate may be *unstable*. In that case the adsorbate forms droplets (Fig. 1 b c) rather than a continuous layer [1]. It has been suggested to use this to produce quantum dots [2]

The instability may be due either to a chemical, short range effect, or to elasticity. In the former case, the adsorbed atoms do not like to form chemical bonds with the substrate and prefer to bind between themselves. In the latter case, the elastic stress exerted by the substrate on the adsorbate is so strong that the elastic energy is lowered by splitting the adsorbate into pieces. The chemical effect is often called "capillary" and is related to the surface tension of the various interfaces (adsorbate-substrate, adsorbate-vapour and substrate-vapour). The elastic effect is long ranged, and the atomic displacement $\vec{u}(\vec{r})$ at a distance $\vec{r}$ of an adatom at the surface of an isotropic solid is proportional to $\vec{r}/r^3$ [3]. This is the same behaviour as the electric field produced by a point charge, and actually the equation of elasticity is analogous to the Laplace equation. The droplets observed experimentally show a fairly regular array (1,2), and this suggests a long range interaction between droplets, which should be due to elasticity. The present article is focussed on instabilities due to elasticity. Using a classical argument [3], it will be shown in the next section that the surface cannot be flat at equilibrium in a continuum description. In section 3, this result will be found to hold also when the atomic nature of matter is taken into account, but the existence of an activation energy barrier will be demonstrated in the case of a singular surface. In section 4, the M.B.E. growth of a stepped crystal surface will be investigated. Elasticity will be found to lead to step-bunching in certain cases only. Other instabilities will be briefly described in Section 5.

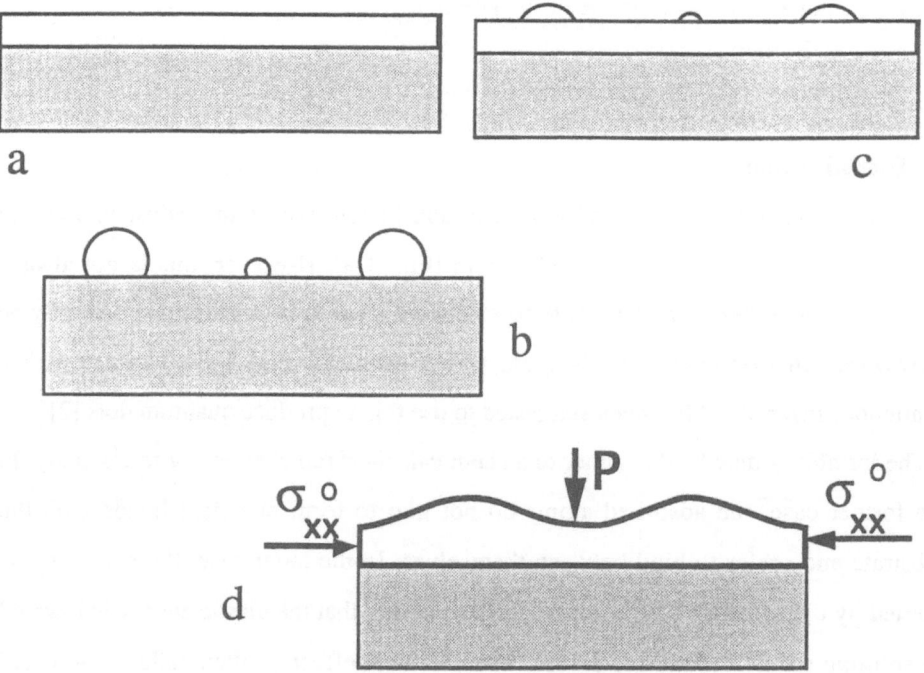

Figure 1. a) Layer by layer growth of an adsorbate (clear) on a substrate (dark). b) The adsorbate forms droplets (Volmer-Weber growth of Pb on graphite). c) The adsorbate forms a finite number of layers, and then droplets (Stranski-Krastanov growth of Pb on Ge(111)). d). The Grinfeld instability. The effect of the substrate is equivalent to a stress σ_o.

2. The Asaro-Tiller-Grinfeld instability   [4,5,6]

Assume that, by some experimental trick, one has been able to prepare a regular adsorbed layer on a foreign substrate (Fig. 1 a). Moreover, the adsorbate is assumed to be *commensurate* with the substrate, and therefore subject to a strong elastic stress, since the two natural atomic distances are never exactly compatible. As argued above, the system may be expected to be unstable with respect to droplets formation. One might expect that the instability does not always occur, or that there is some potential barrier to overcome. Asaro and Tiller [4], and Grinfeld [5], have independently proved that, *within continuum elasticity theory,* it is not the case: the instability always occurs (after a possibly very long time) and there is no activation barrier. The method consists in writing the free energy (or the free enthalpy if pressure is taken into account) and showing that it is lower for a slight sinusoidal modulation (Fig. 1d) of the thickness, than for a flat surface (Fig. 1 a). The sinusoidal modulation may be viewed as the *incipient state* of a process which eventually leads to droplets, and requires diffusion of matter on the surface -a slow process at usual temperature on an usual solid.

The intuitive reason of the instability is the same as that of droplet formation: the top atoms are free to change their distance which thus becomes closer to the distance in the bulk material. However, the modulation costs chemical bonds, and therefore capillary energy. It turns out that, for sufficiently long wavelengths, the capillary energy loss is overcompensated by the elastic energy gain.

The substrate will be assume to impose to the the adsorbate a fixed average strain $\varepsilon^o_{\alpha\gamma}$. Neglecting the pressure, the free energy to be minimized is the following integral on the volume V of the adsorbate:

$$F = \frac{1}{2} \int_V d^3r \sum_{\alpha\gamma\xi\zeta=x,y,z} \Omega^{\xi\zeta}_{\alpha\gamma} \left[\varepsilon_{\alpha\gamma}(\bar{r}) + \varepsilon^o_{\alpha\gamma} \right] \left[\varepsilon_{\xi\zeta}(\bar{r}) + \varepsilon^o_{\xi\zeta} \right] \qquad (1)$$

where $\varepsilon_{\alpha\gamma}(\bar{r})$ is the local strain counted from the non-modulated state. Thus, $\varepsilon_{\alpha\gamma}(\bar{r}) + \varepsilon^o_{\alpha\gamma}$ is the local strain counted from the unstrained adsorbate. For an isotropic solid, the elastic constants are

$$\Omega_{\alpha\gamma}^{\xi\zeta} = \lambda\delta_{\alpha\gamma}\delta_{\xi\zeta} + \mu\delta_{\alpha\xi}\delta_{\gamma\zeta} + \mu\delta_{\alpha\zeta}\delta_{\gamma\xi} \tag{2}$$

The Euler-Lagrange equations which minimize (1) are

$$\sum_\gamma \sum_{\xi\zeta} \Omega_{\alpha\gamma}^{\xi\zeta}\partial_\gamma\left[\partial_\zeta u_\xi + \partial_\xi u_\zeta\right] = 0 \qquad \text{in the bulk} \tag{3 a}$$

and $\qquad \sum_{\gamma\xi\zeta} \Omega_{\alpha\gamma}^{\xi\zeta}\left[\partial_\zeta u_\xi + \partial_\xi u_\zeta\right]n_\gamma = 0 \qquad$ on the surface. $\tag{3 b}$

where $\bar{u}(\bar{r})$ is the local atomic displacement, so that $\varepsilon_{\alpha\gamma}(\bar{r}) = \left[\partial_\alpha u_\gamma(\bar{r}) + \partial_\gamma u_\alpha(\bar{r})\right]/2$, and $\bar{n}(\bar{r})$ is the unit normal. Formula (1) may be written $F = F_1 + F_2 + \text{Const}$, where

$$F_1 = \frac{1}{2}\int_V d^3r \sum_{\alpha\gamma\xi\zeta=xyz} \Omega_{\alpha\gamma}^{\xi\zeta}\varepsilon_{\alpha\gamma}(\bar{r})\varepsilon_{\xi\zeta}(\bar{r})$$

and $\qquad F_2 = \int_\Sigma d^3r \sum_{\alpha\gamma=xy} \sigma_{\alpha\gamma}^o\varepsilon_{\alpha\gamma}(\bar{r})$

where $\qquad \sigma_{\alpha\gamma}^o = \sum_{\xi\zeta=x,y}\Omega_{\alpha\gamma}^{\xi\zeta}\varepsilon_{\xi\zeta}^o \tag{4 a}$

may be viewed as an external stress. It is also possible to impose this stress mechanically in the absence of a substrate, and this is the appropriate way to observe the Grinfeld instability experimentally [7].The stress (4 a) exists in the whole adsorbate, but it is effective only within the thickness h of the modulation because its integral on the rest of the adsorbate is a constant. For an isotropic solid, (2) and (4 a) yield

$$\sigma_{\alpha\gamma}^o = \sigma_o\delta_{\alpha\gamma}\left(\delta_{\alpha x} + \delta_{\alpha y}\right) \tag{4 b} \qquad \text{with} \qquad \sigma_o = -\frac{\delta a}{a}\frac{E}{1-\hat{\sigma}} \tag{4 c}$$

where E is the Young modulus and $\hat{\sigma}$ the Poisson coefficient. The misfit $\delta a = a' - a$ is the difference between the atomic distances a and a' in the substrate and the free adsorbate, respectively. A positive value of δa corresponds to adsorbate atoms bigger than substrate atoms.

Solving the system (3) for a sinusoidal surface modulation of wave vector k yields an elastic perturbation which decays like $\exp[-k|z|]$ with the distance $|z|$ to the surface. The local energy density at long distance $|z|$ is proportional to $Ek^2\varepsilon^2(0)\exp[-k|z|]\sin^2(kx)$, where $\varepsilon(0)\sin(kx)$ is the strain at the surface. Integration yields $F_1 \approx ES\varepsilon^2(0)/|k|.$, where S is the surface area. On the other hand, if h is the amplitude of the modulation, F_2 is of order $Sh\sigma_o\varepsilon(0)$. Minimisation of $F_1 + F_2$ with respect to $\varepsilon(0)$ yields the total elastic free energy due to the modulation:

$$F_{el} = -S\frac{1-\hat{\sigma}^2}{2E}|k|h^2\sigma_o^2 \tag{5}$$

where the coefficient follows from a careful but standard calculation of elasticity [6].

On the other hand, the capillary free energy is easily seen to be

$$F_{cap} = S\gamma k^2 h^2 / 2 \tag{6}$$

where γ is the surface stiffness. The sum $F_{el} + F_{cap}$ is therefore a negative, decreasing function of h if k is small enough, and this proves the instability. The Grinfeld instability has been observed in solid He [7].

The linear stability analysis described above can be complemented by a nonlinear calculation. Grilhé [8] has found that, if one starts from a flat surface, the instability should ultimately result in holes which are the origins of misfit dislocations.

3. The Grinfeld instability on a singular surface

A singular surface is a surface where formula (6) does not apply, but instead should be replaced (as justified below) by

$$F_{cap} \approx 2Sg|k|h / (\pi a) \qquad (7)$$

where g is the line tension of atomic steps and a is the atomic distance. High symmetry surfaces of usual materials at usual temperatures are generally singular. An equivalent statement is that they are below their roughening temperature T_R. Above T_R, g is negative and (7) would be meaningless. Formula (7) results from the fact that the total number of steps in a period is $4h/a$, the number of periods is $L_x k/2\pi$ and the length of a step is S/L_x, where L_x is the length in the direction x parallel to steps.

Comparison of (7) with (5) suggests that the total free energy $F_{el}+F_{cap}$ is now positive for any k if h is small. This conclusion would be incorrect, because not only formula (6), but also the formula (5) of continuous elasticity theory should also be modified when dealing with a singular surface.

A sinusoidal modulation is not a convenient thing to consider if the thickness of interest is of the order of magnitude of the atomic distance. Therefore we shall investigate the stability with respect to stripe formation (Fig. 2). The last atomic layer will be supposed to form stripes of width ℓ and distance $(L-\ell)$. We want to calculate the value of L and ℓ which minimizes the free energy. The elastic free energy is still given by (1) and can be explicitly calculated as a function of the height h and the half-wavelength ℓ by calculating the Fourier components of the profile and summing all contributions of type (5):

$$F_{el} \cong -4S\frac{1-\hat{\sigma}^2}{\pi EL}h^2\sigma_o^2 \sum_{n=1}^{L/a}\frac{1}{n}\left|\sin^2(\pi\ell / L)\right|$$

or

$$F_{el} \cong -2S\frac{1-\hat{\sigma}^2}{\pi EL}h^2\sigma_o^2 \ln\left[\frac{L}{a}\left|\sin(\pi\ell / L)\right|\right] \qquad (8)$$

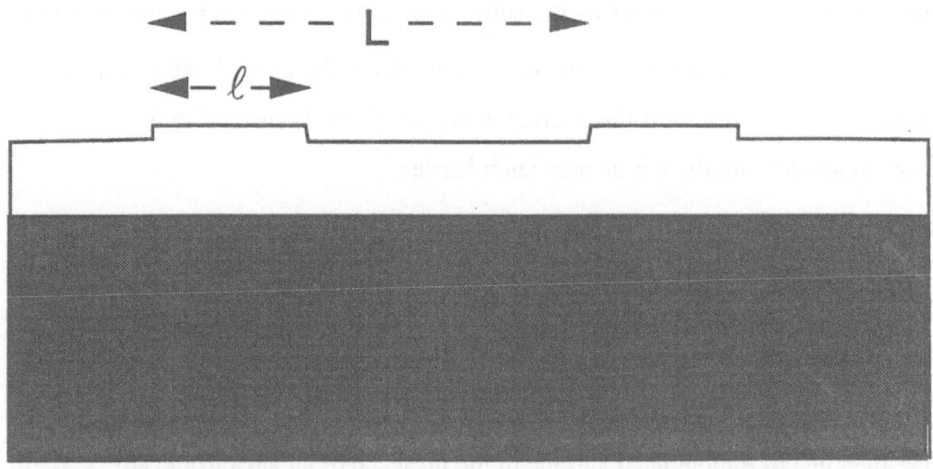

Figure 2 An adsorbate (white) on a substrate. A shorter period L costs more step energy, but less elastic energy.

The cut off L in the sum is due to the fact that wavelengths shorter than the atomic distance a would be unphysical. For a fixed period L, the free energy (8) is minimal for $\ell=L/2$ and is dominated by a logarithmic divergence in L:

$$F_{el} \cong -2S\frac{1-\hat{\sigma}^2}{\pi EL}h^2\sigma_o^2 \ln\left(\frac{L}{a}\right) \cong S\frac{1-\hat{\sigma}^2}{\pi^2 E}kh^2\sigma_o^2 \ln(ka) \qquad (9)$$

A very similar logarithmic dependence on the period has been obtained by Alerhand et al in their study of Silicon (001) surface [9]. The flat surface is unstable with respect to stripe formation with any period $L=2\pi/k$ such that the sum of (7) and (9) is negative. For fixed h, the free energy is minimal for a period L equal to

$$L_c \cong a\exp\left[\frac{2\pi gE}{ah\sigma_o^2}\frac{1}{1-\hat{\sigma}^2}\right] \qquad (10)$$

In practice, the stress σ_o cannot be very large since the adsorbate has been assumed to be commensurate with the substrate. For a thickness h=1 atomic distance and a natural misfit $\delta a/a$ of 1% between the substrate and the adsorbate, L_c turns out to be unphysically large. It does become reasonable for a thickness h of a few atomic distances. The conclusion of this section is that the Grinfeld instability does occur on a singular surface, but there is an activation barrier.

4. Dynamics in the step flow regime: step bunching instability

In the last two seccions, the free energy of a planar surface has been shown to be larger than that of a modulated surface in the presence of an anisotropic stress σ_o. In practice, the modulation (or bubble formation) generally occurs during the growth. The problem of growth instabilities due to elasticity has been addressed by Spencer et al [10] and by Duport et al [11]. The two approaches refer to two completely different physical situation. Spencer et al consider a solid which is close to equilibrium with its vapour or liquid phase, and describe the growth process from a macroscopic point of view. Duport et al, on the other hand, consider growth by Molecular Beam Epitaxy (MBE) -a situation very far from equilibrium.In the following, we summarize the discussion and results of Duport et al.

The strain due to a big cluster of adatoms has been discussed in the previous section. It is clear that an isolated adatom should produce also a strain. However, the mechanism is quite different! As seen from (4 c), a large cluster of adatoms produces a strain because of the misfit δa between the interatomic distances of the adsorbate and of the substrate. On the other hand, an isolated adatom has no interatomic distance. It acts through another mechanism, which may be called the "broken bond mechanism" and is described by Fig. 3 in the case of central, pairwise interactions between nearest and next-nearest neighbours. An isolated adatom is seen to exert forces $\vec{f}_R$ on the other atoms, the location of which is designated by $\vec{R}$. The total force is zero at equilibrium, but the dipole moments

$$m_{\alpha\gamma} = \sum_R R_\alpha f_R^\gamma \qquad (10)$$

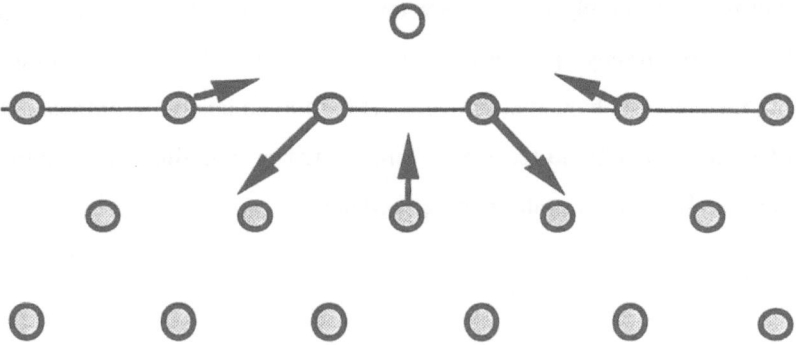

Figure 3. Forces exerted by an atatom on a substrate. The tensor of the moments of the forces does not vanish, and these moments may be identified with an external strain acting on the substate. The effect persists if the substrate is of the same chemical nature as the adatom. Thus, what is called substrate here is called adsorbate in Fig. 4, for instance.

are not zero if $\alpha=\gamma$. If the x and y axes are chosen parallel to the surface, symmetry imposes: $m_{xx} = m_{yy} = m$.

In contrast with the misfit effect, the broken bond effect is not additive: the strain produced by a big cluster of adatoms is not proportional to the number of adatoms; it increases much less rapidly and can be neglected with respect to the misfit effect. Therefore, the interaction force between an adatom and a cluster is proportional to the misfit effect of the cluster and the broken bond effect of the adatom. The local strain resulting from these two effects may have identical or different signs, and therefore the interaction may be attractive or repulsive, depending on the nature of the adsorbate and of the substrate. As a result, elastic interactions may either stabilize or destabilize a planar surface in MBE growth. This is in contrast with Grinfeld's calculation, which always predicts destabilization. As will now be argued, the elastic effect is destabilizing if the stress associated to both mechanisms has the same sign, and stabilizing in the opposite case.

The argument is the following. If the stress associated to both mechanisms has the same sign (Fig. 4) an adatom is attracted to the donward edge of its terrace, where it does not feel the substrate very much. From there, it jumps more easily to the lower terrace. Thus, atoms go preferably to the lower terrace. It results that broad terraces become broader -an obviously destabilizing effect. Conversely, if the stresses associated to both mechanisms have opposite signs (Fig, 4 a) an adatom is attracted to the upper end of its terrace, a broad terrace becomes narrower and this is stabilizing with respect to the step bunching instability described here.

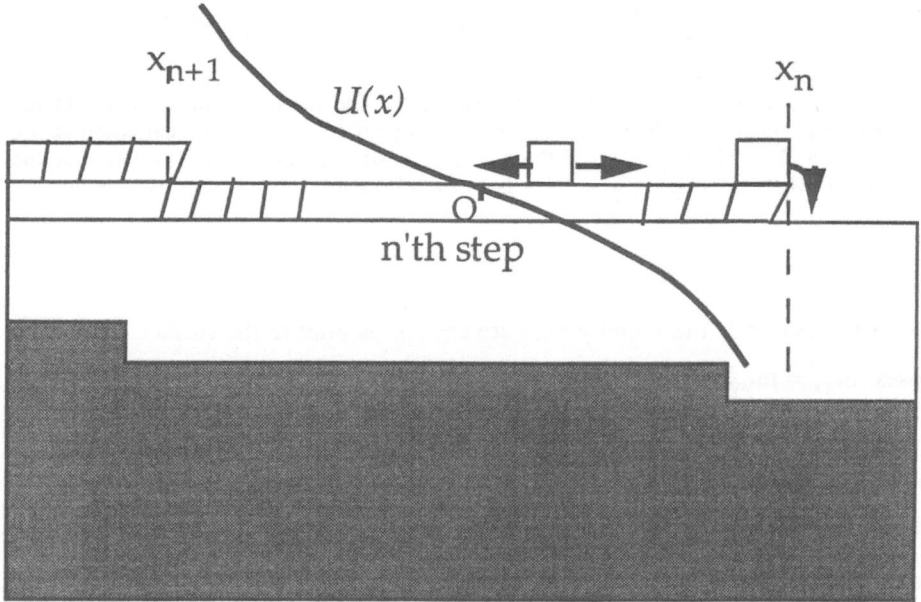

Fig. 4. Schematic picture of strains on a stepped adsorbate in the case $\delta a < O$ .On the uphill side of a step, the atomic distance is closer to its unconstrained value. On the downhill side, it is farther. An adatom is attracted to the upper or lower edge of its terrace, according to the sign of the broken bond effect. If it is attracted to the upper edge, there is an instability. The curve shows the elasticity-mediated potential U(x) seen by an adatom.

5. Other instabilities and other mechanisms

5.1 *The Bales-Zangwill instability*

So far, we have assumed that steps remain straight. Bales and Zangwill [12] have shown that straight steps may become unstable. This happens if adatoms go preferably to the uphill step (i.e. leftward on Fig. 4) than to the lower terrrace. This is the normal situation in M.B.E. and this is called the *Schwoebel effect* [13]. The mechanism of the instability is analogous to that of fractal growth in D.L.A., and the linear stability analysis is reminiscent of the well-known Mullins-Sekerka instability: if a tip develops when a step proceeds, adatoms go preferably to the tip, and the tip become bigger. Thus, the Schwoebel effect, during the growth of a stepped surface, is destabilizing with respect to step meandering. On the other hand, it follows from an discussion analogous to the previous section that the Schwoebel effect is stabilizing with respect to step bunching. Elasticity has certainly an influence on step meandering instabilities, and this effect is currently being studied. However, the Schwoebel effect is generally expected to dominate elasticity.

5.2 Droplet formation

As seen in Section 1, elasticity can produce droplets instead of step bunches. Droplets are expected to form if the growth rate is fast, or if the distance between steps is long, so that adatoms spend a long time before reaching a step and have more chance to meet colleagues willing to form droplets. A recent study of Ratsch and Zangwill [14] may be considered as an investigation of the early stage of droplet formation. In principle, droplets may also form in the opposite limit, when the growth rate is so slow that adatoms have time to detach from a step before being buried by another step. However, diffusion is often so slow that the actual observation of this effect is not frequent. Crack formation in polymerised polydiacetylene has been attributed to the Grinfeld instability [15]. The role of diffusion is, however, dubious in that case, and one can wonder whether the dynamics do not result from anharmonic motions of local character.

In conclusion, we have given a qualitative description of elasticity-driven instabilities. The mechanism described by Asaro and Tiller, and by Grinfeld, has been seen to be less efficient on a high symmetry (i.e., in practice, singular) crystal surface, because a free energy barrier has to be overcome. On the other hand, elasticity-driven instabilities may be suppressed in the MBE growth of a stepped surface, due to the Schwoebel effect, which actually is generally the dominating effect [11].

REFERENCES

1. Chamberod, A., Hillairet, J., ed "Metallic Multilayers" (Material Science Forum, Vol. 59-60, Zürich, 1990)

2. R. Nötzel, J. Temmio, T. Tamamura. Nature **369**, 131 (1994)

3. Lau, K.H., Kohn, W., Surface Science **65**, 607 (1977)

4. Asaro, R.J., Tiller, W.A., Metallurgical Transactions 3, 1789 (1972)

5. Grinfeld, M.A., J. Nonlinear Science Vol. **3**, 35 (1983)

6. Nozières, P., in "Solids far from equilibrium" Ed. C. Godrèche (Cambridge Univ. Press, 1991).

7. Thiel, M., Willibald, A., Evers, P., Levchenko, A., Leiderer, P., Balibar, S., Europhys. Lett. **20**, 707 (1992)

8. Grilhé, J., Europhys. Lett. **23**, 141 (1993)

9. Alerhand, O.L., Vanderbilt, D., Meade, R.D., Joannopoulos, J.D., Phys. Rev. Lett. **61**, 1973 (1988)

10. Spencer, B.J., Voorhees, P.W., Davis, S.H., Phys. Rev. Lett. **67**, 3696 (1991)

11. Duport, C., Nozières, Ph., Villain, J., preprint (1994)

12. Bales, G.S., Zangwill, A., Phys. Rev. **B41**, 5500 (1990)

13. Schwoebel, R.L., Journal of Applied Physics **40**, 614 (1969).

14. Ratsch, C., Zangwill, A., Appl. Phys. Lett. **63**, 2348 (1993)

15. Berréhar, J. ,Caroli, C., Lapersonne-Meyer, C., Schott, M., Phys. Rev. B **46**, 13487 (1992)

VAPOR PHASE NUCLEATION: MOLECULAR MECHANISMS FOR EMBRYO DEVELOPMENT

H. Reiss, C. L. Weakliem, and H. M. Ellerby
University of California, Los Angeles

1.0   Introduction

The process of vapor phase nucleation has its beginnings in the density fluctuations that occur in a supersaturated vapor. Each fluctuation can develop into a liquid drop at a rate characteristic of its individual nature, and the rate of nucleation is then the sum of these individual rates. A truly dynamic theory of nucleation would evaluate this sum, after having derived the individual rates as precisely as possible. Unfortunately the theoretical tools required for the conduct of this program are not yet available, and so workers have followed another route in which an *average* fluctuation has been defined and its properties evaluated, after which the *average* rate at which the *average* fluctuation develops into a drop is calculated.

The average density fluctuation is usually treated as a "physical cluster" and in the classical theory of nucleation (1, 2) that cluster is modeled as a liquid drop having the properties of the bulk liquid, and a sharp interface with the vapor where the surface tension is also that of the bulk liquid. The derivation of the rate theory involves the principle of detailed balance so that an "equilibrium distribution" of clusters must also be derived.

These kinds of theories (the classical and more recent ones (3)) are deficient, beyond the use of the above mentioned inversion of the order of averaging, in that (1) they contain no natural means for distinguishing between the overwhelming majority of fluctuations that do not produce embryos for the formation of liquid drops and those relatively few fluctuations that do, and (2) they omit important mechanisms of embryo development. The above mentioned distinction is usually made on the basis of an ad hoc assumption, e.g., embryos are clusters having the density of the bulk liquid or densities that are simple functions of the liquid density.

Recently we have developed a theory that has the potential to avoid these ad hoc assumptions (4 - 10). Although considerable progress has been made, the theory is still incomplete. Furthermore, its unconventional nature, coupled with the fact that its development has evolved throughout seven detailed papers, has undoubtedly made it difficult for all but the most dedicated reader to acquire a thorough understanding of the new approach. Therefore, in the present paper we offer a less detailed and descriptive account of the theory, its progress to date, and suggestions concerning the future course of its development.

2.0   The Fluctuation Cluster

As in the case of previous theories, it is still necessary to invert orders of averaging and to appeal to the principle of detailed balance. In addition, the present theory is an isothermal one such that the heat evolved in the growth of the cluster that represents a density fluctuation is assumed to be removed by a non-condensable supporting gas (e.g., He, Ar, N_2, etc. as in most experiments). The main points of difference are that (1) the average cluster used to model the average density fluctuation, i.e., the "fluctuation cluster" is not defined in an ad hoc manner, e.g., as a liquid drop, (2) it is defined in molecular terms, and (3) it is characterized by a larger number of parameters than molecular content alone.

To assure that the correct "equilibrium distribution" of clusters (for the purposes of detailed balance) is obtained, the clusters are defined so that they can be employed as rigorously as possible in the evaluation of the partition function of the supersaturated vapor, and therefore in the specification of its thermodynamic properties.

The cluster which meets these various criteria may be described as follows. In the supersaturated vapor we notice what appears to be a density fluctuation, positive or negative, but it is convenient to think of a positive fluctuation although negative fluctuations are included implicitly within the description. (Figure 1 may be helpful in following the cluster definition.) The temperature of the supersaturated vapor is assumed to be far below the critical temperature and the vapor is assumed to behave

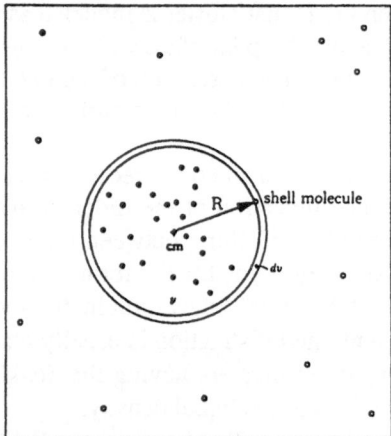

Fig. 1. "Snapshot" of supersaturated vapor containing defined physically consistent cluster. Solid dots are the i particles of the cluster. The shell molecule in the volume dv defines the cluster volume v. Remaining open dots are ideal gas molecules of the supersaturated vapor.

almost as an ideal gas (a condition that is met in most experiments on vapor phase nucleation). Considering a positive density fluctuation, involving $i+1$ molecules, we can find the molecule within it that lies furthest from its center of mass. This is a molecule that by some reasonable criterion is far enough from all other molecules, not

3.0   Characteristics of $W_i(v)$

Once the intermolecular potential u_i is available A_{iv}^*, and therefore $W_i(v)$, can be evaluated by several means. An example, for argon vapor, using the appropriate Lennard-Jones potential in Monte Carlo simulation, appears in figure 2 where $W_i(v)/kT$ is plotted as a surface above the i,v plane. Surfaces computed in this

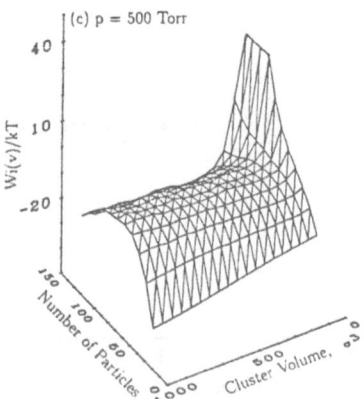

Fig. 2. Argon free energy surface as a function of cluster particle number, i, and volume, v, from computer simulations. $T = 70K$, $V = 1.0$ cm3, and (a) $p = 298$ Torr; (b) $p = 400$ Torr; (c) $p = 500$ Torr.

manner exhibit two effects. First, the surfaces decrease in height as the pressure is increased, i.e., the values of $W_i(v)kT$ are lower for given i and v values, and second, the location of the ridge on the barrier is shifted towards smaller values of i.

By considering figure 2, where $T = 70K$ and $p= 500$ Torr, we can examine all the features of the surface for the range of i and v values for which calculations were made. In that figure, the following features stand out. At small values of v/σ^3 (at the right) there is a steeply rising "wall." This, of course, corresponds to the sharp increase of potential energy as the electron shells of the argon atoms are forced to overlap. For later use, we refer to this wall as the "incompressibility boundary." The next most prominent feature on the surface is a "ridge" that runs not quite parallel to the v/σ^3 axis. This corresponds to the barrier which clusters must overcome on their way to becoming drops. It is not visually obvious from the figure, but this ridge descends in elevation toward the left, i.e., toward larger values of v/σ^3. In fact, extension of the surface with more calculations would show it reaching a "global minimum" at values of v such that i/v approaches the average density of the supersaturated vapor. This is associated with the fact that i/v clusters of this type (if not excluded from the distribution) would be the most populous.

Perhaps, the most important feature of the surface is the existence of a "valley" whose axis runs not quite parallel to the i axis on the far side of the hill. This valley cannot be seen in the figure, but it is clearly present in the $W_i(v)$ data. It appears

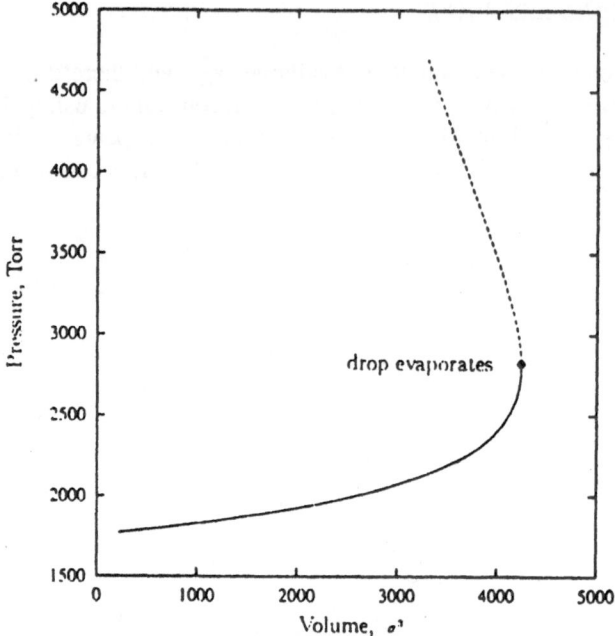

Fig.3. Pressure vs. volume isotherm for argon from the modified liquid drop theory for $T = 85$ K and $i = 80$.

This can be proved to be so even though the lower branch possesses a negative compressibility. At the point separating the two branches, and marked by "drop evaporates," the drop on the lower branch, though small, is still present! However, the slightest increase in v will cause this residual drop to evaporate so that only i molecules of ideal vapor will remain within v. We refer to this point as lying on the "evaporation boundary."

The full pressure p within the container is not given by p^* alone since the wandering drop may also exchange momentum with the wall, R, and since the drop is approximated as an ideal gas molecule the pressure it exerts is added to p^* to yield p.

Then the Helmholtz free energy of the cluster can be evaluated from the standard thermodynamic formula

$$A = -\int_\infty^v p\,dv, \tag{6}$$

where for the range of volumes beyond the evaporation boundary $p = ikT/v$, while below this boundary it is specified as above, making use of the stable lower branch of the curve in figure 3.

However, A specified by eq (6) is not A_{iv}^* since the latter quantity is evaluated with the center of mass fixed, whereas no such restriction has been placed on the MLDM model. Fortunately a term can be added to A to convert it exactly to A_{iv}^* (13).

When the MLDM is used to determine A_{iv}^* in this manner, $W_i(v)$ can be evaluated using eq (5). Figure 4 is a contour map of a $W_i(v)/kT$ surface obtained in this way for argon vapor. It exhibits all of the features displayed by the Monte Carlo

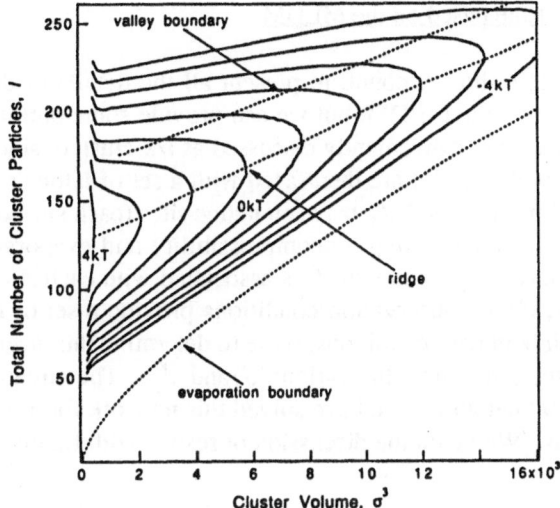

Fig. 4. $W_i(v)$ contour map (MLDM) for argon, 85K, 1419 Torr, 1.0 cm^3.

results of figure 2, e.g., the ridge, the valley on the far side of the ridge but not on the near side, the incompressibility boundary, and the downward slope of the surface (outside the valley) towards increasing v. Some of these features are marked by dashed lines in the figure. An additional feature, the "evaporation" boundary, defined in figure 3, is also shown.

4.0   Rate Theory

Within the theoretical framework, thus far developed, the nucleation rate must be evaluated in terms of how the various i/v clusters are inter-converted, i.e., in terms of the cluster fluxes in both the i and v directions. The significance of the i-component of flux is reasonably clear, but the v component requires some discussion. Consider a particular cluster in which the shell molecule moves to a greater distance from the center of mass. Does this constitute a conversion of the cluster to one having a larger v ? The answer is "not necessarily" because (1) an i/v cluster is defined in terms of the collection of configurations available to the i molecules within v and (2) the shell molecules might move so far from the other molecules that it will have "escaped" from the cluster and the process will then correspond to an i-component of flux in which i goes to $i-1$. The second point shows that the "range" criterion employed to distinguish the cluster from the surrounding ideal gas must be used to show that an apparent v component is actually an i-component, i.e., if the shell molecule escapes to beyond this range, the process is an i-component.

   As far as the bonafide v-process is concerned, it cannot be discussed (because all configurations of the i molecules are involved) in anything but an ensemble sense, once again emphasizing the fact that we must still appeal to an inversion of the order of averaging. What we are really attempting to evaluate is the following. In time dt, what is the net conversion of i/v clusters to $i/(v+dv)$ clusters? The net rate of conversion will involve the loss experienced by the entire set of i/v clusters in *all*

5.0   Nucleation Rate based on the MLDM

It is impossible, in this short account, to present all the details of the calculation of nucleation rate based on the MLDM, but we can provide some outline. The now discrete i/v clusters are more conveniently discussed as i/s clusters since s is in one to one correspondence with v. We are then faced with a set of lattice sites having coordinates i,s, and net fluxes J_i and J_s between lattice sites (each site corresponding to a cluster). Except at boundaries (e.g., incompressibility and evaporation boundaries) there are four fluxes, two J_i's and two J_s's associated with each site. For the steady state of i/s clusters, these conservation conditions provide a set of linear equations, which, together with boundary conditions, serve to determine the steady concentrations of clusters and, at the same time, the various J_i and J_s. The situation is analogous to that of an electrical network. We have solved this network for argon vapor under a variety of conditions. We begin the discussion of results with figure 5.

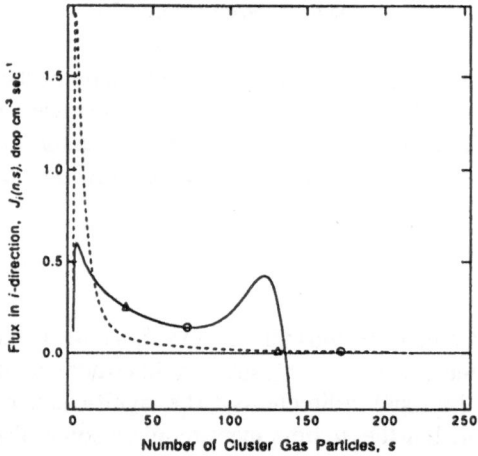

Fig. 5. i-flux for argon (MLDM), 85K, 1417 Torr, 1.0 cm^3, solid curve, $i = 200$, dashed curve, $i = 300$.

Figure 5 shows the values of J_i as a function of s for two different values of i, namely $i = 200$ (solid line) and $i = 300$ (dashed line). The curves correspond to 1.0cm^3 of argon vapor at T = 85K and $p = 1417$Torr. The triangle on each curve represents the point that marks the right boundary of the valley. The circles represent the points where the $i = 200$ and $i = 300$ curves intersect the ridge, such that i,s clusters whose rates appear to the left of the circles have crossed the ridge but those whose rates appear to the right of the circles have not. As can be seen from the plot, for $i = 200$, the flux in the i-direction is bimodal. Clusters are being funneled through the valley at the left while there is also a significant flux at larger values of s. However, the latter flux involves clusters which have not crested the ridge, i.e., they occur to the right of the circle which marks the intersection of $i = 200$ curve with the ridge. In contrast, for the $i = 300$ curve, most of the flux in the i-direction is found in the valley, with negligible *i*-flux anywhere else. In addition, the flux within the valley is concentrated along its floor.

Figure 6 shows the corresponding plot for J_S. The conditions are the same as those

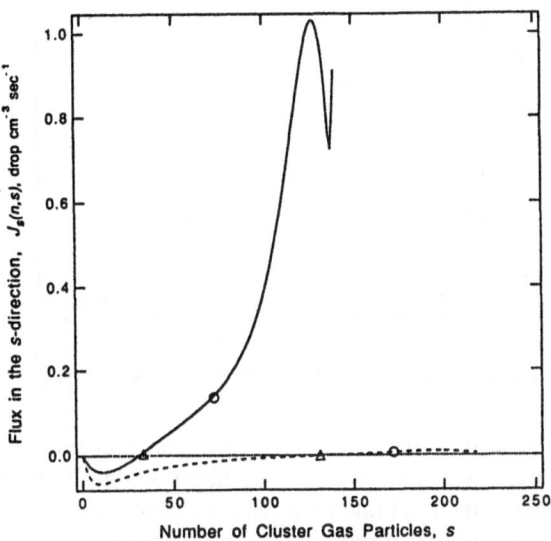

Fig. 6. s-flux for argon (MLDM), 85K, 1417 Torr, 1.0 cm^3, solid curve, $i = 200$, dashed curve, $i = 300$.

in Figure 5. The solid line represents J_S and a function of s for $i = 200$, while the dashed line plots J_S for $i = 300$. Again, the triangles indicate the positions of the right boundary of the valley, while the circles represent the intersection of the constant i curve with the ridge. A negative J_S indicates a net flux to the left, i.e., in the direction of decreasing s so that there is net condensation rather than evaporation, while a posi- tive J_S indicates a net flux toward the right, i.e., toward the evaporation boundary. As can be seen for both curves, the s-fluxes within the valley are negative, indicating that valley clusters are channeled toward higher densities, i.e., to smaller s. The $i = 300$ curve also shows a negative J_S persisting beyond the right boundary of the valley. This corresponds to an entropic s-flux into the valley. Looking at the curve for $i = 200$, one immediately sees, toward the right, the relatively large positive s-flux repre- senting the flow of density fluctuations toward the evaporation boundary and the sub- sequent return of the clusters to the vapor. Most notably, the positive values are found to the left of the circle which demonstrates that while clusters may have crossed the ridge, that fact alone does not insure that they enter the valley and become drops. Therefore we see that the s-flux provides the natural means of identifying those density fluctuations (the overwhelming majority) that return to the vapor and those that enter the valley and go on to form drops. On the $i = 300$ line, the low values of J_S indicate that the separation of clusters into embryos and those that have already been lost to the vapor is almost complete.

Figure 7 plots J_i summed across the valley, i.e., J_T as a function of i. We see that J_T increases with increasing i towards a limit that marks the rate of nucleation. This

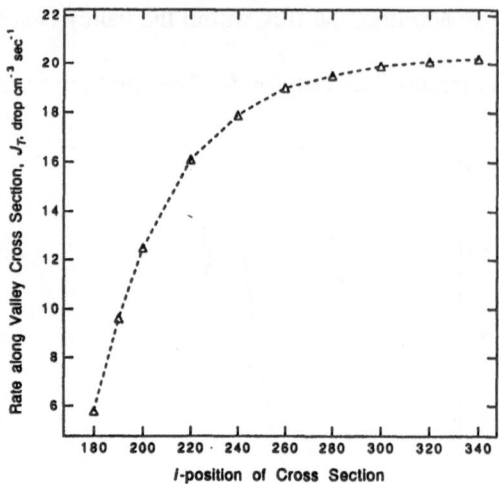

Fig. 7. J_T versus i for argon (MLDM), 85K, 1417 Torr, 1.0 cm³.

limited increase in J_T reflects the s-flux (decreasing with i) into the valley from the right. In the present case the limit is in the neighborhood of 20 drops cm⁻³s⁻¹. There is 90.5 percent change in the total rate between $i = 320$ and $i = 340$. Thus the rate of nucleation is approximately 20 drops cm⁻³s⁻¹. Under the conditions that prevail in Figure 7, classical nucleation theory predicts a rate of 1.0 drops cm⁻³s⁻¹.

In the manner just described we have calculated the nucleation rate at three temperatures and a range of argon pressures such that the rate varies over six orders of magnitude. The results are shown in Figure 8 as plots of the logarithm of the rate derived from the present theory versus the logarithm of the rate predicted by the classical theory. The dashed line without symbols is a line of unit slope passing through the origin, and would be the curve generated if the present theory agreed

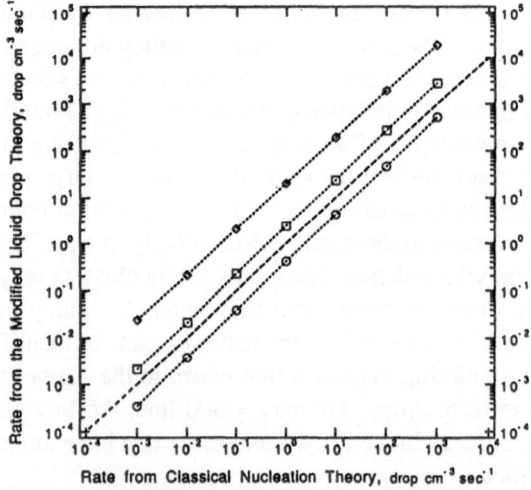

Fig. 8. MLDM theory compared to classical rate theory for argon, 1.0 cm³, diamonds are 85K, squares 90K, circles 95K.

exactly with the classical. Although our results do not fall on this line the data for each temperature fall on lines approximately parallel to it. If J_{cl} denotes the classical rate, this behavior requires the relation, $J_T = K(T)J_{cl}$, where K depends on temperature T alone. Because we can find a temperature where $K(T) = 1.0$ we can describe this situation by saying that the MLDM theory, predicts the same dependence of rate on supersaturation as the classical theory, but a different dependence on temperature. At the two lower temperatures, the MLDM theory yields rates higher than those predicted by classical theory. These converge on the classical rate as the temperature increases, and at the highest temperature, the MLDM rate is lower than those predicted by classical theory. While experiments are not available for argon, those for other vapors (15-17) have shown that, at lower temperatures, classical theory underestimates the rate while at higher temperatures, it overestimates it. In this comparison with classical theory, we see that our results agree with those trends found in experiments.

A study by Zeng and Oxtoby (18) has shown similar results, e.g., a supersaturation dependence in agreement with the predictions of classical theory and a temperature dependence in disagreement, but with the trend found for experimental rates. However, at T = 85K the MLDM rates are approximately 3 orders of magnitude smaller than those calculated by Zeng and Oxtoby, and ~ 4 orders of magnitude smaller at T = 90K and T = 95K. Unfortunately, for argon vapor, there are no reliable measurements for the rate of nucleation. There are some measurements (19) for the onset of nucleation (critical supersaturations) where a large error in the calculated rate can still allow reasonable agreement between theory and experiment. With this qualification the MLDM rate is consistent with experiment, but the same is true of the theory of Zeng and Oxtoby.

6.0   Summary

The special features that distinguish the present theory from previous ones are (1) its molecular formulation such that the physical clusters upon which it is based can be used to accurately evaluate the partition function of the full supersaturated vapor, and (2) the characterization of clusters by more parameters than molecular content i alone. Only one additional parameter is used, namely v but this is enough to allow a natural distinction (the valley and the evaporation boundary) between clusters that serve as embryos and those that do not. Furthermore the v parameter reveals the existence of an important additional mode of cluster variation, i.e., the collective relaxation of its radial density distribution toward the evaporation boundary. If the valley existed on both sides of the ridge this mode would be unimportant since the entire nucleation process, on both sides of the ridge, would be channeled along the axis of the valley, and a characterization of the cluster by i alone would be almost sufficient.

Unfortunately, the nonexistence of the valley on the near side of the ridge has been documented by both Monte Carlo simulation and the MLDM method. Its absence on the near side has also been demonstrated by Talanquer and Oxtoby (20) using a density functional method.

Current and future work is and will be aimed at exploiting the full molecular nature of the theory.

Acknowledgment: This work was supported by the National Science Foundation under NSF grant #CHE93-19519.

7.0   References

1.   R. Becker and W. Doring, Ann. Phys. **2 4**, 719 (1935).
2.   F. F. Abraham, "Homogeneous Nucleation Theory" (Academic Press, New York, 1974).
3.   A. Dillman and G. E. A. Meir, Chem. Phys. Lett. **1 6 0**, 71 (1989).
4.   H. Riess, A. Tabazadeh, and J. Talbot, J. Chem. Phys. **9 2**,1266 (1990)
5.   H. M. Ellerby, C. L. Weakliem and H. Reiss, J. Chem. Phys. **9 5**, 9209 (1991).
6.   H. M. Ellerby and H. Reiss, J. Chem. Phys. **9 7**, 5766 (1992).
7.   C. L. Weakliem and H. Reiss, J. Chem. Phys. **9 9**, 5374 (1993).
8.   C. L. Weakliem and H. Reiss, J. Chem. Phys. in press.
9.   H. M. Ellerby, Phys. Rev. B, in press.
10.   C. L. Weakliem and H. Reiss, J. Phys. Chem. in press.
11.   J. K. Lee, J. A. Barker, and F. F. Abraham, J. Chem. Phys. **5 8**, 3166 (1973).
12.   G. N. Lewis and M. Randall, "Thermodynamics", revised by K. S. Petzer and L Brewer, Ch. 29 (McGraw-Hill, New York, 1961).
13.   Reference 7 and H. Reiss, J. L. Katz, and E. R. Cohen, H. Chem. Phys. **4 8**, 5553 (1968).
14.   D. A. McQuarrie, "Statistical Mechanics" Ch. 20 (Harper and Row, New York, 1973).
15.   P. E. Wagner and R. Strey, J. Chem. Phys. **8 0**, 5266 (1984).
16.   G. W. Adams, J. L. Schmitt, and R. A. Zalabsky, J. Chem. Phys. **8 1**, 5074 (1984).
17.   C. H. Hung, M. J. Krasnopoler, and J. L. Katz, J. Chem. Phys. **9 0**, 1856 (1989).
18.   X. C. Zeng and D. W. Oxtoby, J. Chem. Phys. **94**, 4422 (1991).
19.   B. J. C.Wu, P. P. Wegener, and G. D. Stein, J. Chem. Phys. **6 9**, 1776 (1978).
20.   V. Talanquer and D. W. Oxtoby, J. Chem. Phys. **1 0 0**, 5190 (1994).

CHAOS IN LORENTZ LATTICE GASES

M.H. Ernst
Institute for Theoretical Physics
University of Utrecht, The Netherlands

and

J.R. Dorfman
Institute for Physical Science and Technology
University of Maryland, College Park, MD 20742, USA

Abstract

Lorentz lattice gases belong to the category of dynamical systems with positive Lyapunov exponents, and are therefore chaotic. We show using techniques from the kinetic theory of gases that these dynamical quantities can be computed explicitly.

1 Introduction

The chaotic behavior of dynamical systems underlies the foundations of statistical mechanics through ergodic theory. Here we try to make these ideas more concrete, and show how to quantify some of the chaotic properties that are of interest for statistical mechanics. Relevant questions are: is there at a more practical level any relationship between chaos theory on the one hand, and non-equilibrium statistical mechanics and kinetic theory on the other hand? More specifically, is it feasible to calculate the basic quantities, such as Lyapunov exponents, Kolmogorov-Sinai (KS)-entropies and dynamical partition functions from kinetic theory, and relate them perhaps to transport properties or time correlation functions?

In the last few years such relationships have been derived for a fluid-type model under constant shear, coupled to a Gaussian thermostat by Evans et al. [1], and for many-particle systems standardly used in statistical mechanics, such as the Lorentz gas by Gaspard and Nicolis [2] and classical fluids by Dorfman and Gaspard [3]. In all cases the relevant transport coefficients have been related to Lyapunov exponents and dynamical entropies. Also some analytic studies [4] and computer simulations [5] in deterministic cellular automata have appeared.

Here we will discuss an even simpler system, namely a stochastic Lorentz lattice gas, interpret it as a dynamical system, and show how to calculate dynamical quantities from kinetic theory. We restrict ourselves to closed systems in which the total number of particles is conserved. The concepts of Lyapunov exponents and KS-entropies are introduced and calculated for random walks and Lorentz lattice gases using kinetic theory.

It is very attractive to use the latter set of systems because transport coefficients and time correlation functions have been extensively studied and the results consolidated using the methods of nonequilibrium statistical mechanics and kinetic theory [6, 7] and computer simulations [8]. The discussion of the Lorentz lattice gas, presented here, is based on the publications of Dorfman, Ernst and Jacobs [9], and we refer to these papers for a more extensive and detailed account.

2 Chapman-Kolmogorov Equation

It is instructive to start with a random walk model, the *persistent* random walk. The Lorentz lattice gas, to be discussed below, contains this random walk model as a special limiting case.

Consider a persistent random walker (RW) on a site r of a one-dimensional lattice, which takes at discrete times $t = 0, 1, 2, \ldots$ a step to one of its nearest neighbor sites. With probability p this step occurs in the same direction as in the previous move, and with probability $q = 1 - p$ its direction of motion is reversed. The lattice contains L sites, and it is convenient to use periodic boundary conditions.

The probability that the RW is in state $x = \{r, c\}$ at time t is denoted by $p(x, t) = p(rc, t)$, where r is its lattice position at time t and $c = \pm 1$ is its incoming or pre-transition velocity, indicating that the RW came from site $r - c$ in the previous time step. The probability for the persistent RW evolves in time according to the Chapman-Kolmogorov equation,

$$p(x, t+1) = \sum_y w(x|y) p(y, t). \tag{1}$$

The matrix of transition probabilities,

$$w(x|y) = w(rc|r'c') = W_{cc'} \delta(r, r' + c), \tag{2}$$

is a stochastic matrix that satisfies the normalization condition,

$$\sum_x w(x|y) = 1. \tag{3}$$

Here δ denotes a Kronecker delta function, c and c' label the nearest neighbor lattice vectors $c = \pm 1$, and $W_{++} = W_{--} = p$ and $W_{+-} = W_{-+} = q$.

3 Lorentz Lattice Gas

In the random walk models all lattice sites are scattering sites. In a lattice Lorentz gas only a fraction ρ of all sites, chosen at random (with $0 < \rho \le 1$) are scattering sites. At a scattering site the transition probabilities are given by (2). At a non-scattering site the particle moves like a free particle, i.e. $w(x|y) = \delta(c, c')\delta(r, r'+c)$. The matrix of transition probabilities depends therefore on the *fixed* or *quenched* configuration of scatterers, and can be written conveniently as (compare (2))

$$w(x|y) = \{(1 - \hat{\rho}(r'))\delta(c, c') + \hat{\rho}(r')W_{cc'}\}\delta(r, r' + c), \tag{4}$$

where the Boolean variable $\hat{\rho}(r)$ takes the value $\hat{\rho}(r) = 1$ (scattering site) with probability ρ and the value $\hat{\rho}(r) = 0$ (non-scattering site) with probability $1 - \rho$.

In the Lorentz lattice gas the probability distribution $p(x, t|x_0, \hat{\rho})$ for the moving particle to be in state x at time t, depends on the initial state x_0 and on the quenched distribution of scatterers $\hat{\rho} = \{\hat{\rho}(r)\}$. Its time evolution is again described by the Chapman-Kolmogorov equation (1), and the transition matrix (4) satisfies the normalization (3). In case *all* lattice sites are scattering sites, the stochastic Lorentz lattice gas reduces to the persistent random walk.

We quote some basic properties of the Chapman-Kolmogorov equation. On account of (3) it conserves probability, $\sum_x p(x, t) = 1$. For a given initial condition x_0 and a fixed configuration of scattering sites $\hat{\rho}$ the formal solution of equation (1) is given by the t-th power of the transition matrix, defined as

$$p(x, t|x_0, \hat{\rho}) = w^t(x|x_0) \equiv \sum_{x_1}\sum_{x_2}\cdots\sum_{x_{t-1}} w(x|x_{t-1})\ldots w(x_2|x_1)w(x_1|x_0). \tag{5}$$

Both for the persistent random walk and the Lorentz lattice gas the stationary solution of (1) is spatially uniform, i.e.

$$p(x, \infty|x_0, \hat{\rho}) = (bV)^{-1}, \tag{6}$$

where $V = L$ is the number of sites in the lattice, and $b = 2$ is the number of nearest neighbor sites. This is so, because there are no excluded sites in these models. The extension of these models to higher-dimensional lattices with $V = L^d$ sites ($d = 2, 3, \ldots$) and with different coordination number b is obvious.

The solution $p(x, t|x_0, \hat{\rho})$ of the Chapman-Kolmogorov equation will in general depend sensitively on the microscopic distribution of scatterers $\hat{\rho}$, and possibly also on the initial state x_0 of the moving particle. To calculate macroscopic properties of the system one needs the single particle distribution function

$$f(x, t) = \langle p(x, t|x_0, \hat{\rho})\rangle_\rho, \tag{7}$$

which involves an average over the initial distribution $p(x_0, 0|\hat{\rho})$ of the moving particle in a fixed configuration of scatterers, and a subsequent average over the configuration of scatterers, as described below (4). If one is interested in non-equilibrium properties one may choose an adequate initial distribution; for instance $p(x_0, 0) = \delta(c, c_0)\delta(r, r_0)$. To calculate equilibrium properties one should average over the stationary distribution of the Chapman-Kolmogorov equation, as given in (6).

The Chapman-Kolmogorov equation (1) with transition matrix (4) forms the starting point of the *kinetic theory* for Lorentz lattice gases, where one derives kinetic equations for the single particle distribution function $f(x, t) = \langle p(x, t) \rangle_\rho$, and possibly higher order correlation functions $\langle p(x, t)\hat{\rho}(r') \rangle_\rho$ of the moving particle and a scatterer.

The kinetic theory is well developed and yields the Boltzmann or mean field equations, as well as higher order ring equations [7]. The diffusion coefficient and the algebraic long time behavior of time correlation functions have been studied as a function of the density of scatterers ρ. The most interesting conclusion from these studies is that the Boltzmann approximation is *not a valid* low density approximation in Lorentz lattice gases with a non-vanishing reflection probability q. Exact enumeration methods or higher order kinetic equations have to be used for an analytical evaluation of the diffusion coefficient [7].

4 Deterministic Map

Lyapunov exponents describe properties of *deterministic* trajectories of dynamical systems in phase space. In order to treat the *stochastic* persistent random walk and Lorentz lattice gas as a dynamical system it is necessary to have deterministic equations of motion. This can be done by replacing the above stochastic models by an ensemble of non-interacting particles. Each particle is characterized by a *continuous* coordinate x_t, which is distributed uniformly over the interval $(0,1)$ at the initial time $t = 0$, and which satisfies a deterministic equation of motion. The relationship between the time dependent variable x_t and the actual position of the moving particle on the discrete lattice will be explained below.

Once the discrete time dynamics, $x_t \rightarrow x_{t+1}$, has been represented by a deterministic map,

$$x_{t+1} = M(x_t), \tag{8}$$

the concept of exponential separation of trajectories can be applied directly to our stochastic models. Consider as an example the map $M(x)$, illustrated in Fig. 1, which maps the unit interval onto itself. Let $dx_0 = x_0^{(1)} - x_0^{(2)}$ be the initial separation of two nearby trajectories. The separation after one time step is $dx_1 = |M'(x_0)|dx_0$.

The initial separation dx_0 after applying the map t times has increased by a factor

$$\left| \frac{dx_t}{dx_0} \right| = \prod_{\tau=0}^{t-1} \left| \frac{dx_{\tau+1}}{dx_\tau} \right| = \prod_{\tau=0}^{t-1} |M'(x_\tau)|, \tag{9}$$

where $M'(x)$ is the slope of the map at x. The above equation follows by applying the chain rule of differentiation. If the separation factor (9) increases exponentially with the number of time steps, i.e. $\sim \exp(\lambda t)$, then the trajectory has a

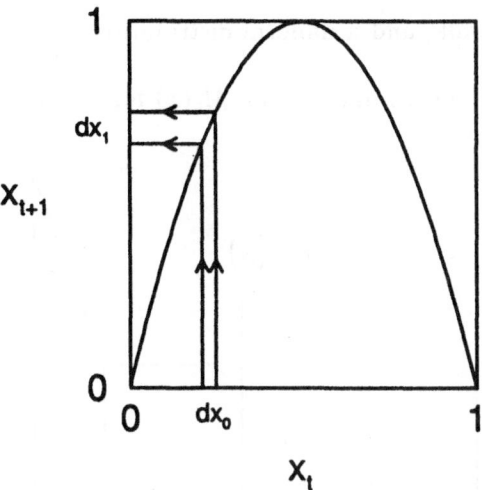

Fig. 1. Example of a deterministic map $x_{t+1} = M(x_t)$.

non-vanishing *positive* Lyapunov exponent λ, given by

$$\lambda = \lim_{t \to \infty} \frac{1}{t} \sum_{\tau=0}^{t-1} \log M'(x_\tau) = \langle \log M'(x) \rangle. \tag{10}$$

The time average can be replaced by an ensemble average over the x-interval $(0,1)$ if the system is ergodic. For Markov processes, like persistent random walks, this property is well established [10]. Quantities satisfying the second equality in (10) are called self-averaging.

5 Lyapunov Exponents for Random Walks

We now turn to the stochastic models, where $\{x_t, c_t\}$ represents the dynamic phase variable, and construct a deterministic map,

$$x_{t+1} = M_{c_t}(x_t), \tag{11}$$

that represents the time evolution of a single independent particle in the ensemble. The map will depend on the value $c_t = \pm 1$ of the incoming or pre-transition velocity at time t. The component x_t is a continuous variable defined on the real axis, whose integer part $[x_t] = r_t$, represents the position of the moving particle on the lattice. The remainder $\tilde{x}_t = x_t - [x_t]$ is a continuous variable on $(0,1)$ that uniquely defines the entire future of the trajectory, i.e. there exists a one-to-one relationship between the ensemble of all possible trajectories,

$$\omega(t) = \{r_0 c_0, r_1 c_1, \ldots r_t c_t, \ldots (t \to \infty)\}, \tag{12}$$

of the persistent random walk, and a uniform distribution of points $\tilde{x}_t$ on the interval $(0,1)$ [10].

The explicit form of the deterministic map $M_c(x)$ for an incoming velocity $c = \pm 1$ is given by

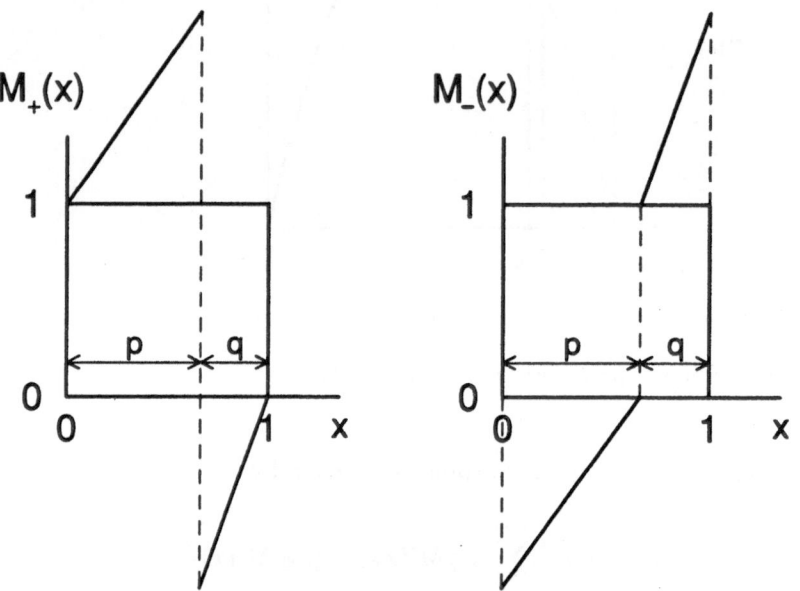

Fig. 2. Map $M_c(x)$ for particle with incoming velocity $c = \pm 1$ at a scattering site.

$$M_c(x) = \begin{cases} [x] + c + (1/p)\tilde{x} & \text{if } \tilde{x} \, \epsilon \, (0,p) \\ [x] - c + (1/q)(\tilde{x} - p) & \text{if } \tilde{x} \, \epsilon \, (p,1), \end{cases} \qquad (13)$$

and illustrated in Fig.2.

Inspection of Fig. 2 and (13) shows that the relation $[x_{t+1}] = [x_t] + c_t$ holds for a fraction p of particles in the ensemble, and corresponds to an outgoing velocity $c_t^* = c_t$. For the remaining fraction q of particles in the ensemble one has the relation $[x_{t+1}] = [x_t] - c_t$ corresponding to an outgoing velocity $c_t^* = -c_t$. Note that the fraction of independent particles in the ensemble moving to the right or left corresponds to the transition probability p or q for the persistent random walker to be transmitted or reflected at site $[x_t] = r_t$.

Note further that the map stretches the intervals of length p and q by factors $1/p$ and $1/q$ respectively, so that the dynamic variable $\tilde{x}_{t+1}$ is again uniformly distributed over the interval $(0,1)$. This stretching, or equivalently the property $|M'(x)| > 1$, is the essential reason why the persistent random walk has a *positive* Lyapunov exponent, so that initially nearby trajectories are exponentially diverging according to (9)-(10). The persistent random walk is therefore *chaotic* in the sense of dynamical systems theory.

Next we consider a subsequent application of the map. Consider for instance those particles with $x_{t+1} \epsilon (1,2)$, which therefore have $c_{t+1} = +1$. If $x_{t+1} \epsilon (1 + p, 2)$, then $x_{t+2} \epsilon (0,1)$ and the corresponding particle is reflected, and is mapped again onto the interval $(0,1)$, etc.

Having completed the description of how to map a *stochastic* random walk on an *ensemble* of independent particles with a *deterministic* time evolution, we are able to calculate the Lyapunov exponent for the *persistent* random walk using equations (9)-(10). In fact we have to count the number of times, $N_p(t)$, that a particle is transmitted at a given site, and $N_q(t)$, the number of times it is reflected in t steps. For each transmission there is a factor $1/p$ in dx_t/dx_0, and for each reflection there is a factor $1/q$. Thus we compute the stretching factor on a trajectory of t time steps, with $N_p + N_q = t$, as,

$$\left| \frac{dx_t}{dx_0} \right| = \left(\frac{1}{p} \right)^{N_p(t)} \left(\frac{1}{q} \right)^{N_q(t)}. \tag{14}$$

The Lyapunov exponent λ for the persistent random walk is obtained by taking the logarithm of (14). For sufficiently long times, the number of $p-$ and q-collisions per unit time approaches an average value, i.e.

$$\lambda(t|x_0) = \frac{1}{t} \langle N_p(t|x_0) \rangle \log \frac{1}{p} + \frac{1}{t} \langle N_q(t|x_0) \rangle \log \frac{1}{q}, \tag{15}$$

where the average is taken over all possible trajectories $\omega(t)$, defined in (12). As $t \to \infty$ the number of transmissions and reflections per unit time approaches the averages $\langle N_p(t) \rangle / t = p$ and $\langle N_q(t) \rangle / t = q$, so that

$$\lambda = \lim_{t \to \infty} \lambda(t|x_0) = -p \log p - q \log q. \tag{16}$$

In the special case of a simple random walk, where $p = q = 1/2$, the Lyapunov exponent is given by $\lambda = \log 2$.

The persistent random walk on a line was described in terms of a single independent random variable with transition probability $p = 1 - q$. On higher dimensional lattices with coordination numbers b ($b > 2$) there are more independent random variables, described by a matrix of transition probabilities $W_{cc'}(c, c' = 1, 2, \ldots b)$ with $\sum_c W_{cc'} = 1$ and the above result can be generalized to

$$\sum_n^{(+)} \lambda_n = -\frac{1}{b} \sum_{cc'} W_{cc'} \log W_{cc'}, \tag{17}$$

where the left hand side represents the sum of all positive Lyapunov exponents.

6 Links between Chaos and Kinetic Theory

In order to establish the connection between dynamical systems theory and non-equilibrium statistical mechanics it is essential to make the link between the separation rate of nearby trajectories in (14) and the *inverse* of the probability $P(\omega, t|x_0)$ on a trajectory $\omega(t)$ of t time steps starting at x_0 with $N_s(t|x_0)$ collisions of type s with $s = \{p, q\}$, i.e.

$$\left|\frac{dx_0}{dx_t}\right| = P(\omega, t|x_0) = p^{N_p(t)} q^{N_q(t)}. \tag{18}$$

The total probability is of course normalized as

$$\sum_\omega P(\omega, t|x_0) = \sum_{N=0}^{t} \binom{t}{N} p^N q^{t-N} = 1. \tag{19}$$

Consequently, the Lyapunov exponent in (15) can also be expressed as

$$\lambda(t|x_0) = -\frac{1}{t} \sum_\omega P(\omega, t|x_0) \log P(\omega, t|x_0). \tag{20}$$

If more independent random variables are involved, the right hand side of (20) should be replaced by the *sum* of all positive Lyapunov exponents, as in (17). Moreover the right hand side of (20) is by definition equal to the Kolmogorov-Sinai (KS) entropy per unit time $h_{KS}(t|x_0)$, so that

$$\lambda(t|x_0) = h_{KS}(t|x_0) \tag{21}$$

for the persistent random walk models. In the long time limit the right and left hand side of (21) approach their equilibrium values λ and h_{KS} respectively. In chaos theory the equality, $\lambda = h_{KS}$, is referred to as Pesin's theorem [10]. It applies to *closed* systems without escape where the total probability of trajectories is conserved (see (19)).

Of course these results are well known for Markov processes [10], but have been presented here to facilitate the transition to stochastic Lorentz lattice gases. In the latter models one has to distinguish between *scattering* sites and *non-scattering* sites, as was done in equation (4). At a scattering site, the deterministic map is the same as in the random walk (see Fig. 2 and (13)). At a *non-scattering* site the particle is in free motion, and the map for a particle with incoming velocity c_t is simply given by

$$x_{t+1} = M_{c_t}(x_t) = x_t + c_t. \tag{22}$$

As the slope of this map $M_c'(x) = \pm 1$, a non-scattering site does not produce any stretching of the distance between two nearby trajectories. This implies that equations (14) and (15) are still valid for a Lorentz lattice gas. Now the total number of p– and q–collisions on a trajectory of length t is less than t. Moreover, the average number of s-collision per unit time $\langle N_s(t|x_0, \hat{\rho}) \rangle / t$ with $s = \{p, q\}$ does not only depend on the initial state x_0, but also on the quenched configurations of scatterers.

It is now assumed that the Lyapunov exponent for the Lorentz lattice gas is obtained by averaging the corresponding quantity $\lambda(t|x_0, \hat{\rho})$ in (15) over the initial state x_0 of the moving particle, and over the quenched configurations of scatterers, i.e.

$$\lambda(t) = \langle \lambda(t|x_0, \hat{\rho}) \rangle_\rho$$
$$= -\frac{1}{t} \langle N_p(t) \rangle_\rho \log p - \frac{1}{t} \langle N_q(t) \rangle_\rho \log q. \qquad (23)$$

To obtain the equilibrium value of $\lambda(t)$ we average in $\langle \dots \rangle_\rho$ over the stationary distribution $p(x_0, \infty|\hat{\rho})$ as defined in (6). Then $\langle N_p(t) \rangle_\rho / t$ is proportional to the fraction of scattering sites, so that

$$\langle N_p(t) \rangle_\rho / t \simeq \rho p, \qquad (24)$$

and the Lyapunov exponent $\lambda(t)$ for a one-dimensional Lorentz lattice gas approaches its equilibrium value

$$\lambda = -\rho p \log p - \rho q \log q. \qquad (25)$$

The exponents $\lambda(t)$ and λ are *positive*, so that a Lorentz lattice gas is also *chaotic* in the sense of dynamic systems theory.

The above results can be generalized directly to higher dimensional Lorentz lattice gases, and the relation analogous to (17) is

$$\sum_n^{(+)} \lambda_n = -\frac{\rho}{b} \sum_{cc'} W_{cc'} \log W_{cc'}. \qquad (26)$$

7 Kinetic Theory of Closed Systems

In the previous section we have considered only the equilibrium value of the (sum of all) positive Lyapunov exponents. In order to study the approach of $\lambda(t)$ in (23) to equilibrium, we need to consider the number of p-collisions $N_p(t)$ on a trajectory starting in a non-stationary initial state $p(x_0, 0) = \delta(c, c_0)\delta(r, r_0)$.

The average number of (cc')-collisions per unit time on a trajectory of length t, starting in x_0, averaged over all configurations of scatterers, can be expressed as [9],

$$t^{-1} \langle N_{cc'}(t) \rangle_\rho = \frac{1}{t} \sum_{\tau=0}^{t-1} \sum_r \langle \hat{\rho}(r) p(rc', \tau) \rangle_\rho W_{cc'}. \qquad (27)$$

The product $\hat{\rho}(r) p(rc', \tau)$ represents the probability in a quenched configuration of scatterers that the moving particle has at time τ the precollision velocity c' and is at the scattering site r. A fraction $W_{cc'}$ of these particles will exit from this scattering site in the direction c. Summing this quantity over all sites r and all time steps τ, and performing finally an average over the quenched configuration of scatterers gives the desired average in (27).

The *pair correlation* function $\langle \hat{\rho}(r) p(rc', \tau) \rangle_\rho$ represents the probability that the moving particle at time τ is at a scattering site r with incoming velocity c'. This quantity can be calculated using standard techniques of kinetic theory.

The non-equilibrium expression for the (sum of all) positive Lyapunov exponents in terms of the non-equilibrium pair function becomes then

$$\lambda(t) = -\frac{1}{t} \sum_{\tau=0}^{t-1} \sum_{r} \sum_{cc'} \langle \hat{\rho}(r) p(rc', \tau) \rangle_\rho W_{cc'} \log W_{cc'}. \tag{28}$$

In Ref.[9] this quantity has been calculated by means of kinetic theory, and yields for long times

$$\lambda(t) \simeq \lambda + \mathcal{O}(t^{-1}) + \mathcal{O}(t^{-\frac{d}{2}-1}). \tag{29}$$

The terms of $\mathcal{O}(t^{-1})$ result from rapidly decaying terms in the summand of (27), and the terms of $\mathcal{O}(t^{-d/2-1})$ result from ring kinetic equations (mode coupling effects). All coefficients can be calculated explicitly.

8 Conclusions and Perspectives

In the preceding sections we have only explicitly discussed the behavior of the Lyapunov exponents and the Kolmogorov-Sinai (KS) entropy in *closed* systems, and we have shown how the basic quantities from chaos theory can be calculated for stochastic Lorentz lattice gases, using kinetic theory.

The analysis can also be extended to open systems, from where the moving particles can escape, i.e. escape from a finite domain D with a random array of scatterers. Then one is interested in the set of unstable trajectories, that stay inside D as $t \to \infty$ (repeller). We have studied the survival probability $Q(t|x_0, \hat{\rho})$, and the average escape rate for Lorentz lattice gases,

$$\gamma(t) = -\frac{1}{t} \langle \log Q(t|x_0, \hat{\rho}) \rangle_\rho. \tag{30}$$

For large systems ($V = L^d \to \infty$) and large times ($t \to \infty$) the escape rate approaches $\gamma \simeq Dk_0^2$, where D is the diffusion coefficient and $k_0 \simeq \pi/L$ is the wave number of the slowest decay mode in the system.

On the other hand, in an open system the escape rate formalism provides the relationship for an Lorentz lattice gas, i.e.

$$\gamma(t) \simeq \lambda(t) - h_{KS}(t), \tag{31}$$

where $\lambda(t)$ is the (sum of all) positive Lyapunov exponents and $h_{KS}(t)$ the KS-entropy for the stochastic Lorentz lattice gas, averaged over the configurations of scatterers. The precise definitions of these quantities can be found in Ref.[9]. Methods are presently being developed for an analytic evaluation of the averages in (30) and (31) for open systems.

The above results can also be extended to the so-called thermodynamic formalism, where the dynamic partition function is defined as [11]

$$Z_\beta(t|x_0, \hat{\rho}) = \sum_\omega (P(\omega, t|x_0, \hat{\rho}))^\beta. \tag{32}$$

(compare (20)) and the corresponding dynamic entropy or pressure can be calculated from

$$\psi_\beta(t) = \frac{1}{t} \langle \log Z_\beta(t|x_0, \hat\rho) \rangle_\rho. \tag{33}$$

Moreover, computer simulations provide an important tool to study the chaos quantities discussed in this paper. For open systems this may be achieved by using two different methods:

(i) For a given realization of a distribution of scatterers one determines numerically the largest eigenvalue $\Lambda(\hat\rho)$ of the transition matrix $w(x|y)$ for a Lorentz lattice gas (compare (4)) with escape, where $\sum_x w(x|y) < 1$. Then the long time limit of the escape rate for the Lorentz lattice gas should be averaged over different configurations of scatterers, and is given by $\gamma = -\langle \log \Lambda(\hat\rho) \rangle_\rho$. Note that $w(x|y)$ is a *large* matrix, typically of size $2L \times 2L$.

(ii) We have derived Chapman-Kolmogorov type equations (compare (1)), possibly with source terms, for the Lyapunov exponent in (20), the survival probability $Q(t|x_0, \hat\rho)$, the dynamic partition function $Z_\beta(t|x_0, \hat\rho)$, etc., and we have developed an exact enumeration method by solving these equations numerically for a quenched configuration of scatterers, and average the results over different runs according to (23), (31) or (33).

In summary, our preliminary results for stochastic Lorentz lattice gases with or without escape indicate that it is feasible to calculate the basic quantities of chaos theory in the approach to stationary values on the basis of nonequilibrium statistical mechanics and chaos theory.

Acknowledgements

The authors want to thank D. Jacobs, P. Gaspard, H. van Beijeren and R. Brito for their continuing help and stimulating discussions.

References

1. D.J. Evans, E.G.D. Cohen and G. Morris, Phys. Rev. A42, 5990 (1990); N.I. Chernov, G.L. Eyink, J.L. Lebowitz and Ya.G. Sinai, Phys. Rev. Lett. 10, 2209 (1993); Comm. Math. Phys. 154, 569 (1993).
2. P. Gaspard and G. Nicolis, Phys. Rev. Lett. 65, 1693 (1990); P. Gaspard and F. Baras, in: *Microscopic Simulations of Complex Hydrodynamic Phenomena*, M. Maréschal and B.L. Holian, Eds. (Plenum Press, New York, 1992) p. 301.
3. J.R. Dorfman and P. Gaspard, Preprint, May 1994.
4. M.A. Shereshevsky, J. Nonlinear Science 2, 1 (1992).
5. F. Bognoli, R. Rechtman, and S. Ruffo, Physics Letters A172, 34 (1992); M. Cieplak, U. d'Ortona, D. Salin, R.B. Rybka, J.R. Banavar, Comput. Material Science 1, 87 (1992).
6. G.A. van Velzen, *Lattice Lorentz Gases*, Ph.D. Dissertation, University of Utrecht (1990); M.H. Ernst, in: *Ordering Phenomena in Condensed Matter Physics*; Z.M. Galasiewicz and A. Pekalski, Eds. (World Scientific, Singapore 1991), p. 291.

7. M.H. Ernst and G.A. van Velzen, J. Phys. A: Math. Gen. 22, 4327 (1989); J. Stat. Phys. 57, 455 (1989); A.J.H. Ossendrijver, A. Santos and M.H. Ernst, J. Stat. Phys. 71, 1015 (1993); H. van Beijeren and M.H. Ernst, J. Stat. Phys. 70, 793 (1993).

8. D. Frenkel, F. van Luyn and P.M. Binder, Europhys. Lett. 20, 7 (1992); P.M. Binder and D. Frenkel, Phys. Rev. A42, 2463 (1990); P.M. Binder, Phys. Rev. E49, R3565 (1994).

9. J.R. Dorfman, M.H. Ernst and D. Jacobs, preprint, June 1994.

10. E. Ott, *Chaos in Dynamical Systems*, (Cambridge Univ. Press, 1993); P. Walters, *An Introduction to Ergodic Theory*, (Sringer Verlag, Berlin 1992).

11. C. Beck and F. Schlögl, *Thermodynamics of Chaotic Systems: An Introduction*, (Cambridge University Press, Cambridge 1993).

Dynamics of Reptation

J.D. Balkenende, J.A. Leegwater and J.M.J. van Leeuwen

Instituut-Lorentz, Rijksuniversiteit Leiden, Postbus 9506, NL–2300 RA LEIDEN, Netherlands

Abstract: Discrete models for reptation are discussed. It is shown that in the weak field limit a formula for the drift velocity exists, which is applicable to a wide class of dynamical models. In the long polymer limit the behavior is dominated by the existence of a permanent head and tail of the polymer which influences the statistics and the expressions for drift and mobility in a fundamental way.

Introduction

The dynamics of a long polymer has been the subject of sustained studies for decades. While a realistic description is at best achievable for relatively short chains by molecular dynamics simulations, more analytic investigations are limited to extremely long chains where some average description becomes possible. Recently the study of polymer dynamics has received an impetus due to the introduction of stochastic lattice models by Rubinstein [1] and Duke [2]. In these lattice models the idea of reptation, as coined by de Gennes [3], is implemented in a simple algorithm describing the motions. This has opened the way for new analytic and simulation research.

Such models are particularly relevant for gel electrophoresis, where the polymer is embedded in a gel, which is a fixed porous structure. The pores of the gel are modeled as the cells of a lattice and the dynamics is schematized by keeping track of the polymer segments occupying the cells of the lattice (See ref. [4] and [5] for details). These polymer segments are called reptons. They may be viewed as Kuhn's units which are sufficiently large such that the motions of subsequent reptons have little correlations. Then the motion of the reptons is taken stochastic with transition probabilities biased by the electric field which drives the polymer through the gel. The possible moves of the reptons are severely restricted by the requirement that the polymer remains connected, i.e. successive reptons are either in the same or in neighboring cells. Thus following the reptons along the polymer chain, one traces out a path of occupied cells in the lattice. This path is only changed by the motions of the end reptons, which can either occupy a new cell or retreat from an occupied cell, thereby enlarging viz. shortening the path. As the connectivity of the polymer chain requires to have at least one repton in each cell of the path we may focus on the extra reptons (or extrons) in the cells. The extrons are created at the new end-cell when the repton retreats inwardly or annihilated in the (old) end-cell when the end repton embarks on a new cell. The description in terms of the extron motion visualizes the reptation as a diffusion of stored length. The amount of stored length in a cell is simply the number of extrons in that cell.

The main emphasis in this paper is on the behavior of the drift velocity (or the associated mobility) as a function of the strength E of the driving electric field and the number N of reptons of the chain (or equivalently its length). We will always work with long chains ($N \gg 1$) and small fields. Measuring the field E in terms of a dimensionless parameter ε we require that $\varepsilon \ll 1$, which means that on the level of a simple repton move the bias for motions in the direction of the field is weak. In our simulations we have observed that for long polymers, keeping ε constant, we are in a regime which is characterized by the fact that the polymer develops a definite head and tail, where the head is the end which grows dominantly and the tail is the side where the polymer shrinks most of the time. On the other hand, taking the limit $\varepsilon \to 0$ while keeping N constant we arrive in the regime where head and tail may frequently interchange positions at the ends of the polymer. The crossover between these two regimes is characterized by a certain scaling combination, $x = \varepsilon N^\sigma$, where the value of σ is one of the most important quantities to be found.

The first part of the paper is devoted to the regime $x \ll 1$, which we name the weak field limit, whereas the second part is mainly devoted to $x \gg 1$, denoted as the long chain limit. We show that in the weak field limit the drift velocity becomes, for general circumstances, linear in ε, with a coefficient which is inversely proportional to a power of N. The proportionality coefficient can be calculated explicitly in the lattice models.

The more intriguing long chain limit shows quite different behavior. The polymer develops a definite head and tail, separated by a middle part, which contains the bulk of the chain. The drift velocity is restricted by the motions along this middle part and becomes independent of the length of the polymer. Our analysis goes hand in hand with simulations and the regime $x \gg 1$ is obviously very difficult to simulate. We benefit from a simplified model where we take the internal motion of the reptons fast with respect to the end point motion. Consequently, the extron distribution becomes a static problem which allows a much more detailed analysis and simulations of much longer chains. The paper closes with a discussions of the results and a comparison with related work.

1 The Master Equation

Our polymer consists of a set of occupied cells forming a path on a lattice [4]. We take a d-dimensional cubic lattice to start with and discuss later on random spatial structures. The cells are labeled $1, \ldots, L$ and the extron occupation is $n_1, \ldots, n_L$ such that the total number of reptons N constrains the occupation by the condition

$$N = L + \sum_{i=1}^{L} n_i . \tag{1.1}$$

For the parameters specifying the path (consisting of $L - 1$ segments) we take the sequence $s_{\frac{3}{2}}, \ldots, s_{L-\frac{1}{2}}$, where $s_{l+\frac{1}{2}}$ locates cell $l + 1$ with respect to cell l. The $s_{l+\frac{1}{2}}$ should be d-dimensional nearest neighbor vectors but we can do for the moment with the projections along the field directions which we take along the body diagonal. This has the advantage that $s_{l+\frac{1}{2}}$ is either positive or negative and by taking suitable units we can restrict ourselves to $s_{l+\frac{1}{2}} = \pm 1$. The complete state of the polymer is denoted by (s, n) where $s = s_{\frac{3}{2}}, \ldots, s_{L-\frac{1}{2}}$ and $n = n_1, \ldots, n_L$.

The probability $P(s, n)$ on (s, n) is the (stationary) solution of the master equation

$$\sum_{(s', n')} W(s, n | s', n') P(s', n') = \sum_{(s', n')} W(s', n' | s, n) P(s, n) . \tag{1.2}$$

$W(s, n | s', n')$ is the transition rate from (s', n') to (s, n). These are biased by the electric field through a Boltzmann factor

$$B = e^{\epsilon/2} = e^{qaE/2k_B T} ,$$

ϵ being the dimensionless measure for the electric field E. The hops of the extrons through a link $s_{l+\frac{1}{2}}$ thus have a transition rate

$$W = w B^{s_{l+\frac{1}{2}}}$$

where w is an overall transition rate.

We are interested in the stationary state which is drifting with a certain velocity along the electric field. As this is the stationary state rather than the equilibrium state, no general solution of Eq. (1.2) exists and detailed balance does not hold.

We distinguish between internal and external moves. In an internal move an extron hops between two neighboring cells, thereby changing n but leaving s invariant. The external moves involve the end reptons. They either enlarge the path from L to $L + 1$ by adding a new cell (and a corresponding s) and annihilating an extron, or they shrink the path from L to $L-1$ by retreating from an end cell (and omitting the s) and creating an extron in the new end cell. The distinction between internal and external moves is made more explicit in the (s, n) notation.

As a side remark for this paper we note that one can simplify the model in an essential way by imposing periodic boundary conditions [5]. In this model it means that both ends move synchronously such that when an element $s_{L-\frac{1}{2}}$ is annihilated at the end it is simultaneously created as $s_{\frac{3}{2}}$ at the beginning and vice versa. This leaves the sequence s invariant, apart from a cyclic renumbering, or one could say all moves become internal. The master equation Eq. (1.2) is then an equation in n-space for a fixed s. This master equation can be solved generally and the drift velocity can be calculated for any path s. Averaging over s as over unbiased random walkers leads to a drift velocity $v \sim \varepsilon/N$ with a lattice dependent proportionality coefficient. We show in this paper that this result is also pertinent for the present case with open ends, provided we stay in the limit $x \ll 1$.

We derive now from Eq. (1.2) formally a master equation for the reduced (path) distribution function $P(s)$

$$P(s) = \sum_n P(s, n)$$

where the sum over n runs over all allowed (see Eq. (1.1)) extron distributions. In deriving the equation for $P(s)$ we have to consider only the external moves which change s. For this moment we arbitrarily name the ends head and tail, respectively; later on the head and tail get a dynamic meaning. We then have that s can change into four new possible paths $s_h^\pm, s_t^\pm$

$$
\begin{aligned}
\text{longer head}: \quad & s_h^+ = & s_{\frac{3}{2}}, \ldots \ldots, s_{L-\frac{1}{2}}, s_{L+\frac{1}{2}} \\
\text{shorter head}: \quad & s_h^- = & s_{\frac{3}{2}}, \ldots, s_{L-\frac{3}{2}} \\
\text{longer tail}: \quad & s_t^+ = s_{\frac{1}{2}}, s_{\frac{3}{2}}, \ldots \ldots, s_{L-\frac{1}{2}} \\
\text{shorter tail}: \quad & s_t^- = & s_{\frac{3}{2}}, \ldots, s_{L-\frac{1}{2}}.
\end{aligned}
\tag{1.3}
$$

At the tail we prefer to add $s_{\frac{1}{2}}$ or omit $s_{\frac{3}{2}}$ without renumbering to keep the notation simple. Averages over the extron occupation are denoted by $\langle \cdot \rangle$, e.g., the probability that at least one extron is present in cell l is given by

$$\theta_l(s) \equiv \langle \theta(n_l) \rangle = \sum_{n_1,\ldots,n_l} P(s, n)\theta(n_l),$$

where θ is the step function, i.e. $\theta(0) = 0$ and $\theta(n \geq 1) = 1$. For the end cells we call them θ_t and θ_h at the tail and head side, respectively. Summing Eq. (1.2) the internal moves cancel between left and right hand side and the moves Eq. (1.3) survive. So we arrive at the equation

$$\sum_{s_{\frac{1}{2}}} B^{s_{\frac{1}{2}}}\left(1 - \theta_t(s_t^+)\right)P(s_t^+) + B^{-s_{\frac{3}{2}}}\theta_t(s_t^-)P(s_t^-)$$

$$+ B^{s_{L-\frac{1}{2}}}\theta_h(s_h^-)P(s_h^-) + \sum_{s_{L+\frac{1}{2}}} B^{-s_{L+\frac{1}{2}}}\left(1 - \theta_h(s_h^+)\right)P(s_h^+)$$

$$= \Big[(B + B^{-1})(\theta_t(s) + \theta_h(s))$$

$$+ B^{s_{\frac{1}{2}}}\left(1 - \theta_t(s)\right) + B^{-s_{L-\frac{1}{2}}}\left(1 - \theta_h(s)\right)\Big]P(s). \qquad (1.4)$$

The averages over the $\theta(n_l)$ have to be inserted to allow for the possible moves, e.g. the longer tail path s_+^t can only reduce to s when cell 0 does not carry an extron, etc. We note that the averages θ_l require knowledge of the full $P(s, n)$. Thus in order that Eq. (1.4) is useful this knowledge has to be supplied. Note also that Eq. (1.4) yields an identity when summed over all s, which is nothing else than the conservation of probability.

A third form of the master equation results by reducing the description further such that only the length L of the path remains. Defining

$$P_L = \sum_s P(s)$$

as the probability of a path of L cells, one derives the equation for P_L from the balance in the steady state

$$W(L|L+1)P_{L+1} + W(L|L-1)P_{L-1} = \left[W(L+1|L) + W(L-1|L)\right]P_L, \qquad (1.5)$$

where the transition rates are defined formally through

$$W(L|L+1)P_{L+1} = \sum_s \Big[\sum_{s_{\frac{1}{2}}} B^{s_{\frac{1}{2}}}\left(1 - \theta_t(s_t^+)\right)P(s_t^+)$$

$$+ \sum_{s_{L+\frac{1}{2}}} B^{s_{L+\frac{1}{2}}}\left(1 - \theta_h(s_h^+)\right)P(s_h^+)\Big] \qquad (1.6)$$

and e.g. the inverse

$$W(L|L-1)P_{L-1} = \sum_s [B^{-s_{\frac{3}{2}}}\theta_t(s_t^-)P(s_t^-)$$

$$+ B^{s_{L-\frac{1}{2}}}\theta_h(s_h^-)P(s_h^-)] \qquad (1.7)$$

with similar expressions for the opposite processes. Again we stress that equations like Eq. (1.5) only have a formal meaning since one has to know the probabilities in order to calculate the transition rates from Eq. (1.6) and Eq. (1.7).

2 The Drift Velocity

In order to discuss the drift velocity of the polymer we inspect first the traffic of extrons along a segment $s_{l+\frac{1}{2}}$ (between cell l and $l+1$). The net current along this segment is given by

$$J^c_{l+\frac{1}{2}} = wB^{s_{l+\frac{1}{2}}}\theta_l - wB^{-s_{l+\frac{1}{2}}}\theta_{l+1} \,. \tag{2.1}$$

The superscript c indicates that it is the curvilinear current. The first term describes the traffic from l to $l+1$ and is restricted by θ_l to cases where an extron is available. The second term gives the flow in the opposite direction. The velocity component in the direction of the field due to the motions along segment $l+\frac{1}{2}$ then follows as

$$v_{l+\frac{1}{2}} = \frac{a}{N} J^c_{l+\frac{1}{2}} s_{l+\frac{1}{2}} \,. \tag{2.2}$$

We have multiplied by a distance a/N which is the change in position of the center of mass due to the move. The factor $s_{l+\frac{1}{2}}$ accounts for the sign of v (in the direction of the field). The total drift velocity of the polymer equals

$$v = \sum_{l=1}^{L-1} v_{l+\frac{1}{2}} \,. \tag{2.3}$$

Again we note that all these quantities refer to a fixed value of s.

In the stationary state one expects $J^c_{l+\frac{1}{2}}$ to be the same for all segments $l+\frac{1}{2}$

$$J^c_{l+\frac{1}{2}} = J \,, \tag{2.4}$$

otherwise extrons would accumulate, since $J^c_{l-\frac{1}{2}}$ is the supply from the tail side of cell l and $J^c_{l+\frac{1}{2}}$ the drain to its head side. This relation is not as obvious as one might think intuitively, but assumes that there are no correlations in extron occupation between end cells and the cell l:

$$\langle \theta(n_1) n_l \rangle = \langle \theta(n_1) \rangle \langle n_l \rangle \,, \tag{2.5}$$

which one expects to be true when l is not too close to the end cells. From Eq. (2.3) and Eq. (2.2) we then find

$$v = \frac{a}{N} J S$$

with

$$S = \sum_{l=1}^{L-1} s_{l+\frac{1}{2}}$$

the distance between head and tail measured in the field direction.

We can derive an expression for J in the weak field limit. We use the identity

$$B^s = \frac{1}{2}(B + B^{-1})(1 + st) \tag{2.6}$$

where t stands for the abbreviation

$$t = \frac{B - B^{-1}}{B + B^{-1}} = \tanh \frac{\varepsilon}{2} \tag{2.7}$$

which will be considered as a small quantity. Inserting Eq. (2.6) into Eq. (2.1) we have

$$J(s) = \frac{w}{2}(B + B^{-1})\left[\theta_l(s) - \theta_{l+1}(s) + s_{l+\frac{1}{2}}t(\theta_l(s) + \theta_{l+1}(s))\right]. \tag{2.8}$$

Summing Eq. (2.8) over l yields

$$(L - 1)J(s) = \frac{w}{2}(B + B^{-1})\Big[\theta_t(s) - \theta_h(s)$$
$$+ t\sum_{l=1}^{L-1} s_{l+\frac{1}{2}}\big(\theta_l(s) + \theta_{l+1}(s)\big)\Big]. \tag{2.9}$$

Now we use the weak field limit $t \to 0$. Then, to lowest order, we may use the field free averages in the last term

$$\theta_l(s) \simeq p \tag{2.10}$$

where p is the probability of finding at least one extron in cell l for $\varepsilon = 0$. This p can be calculated from the equilibrium distribution for $\varepsilon = 0$ where all configurations are equally probable. In the same limit the difference of the first two terms is of order t and thus contributes of order t/L to J, which is small compared to the contributions of the last terms. So

$$J \simeq w(B + B^{-1})tpS/L \tag{2.11}$$

which implies for the drift velocity

$$v \simeq \frac{awp}{N}(B - B^{-1})(S^2/L). \tag{2.12}$$

As v contains already a factor of the order ε through $B - B^{-1}$, the average over the paths of the last factor S^2/L may be carried out in the zero field limit to obtain the leading term, giving $v = aw\varepsilon/(2d+1)N$, which is the same as for the periodic chain model where Eq. (2.12) follows strictly, no influence of endpoints being present. For the details we refer to [5].

Going over the derivations again we see that Eq. (2.5) must hold beyond a certain region which does not grow with the size of the polymer chain. The existence of finite correlations at both end regions is a general feature of these reptating polymers, as it emerges from simulations. It is also clear that we first take the limit $\varepsilon \to 0$ such that Eq. (2.11) results and then consider large L. Therefore Eq. (2.12) applies to the regime $x \ll 1$.

3 Generalizations of the Drift Formula

Eq. (2.12), which links the drift velocity of a chain to its length and displacement S in the field directions, is of more general validity than its derivation in the context of a specific lattice structure and simple hopping rates suggests. One finds it in the literature [6] where it is based, through the Einstein relation, on an expression for the diffusion constant which follows from speculations on the renewal time. Here the viewpoint is different as it is seen as an expression for a given arbitrary shape $s = s_{\frac{3}{2}}, \ldots, s_{L-\frac{1}{2}}$.

In the first generalization of Eq. (2.12) we want to show that this expression is largely independent of the spatial structure of the lattice and in the second generalization we want to discuss the influence of the fact that a realistic polymer has to avoid itself in its motions.

In a general spatial structure of the gel we may consider the pores of the gel as randomly shaped cells located at spatial positions r_l, with arbitrary hopping frequencies $w_{l+\frac{1}{2}}$ determining the traffic between cells l and $l+1$. The characterization of the polymer would still be a path of neighboring cells with an occupation of extrons. The difference is that we cannot use the simple variables $s_{l+\frac{1}{2}} = \pm 1$ to characterize the shape of the path, but rather the displacements $r_{l+1} - r_l$. The hopping rates along a link $l + \frac{1}{2}$ will be given by

$$W = w_{l+\frac{1}{2}} e^{\pm \epsilon_{l+\frac{1}{2}}/2}$$

with $\varepsilon_{l+\frac{1}{2}}$ given by

$$\varepsilon_{l+\frac{1}{2}} = \frac{q(r_{l+1} - r_l) \cdot E}{k_B T}.$$

The $+$-sign in the bias refers to the jump $l \to l+1$ and the $-$-sign to the reverse jump.

Following the derivation of section 2 we have for the curvilinear current

$$J^c_{l+\frac{1}{2}} = w_{l+\frac{1}{2}} \left[e^{\epsilon_{l+\frac{1}{2}}/2} \theta_l(s) - e^{-\epsilon_{l+\frac{1}{2}}/2} \theta_{l+1}(s) \right]$$

with the expression for the velocity

$$v_{l+\frac{1}{2}} = \frac{(r_{l+1} - r_l) \cdot \hat{E}}{N} J^c_{l+\frac{1}{2}}.$$

For the same reason as given earlier $J^c_{l+\frac{1}{2}}$ is (nearly) constant along the chain and for the total velocity one has instead of Eq. (2.3)

$$v = \frac{(r_L - r_1) \cdot \hat{E}}{N} J.$$

The calculation of J in the small E limit proceeds as before. We arrive at the equivalent of expression Eq. (2.9) which we write as

$$\frac{(L-1)J}{w_{l+\frac{1}{2}}} = \cosh(\varepsilon_{l+\frac{1}{2}}/2)\left[\theta_t - \theta_h + \sum_{l=1}^{L-1} t_{l+\frac{1}{2}}(\theta_l + \theta_{l+1})\right] \qquad (3.1)$$

where $t_{l+\frac{1}{2}}$ now stands for

$$t_{l+\frac{1}{2}} = \tanh(\varepsilon_{l+\frac{1}{2}}/2).$$

Expansion for small $\varepsilon_{l+\frac{1}{2}}$ than leads to

$$J\sum_{l=1}^{L-1}\frac{1}{w_{l+\frac{1}{2}}} \simeq p\frac{q(\boldsymbol{r}_L - \boldsymbol{r}_1)\cdot \boldsymbol{E}}{k_B T}$$

with p defined again by Eq. (2.10). So we see that defining the average along the chain as

$$\overline{1/w} = \frac{1}{L-1}\sum_{l=1}^{L-1} 1/w_{l+\frac{1}{2}},$$

we obtain for Eq. (2.12)

$$v \simeq \frac{1}{\overline{1/w}}\frac{pq}{Nk_B T}\frac{(\boldsymbol{r}_L - \boldsymbol{r}_1)\cdot \hat{\boldsymbol{E}}\,(\boldsymbol{r}_L - \boldsymbol{r}_1)\cdot \boldsymbol{E}}{L}$$

which may be put in vectorial form as

$$\boldsymbol{v} = \frac{1}{\overline{1/w}}\frac{pq}{Nk_B T}\frac{(\boldsymbol{r}_L - \boldsymbol{r}_1)(\boldsymbol{r}_L - \boldsymbol{r}_1)\cdot \hat{\boldsymbol{E}}}{L}.$$

This formula shows even more clear than Eq. (2.12) that it is the square displacement in the direction of the field which determines the drift velocity. For lowest order of v in the field E all these other quantities have to be determined in zero field.

More delicate is the influence of the self avoidance of the polymer chain. We have two effects: a local effect where the reptons hinder each other in a cell and a global effect which affects the endpoints in choosing new cells. The local effect can be incorporated by assigning a tolerance k to a cell being the maximum number of allowed reptons in a cell. We may take the tolerance as low as $k = 2$ allowing for maximally 1 extron in a cell (for $k = 1$ no motion would be possible). The effect of $k = 2$ has been investigated in the periodic chain model, which yields a general result for the $x \ll 1$ regime. It lowers the constant p but has otherwise no influence.

The global effects of self avoiding walks are more profound. Again we may stick to tolerance $k = 2$. Starting from a path which is not self-intersecting, the polymer will never reach an intersecting configuration. The head may occasionally visit a cell with one repton, but it has to retreat before traffic through that cell becomes possible. So the polymer is effectively self-avoiding. As known, the mean squared displacement is larger for self-avoiding walks:

$$\langle S^2 \rangle \sim L^{2\nu} ,$$

where ν is a critical exponent ($v = 1, \frac{3}{4}, 0.59, \frac{1}{2}$ for $d = 1, 2, 3, 4$). Thus it changes the dependence of v from E/N to $E/N^{2-2\nu}$ as pointed out by Lerman and Frisch [6]. The difference in substantial for $d = 2$. Also the scaling parameter σ is modified due to the self avoidance [7].

4 Long Polymers

Simulations show that, at small but fixed value of the driving field, the longer polymers start to develop a definite head and tail. At the head side the path of the polymer grows predominantly and it shrinks at the tail side. For shorter polymers head and tail frequently interchange role, but for longer polymers it becomes so rare that it does not show up even in long simulations. It means that one can collect arbitrarily accurate statistics in the situations where the symmetry between head and tail is broken. The point where one crosses over to this asymmetric situation depends on the value of the combination $x = \epsilon N^{\sigma}$. Barkema et al. [8] have found strong evidence that $\sigma = 1$ is the appropriate power for crossover to long polymer behavior for models of the type discussed so far. They make however no explicit connections to the asymmetry between head and tail which is here put forward as the main characteristic. In this section we analyze the consequences of a permanent head and tail.

Extrons are created at the tail and they flow in a constant stream through a long middle section towards the head where they are annihilated. So we may split the master equation into two equations: one for the balance at the head and one for the balance at the tail. The occupied cells are numbered from tail (cell 1) to head (cell L). For the tail we collect from Eq. (1.4) the terms referring to the tail side and equate them:

$$\sum_{s_{\frac{1}{2}}} B^{s_{\frac{1}{2}}} \left(1 - \theta_t(s_t^+)\right) P(s_t^+) + B^{-s_{\frac{3}{2}}} \theta_t(s_t^-) P(s_t^-)$$

$$= \left[(B + B^{-1})\theta_t(s) + B^{s_{\frac{3}{2}}} \left(1 - \theta_t(s)\right) \right] P(s) \qquad (4.1)$$

A similar equation holds for the head. If we assume that the dependence of the θ_t on the first few segments is the most important, we may derive an equation for $\langle s_{l+\frac{1}{2}} \rangle$:

$$\langle s_{l+\frac{1}{2}}\rangle = \sum_{s} s_{l+\frac{1}{2}} P(s)$$

by using that for sufficiently large l the correlation with the first few $s_{\frac{3}{2}}, \ldots$ is weak. So multiplying Eq. (4.1) by $s_{l+\frac{1}{2}}$ and summing over all s we find

$$\langle B^{s_{\frac{3}{2}}}\big(1 - \theta_t(s)\big)\rangle\langle s_{l+\frac{3}{2}}\rangle + (B + B^{-1})\langle\theta_t(s)\rangle\langle s_{l-\frac{1}{2}}\rangle$$
$$= \Big[(B + B^{-1})\langle\theta_t(s)\rangle + \langle B^{s_{\frac{3}{2}}}\big(1 - \theta(s)\big)\rangle\Big]\langle s_{l+\frac{1}{2}}\rangle. \qquad (4.2)$$

Eq. (4.2) only holds in the regime where the segment $l + \frac{1}{2}$ is sufficiently far from the tail repton. The structure of Eq. (4.2) in interesting in spite of the fact that it involves many unknown quantities. Eq. (4.2) can be written as

$$\alpha_t\langle s_{l+\frac{3}{2}}\rangle + \gamma_t\langle s_{l-\frac{1}{2}}\rangle = (\alpha_t + \gamma_t)\langle s_{l+\frac{1}{2}}\rangle \qquad (4.3)$$

with α_t and γ_t given by

$$\begin{cases} \alpha_t = \langle B^{s_{\frac{3}{2}}}\big(1 - \theta_t(s)\big)\rangle \\ \gamma_t = (B + B^{-1})\langle\theta_t(s)\rangle \end{cases}$$

Both α_t and γ_t are positive. The asymptotic solution of Eq. (4.3) is given by

$$\langle s_{l+\frac{1}{2}}\rangle = A_t + C_t\Big(\frac{\gamma_t}{\alpha_t}\Big)^l \qquad (4.4)$$

as one sees by substitution. The constants A_t and C_t must in general be obtained from fitting with the values of $\langle s_{l+\frac{1}{2}}\rangle$ for small l which must be calculated separately. When $\gamma_t > \alpha_t$ then $C_t = 0$, otherwise we obtain unacceptable dependence on l. We stress that $\langle s_{l+\frac{1}{2}}\rangle \neq 0$ only when the polymer chain has a definite head and tail. When head and tail interchange frequently the average is zero.

We can make another contact with the growing and shrinking rates at the head and the tail. From the master equation for the P_L as given by Eq. (1.5), we find, by multiplying with L and summing over all L,

$$\sum_{L}[-W(L|L+1)P_{L+1} + W(L|L-1)P_{L-1}] = 0.$$

With the help of Eqs. Eq. (1.6) and Eq. (1.7) this may again be split in head and tail and can be written as

$$R_t + R_h = 0 \qquad (4.5)$$

with R_t the growth rate at the tail

$$R_t = \sum_{s}\Big[(B + B^{-1})\theta_t(s) - B^{s_{\frac{3}{2}}}\big(1 - \theta_t(s)\big)\Big]P(s) = \gamma_t - \alpha_t \qquad (4.6)$$

and a similar expression for the growth rate at the head R_h. Now one sees that $R_t < 0$, as the polymer path shrinks at the tail side. Thus $\gamma_t < \alpha_t$ and both the coefficients A_t and C_t are allowed. On the other hand $R_t < 0$ implies

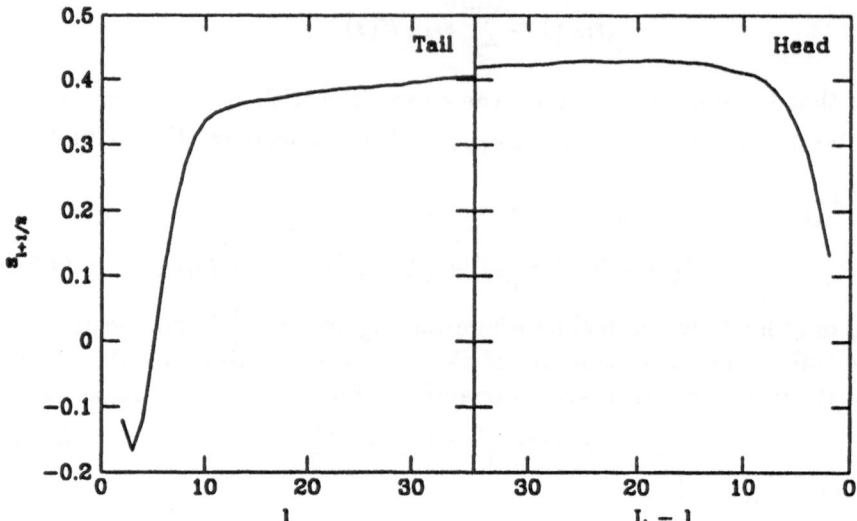

Figure 1. Simulation results of the average orientation of segment $l + \frac{1}{2}$ at tail and head side for $\varepsilon = 0.05$ and $N = 100$.

by Eq. (4.5) that $R_{\rm h} > 0$. Performing the same steps from the head side we find a similar expression as Eq. (4.4) but now $\gamma_{\rm h} > \alpha_{\rm h}$ which excludes a non-vanishing $C_{\rm h}$. This imposes an extra condition from which the value of the average $\langle s_{l+\frac{1}{2}} \rangle$ in the middle of the polymer path follows, because $A_{\rm t} = A_{\rm h}$. A typical simulation plot of $\langle s_{l+\frac{1}{2}} \rangle$ is given in figure 1.

5 The Fast Extron Limit

The considerations of the previous section are not very specific because they involve the unknown quantity $\theta_{\rm t}(s)$, the extron occupation at the tail side. In this section we consider a modification of the model in which we can make a much more specific analysis, thereby clarifying also the general situation. In order to speed up the simulations we considered the case where the internal reptons (the extrons) move much faster than the end reptons. One could see this model as representing polymers with large end-groups inhibiting the motion of the end reptons. Another motive is that this model is complementary to the earlier studied periodic polymers [5] in which the endpoint motion is eliminated where as in this fast extron limit the dynamics is completely limited to the motion of the endpoints.

In the long polymer limit the extrons drift immediately towards the head when they are created at the tail. At the head we have a finite distribution of extrons but at the tail no extrons are present. This simplifies the tail equation

Eq. (4.1) essentially since using

$$\theta_t(s) = 0\,,$$

Eq. (4.1) reduces to

$$\sum_{s_{\frac{1}{2}}} B^{s_{\frac{1}{2}}} P(s_t^+) = B^{s_{\frac{3}{2}}} P(s)\,. \tag{5.1}$$

This is an open hierarchy of equations but we can easily construct a sensible solution

$$P(s) = P_1(s_{\frac{3}{2}}) \prod_{l=2} \left(\frac{1 + s_{l+\frac{1}{2}}\overline{s}}{2} \right) \tag{5.2}$$

or in words, the probabilities factorize and are all equal except for the first cell, which must obey the equation

$$\sum_{s_{\frac{1}{2}}} B^{s_{\frac{1}{2}}} P_1(s_{\frac{1}{2}}) \frac{1 + s_{\frac{3}{2}}\overline{s}}{2} = B^{s_{\frac{3}{2}}} P_1(s_{\frac{3}{2}})$$

as follows from substituting Eq. (5.1) into Eq. (5.2). In this and what follows, $\overline{s}$ is a parameter denoting the average orientation of the polymer in the middle of the chain. Summing this equation over $s_{\frac{3}{2}}$ leads to an identity (reflecting conservation of probability); multiplying it by $s_{\frac{3}{2}}$ and then summing yields

$$\overline{s} \sum_{s_{\frac{1}{2}}} B^{s_{\frac{1}{2}}} P_1(s_{\frac{1}{2}}) = \sum_{s_{\frac{3}{2}}} s_{\frac{3}{2}} B^{s_{\frac{3}{2}}} P_1(s_{\frac{3}{2}})\,.$$

This equation is readily solved when supplemented with the normalization with the result

$$P_1(s_{\frac{3}{2}}) = \frac{1}{2} \left(1 + s_{\frac{3}{2}} \frac{\overline{s} - t}{1 - \overline{s}t} \right)$$

where t is defined in Eq. (2.7).

One sees from Eq. (5.2) that the parameter $\overline{s}$ has the meaning of the average value of $s_{l+\frac{1}{2}}$

$$\langle s_{l+\frac{1}{2}} \rangle = \overline{s} \qquad l = 2, 3, \ldots .$$

It is a constant immediately after the first cell which has the value

$$\langle s_{\frac{3}{2}} \rangle = \frac{\overline{s} - t}{1 - \overline{s}t}\,. \tag{5.3}$$

In figure 2 we see that this remarkable behavior is beautifully confirmed by the simulations. Also one deduces from the analysis that s remains a free parameter as was anticipated in the previous section.

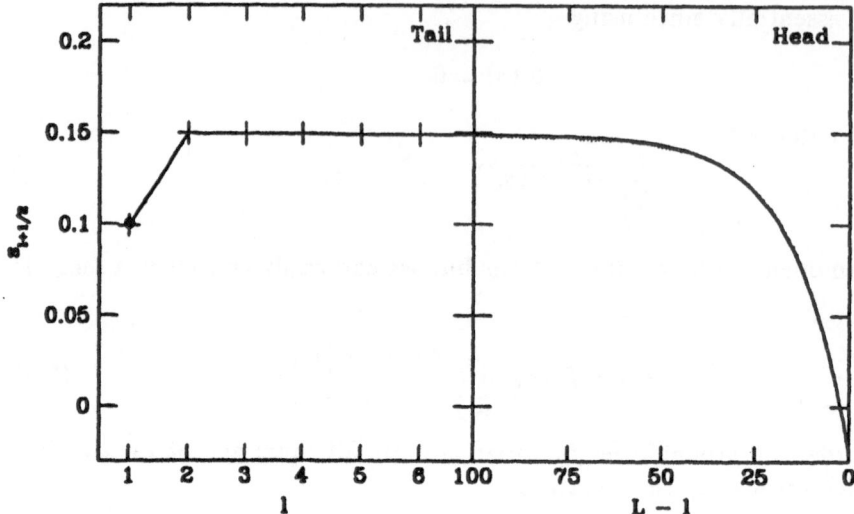

Figure 2. Simulation results of the average orientation of segment $l + \frac{1}{2}$ at the tail side and the head side in the fast extron limit. In the left part of the figure the solid line is a guide to the eye, while in the right part it is an exponential fit. For this plot $\varepsilon = 0.1$ and $N = 1000$.

The discussion of the head is more delicate because $\theta_h(s)$ depends in a non-trivial way on the shape of the polymer path. We refer for a more precise discussion to a paper devoted to the properties of the fast extron limit [9] and restrict ourselves here to a discussion of the principles. So we start by assuming that θ_h is independent of the shape of the path, or rather that its average value over the path may be used. Then the equation for the probabilities at the head side becomes

$$\sum_{s_{L+\frac{1}{2}}} B^{-s_{L+\frac{1}{2}}}(1 - \theta_h)P(s_h^+) + B^{s_{L-\frac{1}{2}}}\theta_h P(s_h^-)$$

$$= \left[B^{-s_{L-\frac{1}{2}}}(1 - \theta_h) + (B + B^{-1})\theta_h \right] P(s) \qquad (5.4)$$

This equation can be solved by assuming that the probabilities factorize. In contrast to Eq. (5.2) we must have now all factors in the probability distribution of equal form since $\theta_h \neq 0$. So we substitute into Eq. (5.4)

$$P(s) = \prod_{n=1} \frac{1 + s_{L-n+\frac{1}{2}}\bar{s}}{2} \qquad (5.5)$$

and the resulting equation for $\bar{s}$ reads

$$(1 - \theta_h)(1 - \bar{s}t)\bar{s} + 2\theta_h t = (1 - \theta_h)(\bar{s} - t) + 2\theta_h\bar{s} \qquad (5.6)$$

which has the solution

$$\bar{s} = \frac{-\theta_h + [\theta_h^2 + (1 - \theta_h^2)t^2]^{1/2}}{(1 - \theta_h)t}. \tag{5.7}$$

One observes from Eq. (5.7) or also directly from Eq. (5.6) that $\bar{s} \to t$ for $\theta_h \to 1$. This means that the average $\bar{s}$ is the usual $t = \tanh \varepsilon/2$ due to the bias when the polymer grows with certainty at the head. Note also that for $t = 1$ (strong bias) the average $\bar{s} = 1$.

The result Eq. (5.7) is not supported by the simulations (see fig. 2) which show a rather large head structure both in the original model and in the fast extron limit. This implies that the replacement of θ_h by its average is incorrect.

We see that at the head-side $\bar{s}$ is indeed fixed by the equation. We still have to compute θ_L from the requirement Eq. (4.5) that the growing rate at the head equals the shrinking rate at the tail. Using Eq. (4.6) for R_t with the probability Eq. (5.3) one has

$$R_t = -\frac{1}{2}(B + B^{-1})\frac{1 - t^2}{1 - \bar{s}t}$$

and the value of R_h is computed with Eq. (5.5) and as

$$R_h = \frac{1}{2}(B + B^{-1})\left[-(1 - \theta_h)(1 - \bar{s}t) + 2\theta_h\right]. \tag{5.8}$$

The solutions for θ_h and $\bar{s}$ are

$$\bar{s} = \frac{1}{t} - \frac{1 - t^2}{t(\sqrt{4 + t^2} - 1)} \tag{5.9}$$

$$\theta_h = \frac{2}{1 + \sqrt{4 + t^2}} \tag{5.10}$$

For small bias t s behaves as $s \sim \frac{5}{4}t$ and $\theta_h \sim \frac{2}{3}$, as expected.

6 Discussion

In this paper we have tried to show that discretized versions of the polymer motion are instructive for the understanding in particular of the notion of reptation. In fact the models discussed here concentrate on reptation, leaving out e.g. collective sliding of the reptons as a means of motion.

The main feature that emerges is the existence of two limits: the weak field and the long polymer limit. In the weak field limit the polymer drifts with a velocity linearly proportional to the driving field. Its mobility is enhanced by self-avoidance effects which stretch the polymer a bit. The polymer has no definite orientation, and the end reptons form alternatingly the head or the tail of the polymer motion. In the long polymer limit the polymer acquires a definite orientation which we characterize by the average $\bar{s} = \langle s_{l+\frac{1}{2}} \rangle \neq 0$ for a link between cells l and $l+1$ deep in the middle of the polymer. At the tail side the polymer shrinks with a rate $|R_t|$ while it grows with equal rate R_h at the head side. The drift velocity is given by

$$v = awR_h\bar{s}.$$

Approximate expressions for $\bar{s}$ and R_L are given by Eq. (5.9) and Eq. (5.8) for the fast extron limit. The remarkable property of v in the long polymer limit is its independence of N. While real polymers are of course finite, the most interesting behavior is to be expected in the scaling limit where $\varepsilon \rightarrow 0$ and $N \rightarrow \infty$, such that combinations εN^σ stay finite. For the original Rubinstein-Duke model it is argued by Barkema et al. [8] that the appropriate power for cross-over behavior is $\sigma = 1$ while for the fast extron limit a more elaborate paper on this limit by two of us [9] suggests $\sigma = 1/2$. In the scaling limit we expect a phase transition between the weak field behavior and the long polymer behavior. The weak field phase has $\bar{s} = 0$ and $R_h = 0$ while the other phase (long polymers) has both quantities non-zero. The nature of the phase transition may be hard to detect as simulations, operating always on finite chains, suggest a smooth crossover for the drift velocity.

An unresolved problem is the behavior of self-avoiding long polymers. Simulations so far suggest that when head and tail do not interchange anymore (long polymer limit) the drift velocity is also zero, at least in $d = 2$.

Acknowledgment

The authors acknowledge a stimulating discussion with B. Widom.

References

1. M. Rubinstein, Phys. Rev. Letters **59**, 1946 (1987).
2. T.A.J. Duke, Phys. Rev. Letters **62**, 2877 (1989).
3. P.G. de Gennes, Scaling Concepts in Polymer Physics (Cornell Univ. Press, Ithaca, 1979).
4. B. Widom, J.L. Viovy and A.D. Desfontaines, J. Phys. (Paris) I 1, 1759 (1991).
5. A. Kooiman and J.M.J. van Leeuwen, J. Chem. Phys. **99**, 2247 (1993).
6. L.S. Lerman and H.L. Frisch, Biopolymers 21, 995 (1982).
7. J.A. Leegwater, submitted to Phys. Rev. Lett.
8. G.T. Barkema, J.F. Marko and B. Widom, Phys. Rev. E **49**, 5303 (1994). submitted (1994).
9. J.A. Leegwater and J.M.J. van Leeuwen, to be published.

Non-Equilibrium Fluctuations in Simple Fluids

James W. Dufty[*][1] *and M. Cristina Marchetti*[**][2]

[1] University of Florida, Gainesville, FL 32611, USA
[2] Syracuse University, Syracuse, NY 13244, USA

1 Introduction

Simple fluids provide a prototype for the study of dynamical processes in many-body systems. Twenty five years ago the focus of both theoretical and numerical efforts was on transport properties of states near equilibrium and fluctuations in the equilibrium state. The advances made since then in the theory of equilibrium time correlation functions and the associated linear kinetic equations are described elsewhere in these proceedings [Co]. At the same time there were parallel developments in the theory and application of stochastic processes. These focused on general properties of fluctuations, without the specific limitation to stationary states, in a variety of physical systems (e.g., electronic devices, lasers, chemical reactions, populations) [La]. The microscopic methods of kinetic theory and the mesoscopic methods of stochastic processes are largely complementary. The former addresses the problem of deriving the proper transport equations for the system, while the latter aims to compute the effects of fluctuations. A synthesis of these two approaches is required to develop a fundamental understanding of both transport and fluctuations for states outside the linear response regime [Kl, EC].

Far from equilibrium the Gibbs state is no longer relevant and characterization of the non-equilibrium state of interest poses a formidable challenge. Still, in many cases it is possible to demonstrate a common structural relationship between the kinetic theory for the non- equilibrium state (the transport equations) and the equations for multi-time fluctuations, without having to solve these equations. The objective here is to present and interpret this structure in the context of a general Markov process, and to illustrate it in more detail using kinetic theory for a low density gas and for a dense fluid of hard spheres.

The method of approach is to apply the many body methods developed to derive transport equations (cluster expansions) to obtain kinetic equations for a more general generating functional from which both transport and fluctuation

[*] Research supported by NSF PHY 9312723.
[**] Research supported by NSF DMR 9112330 and DMR 9217284.

properties can be obtained. Generating functionals are commonly used in applications of stochastic processes, but have seen less use in kinetic theory [Du,MD]. In the next section, the expected general structure of the equations for mean values and fluctuations is summarized. Next, the corresponding description for phase space transport and fluctuations is derived for a low density gas using standard cluster expansion techniques. For the special case of a fluid of hard spheres, an approximation is proposed in section 4 to account for the dominant class of higher density effects. These examples extend the familiar Boltzmann and (revised) Enskog equations for transport to an equivalent description of fluctuations

Fluctuations in non-equilibrium states reveal a wealth of phenomena not present in equilibrium states, such as long range correlations in both space and time, pattern formation and coarsening, instabilities and bifurcations. It is not the purpose to discuss these phenomena here but only to present the tools for their analysis. In the final section, two methods for practical application are noted. In the first, a kinetic model is used to replace the nonlinear Boltzmann equation and derived properties for fluctuations. In the second, the transport and fluctuation equations are considered for a lattice gas cellular automaton.

2 Generating Functional

To illustrate the expected inter-relationship of the transport equations and various fluctuation properties, we review some results for a general Markov process [DBM]. Let $\{a_\alpha(t)\}$ be a set of variables whose averages and fluctuations are of interest. A generating functional is defined in terms of a set of conjugate variables $\{\lambda_\alpha(t)\}$ according to,

$$G[\lambda] \equiv \ell n < U[\lambda] > \quad , \quad U[\lambda] \equiv \exp \int_0^T dt \lambda_\alpha(t) a_\alpha(t). \tag{1}$$

The brackets $< .. >$ denote the average of interest (e.g., over some specified initial state, stochastic degrees of freedom). The square brackets in (1) indicate a functional dependence on $\{\lambda_\alpha(t)\}$, which can be exploited to generate the average values and fluctuations over the interval $0 \leq t \leq T$,

$$A_\alpha(t) \equiv < a_\alpha(t) >= \frac{\delta}{\delta\lambda_\alpha(t)} G[\lambda] \mid_{\lambda=0}, \tag{2}$$

$$C_{\alpha_1\cdots\alpha_n}(t_1...t_n) \equiv < \delta a_{\alpha_1}(t_1)...\delta a_{\alpha_n}(t_n) > \tag{3}$$

$$= \frac{\delta^n}{\delta\lambda_{\alpha_1}(t_1)\cdots\delta\lambda_{\alpha_n}(t_n)} G[\lambda] \mid_{\lambda=0}, \tag{4}$$

with $\delta a_{\alpha_i}(t_i) = a_{\alpha_i}(t_i) - A_{\alpha_i}(t_i)$.

In general, the observables $\{a_\alpha\}$ do not form a complete set for an exact description of the dynamics. For a suitable choice of the $\{a_\alpha(t)\}$ a closed set of transport equations may, however, apply after some initial time interval, when the additional degrees of freedom have relaxed to negligible values. Such equations define the macroscopic dynamics for the system,

$$\frac{\partial}{\partial t} A_\alpha(t) + N_\alpha[t; A] = 0. \tag{5}$$

Here $N_\alpha[t; A]$ is a non-linear functional of $\{A_\alpha(t)\}$. An important special case is that for which $N_\alpha \to N_\alpha(A(t))$ is a time independent function of $\{A(t)\}$; in the following, for simplicity of the presentation, only this latter case is considered.

The additional degrees of freedom not represented by $\{a_\alpha\}$ give rise to fluctuations around the average dynamics governed by Eqs. (4). To obtain a description of such fluctuations, it is useful to relax the condition $\lambda = 0$ in (2) and write the more general transport equations,

$$\frac{\partial}{\partial t} A_\alpha[t; \lambda] + N_\alpha(A[t; \lambda]) = I_\alpha[t; \lambda], \tag{6}$$

$$A_\alpha[t; \lambda] \equiv\; <a_\alpha(t) U[\lambda]>\; /\; <U[\lambda]>. \tag{7}$$

Here $A_\alpha[t; \lambda]$ generalizes $A_\alpha(t)$ to $\lambda \neq 0$, with $A_\alpha(t) = A_\alpha[t; \lambda = 0]$, and $N_\alpha(A)$ are the same functions as in (5). The source terms $I_\alpha[t; \lambda]$(with the property $I[t; \lambda = 0] = 0$) represent the additional dynamics from all other degrees of freedom not described by the transport equations (5) alone. Equations (6) provide a unified basis for fluctuations and transport that makes explicit their interrelationship, as illustrated in the remainder of this section. *Our primary observation is that the kinetic theory methods developed to derive the transport equations (5) can be applied also to obtain the more general equations (6).* Explicit examples are given in sections 3 and 4 below.

The transport equation (5) follows from (6) with $\{\lambda_\alpha \to 0\}$; the fluctuations follow from successive derivatives with respect to $\{\lambda_\alpha\}$ at appropriate times, followed by $\{\lambda_\alpha \to 0\}$. For example, bilinear fluctuations at two different times are found in this way to obey the equations,

$$\frac{\partial}{\partial t} C_{\alpha\beta}(t, t_1) + \mathcal{L}_{\alpha\sigma}(A(t)) C_{\sigma\beta}(t, t_1) = 0, \tag{8}$$

$$\mathcal{L}_{\alpha\sigma}(A) \equiv \frac{\partial}{\partial A_\sigma} N_\alpha(A). \tag{9}$$

Equation (8) applies for $t - t_1 > t_c$, where t_c is a characteristic relaxation time for the additional "fast" degrees of freedom. The contribution from the source term $I_\alpha[t; \lambda]$ is assumed to vanish on this time scale (for the idealized case of a Markov process $t_c \to 0^+$). However, $I_\alpha[t; \lambda]$ *does* give a contribution to the equal time correlation functions which obey the equations,

$$\frac{\partial}{\partial t} C_{\alpha\beta}(t, t) + \mathcal{L}_{\alpha\sigma}(A(t)) C_{\sigma\beta}(t, t) + C_{\alpha\sigma}(t, t) \mathcal{L}_{\sigma\beta}(A(t)) = B_{\alpha\beta}(t), \tag{10}$$

$$B_{\alpha\beta}(t) = \left\{ \frac{\delta}{\delta A_\alpha(t)} I_\beta[t; \lambda] + \frac{\delta}{\delta A_\beta(t)} I_\alpha[t; \lambda] \right\}_{\lambda=0}. \tag{11}$$

Equations (5), (8), and (10) form a closed set of equations for the mean values of $\{a_\alpha(t)\}$ and bilinear fluctuations. Similar equations for all higher order fluctuations are obtained by further functional differentiation of (6) with respect to λ. For example, the three time correlation function obeys the equation,

$$\frac{\partial}{\partial t}C_{\alpha\beta\gamma}(t,t_1,t_2) + \mathcal{L}_{\alpha\sigma}(A(t))C_{\sigma\beta\gamma}(t,t_1,t_2)$$
$$+\mathcal{M}_{\alpha\sigma\nu}(A(t))C_{\sigma\beta}(t,t_1)C_{\nu\gamma}(t,t_2) = 0, \tag{12}$$

$$\mathcal{M}_{\alpha\sigma\nu}(A) \equiv \frac{\partial^2}{\partial A_\sigma \partial A_\nu}N_\alpha(A), \tag{13}$$

with the restriction that all three times are separated by more than the relaxation time t_c; it is straightforward to write down the corresponding equations when two or more of these times are equal, including then contributions from the source term $I[t;\lambda]$.

This is sufficient to illustrate the structure expected for a general theory of transport and fluctuations: a closed set of non-linear equations for the average values, and a closely related set of linear equations for the correlation functions. The generator for the linear dynamics, $\mathcal{L}$, depends on the solution to the transport equations and is given explicitly by the derivative of $N(A)$. Thus the simplest correlation functions obey a linear regression law obtained by linear perturbation of the transport equations about their instantaneous solution. For the special case of equilibrium states this is just Onsager's linear regression assumption (formulated sixty years ago!). In the more general context here it is seen that the linear equations for the correlation functions determine the local (linear) stability of the solution to the transport equation - if the eigenvalues of $\mathcal{L}(A(t))$ vanish for some values of $A(t)$ the corresponding solution to the transport equation is unstable. As expected, fluctuations grow at such an instability. The three time correlation function determined by (12) goes beyond linear stability to introduce a driving source that depends on the local curvature of $N(A)$ as well.

The solution to (8), for given $\{A_\alpha(t)\}$, depends on the initial condition $C_{\alpha\beta}(t_1,t_1)$ which in turn is determined from (10). Here the new feature is the source term $B_{\alpha\beta}(t)$. To interpret the latter, we note that in the case of a stochastic process with additive noise $B_{\alpha\beta}(t)$ is the amplitude of the covariance of the noise. Thus a qualitative interpretation is that it represents a first order measure of the intensity of effects due to the additional degrees of freedom. Near equilibrium, it is related to the temperature and appropriate transport coefficients (fluctuation-dissipation relation); more generally, it depends in detail on the non-equilibrium state considered.

The above structure has been shown to follow exactly for any Markov process [DBM]. This generality suggests that it should follow as well from a suitable kinetic theory analysis, without attempting to identify the specific Markov process involved. The first order problem is a microscopic determination of the generating functional (1). Instead, a more practical procedure is to derive an equation that determines its first functional derivative, equation (6). This can be done by applying the same microscopic methods developed to derive the closely related transport equations. In the next section this is illustrated for the prototype Boltzmann gas, analysed ten years ago, but never published in detail [MD]. Section 4 then suggests the phenomenological generalization of this result to higher densities for a fluid of hard spheres.

3 The Boltzmann Gas

Consider a simple atomic fluid at low density and let $f(x,t)dx$ denote the microscopic number of particles in a small volume in the single particle phase space (x denotes both position and velocity variables). The appropriate transport equation for the mean value $F[x,t;\lambda] =< f(x,t)U > / < U >$ is an approximation to the exact first equation of the BBGKY hierarchy (suppressing the λ dependence for notational simplicity),

$$\left\{\frac{\partial}{\partial t} + \mathbf{v}_1 \cdot \nabla_{r_1}\right\} F[x_1, t] = \int dx_2 \theta(12) F^{(2)}[x_1, x_2, t], \tag{14}$$

$$\theta(12) \equiv -\mathcal{F}(1,2) \cdot \{\nabla_{v_1} - \nabla_{v_2}\}. \tag{15}$$

Here, $F^{(2)}[x_1, x_2, t]$ is the joint distribution in phase space for two particles and $\mathcal{F}(1,2)$ is the force on particle one due to particle two. A closed transport equation is obtained by expressing $F^{(2)}[x_1, x_2, t] \equiv F^{(2)}[12, t]$ as a functional of $F[x_1, t] \equiv F[1, t]$,

$$F^{(2)}[12, t] \rightarrow \Phi[12, t; F]. \tag{16}$$

Formally, the possibility for such a functional relationship follows from the fact that both $F^{(2)}$ and F are functionals of λ, so that elimination of λ gives the desired functional relationship. The form of this functional is discovered by use of formal cluster expansions for both $F[1, t]$ and $F^{(2)}[12, t]$, representing contributions from successively larger correlated groups of particles. These series can be rearranged to express $F^{(2)}[12, t]$ as a functional expansion in $F[1, t]$, with terms ordered as increasing powers of the average density. To lowest order in the density the resulting functional is found to be,

$$\Phi[12, t; F] \rightarrow g[12, t] S_t F[1, t] F[2, t], \tag{17}$$

$$g[12, t] = \exp \int_0^T d\tau \int dx \lambda(x, \tau)[S_{t-\tau} - S_\tau][\delta(x - \bar{x}_1(\tau)) + \delta(x - \bar{x}_2(\tau))]. \tag{18}$$

Here $S_t \equiv S_{-t}(12) S_t^o(12)$ is the two particle scattering operator, where S_t and S_t^o generate the two particle dynamics with and without interactions, respectively. Also, $\bar{x}(t) \equiv S_t^o x$. An important property of S_t is that it reaches a constant limit for $t > t_c$ = collision time. Consequently, $g(12, t) \rightarrow 1$ if both $t - \tau$ and τ are greater than the collision time.

Substitution of (17) into (14) allows identification of the desired kinetic equation for $F[x, t]$ in the form (6)

$$\left\{\frac{\partial}{\partial t} + \mathbf{v}_1 \cdot \nabla_{r_1}\right\} F[1, t] = J[F, F] + I[1, t], \tag{19}$$

$$J[F, F] \equiv \int dx_2 T(12) F[1, t] F[2, t], \tag{20}$$

$$I[1, t] = \int dx_2 \theta(12) [g(12, t) - 1] S_\infty F[1, t] F[2, t]. \tag{21}$$

The limit of $t > t_c$ has been taken, and $T(12) \equiv \theta(1, 2) S_\infty$ is the two particle scattering operator. By standard transformations, $J[F, F]$ is recognized as the non-linear Boltzmann (-Bogoliubov) collision operator.

The source term $I[1, t]$ depends on $g(12, t) - 1$ which clearly vanishes for $\lambda = 0$. As expected, the macroscopic transport equation in this case is the non-linear Boltzmann equation. The source term also vanishes for $t - \tau > t_c$ and $\tau > t_c$ due to the limiting property of S_t in (18). Consequently, the functional derivative of $I[x, t]$ with respect to $\lambda(x', \tau)$ will vanish for $t - t_c > T$. For example, the two time correlation function equations become,

$$\frac{\partial}{\partial t_1} C(1, t_1; 2, t_2) + \mathcal{L}[13; F(t)] C(3, t_1; 2, t_2) = 0, \tag{22}$$

$$\mathcal{L}[12; F] \equiv \mathbf{v}_1 \cdot \nabla_{r_1} \delta(12) - \frac{\delta}{\delta F[2, t]} J[F, F]. \tag{23}$$

This same equation applies for all multi-time correlation functions with times differing by more than the collision time. However, the source term does contribute to correlation functions with one or more times equal. For example, the two time correlation function is found to be [MD],

$$\frac{\partial}{\partial t} C(1, t; 2, t) + \mathcal{L}[13; F(t)] C(3, t; 2, t) + \mathcal{L}[23; F(t)] C(1, t; 3, t) = B(12, t), \tag{24}$$

$$B(12, t) \equiv \mathcal{L}[13; F(t)] \delta(32) F[2, t] + \mathcal{L}[23; F(t)] \delta(13) F[1, t]$$
$$+ T(12) F[1, t] F[2, t] + \delta(12) J[F, F]. \tag{25}$$

It is straightforward to write down the corresponding equations for multi-time correlation functions at arbitrary times, equal or not.

Equation (19) is a compact representation of all transport and fluctuation properties in a low density gas; the derivation extends the Boltzmann approximation for transport to a corresponding approximation for fluctuations. The source term, $I[1, t]$, represents "noise" in the transport equations due to degrees of freedom other than the phase space densities, $f(x, t)$. At low density this noise arises from correlations induced by a single pair collision: uncorrelated pre-collision states become correlated post-collision states on a time scale of t_c. In (24) this noise appears in the source term B(12,t).

To understand the content of (24) consider first the equilibrium state. Then the last two terms on the right side of (25) vanish (detailed balance), and (24) is recognized as the fluctuation-dissipation relation for the Boltzmann gas [BZ,FU,BDD]. The same combination of linear operators, $\mathcal{L}$, occurs on the left and right sides of (24) and the solution is immediately recognized as $C(12) = \delta(12) F(1)$. The properties of $\mathcal{L}$ play no essential role in determining $C(12)$ at equilibrium. The situation is quite different for nonequilibrium stationary states. Then the last two terms of (25) do not vanish since detailed balance is violated by boundary conditions, non- conservative forces, etc. needed to produce the stationary state. The equal time correlation function $C(12)$ now depends sensitively on the spectrum of $\mathcal{L}$. It follows directly from (23) that this spectrum includes the hydrodynamic modes at long wavelengths. Also, it is easily verified that B(12) is orthogonal to the summational invariants $(1, \mathbf{v}, v^2)$. These facts lead to the conclusion that $C(12)$ has long range spatial correlations in nonequilibrium stationary states that are not present in equilibrium [KCD]. For similar reasons, the approach to the stationary state has a slow algebraic dependence in time. The origin of this behavior is the coupling of two hydrodynamic modes (the two $\mathcal{L}$ operators in (24)) by the source term $B(12)$. These effects occur in the kinetic theory of the transport equation only beyond the Boltzmann approximation. Here we see that they occur in the equal time correlation function equation even at the Boltzmann level of approximation.

4 The Hard Sphere Enskog Fluid

The systematic generalization of the Boltzmann equation to higher densities is fraught with difficulties [Co]. For the special case of hard spheres there is, however, a phenomenological kinetic equation that is remarkably accurate even in the dense fluid phase. The transport equation in its most refined form is called the revised Enskog equation [vBE]. As in the previous section, the "derivation" of this equation can be extended to include the test functions $\lambda(x, t)$ for a description of fluctuations in the same approximation.

The Enskog approximation starts with the first BBGKY hierarchy equation (14), and assumes that the functional dependence of $\Phi[12, t; F]$ on $F[1, t]$ can be represented by its form at $t = 0$. For initial non-equilibrium states without velocity correlations, this functional dependence can be determined in terms of the equilibrium free energy density functional for an inhomogeneous fluid; the density field of this functional is that for the true nonequilibrium state. An advantage of this approximation is that it is exact at $t = 0$, makes no apriori limitations on density or space scale, and entails only two particle dynamics. Its justification lies in the fact that the collision time for hard spheres is arbitrarily small (instantaneous momentum transfer for contact configurations), so that this short time approximation in fact represents times long compared to a collision time. As noted above, it works well in practice [AAY].

To describe fluctuations it is necessary to generalize the analysis by including the functional $U[\lambda]$. This leads to both velocity correlations and multi-particle dynamical effects even in the initial state at $t = 0$. We therefore extend the Enskog approximation to this case by neglecting all dynamical effects beyond two particles and all velocity correlations. The result is then,

$$\Phi[12, t; F] \rightarrow g[12, t]\chi[\mathbf{r}_1\mathbf{r}_2; n]\mathcal{S}_t F[1, t]F[2, t], \tag{26}$$

$$n(\mathbf{r}, t) = \int d\mathbf{v} F[x, t]. \tag{27}$$

This is the same as the low density Boltzmann result, except for the additional factor of $\chi[\mathbf{r}_1\mathbf{r}_2; n]$. In detail, this factor is the equilibrium pair correlation function of an inhomogeneous fluid whose density field is that for the exact nonequilibrium state. Thus the functional dependence on this nonequilibrium density is known from *equilibrium* statistical mechanics (it is determined from the second functional derivative of the equilibrium free energy density functional, for which good approximations are known even at liquid densities). Substitution of this functional into the BBGKY hierarchy equation (14) gives the revised Enskog kinetic theory,

$$\left\{\frac{\partial}{\partial t} + \mathbf{v}_1 \cdot \nabla_{r_1}\right\} F[1,t] = J_E[F,F] + I_E[1,t], \tag{28}$$

$$J_E[F,F] \equiv \int d x_2 T(12)\chi[\mathbf{r}_1\mathbf{r}_2; n]F[1,t]F[2,t], \tag{29}$$

$$I[1,t] = \int d x_2 T(12)[g(12,t) - 1]\chi[\mathbf{r}_1\mathbf{r}_2; n]F[1,t]F[2,t]. \tag{30}$$

Here $T(12)$ is the hard sphere scattering operator. Actually, there are several different scattering operators (forward or backward in time, and their adjoints) so the notation is not precise. Some care in this regard is required to derive the results quoted below.

The results (28)-(30) provide a description of transport and fluctuations in the dense hard sphere fluid. Setting $\lambda = 0$ in (28) gives the nonlinear revised Enskog transport equation. Functional differentiation with respect to λ gives correlation function equations or arbitrary order which are depend on the transport equation solution. For example, the two time non-equilibrium fluctuations are determined from the equation (22), with the linear operator $\mathcal{L}$ now given by,

$$\mathcal{L}_E[12; F] = \mathbf{v}_1 \cdot \nabla_{r_1}\delta(12) - \frac{\delta}{\delta F[2,t]}J_E[F,F]$$

$$\equiv \mathbf{v}_1 \cdot \nabla_{r_1}\delta(12) + \Lambda[12; F] + L_E[12; F], \tag{31}$$

$$L_E * C = -\int d x_2 d x_3 T(12)\chi[12; n]F(2,t)\{C(1) + C(3)\}, \tag{32}$$

$$\Lambda * C = -\int d x_2 d x_3 T(12)F(1,t)F(2,t)C(3)\frac{\delta}{\delta n(\mathbf{r}_3)}\chi[12; n]. \tag{33}$$

Here $L_E[12; F]$ is the Enskog collision operator linearized about the state, $F(1,t)$. In addition to this and the free streaming operator, $\mathcal{L}_E[12; F]$ also has a mean field contribution Λ coming from the functional derivative of χ with respect to the density. This represents an effective mean force due to the correlation with other particles, and is absent in the low density Boltzmann limit.

The initial condition for the two time correlation function requires the equal time correlation function. This is given by (24) with the Enskog form for $L[12; F]$ above, and with the Enskog "noise" term,

$$B(12,t) \equiv L_E[13; F(t)]\delta(32)F[2,t] + L_E[23; F(t)]\delta(13)F[1,t]$$

$$+ T(12)\chi[12; n]F[1,t]F[2,t] + \delta(12)J[F,F]. \tag{34}$$

As expected, only the collisional part of $\mathcal{L}_E[12; F]$ contributes to the noise. For the equilibrium state, this result agrees with that of [BDD].

The revised Enskog approximation is quite remarkable. For the equilibrium state, it allows quantitative calculation of fluctuations over all frequencies and wavevectors, with good accuracy from low density almost to freezing [AAY]. Its application to states far from equilibrium has not been exploited since solutions to the transport equation are limited. An exception is the case of uniform shear flow. Approximate "hydrodynamic" modes associated with $\mathcal{L}_E[12; F]$ for this case have been studied [LD, KN] to indicate an instability leading to a disorder-order transition at large shear rates.

For equilibrium fluctuations (22) with (31) is exact in the short time limit. However, for non-equilibrium states it is exact in the same limit only for states without velocity correlations. This might suggest a severe limitation in applications. However, the class of initial states considered in the Enskog approximation can be understood as an approximation to more general non-equilibrium states in the sense of information entropy. Consider a general state leading to the reduced single particle density $F(1, t)$. An approximate ensemble can be constructed such that the same single particle density is implied, by maximizing the information entropy functional subject to appropriate constraints. This leads to an ensemble with spatial correlations as in equilbrium, but with no velocity correlations. Thus the Enskog approximation should be applicable to those non-equilibrium states that can be represented approximately as functionals of their one particle reduced distribution function alone. Of course, this means that the velocity correlations should be small.

It is possible to extend the Enskog approximation to include velocity correlations in the initial state exactly. Then $\chi[\mathbf{r}_1\mathbf{r}_2; n] \rightarrow \chi[x_1, x_2; F]$ is the exact pair correlation function for the non-equilbrium state considered, and equation (22) with (31) is exact in the short time limit. In fact, this is the form used in [LD] while the above form is used in [KN]. However, in general the pair correlations in the non-equilbrium state are not known. They can be determined from (24) and (34). If the exact pair correlations are used to determine χ then these equations become identically the stationary BBGKY equations. If velocity correlations are neglected in χ, then (24) and (34) define approximate equations to determine these correlations. It is not known yet if these equations have solutions (except at equilibrium) or what is their relative accuracy.

5 Practical Applications

The above examples show that kinetic theory methods can be applied to give a unified description of transport and fluctuations far from equiliibrium. The content of these equations is still hidden, however, by the difficulty of solving the non-linear transport equation (e.g., the non-linear Boltzmann or Boltzmann-Enskog equation). This is the primary problem since the correlation function equations, while linear, are functionals of the solution to the transport equation.

We indicate here two ways in which the transport and fluctuation equations can be modelled for practical applications.

5.1 Kinetic Model

There are very few solutions to the non-linear Boltzmann equation, particularly for spatially inhomogeneous states. However, some progress has been possible by replacing the Boltzmann collision operator with a single relaxation time kinetic model [BGK],

$$J[F, F] \rightarrow -\nu \left\{ F(x, t) - F_\ell(x, t) \right\}, \qquad (35)$$

$$\left\{ \frac{\partial}{\partial t} + \mathbf{v}_1 \cdot \nabla_{r_1} \right\} F(1, t) = -\nu \left\{ F(x, t) - F_\ell(x, t) \right\}. \qquad (36)$$

Here ν is an effective collision frequency and $F_\ell(x, t)$ is a local equilibrium distribution depending on five local fields. The latter are fixed by requiring that the integral of $J[F, F]$ times $1, \mathbf{v}$, and v^2 vanish. This assures that the kinetic model retains the conservation laws, and consequently describes as well the rich hydrodynamic behavior derived from them. The local equilibrium distribution then becomes a functional of F through these fields. In recent years several exact solutions to the Boltzmann equation with this approximate collision operator (BGK equation) have been obtained [Du2]. They represent a variety of problems involving diffusion, and heat and momentum transport far from equilibrium.

The extension of the BGK kinetic model to fluctuations follows from the structural relations implied by the generating functional method [DLB]. The non-equilibrium pair fluctuations obey equations (22) and (24), but with the kinetic models for $\mathcal{L}[1; F]$ and $B(12)$,

$$\mathcal{L}[1; F] \rightarrow \left\{ \mathbf{v}_1 \cdot \nabla_{r_1} + \nu(1 - P_{1\ell}) \right\} \delta(12), \qquad (37)$$

$$B(12) \rightarrow \nu(1 - P_{1\ell})(1 - P_{2\ell}) \left\{ \delta(12) F(1, t) + \delta(12) F_\ell(1, t) \right\}. \qquad (38)$$

Here P_ℓ is the local equilibrium projection operator onto the conserved densities,

$$PX(x) = F_\ell(x, t) \psi_\alpha(\mathbf{v}) g_{\alpha\beta}^{-1}(\mathbf{r}, t) \int d\mathbf{v}_1 \, \psi_\beta(\mathbf{v}_1) X(\mathbf{r}, \mathbf{v}_1), \qquad (39)$$

$$g_{\alpha\beta}(\mathbf{r}, t) = g_{\alpha\beta}(\mathbf{r} | F(t)) \equiv \int d\mathbf{v} \, F_\ell(x, t) \psi_\alpha(\mathbf{v}) \psi_\beta(\mathbf{v}), \qquad (40)$$

and $\psi_\alpha(\mathbf{v}) \equiv \left\{ 1, \mathbf{v}, v^2 \right\}$.

These kinetic model equations retain all the relevant physical properties of the Boltzmann description, but allow more practical applications. For example, it is straightforward to include boundary conditions or external forces driving the non-equilibrium state. Finding the solution to (36) is still non-trivial in general, but considerably simpler than the Boltzmann equation itself. Once a particular solution is found, the analysis of the linear fluctuation equations is straightforward. Application of this kinetic model for transport and fluctuations in uniform shear flow far from equilibrium is in progress.

5.2 Lattice Gas Cellular Automata

A second example for practical applications is a lattice gas cellular automaton (LGCA). Here both space and time are discretized. The microscopic phase point becomes $x \rightarrow (\mathbf{r}, \mathbf{c}_i)$, where $\mathbf{r}$ label lattice sites and $\mathbf{c}_i$ label velocity states. Then $f(x, t) \rightarrow f_i(\mathbf{r}, t)$ is the microscopic occupation number (one or zero) for the velocity state $\mathbf{c}_i$ at node $\mathbf{r}$. The microdynamics is a discrete map of the set of these occupation numbers onto themselves at each time step. This map is specified by a propagation step and a transition matrix for velocity changes ("collisions"). Here we consider only the case of local collision rules, for which the transition matrix factorizes over all sites (i.e., independent transitions at each site). It is then possible to discuss the transport equation for the dynamics of the average occupation numbers, $F_i(\mathbf{r}, t) \equiv < f_i(\mathbf{r}, t) >$, and fluctuations in these occupation numbers. As for continuous fluids, a lattice gas Boltzmann equation is obtained when all fluctuations are neglected,

$$F_i(\bar{\mathbf{r}}, t+1) - F_i(\mathbf{r}, t) = J_i[F]. \tag{41}$$

The lattice site $\bar{\mathbf{r}} \equiv \mathbf{r} + \mathbf{c}_i$ results by propagation from $\mathbf{r}$ when the velocity state $\mathbf{c}_i$ is occupied. Thus the left side of (41) is the analog of that in (19). The right side represents the collisions. Due to a "Fermi exclusion rule" whereby no more than one particle is allowed in a given velocity state at the same node, there is a higher order non-linear dependence of $J_i[F]$ on F than for continuous fluids. Still, the lattice Boltzmann equation is a discrete equation that admits efficient numerical solution. For fluid-type LGCA the lattice Boltzmann equation yields complex hydrodynamics behavior on sufficiently long space and time scales. Study of non-equilibrium states via numerical solution of (41) is currently a very active field.

The equations for the pair fluctuations at one and two times is analogous to those for the continuous fluid case [BED],

$$C_{ij}(\bar{\mathbf{r}}_1, t_1 + 1; \mathbf{r}_2, t_2) - (1 - L[F])_{ik} C_{kj}(\mathbf{r}_1, t_1; \mathbf{r}_2, t_2) = 0, \tag{42}$$

$$L_{ij}[F] = -\frac{\partial}{\partial F_j} J_i[F], \tag{43}$$

$$C_{ij}(\bar{\mathbf{r}}_1, t+1; \bar{\mathbf{r}}_2, t+1) - (1 - L[F])_{ik} (1 - L[F])_{jl} C_{kl}(\mathbf{r}_1, t; \mathbf{r}_2, t)$$
$$= B_{ij}(\mathbf{r}_1, \mathbf{r}_2, t). \tag{44}$$

These are the discrete space-time analogues of equations (22) - (24). The form of $B_{ij}(\mathbf{r}_1, \mathbf{r}_2, t)$ is complex and will not be given in detail here (see [BED]) It has the same interpretation as (24); there is a part expected from the equilibrium fluctuation-dissipation relation, plus terms that contribute only for non-equilbrium states. Recently, these equations have been applied to a LGCA with local collision rules that violate detailed balance. The existence of a non-Gibbs stationary solution was established numerically, and the velocity correlations calculated from (44). Excellent agreement with direct computer simulation of the microdynamics was found [BED]. This example illustrates the practical utility of LGCA for the study of complex non-equilibrium states.

6 Discussion

We have reviewed a means to address transport and fluctuations in simple fluids from a unified point of view. The idea is to study the generating functional from which all properties of interest can be derived. A direct microsopic analysis of this generating functional is quite difficult. However, its first functional derivative with respect the test fields obeys an equation similar to the macroscopic transport equation. There are well-developed many-body methods available to derive such equations that can be brought to bear on this more general functional. Thus, instead of attempting to determine the generating functional itself, an equation that determines its first functional derivative is sought. While the solution to this equation is often difficult, it is not needed to obtain corresponding equations for the fluctuations. In this way a closely related set of transport and fluctuation equations is obtained, which form the basis for subsequent calculation of properties of interest. In the last section, we have indicated some practical methods that can be used to solve these equations.

Attention here has been restricted to simple fluids. The Boltzmann approximation has been extended to discuss fluctuations of arbitrary order. Non-trivial applications of this result include the calculation of non-linear response functions [DM]. For comparison with computer simulations, a corresponding theory of fluctuations at higher density is required. The Enskog approximation presented here for the hard sphere fluid appears to be the only case for which higher order fluctuations can be calculated. Few applications however have been attempted to date.

The method of analysis illustrated by these examples can be extended to other systems as well. For example, the cluster expansion method can be adapted to degenerate quantum gases. The corresponding quantum Boltzmann equation can be extended to obtain the generator for quantum fluctuations, just as in section 3 above. In another direction, the Enskog approximation can be extended to the crystal phase as well as the fluid phase. It was noted recently [KDEP] that the Enskog transport equation includes both fluid and crystal equilibrium solutions, if a sufficiently accurate model for the functional $\chi[\mathbf{r}_1\mathbf{r}_2; n]$ is used. Of particular interest would be to study the dynamics of fluctuations in metastable fluid states, with asymptotic approach to the stable crystal phase.

The description considered here is mesoscopic, at the level of the single particle phase space. These equations imply a corresponding macroscopic hydrodynamic description [MD]. Thus the stucture of transport and fluctuations can be contracted to this level whenever the phenomenon occur on sufficiently long space and time scales.

The formalism is consistent with that for a Markov process, as indicated in section 2. However, the analysis is based on appropriate approximations to the underlying Liouville dynamics and it is not necessary to specify or justify a particular process. In fact, the limitations on time scales indicated in section 3 shows that the phenomena are not strictly Markov. Also, higher order equal time correlations are generated from the source term in the transport equation. There is no a priori assumption of Gaussian, Poisson, or other specific statistics.

The "noise" is fixed by the approximation leading to the transport equation. In both the Boltzmann and Enskog cases here, all properties of the noise are due to correlations induced by a single binary collision.

The examples considered here represent transport equations in a "mean field" approximation. In general the fluctuations determined at this same level of approximation lead to renormalization of the transport equations. In some cases it is possible to write a self-consistent set of equations whereby such fluctuation effects on the transport equations is included [EC, BED].

References

[AAY] W. Alley, B. Alder, and S. Yip, Phys. Rev. **A27** (1983) 3174.

[vBE] H. van Beijeren and M. Ernst, Physica *68* (1973) 437; J. Stat. Phys. **21** (1979) 125

[BGK] P. Bhatnager, E. Gross, abd M. Krook, Phys. Rev. **94** (1954) 511.

[BDD] M. Bixon, J. R. Dorfman, and J. Dufty, J. Phys. Chem. **93** (1989) 7019.

[BZ]   M. Bixon and R. Zwanzig, Phys. Rev. **187** (1969) 267.

[BED] H. Bussemaker, M. Ernst, and J. W. Dufty, J. Stat.Phys. (to be published).

[Co]   E. G. D. Cohen (this volume).

[Du]   J. W. Dufty, Phys. Rev. **A13** (1976) 2299.

[Du2] For a recent review, see J. W. Dufty in *Lectures on Thermodynamics and Statistical Mechanics*, M. Lopez deHaro, ed. (World Scientific, N.J., 1990).

[DBM] J. W. Dufty, J. J. Brey, and M. C. Marchetti, Phys. Rev. **A13** (1986) 4307.

[DLB] J. W. Dufty, M. Lee, and J. Brey, Phys. Rev. E (to be published).

[DM]   J. W. Dufty and M. Marchetti, J. Chem. Phys. **75** (1981) 422.

[EC]   For extensive references see M. H. Ernst and E. G. D. Cohen, J. Stat. Phys. **25** (1981) 153.

[FU]   R. Fox and G. Uhlenbeck, Phys. Fluids **13** (1970) 1893.

[KCD] T. Kirkpatrick, E. G. D. Cohen, and J. R. Dorfman, Phys. Rev. **26** (1982) 950.

[KDEP] T. Kirkpatrick, S.Das, M. Ernst, J. Piasecki, J. Chem. Phys. **92** (1990) 3768.

[KN]   T. Kirkpatrick and J. Nieuwoudt, Phys. Rev. Lett. **56** (1986) 885.

[Kl]   A synthesis of kinetic and stochastic methods is accomplished in the Klimontivich formulation; see Yu. Klimontovich, *The Statistical Theory of Nonequilibrium Processes in Plasma* (Pergamon, Oxford, 1967).

[La]   See the series on classical and quantum noise at this time by M. Lax; of particular interest here is M. Lax, Rev. Mod. Phys. **38** (1966) 359.

[LD]   J. Lutsko and J. Dufty, Phys. Rev. Lett. **56** (1986) 1571.

[MD]   M. C. Marchetti and J. W. Dufty, Physica **118A** (1983) 205; M. C. Marchetti, PhD thesis (1982) University of Florida.

Long time tails in stress correlation functions

M. H. J. Hagen   C. P. Lowe

D. Frenkel

FOM Institute for Atomic and Molecular Physics

Kruislaan 407

1098 SJ Amsterdam

The Netherlands

Abstract

The algebraic long time tail of the stress correlation function is observed in a simple lattice Boltzmann model. The amplitude of this tail is compared with the mode coupling expression for the long time tail in the stress correlation function. Agreement is found between mode coupling theory and simulation in both two and three dimensions.

1   Introduction

In 1970 Alder and Wainwright [1] reported the results of a computer simulation study of the decay of velocity fluctuations in a hard-sphere fluid. These simulations revealed that velocity fluctuations do not decay exponentially, as had been previously assumed, but algebraically. This observation was of great importance because non-exponential decay of the velocity autocorrelation function (velocity ACF) is not compatible with Boltzmann's 'molecular chaos' hypothesis, i.e. the assumption that there is no correlation between the velocity of a particle at time t and the velocity of its collision partners at any later time.

Subsequently, mode coupling [2] and kinetic theories [3] were developed to provide a theoretical framework for the description of long time tails in correlation functions. Both classes of theory reproduce the algebraic decay of the velocity ACF $\phi(t) \sim t^{-d/2}$, where d is the dimensionality of the fluid, and t the time. In addition, the same theories also predict an algebraic long time tail in the stress correlation function. The mode-coupling theory prediction for the asymptotic form of the stress autocorrelation function is [2]:

$$\phi_{xy}(t) = \frac{1}{\rho d(d+2)} \left(\frac{d^2 - 2}{(8\pi\nu t)^{d/2}} + \frac{1}{(4\pi\Gamma t)^{d/2}} \right) \equiv \frac{d_0}{t^{d/2}} \qquad (1)$$

In this equation, $\phi_{xy}(t)$ is the correlation function for the xy-component of the stress-tensor, ρ is the number density, ν is the "bare" kinematic viscosity, and Γ is the sound wave damping coefficient. Unlike the long-time tail in the velocity ACF, the algebraic tail in the stress correlation function has thus far not been observed directly neither in simulations nor in experiment, except in a very simple one dimensional model [4] that does not really correspond to a fluid.

In the case of the velocity autocorrelation function, the most accurate numerical results were obtained in simulations of a simplified model for an atomic fluid, namely a lattice-gas cellular automaton of the type introduced by Frisch, Hasslacher and Pomeau [5]. By exploiting some of the special features of the lattice gas, Frenkel and Ernst [6] computed the velocity ACF of a tagged particle with an accuracy that was at least four orders of magnitude better than was hitherto possible. For the velocity ACF of a tagged particle in a lattice gas, it proved possible to perform an average over *all* possible labelings of the tagged particle. In contrast, no such averaging can be performed in the case of the stress, which is a collective, rather than a single-particle property. As a consequence, the stress correlation function is very noisy. It would seem attractive to try to improve the statistics of the stress ACF by performing some kind of pre-averaging that does reduce the statistical fluctuations but not the way in which stress decays in the lattice-gas fluid. A natural pre-averaged version of a lattice gas cellular automaton fluid is the so-called lattice-Boltzmann model introduced by McNamara et al. [7, 8]. The advantage of the lattice Boltzmann model is that one can study the decay of an initial perturbation of the stress without any statistical noise. The disadvantage is that, due to the pre-averaging, it is no longer a truly atomistic model. Moreover, the pre-averaging has killed all *spontaneous* fluctuations. Hence the way to study the stress ACF is not to watch the decay of spontaneous fluctuations in the stress (there are none), but to make use of Onsager's regression hypothesis and study the decay of an *imposed* perturbation of the stress. In this paper, we report calculations of the stress ACF, using a lattice-Boltzmann model.

At first sight, it may seem strange to look for long-time tails in a Boltzmann model. After all, in the Boltzmann equation that determines the time evolution of this lattice model, one ignores the correlations between successive collisions that, in the kinetic theory description, give rise to long-time tails. Yet, the lattice-Boltzmann model does reproduce the hydrodynamic behavior of a fluid. In the mode-coupling theories of long-time tails in simple fluids, it is precisely the slow decay of hydrodynamic modes that is responsible for the appearance of long-time tails (that are, for this reason, often referred to as *hydrodynamic* long-time tails).

2 Lattice-Boltzmann model

The lattice-Boltzmann model is a pre-averaged version of a lattice-gas cellular automaton (LGCA) model of a fluid. In lattice-gas cellular automaton the state of the fluid at any (discrete) time is specified by the number of particles at every lattice site and their velocity. Particles can only move in a limited number of directions (towards neighboring lattice points) and there can be at most one particle moving on a given 'link'. The time evolution of the LGCA consists of two steps - 1. Propagation: every particle moves in one time step, along its link to the next lattice site. 2. Collision: at every lattice site particles can change their velocities by collision, subject to the condition that these collisions conserve number of particles and momentum (and retain the full symmetry of the lattice). In the lattice-Boltzmann method (see e.g. [9]) the state of the fluid system is no longer characterized by the number of particles that move in direction c_i on lattice site r, but by the *probability* to find such a particle. The single-particle distribution function $n_i(r,t)$, describes the average number of particles at a particular node of the lattice r, at a time t, with the discrete velocity c_i. The hydrodynamic fields, mass density ρ, momentum density j, and the momentum flux density Π are simply moments of this velocity distribution:

$$\rho = \sum_i n_i, \qquad j = \sum_i n_i c_i, \qquad \Pi = \sum_i n_i c_i c_i. \tag{2}$$

The lattice model used in this work is the four dimensional Face-Centered Hyper Cubic (FCHC) lattice. A two or three dimensional model can then be obtained by projection in the required number of dimensions. This FCHC model is used because three-dimensional cubic lattices do not have a high enough symmetry to ensure that the hydrodynamic transport coefficients are isotropic.

The time evolution of the distribution functions n_i is described by the discretized analogue of the Boltzmann equation [10]:

$$n_i(r + c_i, t + 1) = n_i(r, t) + \Delta_i(r, t), \tag{3}$$

where Δ_i is the change in n_i due to instantaneous molecular collisions at the lattice nodes. The post-collision distribution $n_i + \Delta_i$ is propagated in the direction of the velocity vector $\mathbf{c}_i$. A complete description of the collision process is given in [11]. The main effect of the collision operator $\Delta_i(\mathbf{r}, t)$ is to relax the non-equilibrium part of the momentum flux. The full, non-linear expression for equilibrium part of the local momentum flux density $\mathbf{\Pi}^{eq}$ is given by:

$$\mathbf{\Pi}^{eq} = p\mathbf{I} + \rho\mathbf{uu}, \tag{4}$$

with p the local pressure, $\mathbf{I}$ the unit tensor, and $\mathbf{u}$ the local fluid velocity. In the linearized version the equilibrium part of the momentum flux density is given by:

$$\mathbf{\Pi}^{eq} = p\mathbf{I}. \tag{5}$$

The rate of stress relaxation, or equivalently, the kinematic viscosity ν, can be chosen freely. In the linear lattice-Boltzmann model (eq. 5) Π_{xy} can only decay exponentially. To observe the long time behavior of the stress ACF, a coupling to the momentum is essential. The second term in eq. (4) which is usually only taken into account to study high Reynolds number flow, does exactly this. In order to observe the long time tail in the stress ACF, the full non-linear stress tensor had to be used in the simulation.

As the lattice-Boltzmann model is purely dissipative, microscopic fluctuations in the fluid are not included. Such fluctuations can be incorporated in the lattice-Boltzmann model by adding a suitable random noise term to the stress [12]. However, for the present work, such fluctuations are not essential for the phenomenon under study yet would seriously deteriorate the statistical accuracy of our calculations.

The stress in the system, which is a collective property is given by:

$$\Sigma = \sum_{\mathbf{r}} (\mathbf{\Pi} - p\mathbf{I}). \tag{6}$$

For the sake of convenience, we consider only one component of the traceless symmetric part of the stress tensor *viz.* the xy component. Other components give rise to the same correlation functions. We compute the stress ACF by correlating the initial perturbation of the stress with the stress at some later time t:

$$\phi_{xy}(t) = \frac{\langle \Delta\Sigma_{xy}(0)\Delta\Sigma_{xy}(t)\rangle}{\langle(\Delta\Sigma_{xy}(0))^2\rangle}, \tag{7}$$

where $\Delta\Sigma_{xy} = \Sigma_{xy}(t) - \Sigma_{xy}(\infty)$. It is important to subtract the steady state ($t = \infty$) value of the stress tensor because, in a finite system, the initial stress perturbation will relax to a uniform velocity field with an associated stress given by eq. (4):

$$\Sigma_{xy}(\infty) = \frac{\Sigma_{xy}(0)}{\rho V}, \tag{8}$$

where V is the volume of the fluid. In a LGCA, where the stress is purely kinetic in origin, the stress at site r at time t is uncorrelated to the stress at that same time at any other lattice point. In our calculations we have therefore chosen to consider the simplest possible initial condition *viz.* a small perturbation of the stress *at one lattice site only.*

3   Results

Having set up the system with an initial local stress perturbation, we followed the time-evolution of the total stress of the system using the dynamics of the lattice-Boltzmann model. In fact, we did the simulations both for the linearized and the, non-linear expression for the stress tensor.

In order to be able to compare the tail amplitude as obtained from the simulations with the theoretical expression (Eq. (1)) we need to know the sound damping coefficient Γ. At the Boltzmann level, this quantity is given by: $\Gamma = 2(d-1)\nu/d + \zeta$ with ζ the kinematic bulk viscosity. ν and ζ were 'measured' by setting up a sound wave in the system and measuring the decay of that wave in the long wavelength limit [14].Γ, ζ and ν were computed for a range of imposed kinematic viscosities between 0.01 and 0.50.

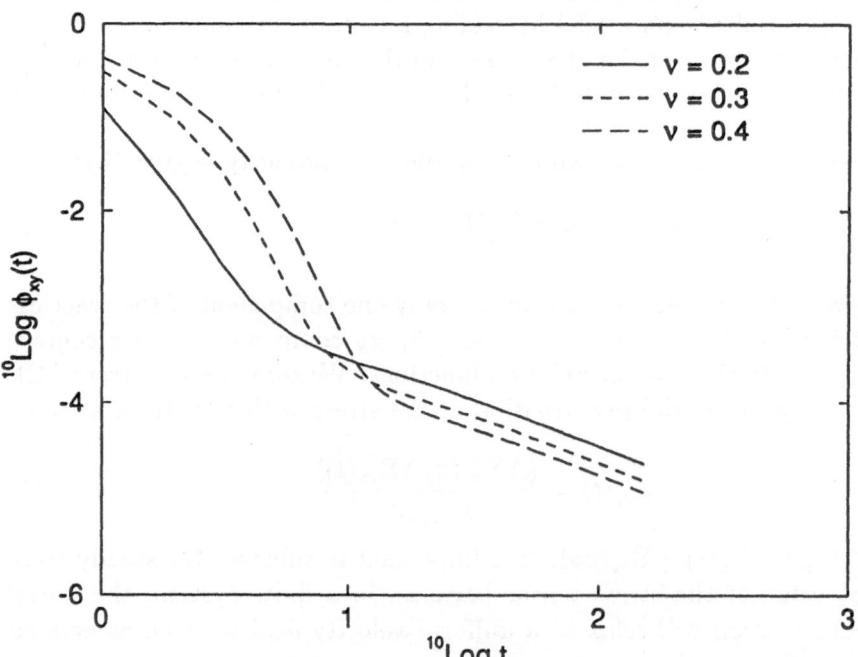

Figure 1: The stress autocorrelation function of a two dimensional lattice-gas fluid.

The simulations in two dimensions were performed on a system of 250 × 250 lattice sites. For this size of simulation box we followed the stress ACF for 140 time steps. This upper limit was chosen because, after this time, interference occurred due to sound waves that cross the periodic system. In figure 1, we show the stress ACF of the lattice Boltzmann model for several different values of the kinematic viscosity and the non-linear expression for the stress. For this model, we do indeed observe a clear algebraic decay of the stress ACF. As expected for a two-dimensional fluid, the exponent of the algebraic long-time tail was − 1. In contrast, *no* algebraic tail is observed if the non-linear terms in the stress are ignored. This is understandable because in the linearized model there is no mechanism by which the different modes can couple.

The limiting value d_0 was determined by plotting $t\phi_{xy}(t)$ as a function of $1/t$. The intercept for $1/t=0$ yields the desired amplitude. The results of this analysis are shown in figure 2. In this figure, we also show the theoretical tail coefficient given by Eqn. (1). Figure 2 shows that the mode-coupling predictions of the tail coefficient are in almost quantitative agreement with

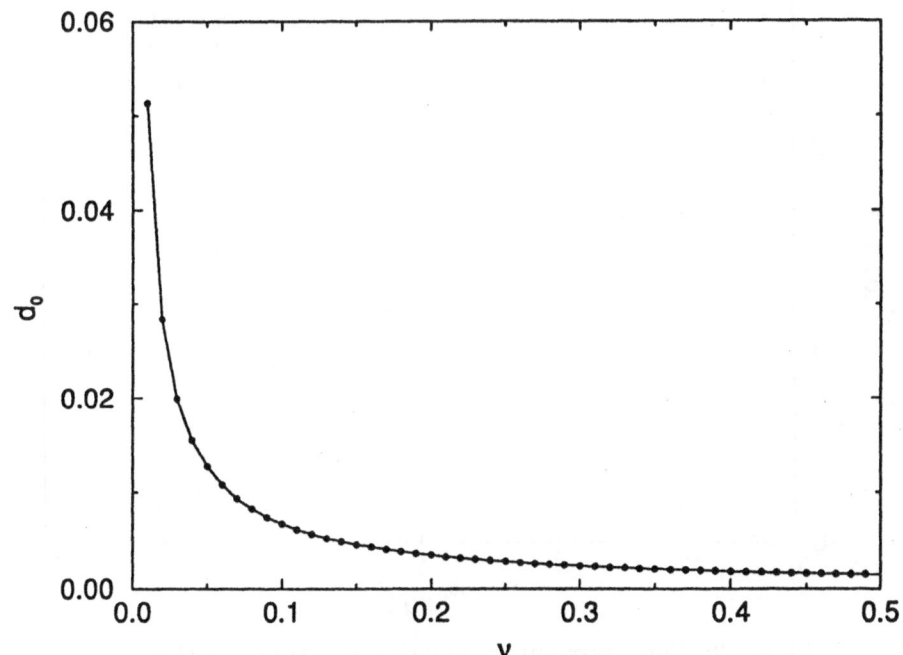

Figure 2: The tail coefficient for a two dimensional lattice gas fluid, as a function of the kinematic viscosity ν. The points are the results of simulations of the lattice-Boltzmann model, while the drawn curve corresponds to the prediction of mode coupling theory.

the simulation results. In fact, there is a very small discrepancy between the mode-coupling theory predictions and the simulation results. However, this discrepancy is consistent with inaccuracy in the determination of d_0.

The simulations in three dimensions were done for a system 90x90x90 lattice sites. For this size simulation box we followed the stress ACF for times up to $t = 50$. After this time interference due to the round-trip of sound waves occurred. In the three-dimensional fluid, we also find algebraic decay of the stress ACF in the non-linear lattice-Boltzmann model only. The algebraic tail is characterized by an exponent -1.5, as expected. Figure 3 shows the stress ACF of the $3D$ lattice Boltzmann fluid for several values of the kinematic viscosity.

We performed almost the same extrapolation procedure as described above to determine the amplitude of the long-time tail. Specifically, we plotted $t^{3/2}\phi_{xy}(t)$ as a function of $1/t$. As before, the amplitude of the algebraic tail d_0 is obtained from the intercept of $t^{3/2}\phi_{xy}(t)$ in the limit $1/t \to 0$.

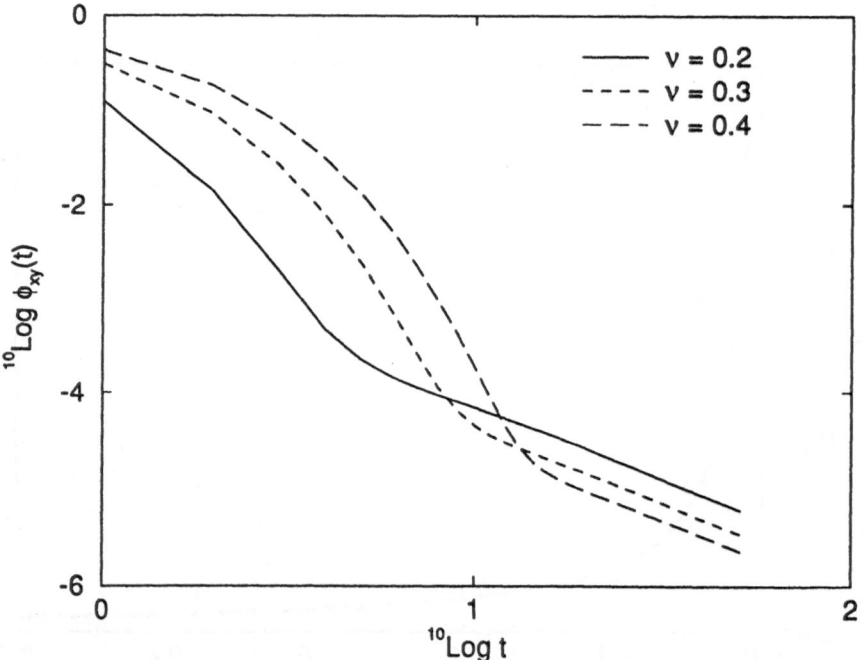

Figure 3: The stress autocorrelation function of a three-dimensional lattice-gas fluid.

Figure 4 shows a comparison of the tail coefficient obtained from the simulations, with the corresponding mode-coupling prediction (Eqn. (1)). As can be seen from this figure, there is again almost quantitative agreement between mode-coupling theory and the simulation results. The small discrepancy is caused by the short simulation time and is expected to disappear if longer simulations in a larger system could be performed.

In the three dimensional system we have also computed to what extent the long-time tail in the stress ACF changes the "bare" kinematic viscosity ν which was computed at the Boltzmann level. This is done by using the Green-Kubo formula for the viscosity [15]:

$$\nu_{\text{hydro}}(t) \sim \frac{1}{2}\phi_{xy}(0) + \sum_{t'=1}^{t} \phi_{xy}(t').$$ (9)

Asymptotically, $\nu_{\text{hydro}}(t) \sim t^{-1/2}$ (from Eq. (1)), and in this way extrapolation was performed to find $\nu_{\text{hydro}} = \lim_{t\to\infty} \nu_{\text{hydro}}(t)$.

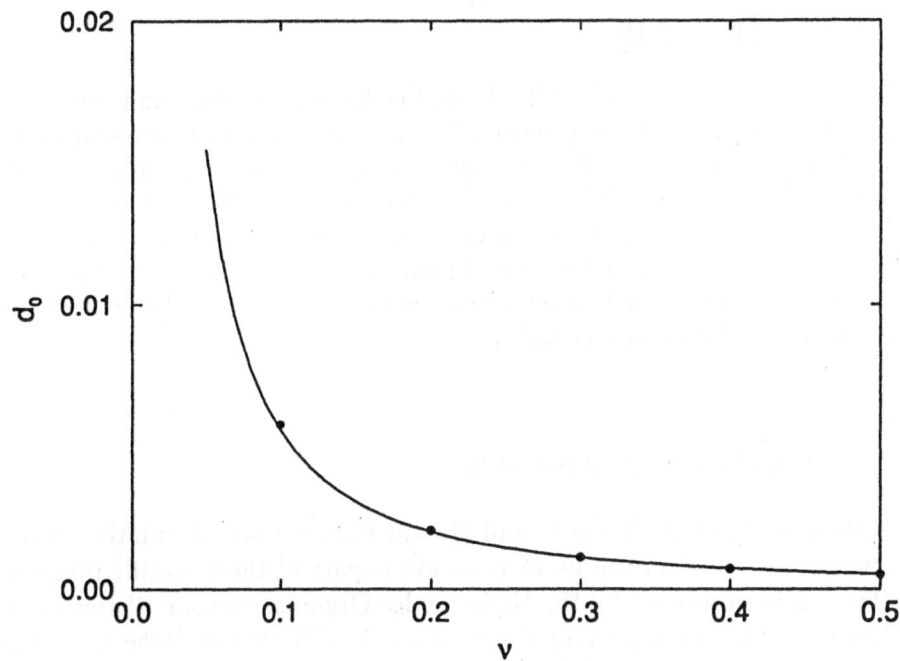

Figure 4: The tail coefficient for a three dimensional lattice-gas fluid, as a function of the kinematic viscosity ν. The points are the results of simulations of the lattice-Boltzmann model, while the drawn curve line corresponds to the prediction of mode coupling theory.

The result of this calculation is shown in table 1. Note that the algebraic

ν	$(\nu_{\text{hydro}} - \nu)/\nu$
0.1	0.171
0.2	0.079
0.3	0.053
0.4	0.041
0.5	0.033

Table 1: The relative effect of the hydrodynamic long time tail on the viscosity in three dimensions.

long-time tail results in a small re-normalization of the viscosity. The same calculation was not performed in two dimensions, because the re-normalized kinematic viscosity diverges in that case.

4   Conclusions

We have computed the stress ACF of a lattice Boltzmann fluid and compared the results with mode coupling theory. We find that both the exponent of the algebraic long time tail $(-d/2)$ and the its amplitude (d_0) are in essentially quantitative agreement with mode coupling theory. The computation of the time dependent viscosity in three dimensions shows that, at least for the simple lattice-gas model studied in this work, the hydrodynamic long-time tail of the stress ACF results in a small correction to the Boltzmann prediction of the kinematic viscosity.

5   Acknowledgements

We gratefully thank M.H. Ernst and H. van Beijeren for stimulating correspondence. The work of the FOM Institute is part of the scientific program of FOM and is supported by the Nederlandse Organisatie voor Wetenschappelijk Onderzoek (NWO). Computer time on the CRAY-C98/4256 at SARA was made available by the Stichting Nationale Computer Faciliteiten (Foundation for National Computing Facilties).

References

[1] B. J. Alder and T. E. Wainwright, Phys. Rev. A. **1**, 18 (1970).

[2] M. H. Ernst, E. H. Hauge and J. M. J. van Leeuwen, Phys. Rev. Lett. **25**, 1254 (1970).

[3] J. R. Dorfman and E. G. D. Cohen, Phys. Rev. Lett. **25**, 1257 (1970).

[4] T. Naitoh and M. H. Ernst, Molecular Simulation. **12**, 197 (1994).

[5] U. Frisch, B. Hasslacher and Y. Pomeau, Phys. Rev. Lett. **56** 1505 (1986).

[6] D. Frenkel and M. H. Ernst, Phys. Rev. Lett. **63** 2165 (1989).

[7] G. R. McNamara and G. Zanetti, Phys. Rev. Lett. **61**, 2332 (1988).

[8] F. Higuera, S. Succi, and R. Benzi, Europhys. Lett. **9**, 345 (1989).

[9] G. R. McNamara and B. J. Alder, in *Microscopic Simulation of Complex Hydrodynamic Phenomena*, edited by M. Mareschal and B. L. Holian (Plenum, New York, 1992).

[10] U. Frisch, D. d'Humières, B. Hasslacher, P. Lallemand, Y. Pomeau, and J. -P. Rivet, Complex Systems **1**, 649 (1987).

[11] A. J. C. Ladd, J. Fluid Mech. **271**, 285 (1994).

[12] A. J. C. Ladd, Phys. Rev. Lett. **70**, 1339 (1993).

[13] M. A. van der Hoef and D. Frenkel, Phys. Rev. Lett. **66** 1591 (1991).

[14] D. d'Humières and P. Lallemand, Complex Systems **1**, 599 (1987).

[15] J. W. Dufty and M. H. Ernst, J. Phys. Chem. **93**, 7015 (1989).

[16] M. H. Ernst and H. van Beijeren (private communication).

New Advances in Laplacian Growth Models

F. Guinea and O. Pla[1], E. Louis[2] and V. Hakim[3]

[1] Instituto de Ciencias de Materiales, CSIC, Cantoblanco, E-28049 Madrid,Spain
[2] Departamento de Física Aplicada, Universidad de Alicante, Apdo. 99, E-09080 Alicante, Spain
[3] Laboratoire de Physique Statistique, Ecole Normale Superiéure, 24 Rue Lhomond, 75231 Paris CEDEX 05, France

1 Introduction

In 1983, the model called Diffussion Limited Aggregation was first introduced in the scientific literature[1, 2], in order to study the formation of very tenuous structures, such as soot and dust. Since its conception, a large number of variations have been discussed, and applied to phenomena as varied as lightning or electrolytic deposition. The underlying feature which can be used to unify such diverse growth phenomena is the existence of a field which satisfies Laplace's equation, and which determines the evolution of the system.

The vast potentiality of Laplacian growth models was developed in a series of succesive steps. Among others, it is worth mentioning:

- The generalization to models unrelated to aggregation proper, but controlled by a Laplacian field: the Dielectric Breakdown Model[3].
- The understanding of its relation to the growth process par excelence, solidification of an undercooled liquid and dendritic formation[4].
- The connexion of DLA to other 'classic' problem, viscous fingering[5].
- The generalization of DBM to elastic problems and crack formation[6].
- The application to the varied morphology found in electrolitic deposition[7].

While the advances in relating Laplacian growth to different physical processes has been rather spectacular, theoretical understanding progresses at a much slower pace. Some basic features were quickly identified, as crucial in the formation of the fractal patterns found in numerical simulations: the long range effects and screening mediated by laplacian field, and the tip splitting instability of the branches of the patterns. The latter instability dissappears in he presence of anisotropy and in the absence of noise. These are always the conditions when the growing pattern is a crystal. Stable branch tips lead to dendritic growth. It is also assumed that the global shape is determined by some kind of competition between the various branches which from it.

The DLA model itself has been the object of extensive numerical simulations, and the resulting patterns have been characterized in great detail[8]. Numerical work in many of the related models is significantly more difficult, as it requires the calculation of the laplacian field throughout the entire space[9].

The model has also proven to be quite hard to tackle by the standard methods of statistical physics. Different attempts to define dome kind of mean field approach have been tried, with limited success[10]. A variety of renormalization group approaches have been used. The most promising schemes focuses on the details of the growth of a particular branch, which then are 'blown up' to the entire aggregate[11]. While this technique describes well the global features, the way in which the scale invariance of the model is generated is left unresolved. The extensive mathematical knowledge of the 2D Laplace equation has also been applied. It was found that, by means of conformal mapping methods, the model allows for an elegant formulation. A number of general properties, conserved quantities and even exact solutions have been found in this manner[12]. It is yet unclear, however, whether the average global features can be obtained by this technique.

Thus, laplacian growth remains a challenging theoretical problem, and its elucidation would surely clarify many issues in the physics of systems evolving far out of equilibrium. To show its broad range of applicability, we show in figures (1) and (2) a pattern obtained by electrolytic deposition (courtesy of J. M. Pasror, M. A. Rubio and E. Crespo (UNED, Madrid), and a similar pattern found in the fossil record (from A. G. Checa and J. M. García-Ruiz, **Ammonoids paleobiology**, N. Chapman ed. Plenum, 1994).

It is impossible in a short article to review in full the many new developments which are being proposed. In the following sections, we will address two topics which are representative of the many approaches to the problem available in the literature. We discuss:

- The variation in shapes of the Dielectric Breakdown Model as function of the growth law. The goal is to find 'simple' limits which allow us to use analytical perturbative schemes. While this objective has not been achieved, a variety of

fdd13194 - 2

Fig. 1. Patterns obtained by electrolytic deposition.

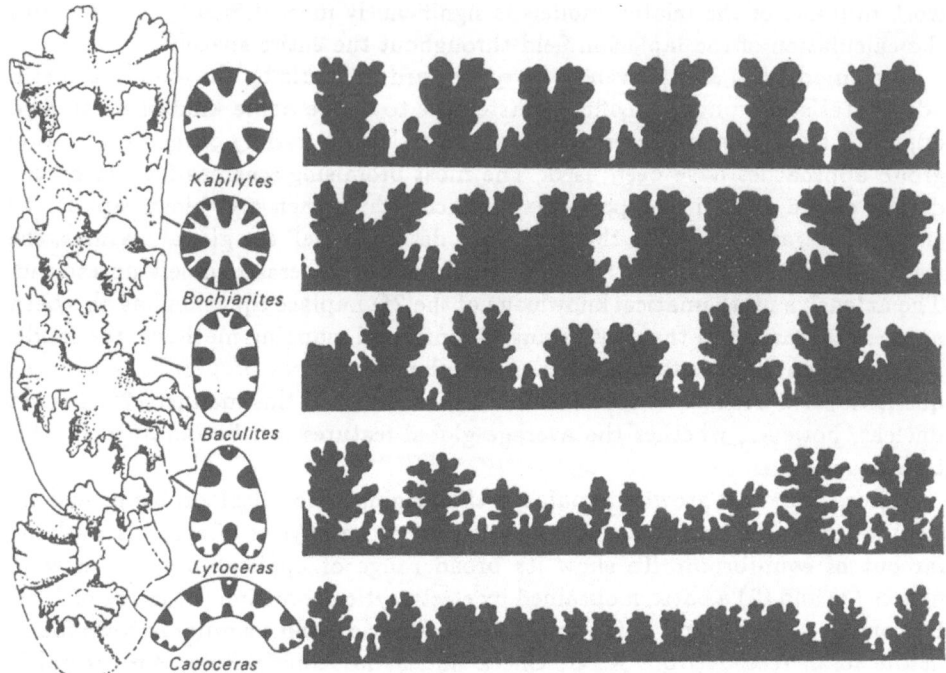

Fig. 2. Patterns in fossil ammonid shells.

interesting results, on the tip splitting instability and branching in general, have been found.

- Pattern formation in the growth of cracks in brittle materials under thermal stresses. Recent experiments[13] show a variety of growth instabilties, such as the sinusoidal modulation of initially straight cracks, or the branching of a single crack into many. As discussed below, this rich morphology can be understood in terms of the laplacian growth models described here.

2 Morphologies in the Dielectric Breakdown Model

As mentioned in the Introduction, the Dielectric Breakdown Model is the simplest modification of DLA. It describes the growth of an aggregate in the presence of a field, ϕ, which obeys the Laplace equation outside it. The model is completely defined by determinig the boundary conditions for the field and the dependence of the velocity of growth at each point of the boundary of the aggregate on the laplacian field. We take $\phi(x) = 1$, for ponits x on an external boundary far away from the aggregate, and $\phi(x) = 0$ when x lies on the boundary of the aggregate. The velocity of growth is taken to be $v(x) \sim |\nabla \phi(x)|^\eta$. η is the only parameter which can be modified to change the properties of the pattern.

It is interesting to note that this generic model can be used for the study of more complex growth laws. Typically, a crossover takes place between the different effective η's by which the growth law can be approximated[14].

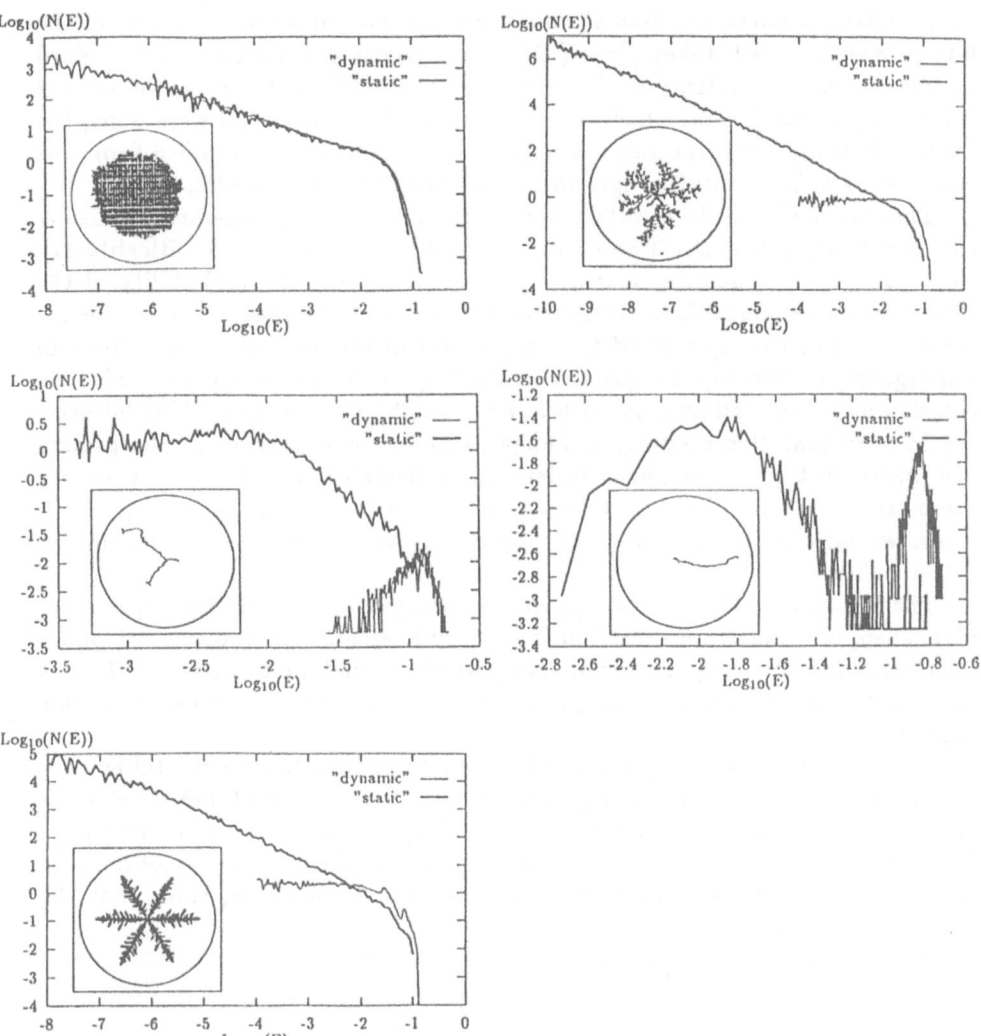

Fig. 3. Patterns, and 'catastrophe distribution " in Dielectric breakdown models for different values of η.

Typical patterns for different values of η are shown in fig. (3). As η increases, the probability of growth becomes more and more concentrated at the tips. The splitting instability characteristic of DLA disappears first, and, eventually, branching ceases altogether. This last feature is in agreement with a simple 'star' model which can be solved by conformal transformations[15]. Within this scheme, used to study the competition between branches, only a single branch survives for $\eta > 6$. An estimation of the value of η at which the fractal dimension goes to one can also be obtained from the analysis of Turkevich and Scher [16]. These authors, by relating the fractal dimension to the field singularities, arrived to a relation between the fractal dimension and an angle characteristic of the lattice where growth takes place (α). This analysis can be easily generalized to the case of an arbitrary η. The result is, $D = \eta(\frac{\pi}{\alpha} - 1) + 2$. In the case of the triangular lattice, on which the simulations of Reference [9] were done, the angle α is $4\pi/3$. Thus, the value of η for which D becomes 1, obtained from this equation, is 4, in qualitative agreement with the numerical results.

The complexity in the growth process also decreases as η increases. We define the complexity within thw framework developed fo self organized criticality[17]. Each growth event induces a rearrangement in the growth probability of the sites close to the one where the growth takes place. The magnitude of this rearrangement can be considered the counterpart of the 'catastrophes' defined in self organized criticality. Irrespective of the way of computing the sizes of these catastrophes, they follow a power law, as in the SOC hypothesis. It is interesting to note that this property is independent of some changes in the growth law, provided that η remains well defined, as depicted in figure (3). Thus, we use noise reduction algorithms to move from DLA to dendritic growth (in both cases, we have $\eta = 1$) without an appreciable change in the distribution of catastrophes[18].

The complexity of the growth processes changes significantly for large values of η. Then, the catastrophe distribution is very irregular, and most catastrophes cluster around the largest possible size, which is that of the cluster. This is consistent with the fact screening is reduced as the patterns become more one dimensional.

To complete this investigation, we have analyzed in detail the fractal properties of the aggregates as η is changed[9]. The results for the fractal dimension are plotted in figure (4). The curve comes very close to one for $\eta \sim 6$, and remains flat at larger values of η. This is highly suggestive of the existence of a phase transition at a critical value of η. Simple arguments, however, show that the

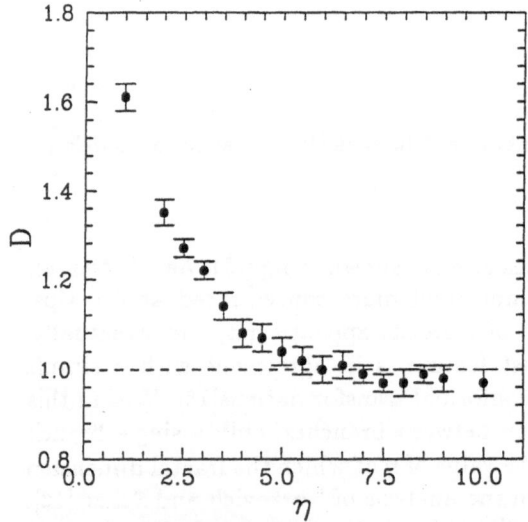

Fig. 4. Fractal dimension as function of η in dielectric breakdown.

simplest model which can describe the large η regime, a directed random walk, does not apply to these Laplacian random walks[19].

In conclusion, the dielectric breakdown model leads to a rather non trivial and interesting extension of DLA. It gives a new way of studying the latter model, if suitable expansions in η can be found.

3 Fracture by thermal stresses

The phenomenology of fracture is very rich, and many of the underlying processes are still poorly understood. Because of it, almost no attempt has been made to clasify crack patterns. Recent experiments, however[13], show that, under controlled circumstances, a number of well defined patterns can be generated, ranging from simple linear cracks to multibranched shapes. Fracture takes place in a brittle glass plate, driven by the existence of a temperature gradient. The basic control parameter is the velocity of motion of the thermal distribution in the sample.

The variety of patterns found in experiments is well reproduced in numerical simulations, using the mechanical breakdown model mentioned in the first section of this work[20, 21], as shown in figure (5). The brittle plate is modeled by an array of elastic springs, whose restoring force is set to zero in the region occupied by the growing crack. For a given thermal distribution, all springs neighboring to the existing crack, and with stresses over a given threshold, break. Once the system achieves equilibrium in the new, fully relaxed situation, the gradient is displaced, and new springs are broken. This separation of time scales is consistent with the conditions at which the experiment is performed.

The fact that the simulations suffice to describe well the variety of experimental patterns implies that the system behaves as a mechanical breakdown model, albeit with a non trivial dependence of the growth velocity on the stresses at the boundaries of the crack (as in DBM).

The instabilities observed in the experiments are sufficently simple, and it is imaginable that a complete understanding of the process will be achieved. Obviously, they are related to the distribution of stresses induced by the thermal gradient, the crack and the free boundaries of the plate. The problem, in the presence of a single straight crack is amenable to an analytical treatment[22]. In an analogous way to the solution of problems in electrostatics by means of fictitious surface charges, the stresses can be seen as generated by a distribution of dislocations at the edges of the plate, and at the crack itself, with the Burgers vector normal to the main axis of the plate.

Numerical results of the stress distribution for various types of cracks are also shown in figure (5). A complete analysis of the changes near the instabilities has not been yet performed, but some salient features can already be identified:

- The crack reduces stresses behind it. This is an expected screening effect, as in electrostatic models.

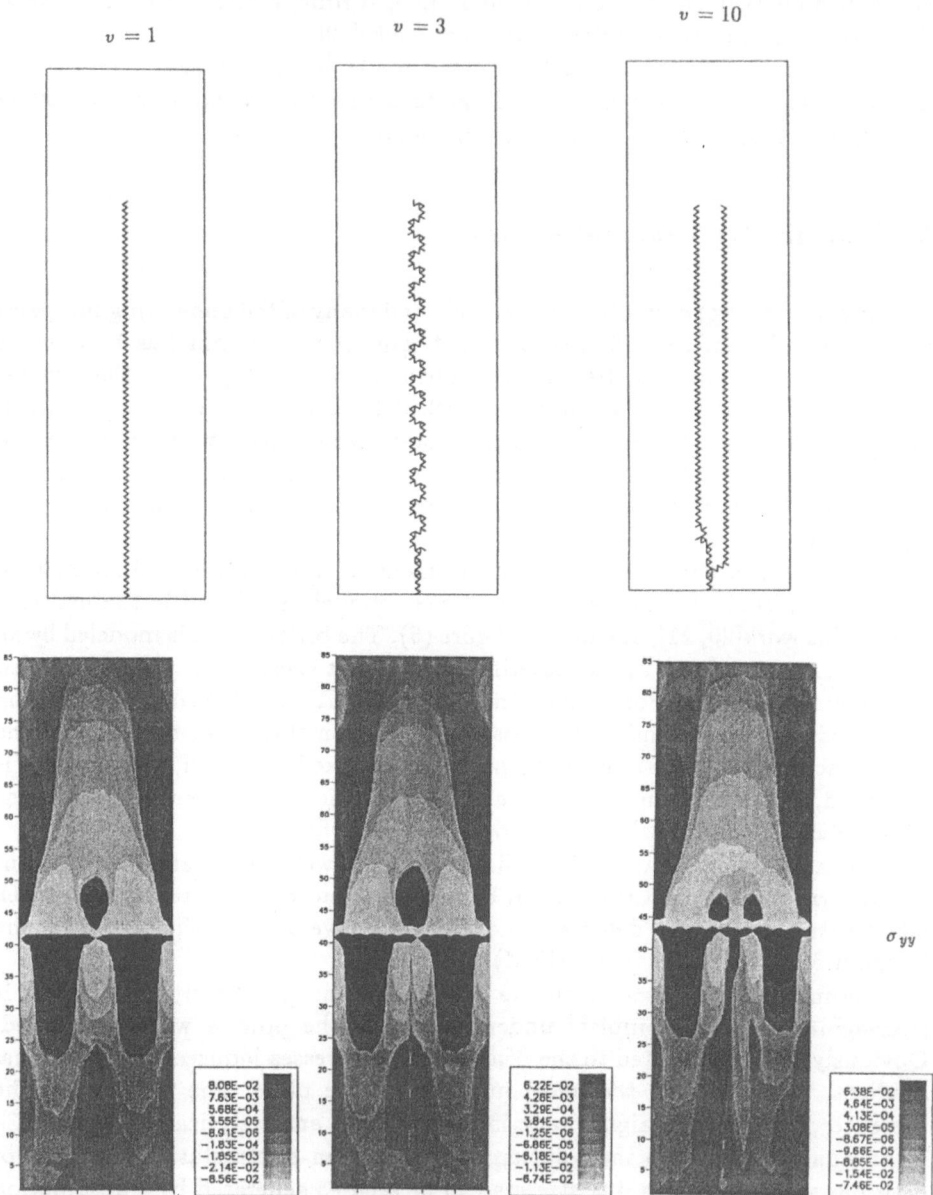

Fig. 5. Crack patterns and stress distributions obtained by a simulation of thermal fracture propagation.

- The leading effect of the crack is a substantial change of the stress distribution in the neighborhood of the crack tip. This effect can be approximately described by imagining a single dislocation placed at the tip itself.
- The instability is preceded by the fission of the bump described in the previous paragraph.

Finally it is interesting to note that a linear stability analysis of a circular crack growing under a constant thermal gradient gives for the ratio between the instantaneous rates of growth of a perturbation of wave number m and that of the circular crack, $\alpha_m = 2m - 1$. This result is similar to that obtained for cracks growing under constant strain or pressure [23], although in this case the tendency to instabilities is slightly stronger. One should note, however, that this result may depend on the actual thermal gradient to which the medium is subjected.

4 Conclusions

In the preceding sections we have tried to give a brief overwiew of some current work in one of the most challenging problems in statistical mechanics: pattern formation out of equilibrium.

The canonical model introduced for its study, Diffussion Limited Aggregation, has proven to be much harder to tackle than expected, despite its apparent simplicity. We have discussed two aspects of the problem which seem likely to lead to new insights and perspectives. We have presented a detailed study of the, more general, Dielectric Breakdown Model, which points out to the existence of a novel type of phase transition. We have also presented recent experimental work on pattern formation in fracture. We have shown that it can be included into the framework of DLA related models. The simplicity of the observed instabilities makes likely that they will be be understood satisfactorily in the near future.

References

1. T. A. Witten and L. M. Sander, Phys. Rev. Lett. 47, 1400 (1981). T. A. Witten and L. M. Sander, Phys. Rev. B 27, 2586 (1983).
2. L. Sander, Scientific American, 255, 94 (1987)
3. L. Niemeyer, L. Pietronero, and H. J. Wiesmann, Phys. Rev. Lett. 52, 1033 (1984).
4. J. Langer, Rev. Mod. Phys. 52, 1 (1980).
5. See D. Bensimon, L. P. Kadanoff, S. Liang, B. I. Schraiman and C. Tang, Rev. Mod. Phys. 58, 977 (1986) and references therein.
6. E. Louis and F. Guinea, Europhys. Lett. 3, 871 (1987).
7. D. G. Grier, D. A. Kessler, and L. M. Sander, Phys. Rev. Lett. 59, 2315 (1987). P. Garik, D. Barkley, E. Ben-Jacob, E. Bochner, N. Broxholm, B. Miller, B. Orr, and R. Zamir, Phys. Rev. Lett. 62, 2703 (1989).
8. P. Meakin, Phase Transitions and Critical Phenomena C. Domb and J. L. Lebowitz eds., Academic Press, London, vol. 12, p. 336, 1988. 34, 234 (1987).
9. A. Sánchez, F. Guinea, L. M. Sander, V. Hakim y E. Louis, Phys. Rev. E 48, 1296 (1993).
10. M. Nauenberg, R. Richter and L. M. Sander, Phys. Rev. B 28, 1649 (1983). H. Levine and H. Tu, Phys. Rev. E 48, R4207 (1993).
11. L. Pietronero, A. Erzan and C. Everstz, Phys. Rev. Lett. 61, 861 (1988). T. C. Halsey, ibid 72, 1228 (1994).

12. M. B. Mineev-Weinstein and R. Mainieri, Phys. Rev. Lett. **72**, 880 (1994).
13. A. Yuse and M. Sano, Nature **362**, 329 (1993).
14. O. Pla, F. G., E. Louis, G. Li, L. M. Sander, H. Yan and P. Meakin, Phys. Rev. A **42**, 3670 (1990).
15. B. Derrida and V. Hakim, Phys. Rev. A, in press.
16. L.A. Turkevich and H. Scher, Phys. Rev. Lett. **55**, 1026 (1985).
17. P. Bak, C. Tang and K. Wisenfeld, Phys. Rev. Lett. **59**, 381 (1987).
18. O. Pla, F. Guinea and E. Louis, Phys. Rev. A **42**, 6720 (1990). F. Guinea, O. Pla and E. Louis, Europhys. Lett. **24**, 701 (1993).
19. W. Lyklema, C. Everstz, and L. Pietronero, Europhys. Lett. **2**, 77 (1986). W. Lyklema and C. Everstz, J. Phys. A **19**, L895 (1986). N. Majumdar, Phys. Rev. Lett. **68**, 2329 (1992).
20. Y. Hayakawa, Phys. Rev. E **49**, R1804 (1994).
21. O. Pla, F. Guinea, V. Hakim and E. Louis, in preparation.
22. M. Marder, Phys. Rev. E **49**, R51 (1994).
23. E. Louis and F. Guinea, Phys. Rev. E **49**, R994 (1994).

Slow Dynamics and Linear Relaxation

B.U. Felderhof

Institut für Theoretische Physik A
RWTH Aachen
Templergraben 55
D-52056 Aachen, Germany

Abstract

Slow dynamics of linear relaxation systems involves a continuous spectrum of arbitrarily long relaxation times. Examples abound in the theory of diffusion-controlled reactions, motion of particles in a viscous fluid, Brownian motion of interacting particles, viscoelasticity, etc.. Such systems may be described in a common mathematical framework. The Laplace transform of a typical relaxation function is conveniently studied as a function of the square root of frequency, and may be approximated by a rational function of this variable. The approximation may be constructed by the method of N-point Padé approximants.

1 Introduction

In many systems relaxation towards equilibrium is slow in the sense that it involves a superposition of purely decaying exponentials with arbitrarily long relaxation times. As a consequence of the continuous spectrum of rates the relaxation function decays with an algebraic long-time tail. The underlying reason is the occurrence of a diffusion process of particles, heat, momentum, or stress.

As an example, consider the problem of diffusion-controlled reactions, first studied by Smoluchowski [1]. Particles diffuse in the space outside a spherical sink of radius a, and are instantaneously absorbed upon reaching the sink surface. This is described by the diffusion equation for the particle density $c(\boldsymbol{r}, t)$ outside the sink

$$\frac{\partial c}{\partial t} = D_0 \, \nabla^2 \, c \qquad\qquad \text{for } r > a \qquad\qquad (1)$$

with boundary condition $c(\boldsymbol{r}, t) = 0$ for $r = a$. Smoluchowski gave the solution with initial concentration $c(\boldsymbol{r}, 0) = c_\infty$ as

$$c(\boldsymbol{r}, t) = c_\infty \left[1 - \frac{a}{r} \text{erfc} \left(\frac{r - a}{\sqrt{4 D_0 \, t}} \right) \right] . \qquad\qquad (2)$$

This may be written $c(\boldsymbol{r}, t) = c_\infty \, S(t|\boldsymbol{r})$, where $S(t|\boldsymbol{r})$ is the survival probability at time t for a particle starting to diffuse from $\boldsymbol{r}$ at time zero. In the end the survival probability tends to the escape probability $\phi(r) = 1 - (a/r)$. This shows that the sink has effect at very long range. The relaxation is correspondingly slow. If one calculates the current into the sink one finds $J(t) = k(t)c_\infty$ with time-dependent rate coefficient

$$k(t) = k_L + k_L \frac{a}{\sqrt{\pi D_0 t}}, \tag{3}$$

with long-time value $k_L = 4\pi D_0 a$. This is a transport coefficient with a remarkable memory effect. We may rewrite Eq. (3) as

$$k(t) = k_L + \frac{k_L a}{\pi D_0} \int_0^\infty \frac{e^{-\lambda t}}{\sqrt{\lambda}} d\lambda. \tag{4}$$

Evidently there is a continuous spectrum of relaxation rates with accumulation near $\lambda = 0$.

It is of interest to consider also the Laplace transform

$$\hat{k}(s) = \int_0^\infty e^{-st} k(t) dt. \tag{5}$$

It is given by

$$\hat{k}(s) = \frac{k_L}{s} + \frac{k_L a}{\sqrt{D_0 s}}, \tag{6}$$

with a singularity at $s = 0$.

In the following we consider similar relaxation phenomena, characterized by a continuous spectrum of relaxation rates, extending to zero. We shall see that slowly relaxing systems can be treated from a common point of view. We discuss a method, developed in collaboration with B. Cichocki, for calculating a good approximation to the relaxation function [2].

2 Slow dynamics

As a further example of slow dynamics consider the motion of a sphere of radius a in a viscous incompressible fluid of shear viscosity η, mass density ρ. The fluid flow velocity $v(r, t)$ and pressure $p(r, t)$ are assumed to satisfy the linearized Navier-Stokes equations

$$\rho \frac{\partial v}{\partial t} = \eta \nabla^2 v - \nabla p, \qquad \nabla \cdot v = 0. \tag{7}$$

We assume stick boundary conditions at the sphere surface. If the whole system is at rest for $t < 0$, and is perturbed by an external force $E(t) = S \delta(t)$ acting on the sphere, then the sphere velocity for $t > 0$ is

$$U(t) = \frac{1}{m^*} F(t) S, \tag{8}$$

with $F(0+) = 1$ and an effective mass given by $m^* = m_p + \frac{1}{2} m_f$, where m_p is the mass of the sphere and $m_f = 4\pi\rho\, a^3/3$ is the mass of the displaced fluid. The function $F(t)$ is calculated conveniently by Fourier transform. Write

$$U(t) = \int_{-\infty}^{\infty} U_\omega\, e^{-i\omega t}\, d\omega, \qquad\qquad E_\omega = \frac{S}{2\pi}, \qquad (9)$$

then the amplitudes are related by

$$U_\omega = \mathcal{Y}(\omega) E_\omega \qquad (10)$$

with the admittance

$$\mathcal{Y}(\omega) = \frac{1}{6\pi\eta a} \frac{1}{1 + \sqrt{-i\omega\tau_v} - i\omega\tau_M}. \qquad (11)$$

Here $\tau_v = a^2\rho/\eta$ is the viscous relaxation time, and $\tau_M = m^*/6\pi\eta a$ is the mean relaxation time. The admittance has a branch cut starting at $\omega = 0$. In terms of the Laplace variable $z = -i\omega\tau_M$

$$\mathcal{Y}(\omega) = \frac{1}{6\pi\eta a} \frac{1}{1 + \sigma\sqrt{z} + z}, \qquad (12)$$

where $\sigma = \sqrt{\tau_v/\tau_M} = \sqrt{9m_f/2m^*}$. The branch cut can be taken along the negative real z axis. It is convenient to introduce the variable $y = \sqrt{z}$. Then

$$\mathcal{Y}(\omega) = \frac{1}{6\pi\eta a} \frac{1}{1 + \sigma y + y^2}. \qquad (13)$$

This shows that the function is characterized by two conjugate poles $y_\pm$ in the left hand y plane, corresponding to the second Riemann sheet in the variable z.

The inverse Fourier transform can be rewritten by a change of integration contour as [3]

$$F(t) = \int_0^\infty p(u)e^{-u\tau}\, du, \qquad\qquad \tau = t/\tau_M, \qquad (14)$$

with positive spectral density

$$p(u) = \frac{1}{\pi} \frac{\sigma\sqrt{u}}{1 + (\sigma^2 - 2)u + u^2}. \qquad (15)$$

We have normalized such that

$$\int_0^\infty p(u)du = 1, \qquad\qquad \int_0^\infty \frac{p(u)}{u}\, du = 1. \qquad (16)$$

The parameter σ determines the width of the spectrum. For $\sigma \to 0$, corresponding to $m_p \gg m_f$, the distribution tends to a delta-function at relaxation rate

$1/\tau_M$. In the general case the density $p(u) \approx \sigma\sqrt{u}/\pi$ at small u leads to the long-time behavior

$$F(t) \approx \frac{\sigma}{2\sqrt{\pi}} \left(\frac{\tau_M}{t}\right)^{3/2} \qquad \text{as } t \to \infty. \qquad (17)$$

The velocity of the sphere for long times is

$$U(t) \approx \frac{1}{12} \frac{\sqrt{\rho}}{(\pi\eta t)^{3/2}} S. \qquad (18)$$

Remarkably, this long-time tail is independent of the size or mass of the sphere. It corresponds to the long-time tail found by Alder and Wainwright [4] in the velocity autocorrelation function of a hard sphere fluid.

Another example with similar analytic behavior is found in the theory of viscoelasticity. The dynamic viscosity of many supercooled liquids is well described by the phenomenological Barlow-Erginsav-Lamb equation [5]

$$\eta(\omega) = \eta(0) \left[1 + \sigma\sqrt{-i\omega\tau_M} - i\omega\tau_M\right]^{-1} \qquad (19)$$

with mean relaxation time $\tau_M = \eta(0)/G_\infty$, where G_∞ is the high-frequency shear modulus. Many liquids are described by this simple expression with the universal value $\sigma = 2$. Experiments on mixtures yield a different value for the parameter.

The diffusion-controlled rate coefficient $\hat{k}(s)$ is related to the admittance

$$\hat{A}(s) = \left[s V_0 + s \hat{k}(s)\right]^{-1} \qquad (20)$$

where $V_0 = 4\pi a^3/3$ is the sink volume. Substituting Eq. (6) we find

$$\hat{A}(-i\omega) = k_L \left[1 + \sigma\sqrt{-i\omega\tau_M} - i\omega\tau_M\right]^{-1} \qquad (21)$$

with mean relaxation time $\tau_M = V_0/k_L$ and $\sigma = \sqrt{3}$.

Finally we consider self-diffusion in a suspension of interacting Brownian particles [6][7]. The mean-square displacement of a selected particle is

$$W(t) = \frac{1}{6} < [R_1(t) - R_1(0)]^2 >, \qquad (22)$$

where the angle brackets denote an equilibrium average. The time derivative defines the self-diffusion coefficient $D_S(t) = dW/dt$ with short-time value $D_S^S = D_S(0)$ and long-time value $D_S^L = D_S(\infty)$. The diffusion coefficient can be expressed as

$$D_S(t) = D_S^L + \mu_S(t), \qquad (23)$$

with relaxation function $\mu_S(t)$. It follows from the many-body Smoluchowski equation that this is a completely monotonic function, i.e. it takes the form

$$\mu_S(t) = \left(D_S^S - D_S^L\right) \int_0^\infty p_S(u)e^{-u\tau} \, du, \qquad \tau = t/\tau_M, \qquad (24)$$

with positive spectral density $p_S(u)$. The spectrum is normalized as in Eq. (16) if the mean relaxation time is defined as

$$\tau_M = \int_0^\infty \mu_S(t)dt/\mu_S(0). \tag{25}$$

For a semidilute suspension of hard spheres the time-dependent diffusion coefficient may be calculated to first order in density n from the two-sphere Smoluchowski equation. With neglect of hydrodynamic interactions the short-time and long-time values are

$$D_S^S = D_0 = \frac{k_B T}{6\pi\eta a}, \qquad D_S^L = D_0(1 - 2\phi), \tag{26}$$

to first order in volume fraction $\phi = 4\pi na^3/3$. The Laplace transform of the relaxation function is [8]

$$\hat{\mu}_S(z) = \left(D_S^S - D_S^L\right)\left[1 + \sigma\sqrt{z} + z\right]^{-1} \tag{27}$$

with $\sigma = \sqrt{2}$, Laplace variable $z = -i\omega\tau_M$, and mean relaxation time $\tau_M = a^2/D_0$. For other interaction potentials the behavior is well approximated by the same expression with different values for σ and τ_M.

The foregoing shows that in a number of cases the Laplace transform of the relaxation function is given by a simple expression with two poles in the complex $\sqrt{z}$ plane. This suggests more generally that the slow dynamics of diffusive systems can be described by Laplace transformed relaxation functions with a simple pole structure in the $\sqrt{z}$ plane. The conjecture is confirmed by calculations on model systems, as well as by comparison with experimental data.

3 Pole structure

We formulate the conjecture in a more general framework. We consider the time-correlation function $C_A(t) =< A(t)A(0) >$ of a fluctuating variable A with zero mean, $< A >= 0$. The relaxation function $\gamma_A(\tau)$ is defined by

$$C_A(t) = C_A(0)\gamma_A(t/\tau_M) \tag{28}$$

with mean relaxation time

$$\tau_M = \int_0^\infty C_A(t)dt/C_A(0). \tag{29}$$

The relaxation function has initial value $\gamma_A(0) = 1$, and is assumed to be completely monotonic

$$\gamma_A(\tau) = \int_0^\infty p_A(u)e^{-u\tau} \, d\tau, \tag{30}$$

with positive spectral density $p_A(u)$, normalized as in Eq. (16). Hence it follows that the Laplace transform

$$\Gamma_A(z) = \int_0^\infty e^{-z\tau}\, \gamma_A(\tau) d\tau \tag{31}$$

can be written as a Stieltjes integral

$$\Gamma_A(z) = \int_0^\infty \frac{p_A(u)}{u+z}\, du. \tag{32}$$

This shows explicitly that all poles are on the negative real z axis and have positive weight. It is advantageous to introduce the complex variable $y = \sqrt{z}$. Without loss of generality we can assume that the Laplace transform $\Gamma_A(z)$ can be represented as a sum of simple poles in the left hand side of the complex y plane

$$\Gamma_A(z) = \sum_j \frac{R_j}{y - y_j}. \tag{33}$$

Since $p_A(u)$ is positive, the function $\Gamma_A(z)$ has a symmetry with respect to the real z axis. As a consequence the poles occur either in conjugate pairs (y_j, y_j^*) with residues (R_j, R_j^*), or singly on the negative real y axis with real residue. For example, in the two-pole expression (13) the poles and residues are

$$y_\pm = -\frac{1}{2}\sigma \pm \frac{1}{2} i\sqrt{4 - \sigma^2}, \qquad R_\pm = \frac{\mp i}{\sqrt{4 - \sigma^2}}. \tag{34}$$

In the general case the Laplace transform can be written as

$$\Gamma_A(z) = \left[1 + \sigma\, y + y^2 + y^2\, \psi(y)\right]^{-1}, \tag{35}$$

with a function $\psi(y)$ which tends to a constant as $y \to 0$ and to zero as $y \to \infty$. The behavior at small z gives the sum rules

$$\sum_j \frac{R_j}{y_j} = -1, \qquad \sum_j \frac{R_j}{y_j^2} = \sigma. \tag{36}$$

The behavior at large z gives the sum rules

$$\sum_j R_j y_j = 1, \qquad \sum_j R_j = 0. \tag{37}$$

The inverse Laplace transform of Eq. (33) reads

$$\gamma_A(\tau) = \sum_j R_j y_j\, w(-i y_j \sqrt{\tau}), \tag{38}$$

where the w function is related to the error function of complex argument [9]. At long times the relaxation function behaves as

$$\gamma_A(\tau) \approx \frac{\sigma}{2\sqrt{\pi}}\, \tau^{-3/2} \qquad \text{as } \tau \to \infty. \tag{39}$$

Often the coefficient σ can be evaluated exactly. If σ vanishes, then the exponent of the long-time tail is larger than $3/2$.

Our conjecture is that usually $\gamma_A(\tau)$ is dominated by a relatively small set D of dominant poles,

$$\gamma_A(\tau) \approx \sum_{d \in D} R_d y_d w(-i y_d \sqrt{\tau}). \tag{40}$$

To determine the set D we must make sure that the long-time behavior is well reproduced, i.e. we must have

$$\sum_{d \in D} \frac{R_d}{y_d} \approx -1, \qquad\qquad \sum_{d \in D} \frac{R_d}{y_d^2} \approx \sigma. \tag{41}$$

The approximate expression (40) corresponds to an approximation to the Laplace transform,

$$\Gamma_A^{(p+1)}(z) = \left[1 + \sigma y + y^2 + y^2\, \psi_p(y)\right]^{-1}, \tag{42}$$

where $\psi_p(y)$ is a rational approximation to the function $\psi(y)$ in Eq. (35), i.e. it is a ratio of two polynomials, the Padé approximant

$$\psi_p(y) = \frac{A_p(y)}{B_p(y)}. \tag{43}$$

One can find the coefficients of the polynomials A_p and B_p by fitting to known values of the function $\Gamma_A(z)$ on the positive real z axis, that is, in the physical domain. If the fit is made at $p+1$ points $\{z_j\}$, then one can construct the approximant $\Gamma_A^{(p+1)}(z)$ with polynomials A_p, B_p of degree $\frac{1}{2} p$ for p even, and with A_p of degree $\frac{1}{2}(p-1)$ and B_p of degree $\frac{1}{2}(p+1)$ for p odd [10]. The values $\{\Gamma_A(z_j)\}$ can be obtained from a model calculation, or can be inferred from experimental data at real frequency.

For example, we have found an approximate expression for the dynamic shear viscosity of the rubber-like substance polyisobutylene, on the basis of old absorption data [11]. The data give the real part of the dynamic viscosity $\eta(\omega)$ at selected frequencies. We find that the dynamic viscosity is well approximated by a four-pole expression. The poles are at $y_{1,2} = -0.06 \pm 0.49i$ and $y_{3,4} = -0.4 \pm 1.2i$ with residues $R_{1,2} = 0.07 \mp 0.12i$, and $R_{3,4} = -0.07 \mp 0.35i$.

The number of poles necessary for an accurate description depends on the details of the dynamics. We have done model calculations on diffusion of a particle in a three-dimensional potential to study the dependence [2]. The required number of poles increases with the number of barriers.

4 Diffusion in a potential

Specifically we have studied particle diffusion in a three-dimensional radially symmetric potential $v(r)$. Diffusion in a potential is described by the Smoluchowski equation

$$\frac{\partial P}{\partial t} = D_0 \nabla \cdot [\nabla P + \beta (\nabla v) P]. \tag{44}$$

The equation has the time-independent solution

$$g(r) = \exp[-\beta v(r)]. \tag{45}$$

We study the correlation function

$$C_A(t) = \int A(r) P(r, t | A) dr, \tag{46}$$

where $P(r, t | A)$ is the solution of Eq. (44) with initial condition

$$P(r, 0 | A) = A(r) g(r). \tag{47}$$

We choose in particular

$$A(r) = \theta(a - r). \tag{48}$$

Then $C_A(t)$ is proportional to the probability to return at time t to the spherical volume $r < a$, when starting there with Boltzmann distribution at time zero.

We define the relaxation function $\gamma_A(\tau)$ as in Eq. (28), with mean relaxation time τ_M as in Eq. (29). The initial value of the correlation function is here

$$C_A(0) = 4\pi \int_0^\infty A^2(r) g(r) r^2 \, dr. \tag{49}$$

It is convenient to write

$$P(r, t | A) = f(r, t) g(r). \tag{50}$$

Then $f(r, t)$ satisfies the adjoint equation

$$\frac{\partial f}{\partial t} = \mathcal{L} f, \tag{51}$$

with adjoint Smoluchowski operator $\mathcal{L} = D_0 [\nabla - \beta (\nabla v)] \cdot \nabla$ and initial value $f(r, 0) = A(r)$. The calculation of the mean relaxation time τ_M requires solution of the steady-state equation

$$D_0 \left[\nabla^2 f^{(0)} - \beta \frac{dv}{dr} \frac{df^{(0)}}{dr} \right] = -A(r). \tag{52}$$

It may be expressed as

$$\tau_M = 4\pi \int_0^\infty A(r) f^{(0)}(r) g(r) r^2 \, dr / C_A(0). \tag{53}$$

The steady-state equation may be rewritten as

$$D_0 \nabla \cdot [g \nabla f^{(0)}] = -g A. \tag{54}$$

This shows that $f^{(0)}(r)$ may be interpreted as the electrostatic potential generated by the charge density $g(r)A(r)/4\pi D_0$ in a medium with dielectric constant $g(r)$.

If $g(r)$ equals unity for $r > r_c > a$, then the solution $f^{(0)}(r)$ decays as Q_A/r for $r > r_c$, where Q_A is the "total charge"

$$Q_A = \frac{1}{D_0} \int_0^\infty A(r)g(r)r^2 \, dr. \tag{55}$$

It may be shown [2] that the width parameter σ can be calculated from Q_A according to

$$\sigma = 4\pi \sqrt{\frac{D_0}{\tau_M^3}} \, Q_A^2 / C_A(0). \tag{56}$$

Thus the two parameters τ_M and σ necessary for construction of the two-pole approximation can be readily calculated.

Consider as an example the single barrier model characterized by the potential

$$
\begin{aligned}
v(r) &= v_0 \quad \text{for} \quad 0 < r < a \\
&= v_1 \quad \text{for} \quad a < r < b \\
&= 0 \quad \text{for} \quad b < r
\end{aligned} \tag{57}
$$

with v_0 negative and v_1 positive, corresponding to a well for $r < a$ separated by a barrier from free space. For this model the Laplace transform $\Gamma_A(z)$ is easily evaluated explicitly in terms of Bessel functions. It shows an infinite number of poles $\{y_j\}$ in the left hand side of the y plane. In typical cases the spectrum $p_A(u)$ is well approximated by the two-pole spectrum given in Eq. (15). Correspondingly, the relaxation function $\gamma_A(\tau)$ is well represented by the two-pole approximation over a wide range of time. One can improve the description by including two more poles by use of the Padé approximant method.

The situation changes if one adds a further well with potential v_2 and barrier with potential v_3. This introduces another time scale. At least four poles are needed for an approximate description of the spectrum. A very accurate fit is obtained with eight poles. This requires ten values $\Gamma_A(z_j)$ for z_j on the positive z axis, besides the values of τ_M and σ.

5 Discussion

The study of diffusion in a potential well shows that the approximation of the exact Laplace transform $\Gamma_A(z)$ by an approximant $\Gamma_A^{(p+1)}(z)$, which is rational in $\sqrt{z}$, works very well. It yields an accurate representation of the spectral density $p_A(u)$ and the relaxation function $\gamma_A(\tau)$. The N-point Padé approximant method permits the explicit construction of the approximation.

The number of poles required is dictated by the physical situation. Usually a small number of poles is sufficient for an accurate description. For example, in the case of self-diffusion in Brownian suspensions we expect that a two-pole approximation is sufficient in a wide range of density, provided the interaction is predominantly repulsive [12]. Study of a square well model [13] shows that a four-pole approximation is required if the well is sufficiently broad and deep. The dynamic viscosity of colloidal suspensions usually requires at least a three-pole approximation [14]. We expect [15] that the intermediate scattering function $F(q,t)$ of a fluid or suspension near the glass transition is well approximated by a four-pole expression, one pair of poles corresponding to α-relaxation, and the other pair to β-relaxation. The positions and residues of the poles depend of course on wavenumber q, density, and temperature.

The method described above may be regarded as a normal mode analysis of diffusive systems. The dynamics of the system is described by a small number of poles and residues in the complex $\sqrt{\omega}$ plane, rather than in the frequency plane of usual linear response theory. In the time domain the superposition of damped oscillating exponentials of usual linear dynamical systems is replaced by a superposition of w functions in the case of linear relaxation systems. The mathematical correspondence between linear relaxation systems of the type considered above and more general linear passive systems has been stressed by Meixner [16][17][18].

References

1. M. v. Smoluchowski, Phys. Z. 17, 557, 585 (1916).
2. B. Cichocki and B.U. Felderhof, submitted.
3. B.U. Felderhof, Physica A 175, 114 (1991).
4. B.J. Alder and T.E. Wainwright, Phys. Rev. A 1, 18 (1970).
5. G. Harrison, The Dynamic Properties of Supercooled Liquids (Academic, London, 1976).
6. B. Cichocki and B.U. Felderhof, Phys. Rev. A 44, 6551 (1991).
7. B. Cichocki and B.U. Felderhof, J. Chem. Phys. 96, 4669 (1992).
8. B. Cichocki and R.B. Jones, Z. Phys. B 68, 513 (1987).
9. Handbook of Mathematical Functions, edited by M. Abramowitz and I.A. Stegun (Dover, New York, 1965).
10. G.A. Baker, Jr., Essentials of Padé Approximants (Academic, New York, 1975) Chap. 8.
11. F.C. Roesler and J.R.A. Pearson, Proc. Phys. Soc. B 67, 338 (1954).
12. B. Cichocki and B.U. Felderhof, J. Chem. Phys. 96, 9055 (1992).
13. B. Cichocki and B.U. Felderhof, Langmuir 8, 2889 (1992).
14. B. Cichocki and B.U. Felderhof, Phys. Rev. A 46, 7723 (1992).
15. B.U. Felderhof, Proceedings of the Tohwa University Symposium on Slow Dynamics in Condensed Matter, Fukuoka 1991, eds. K. Kawasaki, T. Kawakatsu, and M. Tokuyama (American Institute of Physics, New York, 1992), p. 370.
16. J. Meixner, Arch. Rational Mech. Anal. 17, 278 (1964).
17. E.I. Takizawa and J. Meixner, J. Phys. Soc. Japan 35, 654 (1973).
18. J. Meixner, Arch. Rational Mech. Anal. 54, 148 (1974).

Driven Tunneling: New Possibilities for Coherent and Incoherent Quantum Transport

T. Dittrich, P. Hänggi, B. Oelschlägel, and R. Utermann

Institut für Physik, Universität Augsburg, Memminger Straße 6,
D-86135 Augsburg, Germany

Abstract: We study the conservative as well as the dissipative quantal dynamics in a harmonically driven, quartic double-well potential. In the deep quantal regime, we find coherent modifications of tunneling, including its complete suppression. In the semiclassical regime of the conservative system, the dynamics is dominated by the interplay of tunneling and chaotic diffusion. A strong correlation exists between the tunnel splittings and the overlaps of the associated doublet states with the chaotic layer. With weak dissipation, remnants of coherent behaviour occur as transients, such as the tunneling between symmetry-related pairs of limit cycles. The coherent suppression of tunneling observed in the conservative case is stabilized by weak incoherence. The quantal stationary states are broadened anisotropically due to quantum noise, as compared to the corresponding classical attractors.

1 Introduction

During the last decades one observes a strong tendency, both in experimental and theoretical physics, to shift focus from a global, macroscopic point of view towards the microscopic study of moderately small systems—nanometer-scale electronic devices, molecules, small metallic clusters. For such a system in a nonequilibrium environment, there are typically three components which together make up the essential physics: A coherent driving which represents the macroscopic energy source and can often be described classically, the microscopic system itself, with its dynamics characterized by the simultaneous presence of quantum effects and classical nonlinearity, and an environment comprising a large number of weakly coupled degrees of freedom which serve as a sink both for energy and coherent information.

In the present contribution we intend to give an overview over the interplay of these three components, for a specific example: a bistable system driven by a harmonic force. Bistability is an elementary source of nonlinear behaviour [1–5];

with a periodic driving, it enables an enormously rich dynamical repertoire [6–9]. At the same time, bistable systems provide a paradigm for quantum coherence: tunneling [10]. A harmonic driving captures the essence of a ubiquitous energy source, electromagnetic irradiation, specifically of lasers and their relatives.

We shall approach the complexity of weakly dissipative, semiclassical nonlinear behaviour in three steps: Following an introduction into the model in Sect. 2, its dynamics in the deep quantal regime is outlined in Sect. 3. Section 4 is devoted to the semiclassical regime of the conservative dynamics. In Sect. 5, weak dissipation is brought into play. Section 6 contains a summary and an outlook.

The present work forms a synopsis of results partially published elsewhere [11–18].

2 The model and its symmetries

We formulate the harmonically driven bistable system as a quartic double well with a spatially homogeneous, classical sinusoidal driving force. It is described by the Hamiltonian

$$H_{\mathrm{DW}}(x,p;t) = H_0 + H_1, \quad H_0(x,p) = \frac{p^2}{2} - \frac{1}{4}x^2 + \frac{1}{64D}x^4, \quad H_1(x;t) = x\, S\cos\omega t.$$
(1)

With the dimensionless variables used, the only parameter controlling the unperturbed Hamiltonian $H_0(x,p)$ is the barrier height D. It approximately gives the number of doublets with energies below the top of the barrier. Accordingly, the classical limit amounts to $D \to \infty$.

The symmetry of the Hamiltonian under discrete time translations, $t \to t + 2\pi/\omega$, enables to use the Floquet formalism [19–22], which generalizes most of the conceptual tools of spectral analysis to the present context. Its basic ingredient is the Floquet operator, i.e., the unitary propagator that generates the time evolution over one period of the driving force, $U = \mathsf{T}\exp(-\mathrm{i}\int_0^{2\pi/\omega} \mathrm{d}t\, H_{\mathrm{DW}}(t)/\hbar)$, where T effects time ordering. Its eigenvectors and eigenphases are referred to as *Floquet states* and *quasienergies*, respectively. Being phases, the quasienergies are organized in classes, $\epsilon_{\alpha,k} = \epsilon_\alpha + k\omega$, $k = 0, \pm 1, \pm 2, \ldots$, where each member corresponds to a physically equivalent solution. Therefore, all spectral information is contained in a single "Brillouin zone", $-\omega/2 \leq \epsilon < \omega/2$.

Besides invariance under time translation and time reversal, the unperturbed system possesses the spatial reflection symmetry $x \to -x$, $p \to p$, $t \to t$. For the specific time dependence of a harmonic driving as it is used here, the symmetry $f(t + \pi/\omega) = -f(t)$ restores a similar situation as in the unperturbed case: The system is now invariant against the operation [11, 23] P : $p \to -p$, $x \to -x$, $t \to t + \frac{\pi}{\omega}$, which may be regarded as a *generalized parity* in the extended, three-dimensional phase space spanned by x, p, and time $t \bmod(2\pi/\omega)$. As in the unperturbed case, this enables to separate the eigenstates into an even and an odd subset.

3 Driven tunneling and localization

To give an impression of driven tunneling in the deep quantal regime, we study how a state, prepared as a coherent state centered in the left well, evolves in time under the external force. Since this state is approximately given by a superposition of the two lowest unperturbed eigenstates, $|\Phi(0)\rangle \approx (|\Psi_1\rangle + |\Psi_2\rangle)/\sqrt{2}$, its time evolution is dominated by the Floquet-state doublet originating from $|\Psi_1\rangle$ and $|\Psi_2\rangle$, and the splitting $\epsilon_2 - \epsilon_1$ of its quasienergies.

There are two regimes in the (ω, S)-plane where tunneling is not qualitatively altered by the external force: Both in the limits of slow (adiabatic) and of fast driving, the separation of the time scales of the inherent dynamics and of the external force effectively uncouples the two processes and results in a mere renormalization of the tunnel splitting Δ. Both an analytical treatment and numerical experiments show that the driving always reduces the effective barrier height and thus increases the tunneling rate in these two limits [11].

Qualitative changes in the tunneling behavior are expected as soon as the driving frequency becomes comparable to the internal frequencies of the double well, in particular, to the tunnel splitting and to the so-called resonances $E_3 - E_2$, $E_4 - E_1$, $E_5 - E_2$, ... By spectral decomposition, the temporal complexity in this regime is immediately related to the "landscape" of quasienergy planes $\epsilon_{\alpha,k}(\omega, S)$ in parameter space. Features of particular significance are close encounters of quasienergies: Two quasienergies cross one another without disturbance if they belong to different parity classes, otherwise they form an avoided crossing.

The transition at $\omega = E_3 - E_2$, a single-photon transition in the terminology of quantum optics, is called *fundamental resonance*. For $S > 0$, the corresponding quasienergies $\epsilon_{2,k}$ and $\epsilon_{3,k-1}$ form an avoided crossing, since they have equal parity. Fig. 1a shows the time evolution of the *probability to return*, $P^\Phi(t_n) = |\langle \Phi(0)|\Phi(t_n)\rangle|^2$ at the fundamental resonance. The monochromatic oscillation of $P^\Phi(t_n)$ characteristic of unperturbed tunneling has given way to a more complex beat pattern. Its Fourier transform reveals that it is composed mainly of two groups of three frequencies each (Fig. 1b), which can be associated with the allowed transitions among the Floquet states pertaining to the two lowest doublets.

In contrast, a two-photon transition that bridges the tunnel splitting Δ is "parity forbidden", and thus the quasienergies $\epsilon_{2,k-1}$ and $\epsilon_{1,k+1}$ can form exact crossings. A vanishing of the difference $\epsilon_{2,-1} - \epsilon_{1,1}$ will have a remarkable consequence: For a state prepared as a superposition of the corresponding two Floquet eigenstates only, $P^\Phi(t)$ and all other observables become constants, at least at the discrete times t_n, and thus it is possible that tunneling comes to a standstill. According to an argument going back to von Neumann and Wigner [24], exact crossings should occur along one-dimensional manifolds in the (ω, S) plane. In the present case, there is one such manifold $S_{\text{loc},k}(\omega)$ for each condition $\epsilon_{2,-k} = \epsilon_{1,k}$. Fig. 2a shows $S_{\text{loc},0}(\omega)$: a closed curve, reflection-symmetric with respect to the line $S = 0$. A typical time evolution of $P^\Phi(t_n)$ for a parameter point on the linear part of that manifold is presented in Fig. 2b. It clearly

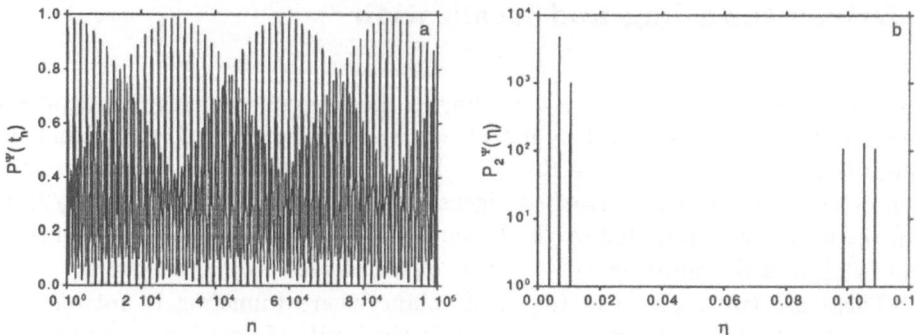

Fig. 1. Driven tunneling at the fundamental resonance $\omega = E_3 - E_2$. (a) Time evolution of $P^{\Phi}(t_n)$ over the first 2×10^5 time steps; (b) local spectral two-point correlation function $P_2^{\Phi}(\eta)$ obtained from (a). The parameter values are $D = 2$, $S = 2 \times 10^{-3}$, and $\omega = 0.876$.

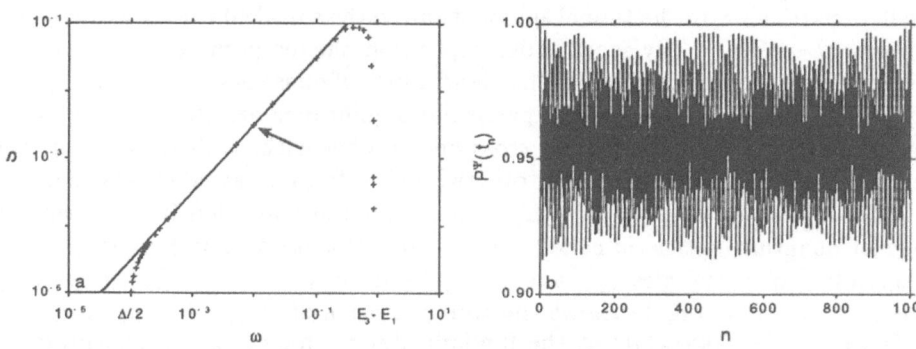

Fig. 2. Suppression of tunneling at an exact crossing $\epsilon_{2,-1} = \epsilon_{1,1}$. (a) The manifold $S_{\mathrm{loc},0}(\omega)$ in the (ω, S) plane where this crossing occurs (data obtained by diagonalization of the full Floquet operator are indicated by crosses, the full line is based on a two-state approximation [13, 25, 26], the arrow indicates the parameter point to which part (b) refers); (b) time evolution of $P^{\Phi}(t_n)$ over the first 10^3 time steps, at $D = 2$, $S = 3.171 \times 10^{-3}$, $\omega = 0.01$.

demonstrates that tunneling is almost completely suppressed. The remaining oscillations of small amplitude can be ascribed to an admixture of higher-lying quasienergy states to the initial state. The suppression of tunneling is an elementary quantum-interference effect, much of which can be understood on basis of a two-state approximation [13, 25, 26].

4 Tunnel splittings and the onset of chaos

In the classical double well, the most significant consequence of the periodic driving is the onset of deterministic chaos [27]. It develops around the hyperbolic fixed point at the top of the barrier and along the separatrix originating there. For sufficiently high S, deterministic diffusion along the separatrix becomes a significant contribution to the classical phase-space transport. Quantum mechanically, it competes with tunneling [28–33]. Applying ideas from Einstein-Brillouin-Keller (EBK) quantization for periodically driven systems [34] and from random-matrix theory for mixed (regular and chaotic) systems [35] to the present context, one obtains the following crude picture of the impact of the chaotic layer on tunneling [29–31]: Even with the driving, the two isolated regular regions within the wells remain related by the generalized parity P. Accordingly, Floquet states residing within these regions form a ladder of tunnel-splitted doublets. For states mainly residing within the chaotic layer, in contrast, random-matrix theory predicts level repulsion. We therefore expect that, as soon as one of the pairs of quantizing tori pertaining to the symmetry-related regular regions resolves in the spreading chaotic layer, the exponentially small splitting of the corresponding doublet breaks up and eventually reaches a size of the order of the mean level separation. As a consequence, the coherent tunneling on an extremely long time scale will give way to a more irregular dynamics on shorter time scales, forming the quantal counterpart of deterministic diffusion along the separatrix. The breakup of the tunnel doublets in the chaotic layer is not a direct consequence of the Lyapunov exponent of the classical dynamics being positive there, but rests on the global condition that diffusive spreading connects all parts of the chaotic layer, bridging the symmetry plane.

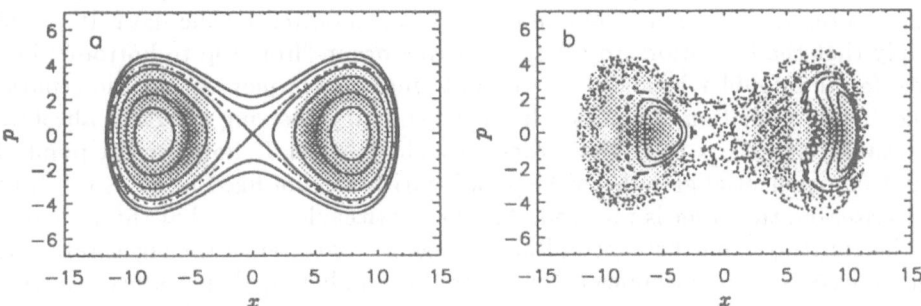

Fig. 3. Husimi distributions (in gray-scale representation) for the quasienergy state $|\phi_7(0)\rangle$, compared with the corresponding classical phase-space portraits, at (a) $S = 10^{-5}$ and (b) $S = 0.2$.

In order to allow for a numerical check, we quantify the distinction between "regular" and "chaotic" eigenstates on the basis of the overlap of the Husimi

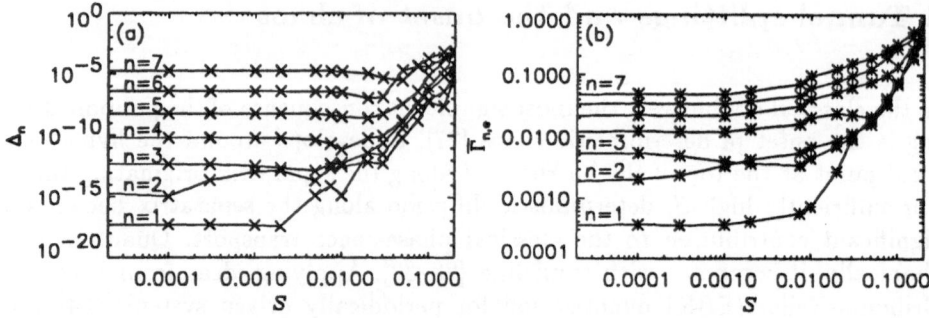

Fig. 4. Tunnel splittings (a) and overlaps with the chaotic layer (b) for the seven lowest tunnel doublets, as functions of the amplitude of the driving.

representation [36] of an eigenstate with the chaotic layer in phase space [17, 18]. To illustrate these concepts, the Husimi distribution for the quasienergy state $|\phi_7\rangle$ is superposed, in Fig. 3, with the corresponding classical phase space portrait, for (a) $S = 10^{-5}$ and for (b) $S = 0.2$, at phase 0 of the driving. In Fig. 4, we compare the S dependencies of the tunnel splittings and the overlaps for the seven quasienergy doublets from $|\phi_1\rangle$, $|\phi_2\rangle$ to $|\phi_{13}\rangle$, $|\phi_{14}\rangle$ [17, 18]. Qualitatively, they are strikingly similar: There is only a weak S dependence, reflecting the influence of the growing first resonance, for $S \lesssim 10^{-3}$. For $S \gtrsim 10^{-3}$, both the tunnel splittings and the overlaps start to grow exponentially, one by one, starting from the lowest doublet, so that the range of these quantities reduces by several orders of magnitude. The S regime of this steep increase coincides with that of the onset of chaotic motion in the classical dynamics. Insofar, the picture sketched above is confirmed. Details, however, need to be revised.

The notion that each splitting widens individually as the corresponding quantizing torus resolves, is not unambiguously corroborated by the data. It would imply that the transitions to a large splitting occur "from top to bottom", i.e., first for the doublet localized on the outermost torus pair within the separatrix. If this transition is assessed from the splittings passing a certain absolute threshold, e.g., $\Delta_\lambda = 10^{-4}$, that order is roughly obeyed. If, however, the point of onset of exponential growth, visible in a logarithmic plot like Fig. 4a, is taken as the criterion, the order is reversed. Another noticeable fact is that the widening of the splittings and the concomitant change in character of the eigenstates, as functions of S, are continuous processes that can only roughly be associated with the decay of a KAM torus, taken as a discrete event. Even doublet states overlapping by 70% with the chaotic layer may still show a relatively small splitting and exhibit the signature of a regular state in their spatial structure and time dependence. It is not clear whether this retarded decay of the tunnel doublets corresponds to the gradual disintegration of classical tori via *cantori* [37] and *vague tori* [38].

5 Driven tunneling with dissipation

In this section, we extend our working model (1), such as to include the influence of dissipation and noise on the microscopic level, following the usual procedure of coupling the central system to a heat bath [39, 40]. Proceeding in a similar way as in Ref. [41], we use the density operator in the Floquet basis, reduced to the double-well degree of freedom, as the basis of our description, and resort to the usual rotating-wave and Markov approximations. This allows us to derive the equation of motion for the density matrix $\tilde{\sigma}$ (in the interaction picture with respect to H_I) in the form of the master equation [41]

$$
\dot{\tilde{\sigma}}_{\alpha,\beta}(t) = \delta_{\alpha,\beta} \sum_\nu (W_{\alpha,\nu} \tilde{\sigma}_{\nu,\nu}(t) - W_{\nu,\alpha} \tilde{\sigma}_{\alpha,\alpha}(t)) +
$$
$$
\frac{1}{2}(1 - \delta_{\alpha,\beta}) \sum_\nu (W_{\nu,\alpha} + W_{\nu,\beta}) \tilde{\sigma}_{\alpha,\beta}(t). \tag{2}
$$

It comprises a closed subset of equations for the evolution of the diagonal elements towards a steady state, and another subset describing the decay of the non-diagonal elements. The coefficients $W_{\alpha,\beta}$ depend on the coupling constants and on the quasienergies; they are given elsewhere [16]. The classical limit of the dissipative quantal dynamics (specifying a linear frequency dependence of the coupling strength with cutoff) leads to a Langevin equation describing a bistable Duffing oscillator [6–9] showing Ohmic damping, with a macroscopic damping constant γ, and fluctuations.

A particularly interesting question is whether the coherent suppression of tunneling observed in the conservative case will survive in the presence of dissipation. In order to obtain an adequate description of this phenomenon on basis of a master equation like (2), the rotating-wave approximation must be avoided. It is valid only if the time scales of the classical relaxation and of the conservative quantal dynamics are clearly separated. However, in the vicinity of the manifolds $S_{\text{loc},k}(\omega)$ where the tunnel splitting vanishes (see Fig. 2a), exceedingly small energy scales and correspondingly large time scales occur in the undamped dynamics. This necessitates to take also quasienergy transitions into account that virtually violate energy conservation. Details of this refinement of the master equation are given in Refs. [16, 42].

Fig. 5a shows the time evolution of $P^\sigma(t_n) = \text{tr}[\sigma(t_n)\sigma(0)]$ (the analogue of $P^\Phi(t_n)$ for a density-matrix representation) at a parameter point very close to, but not exactly on $S_{\text{loc},0}(\omega)$, for $\gamma = 10^{-6}$ and various values of T. For low temperature, $P^\sigma(t_n)$ exhibits a slowly decaying coherent oscillation with a very long period, due to the slight offset from $S_{\text{loc},0}(\omega)$ [16]. Asymptotically, the distribution among the wells is completely thermalized. With increasing temperature, the decay time of the slow coherent oscillation first decreases until this oscillation is suppressed from the beginning (the corresponding part of the graph is not shown in Fig. 5b). After going through a minimum, however, the thermalization time increases again. At a characteristic temperature T^*, this time scale reaches a resonance-like *maximum* where the incoherent processes

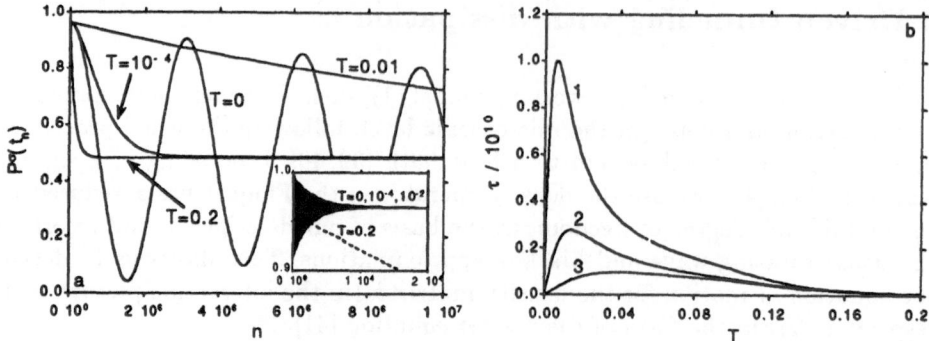

Fig. 5. Coherent suppression of tunneling in the presence of weak dissipation. (a) Time evolution of $P^\sigma(t_n)$ over the first 10^7 time steps, at a parameter point ($D = 2$, $\omega = 0.01$, and $S = 3.171 \times 10^{-3}$) close to $S_{\mathrm{loc},0}(\omega)$ (see Fig. 2a), for $\gamma = 10^{-6}$ and various values of T, starting from a pure, minimum-uncertainty state centered in one of the wells; (b) temperature dependence of the decay time τ of $P^\sigma(t_n)$ for three values of the detuning $\Delta\omega$ from the manifold $S_{\mathrm{loc},0}(\omega)$ (graph 1: $\Delta\omega = -1.4 \times 10^{-7}$, as in part (a), 2: $\Delta\omega = 5.0 \times 10^{-7}$ at $S = 3.1712 \times 10^{-3}$, 3: $\Delta\omega = 1.4 \times 10^{-6}$ at $S = 3.1715 \times 10^{-3}$). The other parameters are as in part (a). The data shown do not extend down to $T = 0$, where $\tau(T)$ diverges, but start only with the rising part of this function.

induced by the reservoir *stabilize* the localization of the wave packet and thus compensate for the detuning introduced deliberately. In Fig. 5b, we present the temperature dependence of the decay time τ (defined by $P^\sigma(t_n) \sim \exp(-n/\tau)$) for three values of the detuning $\Delta\omega$ from the manifold $S_{\mathrm{loc},0}(\omega)$. A variation of γ reveals that the dependence on the damping constant has a similar resonance-like form [42].

This stabilization of the coherent suppression of tunneling by noise is to be distinguished from the trivial localization by strong damping. In fact, it has already been observed in a model analogous to (1), but with the deterministic harmonic driving of the double well replaced by a noisy one, so that the time evolution remained unitary and a damping could not occur [14]. Rather, this phenomenon bears some resemblance to the quantum zeno effect in a bistable system [43], and to the classical stabilization of instable equilibrium states by multiplicative noise [44, 45].

Besides discussing the influence of dissipation on coherence effects, one may conversely ask how the classical dynamics of the driven damped Duffing oscillator [6–9] is modified by quantal interference. A hallmark of the classical dynamics is the existence of attractors of various degrees of complexity, as a function of ω, S, or γ. In Fig. 6 [42], we choose parameter values where classically, there exist five limit cycles, with the frequency of the driving: two symmetry-related pairs with one partner within each well, and a single one encircling the wells. The Husimi distribution in the stationary state at phase 0 is overlayed with the phase-space portrait of the corresponding conservative classical system (in a periodically driven system, the stationary state may still possess a time dependence with the period of the driving, which however is invisible in a stroboscopic plot like this). Both the classical attractors and the maxima of the quantal stationary distribution, while not coinciding exactly, are located near elliptic fixed points of the conservative dynamics and can be associated with the regular regions around the potential minima and the first resonance, respectively (the fifth limit cycle outside the wells is not significantly populated in the quantal stationary state). The quantum noise preferentially broadens the stationary distribution along the limit cycle to which the corresponding classical attractor belongs, the direction in which the classical phase-space flow is least contractive [46].

While the smoothing due to quantum noise is the only quantum effect left in the stationary state, a remnant of coherent tunneling survives in the transient behavior [42]. Fig. 7 shows the time evolution of $P^\sigma(t_n)$ for the same parameter values as above, with the initial states prepared as coherent states at the location of either one of the maxima of the asymptotic distribution (see Fig. 6) within the left well, corresponding to nonresonant motion (a) and to the first resonance (b), respectively. In both cases, we observe a coherent oscillation decaying as the stationary state is approached. Fig. 8 reveals that these oscillations form a remnant of tunneling between the partners of each of the two symmetry-related pairs of regular regions, i.e., tunneling between limit cycles. Thermalization within each pair, and subsequently among the pairs, is reached only on longer time scales of the order of the classical relaxation time.

6 Summary

The present paper is intended to highlight a number of facettes of a generic nonequilibrium system: The nonlinear quantum dynamics in a periodically driven double-well potential, at different stages of the transition from microscopic, coherent to macroscopic, incoherent behavior. (For the effects of periodic driving on the tunneling decay out of a single metastable state, see Ref. [47].) In the deep quantal regime, we find modifications, due to the driving, of the familiar tunneling. They range from a mere acceleration of its rate, in the two extremes of slow and of fast driving, through complex quantum beats near resonances with the unperturbed system frequencies, to an almost complete suppression of tunneling by a coherent mechanism.

Towards the semiclassical limit of the conservative system, we addressed the interplay between coherent transport by tunneling and diffusive transport along the chaotic layer which develops in the vicinity of the separatrix of the undriven system. Eigenstate doublets residing within the symmetry-related pair of regular regions of the classical phase space exhibit exponentially small splittings and thus support tunneling. As the pair of quantizing tori pertaining to such a doublet resolves in the chaotic sea, the splitting widens and tunneling gives way to a more complex dynamics contributing to the quantal counterpart of chaotic diffusion.

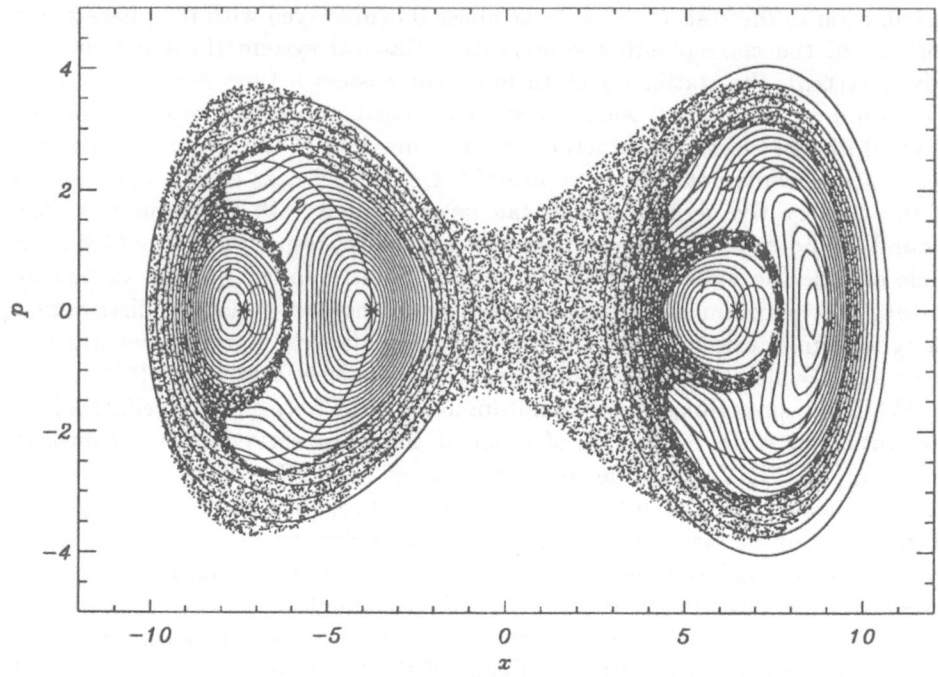

Fig. 6. Quantal stationary state. Contour plot of the asymptotic Husimi distribution at $D = 6$, $S = 0.0849$, $\omega = 0.9$, $\gamma = 10^{-5}$, and $T = 0$, compared to the limit cycles of the corresponding classical system (the positions on the cycles at phase 0 are indicated by asteriscs), and to the phase-space portrait of the corresponding conservative dynamics.

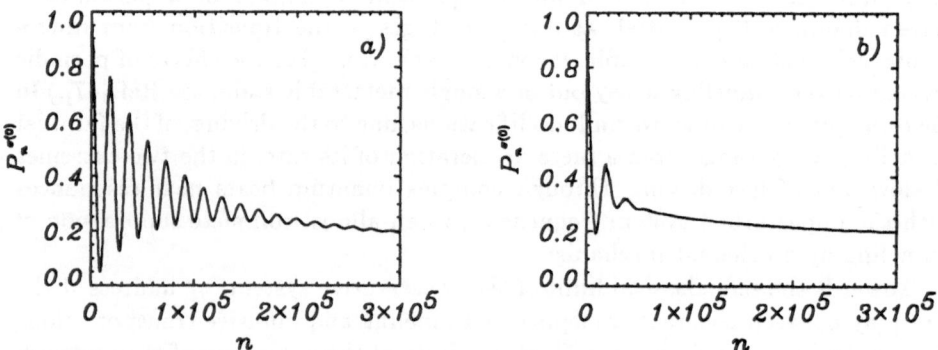

Fig. 7. Transient tunneling between limit cycles, in terms of the time evolution of $P^\sigma(t_n)$ over the first 3×10^5 time steps, at paramater values as in Fig. 6, but with $\gamma = 5 \times 10^{-6}$, for initial states prepared as coherent states located at either one of the maxima of the stationary Husimi distribution in the left well, i.e., at $p = 0$ and (a) $x = -7.5$, (b) $x = -4.2$.

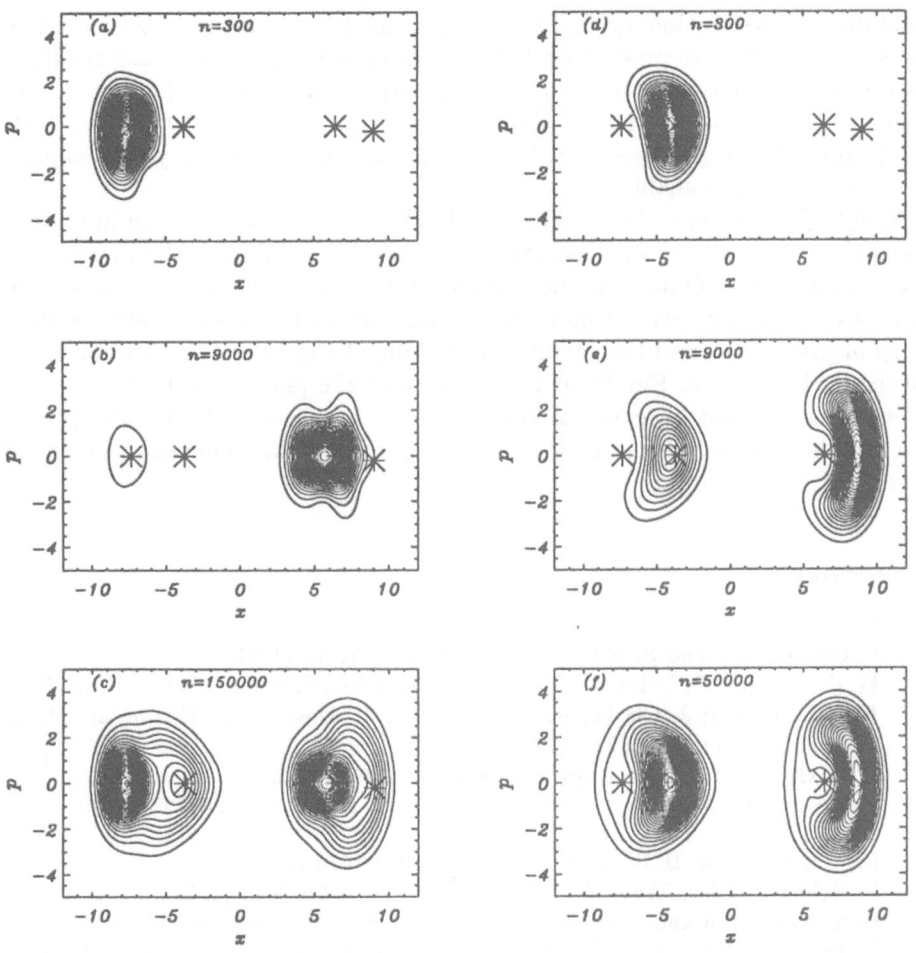

Fig. 8. Transient tunneling between limit cycles, in terms of the Husimi distribution at various times, for parameter values as in Fig. 7, and initial state for panels (a – c) as in Fig. 7a, for panels (d – f) as in Fig. 7b. The positions, at phase 0, on the limit cycles of the corresponding classical dynamics are indicated by asteriscs.

The disturbing effects of the ambient degrees of freedom render the coherence effects observed in the deep quantal regime transients. Surprisingly, the coherent suppression of tunneling is stabilized if damping constant and temperature take specific values, a result akin to the quantum zeno effect and to the classical stabilization of instable equilibria by multiplicative noise. On the time scale of classical relaxation, the dissipative quantum dynamics approaches a stationary state which forms the analogue of the attractors of the corresponding classical dynamics. These stationary states are broadened by quantum noise which is not isotropic, but acts preferentially in the direction where the classical phase-space flow is least contractive, e.g., stronger along limit cycles than transverse to them. On time scales shorter than the classical relaxation, coherence effects such as the tunneling between the basins of attraction of symmetry-related pairs of attractors remain visible.

Many questions have been left open. In the conservative case, an analytical description, in terms of semiclassical concepts, of tunneling in the presence of chaos is not yet available. In the parameter regime of the dissipative system where several strange attractors coexist, both transient tunneling between their basins of attraction, and the quantal smoothing of the fractal basin boundaries [48] should be studied. Finally, it is possible that the phenomenon of stochastic resonance, generated by external classical noise in periodically driven bistable systems [49], can be induced by the inherent quantum noise addressed in Sect. 5 as well.

References

1   S. Chakravarty and S. Kivelson, Phys. Rev. B **32** 76 (1985).

2   M. H. Devoret, D. Estéve, J. M. Martinis, A. Cleland, and J. Clarke, Phys. Rev. B **36** 58 (1987); J. Clarke, A. N. Cleland, M. H. Devoret, D. Estéve, and J. M. Martinis, Science **239** 992 (1988).

3   R. Bavli and H. Metiu, Phys. Rev. Lett. **69** 1986 (1992).

4   J. E. Combariza, B. Just, J. Manz, and G. K. Paramonov, J. Phys. Chem. **95** 10351 (1992).

5   L. M. Sander and H. B. Shore, Phys. Rev. B **3** 1472 (1969).

6   P. Holmes, Philos. Trans. R. Soc. London, Ser. A **292** 419 (1979).

7   B. A. Huberman and J. P. Crutchfield, Phys. Rev. Lett. **43** 1743 (1979).

8   Y. Ueda, J. Stat. Phys. **20** 181 (1979); Ann. N. Y. Acad. Sci. **357** 422 (1980).

9   W. Szemplińska-Stupnicka, Nonlinear Dynamics **3** 225 (1992).

10   F. Hund, Z. Phys. **43**, 803 (1927).

11   F. Grossmann, P. Jung, T. Dittrich, and P. Hänggi, Z. Phys. B **84** 315 (1991); Phys. Rev. Lett. **67** 516 (1991); .

12   F. Grossmann, T. Dittrich, and P. Hänggi, Physica B **175** 293 (1991).

13   F. Grossmann and P. Hänggi, Europhys. Lett. **18** 571 (1992).

14   F. Grossmann, T. Dittrich, P. Jung, and P. Hänggi, J. Stat. Phys. **70** 229 (1993).

15   T. Dittrich, F. Grossmann, P. Jung, B. Oelschlägel, and P. Hänggi, Physica A **194** 173 (1993).

16   B. Oelschlägel, T. Dittrich, and P. Hänggi, Act. Phys. Pol. B **24** 845 (1993); Europhys. Lett. **22** 5 (1993).

17   R. Utermann, T. Dittrich, and P. Hänggi, Phys. Rev. E **49** 273 (1994); Physica B **194-196** 1013 (1994).

18   T. Dittrich, F. Großmann, P. Hänggi, B. Oelschlägel, R. Utermann, in Proc. of the Third Max Born Symposium *Stochasticity and Quantum Chaos*, ed. by Z. Haba (Kluwer Academic, Dordrecht, in print).

19   J. H. Shirley, Phys. Rev. **138B** 979 (1965).

20   H. Sambe, Phys. Rev. A **7** 2203 (1973).

21   N. L. Manakov, V. D. Ovsiannikov, and L. P. Rapoport, Phys. Rep. **141** 319 (1986).

22   S. Chu, Adv. Chem. Phys. **73** 739 (1989).

23   A. Peres, Phys. Rev. Lett. **67** 158 (1991).

24   J. von Neumann and E. Wigner, Phys. Z. **30** 467 (1929).

25   J. M. Gomez Llorente and J. Plata, Phys. Rev. A **45** R6958 (1992).

26   L. Wang and J. Shao, Phys. Rev. A **49** R637 (1994).

27   L. E. Reichl and W. M. Zheng, in *Directions in Chaos*, vol. 1, ed. by H. B. Lin (World Scientific, Singapore 1987), p. 17.

28   M. J. Davis and E. J. Heller, J. Chem. Phys. **75** 246 (1986).

29   O. Bohigas, S. Tomsovic, and D. Ullmo, Phys. Rev. Lett. **64**, 1479 (1990); *ibid.* **65** 5 (1990).

30   S. Tomsovic and D. Ullmo, *Tunneling in the Presence of Chaos*, preprint 1990.

31   O. Bohigas, S. Tomsovic, and D. Ullmo, Phys. Rep. **223** 43 (1993).

32   W. A. Lin and L. E. Ballentine, Phys. Rev. Lett. **65**, 2927 (1990); Phys. Rev. A **45** 3637 (1992).

33   J. Plata and J. M. Gomez Llorente, J. Phys. A **25** L303 (1992).

34   H. P. Breuer and M. Holthaus, Ann. Phys. (N. Y.) **211** 249 (1991).

35   M. V. Berry and M. Robnik, J. Phys. A **17** 2413 (1984).

36   K. Husimi, Proc. Phys. Math. Soc. Jap. **22** 264 (1940).

37   L. E. Reichl, in *The Transition to Chaos: In Conservative and Classical Systems: Quantum Manifestations* (Springer, New York 1992), Chaps. 3.9, 9.5.1, and refs. therein.

38   W. P. Reinhardt, J. Phys. Chem. **86**, 2158 (1982); R. B. Shirts and W. P. Reinhardt, J. Chem. Phys. **77** 5204 (1982).

39   H. Haken, vol. XXV/2c of *Encyclopedia of Physics*, edited by S. Flügge (Springer, Berlin 1970).

40   W. H. Louisell, *Quantum Statistical Properties of Radiation* (Wiley, London 1973).

41   R. Blümel, A. Buchleitner, R. Graham, L. Sirko, U. Smilansky, and H. Walther, Phys. Rev. A **44** 4521 (1991).

42   B. Oelschlägel, Ph. D. thesis, University of Augsburg, unpublished (1993).

43   M. J. Gagen, H. M. Wiseman, and G. J. Milburn, Phys. Rev. A **48** 132 (1993).

44   R. Graham and A. Schenzle, Phys. Rev. A **26** 1676 (1982).

45   M. Lücke and F. Schank, Phys. Rev. Lett. **54** 1465 (1985).

46   T. Dittrich and R. Graham, Europhys. Lett. **4** 263 (1987); Ann. Phys. (N. Y.) **200** 363 (1990).

47   F. Grossmann and P. Hänggi, Europhys. Lett. **18** 1 (1992); Chem. Phys. **170** 295 (1993).

48   C. Grebogi, S. W. McDonald, E. Ott, and J. A. Yorke, Physics Letters **99A** 415 (1983); S. W. McDonald, C. Grebogi, E. Ott, and J. A. Yorke, Physica D **17** 125 (1985).

49   F. Moss, Ber. Bunsenges. Phys. Chem. **95** 303 (1991), and refs. therein.

From the Random Sequential Adsorption to the Ballistic model

P.Schaaf
Ecole Européenne des Hautes Etudes des Industries Chimiques de Strasbourg, 1 rue Blaise Pascal (URA 405, CNRS), 67008 Strasbourg Cedex, France
Institut Charles Sadron(CNRS-ULP), 6, rue Boussingault 67083 Strasbourg Cedex, France

Abstract

Irreversible adsorption processes are often described by the Random Sequential Adsorption (RSA) model, which has received considerable attention from a theoretical point of view. The validity of this model and of its generalizations is discussed in this article from both an experimental and a theoretical point of view. It appears that, due to cancelling between two different effects, hydrodynamic interactions and the diffusion of the particles in the bulk phase during the adsorption process, the RSA is a good first approximation to model the distribution function g(r) of systems of adsorbed particles. On the other hand, for adsorption kinetics no clear conclusions may be drawn, neither experimentally nor theoretically. For deposition of large particles influenced by gravity, the ballistic model is introduced and successfully compared to first experimental results.

I - Introduction

The Random Sequential Adsorption (RSA) model has received considerable attention from a theoretical point of view mainly due to potential applications to irreversible adsorption processes of proteins or particles on solid surfaces. The RSA model is defined by the following rules : (i) Particles are adsorbed randomly and sequentially. (ii) For each adsorption trial, the position of the particle is chosen randomly on the surface. If, at this position, the particle overlaps with an already adsorbed one, the trial is rejected and a new one is started. (iii) Once a particle is adsorbed it can neither diffuse on the surface nor desorb from it.

The model was first introduced by Flory [1] and later extensively investigated by Cohen and Reiss [2] in connection with chemical reactions on polymer chains. Feder and Giaever [3] suggested, some twenty years later, the use of the RSA model to account for their results on the adsorption of ferritin on carbon surfaces. This has been frequently cited as experimental support for studies of the RSA adsorption model from a theoretical point of view. In addition to this, RSA constitutes one of the simplest model that takes irreversibility into account and is therefore particularly inviting. Most of the theoretical results related to RSA have been reviewed recently by Evans [4] with a bibliography which attests to the large number of studies performed in this field.

At this point one may ask whether this model is in fact useful to describe real physical adsorption processes or whether it is not, after all, merely a toy for theoreticians. It is the purpose of this article to try to answer this question from the knowledge presently on hand. We will first discuss this question from an experimental point of view. Some theoretical results that should help to answer the question will be given in a second stage. Finally, generalizations of the RSA model to take gravitational effects into account, proposed over the last two years, will be discussed. The ballistic model, in particular, will be described and compared to recent experimental data.

II - Experimental results

We will discuss here only the RSA model on continuous surfaces and not consider adsorption processes which can be described as occuring on lattice surfaces. Thus, our discussion will not include phenomena such as catalysis. In the limit of continuous surfaces, RSA has been used to describe the adsorption of proteins and colloidal particles. The model was first proposed by Feder and Giaever [3,5] who studied the adsorption of a protein (ferritin) on carbon surfaces by means of electron microscopy. They have investigated the coverage $\theta(\infty)$ of the system in configurations corresponding to the jamming limit (a system reaches the jamming limit when no additional particle can be adsorbed on it). They also discussed the radial distribution function g(r) at the jamming limit. Good agreement was found between the RSA predictions and the experimental findings for both of these aspects. Feder concluded that "Both the saturation coverage and the radial distribution function of the ferritin on carbon using stain is remarkably well described by a random sequential filling model. Clearly more realistic models of protein adsorption with irreversible sticking can be constructed..." [5]. Note that the RSA model was suggested by referring to results obtained purely at the jamming limit. A few years later, Onoda and Liniger [6] studied the adsorption of small colloidal particles on solid surfaces. They determined the jamming limit coverage to be of the order of $\theta(\infty)=0.55\pm0.01$. The distribution function was not analyzed mainly because the observations were done by electron microscopy, a technique that requires drying the sample. Unfortunately, the positions of the particles may change slightly during the drying process. From their result, they also concluded that the RSA might be a good model to describe the process they studied. These two observations raise the following question : **are the jamming limit coverage and the distribution function at the jamming limit signatures of the deposition process?**

One of the most active groups trying to verify the validity of the RSA model is that of Adamczyk who studies the adsorption of latex particles on solid surfaces. Both the adsorption kinetics and the distribution of the particles on the surface have been analyzed. Their experimental results are summarized in a recent review paper [7]. For latex particle suspensions to be stable, the particles usually carry surface charges that induce a repulsion between them. To account for their influence, the experimental results were compared with a generalized RSA model that should, in principle, take interparticle interactions originating from a potential into account. In this model, the adsorption probability of a particle at a position r is proportional to $\exp(-\varphi(r)/kT)$, where $\varphi(r)$ is the potential at the position r of the interaction with the environment and in particular with the already-adsorbed particles, k is the Boltzmann constant and T is the absolute temperature. Since the interactions are usually repulsive between the particles (otherwise the suspension solution would not have been stable), the maximum obtainable coverage is smaller than the limit coverage expected from the "simple" RSA model. It has been proposed that the jamming limit coverage is obtained by renormalizing the RSA jamming limit coverage according to:

$$\theta^{max} = \theta^{max}(RSA)*(2a/d^*)^2 \tag{1}$$

where d^* corresponds to the center/center distance between particle of radius a when their interaction energy is of the order of 10kT (this last value being somewhat arbitrary). This law for the maximum coverage was tested by computer simulations and it is followed reasonably well[8]. The distribution of the particles on the surface was also analysed through the radial distribution function g(r). Here too, good agreement was observed between the experimental g(r) and the distribution function corresponding to the generalized RSA process as found from computer simulations. Good agreement between the adsorption kinetics found experimentally and that determined from computer simulations was observed over a large coverage range. Moreover, Adamczyk argues that the adsorption kinetics measured at long times exhibit a power law[7]:

$$\theta(\infty)-\theta(t) \propto t^{-1/2} \tag{2}$$

as is predicted from the "simple" RSA. In our opinion, this power law is not clearly demonstrated by the experimental data. In addition, it would be fortuitous if a generalized RSA model such as that introduced by Adamczyk would follow such a power law, which originates in geometrical constraints. One can, for example, observe that for particles diffusing toward a perfect adsorbing plane, in the asymptotic regime, and in the absence of hydrodynamic interactions, the adsorption kinetics still follow a power law, but with a power equal to -2/3 intead of -1/2 [9].

The kinetics of protein adsorption have also been tested against the RSA model. In a first article, Ramsden [10] has studied the adsorption of iron-free transferrin on solid surfaces by means of a powerful optical waveguide method. He tested the validity of the available surface function Φ as predicted from the RSA model up to third order in the coverage. He determined adsorption kinetic curves for different bulk concentrations of proteins. Each curve was compared to the RSA prediction using the size of the proteins as an adjustable parameter. The experimental curves were well described by the theoretical predictions but the deduced particle size varied by more then 30% with the bulk protein concentration. This was attributed to relaxation phenomena of the proteins on the surface, a phenomenon which is not taken into account in the basic RSA model. The asymptotic kinetic law was also determined and found to follow a power law that goes like $t^{-1/2}$. Here too, one must ask about the sensitivity of the results to the exponent -1/2. Finally, during these experiments, diffusion of the particles to the interface could have played a role that has not been taken into account. More recently, Kurrat et al. [11] performed adsorption experiments of serum albumin on solid surfaces. They analysed their results by taking the diffusion in the bulk and desorption into account. However, they still used the RSA available surface function expanded up to third order in the coverage even though this model does not take the diffusion into account. Within these approximations they could fit quite accurately their experimental results, the fitting parameters being the adsorption and desorption constants and the size of the particles.

The question that emerges out of these results concerns their sensitivity to the exact form of the available surface function. Is it possible, from the experimental data that are available and due to the great number of additional parameters that come into the problem, to discriminate between different models and in particular to decide whether the RSA model is valid?

RSA has also been used to model the adsorption of synthetic polymers on solid surfaces [12,13]. The experimental results seem to indicate that it describes accurately the adsorption kinetics if, during the adsorption process, the macromolecules undergo a fast structural reconformation at the first contact with the surface. In this configurations the macromolecule then forms a flat carpet with the monomers in immediate contact with the surface. Both for polystyrene adsorbing on silica in tetrachloride and for block copolymers (polyvinyl-4-pyridine/polystyrene) in toluene adsorbing on silica, the RSA kinetics obtained from simulations have been in accordance with the experiments over the entire coverage range, the only adjustable parameter being the number density of adsorbed molecules at the jamming limit.

From this discussion, one may conclude that there exist experimental evidence for the validity of the RSA model as a predictive tool to describe the adsorption kinetics and distributions of particles on surfaces but that most of this evidence is still not very strong due to the uncertainty related to the experimental data and to the uncertain sensitivity of the experimental results to the model. There is need for more experimental data confirming the model. More precisely, the adsorption kinetics should be analysed **over the total coverage range with the same model,** which is not always the case[10]. As far as the surface distribution is concerned, more data are needed with increasing precision to show if, indeed, the RSA model is valid or if it is only a first approximation.

III - Simulation results

One of the rules of the RSA model states that if the particle fails to adsorb during an adsorption trial, it is rejected and a new position chosen **randomly over the entire surface** is selected for a new trial. However, if you observe the adsorption process of particles on a surface through an optical microscope, you immediately realize the importance of the diffusion process of the particles in the vicinity of the surface, before touching it. In particular, it becomes obvious that for a particle wondering near one already adsorbed, the probability of adsorption in the direct vicinity of the adsorbed one is greater than the probability predicted by the RSA rules. The Diffusion RSA (DRSA) algorithm was introduced [14,15] in order to analyse the influence of this diffusion phenomena on the adsorption process. The most significant finding was that DRSA generates jammed configurations whose coverage and structure are, while not rigorously identical to those of the RSA process, indistinguishable (see figure 1) from these at the precision of usual experiments or computer simulations[16]. It implies that the distribution of the particles on the surface at the jamming limit as well as $\theta(\infty)$ are not signatures of the adsorption process. Thus, the results of Feder et al. [3,5] and of Onoda and Liniger [6] cannot serve as strong indices in favor of RSA. On the other hand, at lower coverages, the diffusion significantly modifies the form of the radial distribution function when compared with that of the RSA process (for the same

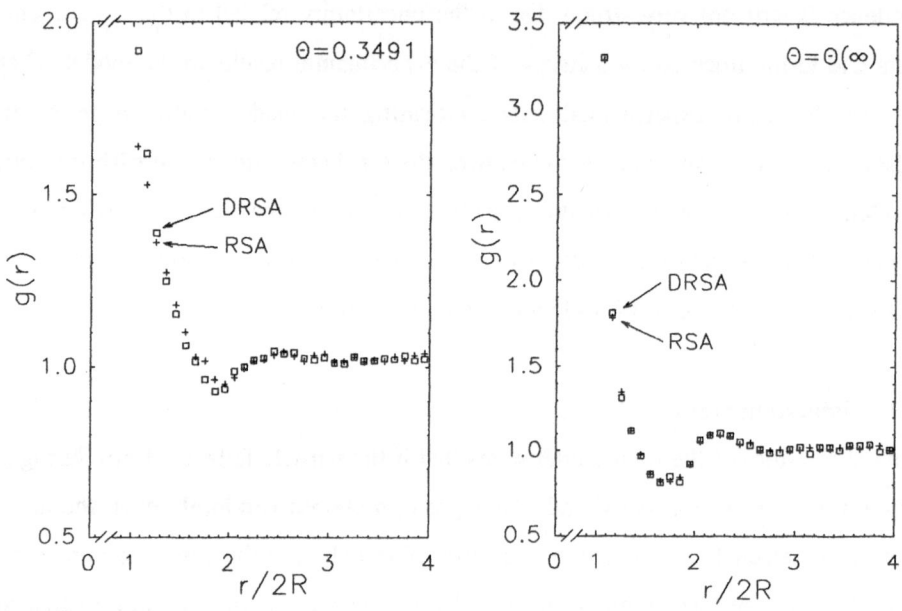

Figrure 1. Comparison of the radial distribution function g(r) of the adsorbed layer resulting from DRSA and RSA for coverage 6 equal to 0.3491 and for the jamming limit 6(∞).

coverage) as can be seen in figure 1. In particular, it is found, as expected, that the first peak in the g(r) is higher in the DRSA case than in the RSA one.

All these models assume that the particles interact through hard core potentials, which is not usually the case in experimental systems. Thus, DRSA simulations which take into account forces that derive from potentials have been performed recently[17]. The main goal of these simulations was to verify if Adamczyk's generalisation of the RSA model[8] i.e. to assume that the adsorption probability is proportional to $\exp(-\varphi/kT)$, is valid. It was found that within the generalized DRSA model, **this adsorption rule is not followed**. In particular, the adsorption probability appears larger then the value predicted by the Boltzman factor near previously adsorbed particles. On the other hand, the ratio of the adsorption probability $p_{pot}(\mathbf{r})$ when the interaction potential is taken into account in the DRSA model with the adsorption probability $p_0(\mathbf{r})$ for "pure" DRSA (in the presence only of hard core interactions) does follow approximately a Boltzmann law:

$$\frac{p_{pot}(\mathbf{r})}{p_o(\mathbf{r})} \propto \exp(-\varphi(\mathbf{r})/kT) \tag{3}$$

where $\varphi(\mathbf{r})$ represents the interaction potential at the position $\mathbf{r}$ at which the particle was adsorbed. This result would be in agreement with Adamczyk's rule if $p_0(\mathbf{r})$ were constant, independant of $\mathbf{r}$ for any position that does not correspond to an overlap between particles. This is the case in RSA but not in DRSA.

Compared to the RSA, the DRSA model already contains essential physical features such as the diffusion of the particles in the bulk. However, due to the fact that the diffusion coefficient is held constant, independant of the position of the particle with respect to the adsorption plane and to the position of the previously adsorbed particles, the hydrodynamic interactions are neglected : DRSA thus corresponds to an adsorption process in "dry water". The introduction of the different hydrodynamic interactions is a difficult task and renders the simulation programs extremely time-consuming. Thus, up to now, no simulations have been performed in which the kinetics and the particle distribution on the surface have been analyzed as a function of the coverage. To our knowledge, the only result available on this problem is that of Bafaluy at al. [18]. The adsorption of one particle in the presence of one or two adsorbed particles on the surface has been studied by taking into account hydrodynamic interactions between the adsorbing particle, the adsorbed ones and the adsorption plane. Thus the results of this study are only valid in the low to intermediate coverage range. Interparticle interactions other than hydrodynamic or hard core were neglected and the hydrodynamic interactions were evaluated by assuming additivity of the different friction tensors. Within these approximations, it has been shown that the particles seem to adsorb almost randomly on the surface without correlation with previously adsorbed spheres. This result originates in the tendency of hydrodynamic interactions to diminish the motion perpendicular to the adsorption plane in comparison to the lateral motion. The lateral diffusion then homogenizes the distribution of the particles. This result implies that, despite the complexity of the adhesion phenomenon, RSA is a suitable model to describe accurately the particle distribution on the surface. Even though these results only apply to coverages lower than or of the order of 30% (for higher coverages three-particle interactions become important [19]) it is likely to remain valid up to the jamming limit. For higher coverages, only small regions in space remain accessible to new particles. Then, during the diffusion of the particles toward the surface, a randomization of the particle in the direction parallel to the surface is likely to occur.

If one assumes that equation (3) remains valid when hydrodynamic interactions are included in the generalized DRSA model that takes interparticle forces deriving from a potential into account, one recovers Adamczyk's rule for generalized RSA. Such a result seems highly plausible due to the fact that after randomization of a particle over a large portion of the surface, one should recover a Boltzmann law. This assumption is also supported by the experimental result of Adamczyk related to the distribution function[7].

One can thus conclude that, due to hydrodynamic interactions, the RSA model or its generalizations that include the presence of interaction potentials between particles seem valid to describe the distribution of the particles on the surface. Computer simulations involving entire surfaces are however not yet available to prove this conclusion which is thus still weak and based only on the result of Bafaluy et al[18]. For the case of adsorption kinetics, no simulation results that include the hydrodynamic interactions are yet available. It is not clear that the presence of hydrodynamic interactions will happen to balance the effect of diffusion during adsorption (knowing the $t^{-1/3}$ at high coverages compared to the $t^{-1/2}$ for the pure RSA model) as it does for particle distributions. It is highly probable that the kinetics depend much more on the precise form of the interactions, hydrodynamic or other, and that general results will be more difficult, even if possible, to obtain. At this time, no prediction can be made for even the asymptotic law of the adsorption kinetics. Several groups have deduced from experimental data coverages increasing as $\theta(\infty)-\theta(t) \propto t^{-1/2}$ near jamming, but one may wonder what result would have been found in the absence of foreknowledge.

IV - From the RSA to the ballistic model

As has been discussed previously, there is need for more experimental results concerning the distribution function $g(r)$. One must however consider the fact that these experiments are most precisely done by means of optical microscopy and thus require the use of particles that are of the order or larger than $1\mu m$ in diameter. In this case gravity starts to play an important role and must be introduced in the deposition model. The DRSA model has been generalized accordingly [20]; it was found that the coverage at the jamming limit varies continuously between 0.546 and 0.61 as the radius of the particles increases up to infinity (see figure 2). In this latter case the adsorption process is said to be ballistic [21,22]. The whole deposition process depends only on a rescaled dimensionless radius R^* given by [20]:

$$R^* = R(4\pi\Delta\rho g / 3kT)^{1/4}$$

where R is the radius of the particles, $\Delta\rho$ the density difference between the particles and water, the other symbols having their usual meaning. One can see from figure 2 that for $R^* < 1.5$ the system behaves according to RSA as far as the jamming limit coverage is concerned, and that it reaches its ballistic value for $R^* > 4$.

Experiments have been carried out for particles of radii $R^* \approx 3.5$ which is close to the ballistic limit [23]. The radial distribution functions of the particle configurations were determined over a wide range of coverages. The experimental distribution functions were compared to the $g(r)$ determined from computer

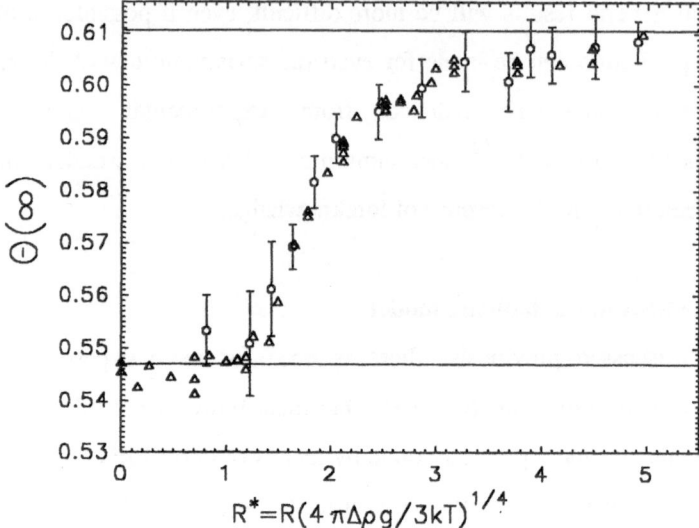

Figure 2. Variation of the coverage at the jamming limit as a function of the reduced sphere radius R* obtained from DRSA simulation taking into account gravitational effects.

simulations. Figure 3 represents g(r) for a coverage equal to θ=58.5% which is close to the jamming limit. Even though the agreement between the two curves is quite good, some discrepancies remain : the simulated distribution functions were lower than the experimental observations for distances slightly above the first and second peaks. This effect was first attributed to some polydispersity in the size of the particles under study. However, by including the polydispersity in the simulations no clear improvement was observed

n comparison to the experimental data. Hydrodynamic interactions not included in the ballistic model, have also been proposed to account for these differences. Recent simulations performed by Pagonabarraga and Rubi [24] for the one dimensional ballistic case including hydrodynamic interactions show that, indeed, the effect of these latter is to decrease the adsoption probability in the close vicinity of already adsorbed particles and that this effect is more pronounced at low to intermediate coverages than near the jamming limit. Computer simulations for the 2 dimensional case are currently under way to verify this result for the experimental systems.

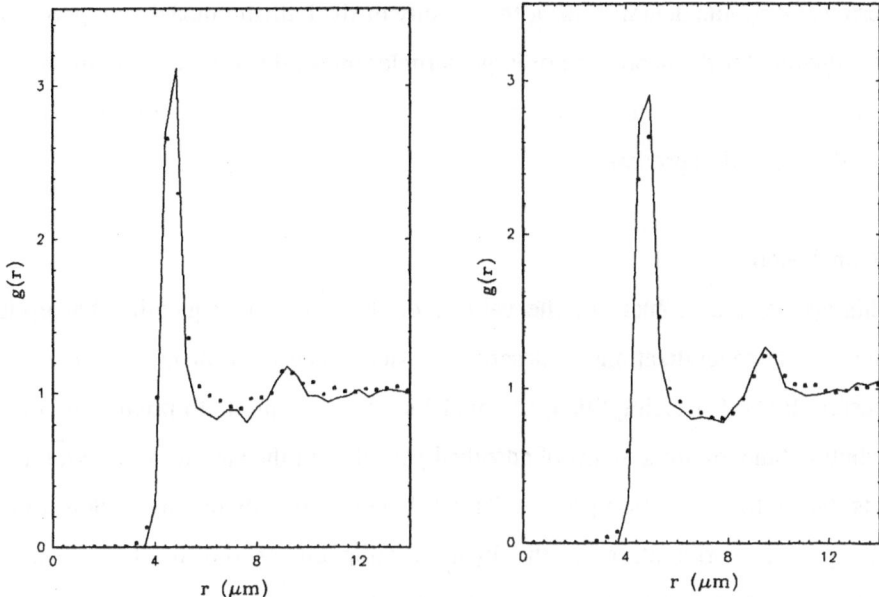

Figure 3. Radial distribution function g(r) for surfaces covered by melamine particles at coverage θ = 35.4% (left) and 58.5% (right). (•) : experimental results, (—) : ballistic deposition simulation .

These experiments also revealed that the fluctuation of the number of particles on surfaces of small area extracted from an infinitely large surface also contains informations about the deposition process[23]. One of the characteristics of the reduced variance $\sigma^2/<n>$ of the deposition processes under gravity is to vary as

$$\sigma^2/<n>=-1-\alpha\theta^3+o(\theta^4)$$

where α is some constant that depends on the process and is equal to 9.612 for the ballistic case. For the RSA case, the reduced variance varies as $1-4\theta+o(\theta^2)$. More generally it can be shown [25] that if the available surface function Φ can be expanded in the coverage θ as $\Phi=1-\alpha_n\theta^n+o(\theta^{n+1})$ then the reduced variance is equal to

$$\sigma^2/<n>=1-\alpha_n\theta^n+o(\theta^{n+1})$$

For the fluctuations as well, good agreement was found between the experiments and the ballistic simulations. Reference 23 constitutes to our knowledge the first experimental results demonstrating the validity of the ballistic model as a good first approximation for the deposition of large particles on solid surfaces. Experiments are currently under way to study, from an experimental point of view, the influence of R^* on the deposition process.

V - Conclusion

In this article, the problem of the validity of the Random Sequential Adsorption model and its generalizations to describe physical adsorption phenomena has been addressed. Even if oversimplified, the model seems appropriate to predict the radial distribution function for systems of adsorbed particles, in the absence of gravitational effects during the deposition process. This behavior is mainly due to randomization of the particles originating in the hydrodynamic effects during the deposition process. No such statement can be made for the adsorption kinetics.

In the presence of strong gravitational effects, the ballistic model seems appropriate as has been demonstrated from an experimental study even if some discrepancies remain in the radial distribution function between experiment and simulation. These are attributed to hydrodynamic interactions and studies are presently under way to investigate this problem.

aknowledgments

The author is indented to E.Mann and B.Senger for stimulating discussions about this subject. This work was supported by the NATO (Grant No 8908872) by the Commission of the European Communities (Contract No SC1-CT-91-0696(TSTS)) and by the NSF/CNRS

[1] P.J. Flory, *J. Am. Chem. Soc.* **61**, 1518 (1939)

[2] E.R. Cohen, H. Reiss, *J. Chem. Phys.* **38**, 680 (1963)

[3] J. Feder, I. Giaever, *J. Coll. Interface Sci.* **78**, 144 (1980)

[4] J.W. Evans, *Rev. Mod. Phys.* **65**, 1281 (1993)

[5] J. Feder, *J. Theor. Biol.* **87**, 237 (1980)

[6] G.Y. Onoda, E.G. Liniger, *Phys. Rev. A* **33**, 715 (1986)

[7] Z. Adamczyk, B. Siwek, M. Zembala, P. Belouschek, *Adv. Colloid Interface Sci.* **48**, 151 (1994)

[8] Z. Adamczyk, M.Zembala, B.Siwek, P.Warszynski, J.Colloid Interface Sci. **140**, 123 (1990)

[9] P.Schaaf, A.Johner, J.Talbot, Phys. Rev. Lett. **66**, 1603 (1991)

[10] J.J. Ramsden, *Phys. Rev. Lett.* **71**, 295 (1993)

[11] R. Kurrat, J.J. Ramsden, J.E. Prenosil, *J. Chem. Soc. Faraday Trans.* **90**, 587 (1994)

[12] A. Elaissari, A. Haouam, C. Huguenard, E. Pefferkorn, *J. Colloid Interface Sci.* **149**, 68 (1992)

[13] C. Huguenard, E. Pefferkorn, *Macromolecules* (submitted)

[14] B. Senger, J.-C. Voegel, P. Schaaf, A. Johner, A. Schmitt, J.Talbot, *Phys. Rev. A* **44**, 6926 (1991)

[15] B. Senger, P. Schaaf, J.-C. Voegel, A. Johner, A. Schmitt, J. Talbot, *J. Chem. Phys.* **97**, 3813 (1992)

[16] B. Senger, J. Talbot, P. Schaaf, A. Schmitt, J.-C. Voegel, *Europhys. Lett.* **21**, 135 (1993)

[17] B. Senger PNAS

[18] J. Bafaluy, B. Senger, J.C. Voegel, P. Schaaf, *Phys. Rev. Lett.* **70**, 623 (1993)

[19] P. Schaaf, J. Talbot, *J. Chem. Phys.* **91**, 4401 (1989)

[20] B. Senger, F.J. Bafaluy, P. Schaaf, A. Schmitt, J.-C. Voegel, *Proc. Natl. Acad. Sci. USA*, **89**, 9449 (1992)

[21] R. Jullien, P. Meakin, *J. Phys. A* **25**, L189 (1992)

[22] A.P. Thompson, E.D. Glandt, *Phys. Rev. A* **46**, 4639 (1992)

[23] P. Wojtaszczyk, P. Schaaf, B. Senger, M. Zembala, J.-C. Voegel, *J. Chem. Phys.* **99**, 7198 (1993)

[24] I. Pagonabarraga, M. Rubi,*Phys. Rev. Lett.* **73**, 114 (1994)

[25] P. Schaaf, P. Wojtaszczyk, E.Mann, B. Senger, J.-C. Voegel, (in preparation)

High Frequency Viscosity of a Dilute Polydisperse Emulsion

R. B. Jones

Department of Physics, Queen Mary & Westfield College, Mile End Road, London E1 4NS, United Kingdom

Abstract: We discuss the effect of polydispersity on the Huggins coefficient of a dilute emulsion of spherical droplets in which all droplets are composed of the same material but differ greatly in size.

1. Introduction

The study of the effective viscosity of a suspension of spherical particles has a distinguished history starting with Einstein's result [1] for a dilute system of identical hard spheres of radius a suspended in a liquid of shear viscosity η,

$$\eta_{\text{eff}} = \eta(1 + \frac{5}{2}\phi) \quad .$$ (1)

Here $\phi = 4\pi a^3 n/3$ is the volume fraction of suspended particles and n is their number density. This virial series expansion approach for dilute suspensions has been extended to treat effects through second order in ϕ [2 -4]. Einstein's first order term arises from the flow perturbation due to the presence of a single sphere while second order contributions arise from two-body interactions and consist of two parts, a short-time (infinite frequency) contribution arising from the instantaneous hydrodynamic interaction of the suspended spheres [3] in a fixed configuration and a long-time (zero-frequency) contribution arising from the change of configuration due to Brownian motion [4]. For simplicity, we shall concentrate on the infinite frequency contribution only, writing the second order expression for the short-time effective viscosity of a monodisperse suspension as

$$\eta_{\infty}^{\text{eff}} = \eta \left(1 + [\eta]\phi + k_{\text{H}}[\eta]^2 \phi^2 \right) \quad ,$$ (2)

where $[\eta]$ is the intrinsic viscosity of the suspension and k_{H} is the Huggins coefficient. Einstein's result was for hard spheres with stick boundary conditions at the surface. However, a linear response calculation of $\eta_{\infty}^{\text{eff}}$ has been presented which enables one to treat other types of boundary conditions and other types of spherical particles such as droplets or porous spheres [5,6]. For some classes of dispersions, however, the condition of monodispersity is badly violated. Emulsions of spherical droplets, which easily break under strong shear fields [7] , can show size polydispersity with great variation of droplet radius; polydisperse suspensions of hard particles may be produced by deliberate mixing of different sized particles [8]. It is of interest to understand the influence of polydispersity on the effective viscosity of such dispersions.

2. Polydisperse Formalism

The viscosity of a suspension differs from that of its pure suspending liquid due to the modification of an incident flow field by the presence of the freely moving suspended particles. Each particle is equivalent to an induced force density [9] acting on the fluid and the contribution of each particle to the effective viscosity of the suspension can be expressed in terms of the induced force dipole moment $\mathcal{F}_{\alpha\beta}^{(2S)}$ which it possesses. To describe a polydisperse suspension we assume that each particle belongs to one of p species indicated by a Greek subscript. Particles of species σ have radius a_σ and number density n_σ. Consider an incident linear flow field of the form

$$v^{\mathrm{inc}}(r) = v_0(r) = g \cdot r \quad , \tag{3}$$

with $g_{\alpha\beta}$ a symmetric traceless constant tensor. If we write

$$\eta_\infty^{\mathrm{eff}} = \eta + \Delta\eta_\infty \quad , \tag{4}$$

then, by suitably generalizing the cluster expansion treatment of a monodisperse suspension [5 , 10], the change in viscosity can be expressed as

$$
\begin{aligned}
2\Delta\eta_\infty\, g_{\alpha\beta} = &- \sum_{\sigma=1}^{p} n_\sigma \mathcal{F}_{\alpha\beta}^{(2S)}(1_\sigma) \\
&- \sum_{\sigma,\tau=1}^{p} \int dR_{2_\tau} \left[n(1_\sigma, 2_\tau) - n_\sigma n_\tau \right] \left[\mathcal{F}_{\alpha\beta}^{(2S)}(1_\sigma|2_\tau) - \mathcal{F}_{\alpha\beta}^{(2S)}(1_\sigma) \right] \\
&- \sum_{\sigma,\tau=1}^{p} \int dR_{2_\tau}\, n(1_\sigma, 2_\tau) \left[\mathcal{F}_{\alpha\beta}^{(2S)}(1_\sigma, 2_\tau) - \mathcal{F}_{\alpha\beta}^{(2S)}(1_\sigma|2_\tau) \right] \quad .
\end{aligned}
$$

$$\tag{5}$$

Here there are one-body and two-body terms and the labels 1_σ and 2_τ refer to particles 1 and 2 of species σ and τ whose centres are located at positions R_1 and R_2 respectively. The number densities of suspended particles of species σ and τ are n_σ and n_τ while $n(1_\sigma, 2_\tau)$ is the pair density. In the first term, $\mathcal{F}^{(2S)}(1_\sigma)$ is the force dipole moment induced on particle 1_σ when it is placed in isolation at R_1 in the incident flow v_0. In the second term of (5) we note that $n(1_\sigma, 2_\tau)$ vanishes when $|R_2 - R_1| < a_\sigma + a_\tau$ so that the integration region includes an overlap of the two particle volumes. For an isotropic system the entire contribution of this term comes from the virtual overlap configuration [11]. The quantity $\mathcal{F}^{(2S)}(1_\sigma|2_\tau)$ is the force dipole moment induced on particle 1_σ when it is placed by itself in the (non-linear) incident field $\boldsymbol{v}^{\text{inc}}(2_\tau)$ which results when particle 2_τ is placed by itself at $\boldsymbol{R}_2$ subject to the linear incident field $\boldsymbol{v}_0$. In the third term, $\mathcal{F}^{(2S)}(1_\sigma, 2_\tau)$ is the force dipole moment induced on particle 1_σ at $\boldsymbol{R}_1$ when it is subject, together with particle 2_τ at $\boldsymbol{R}_2$, to the linear incident field $\boldsymbol{v}_0$. The subtracted term $\mathcal{F}^{(2S)}(1_\sigma|2_\tau)$ in the final integral in (5) removes the long distance parts of $\mathcal{F}^{(2S)}(1_\sigma, 2_\tau)$ so that the integral is absolutely convergent.

The expressions in (5) can be evaluated in terms of complete solutions to the hydrodynamic one- and two-body problems [12 - 14]. To calculate the integrals we use the isotropic low density hard sphere pair density, $n(1_\sigma, 2_\tau) = n_\sigma n_\tau \theta(|R_{2\tau} - R_{1\sigma}| - a_\sigma - a_\tau)$, to give [15]

$$-\sum_{\sigma=1}^{p} n_\sigma \mathcal{F}_{\alpha\beta}^{(2S)}(1_\sigma) = \frac{8\pi\eta}{3} \sum_{\sigma=1}^{p} n_\sigma A_{20}(1_\sigma) g_{\alpha\beta} \quad ,$$

$$-\sum_{\sigma,\tau=1}^{p} \int dR_{2\tau} \left[n(1_\sigma, 2_\tau) - n_\sigma n_\tau \right] \left[\mathcal{F}_{\alpha\beta}^{(2S)}(1_\sigma|2_\tau) - \mathcal{F}_{\alpha\beta}^{(2S)}(1_\sigma) \right] =$$

$$\frac{4\eta}{5} \left(\frac{\sum_{\sigma=1}^{p} 4\pi n_\sigma A_{20}(1_\sigma)}{3} \right)^2 g_{\alpha\beta} \quad ,$$

$$-\sum_{\sigma,\tau=1}^{p} \int dR_{2\tau}\, n(1_\sigma, 2_\tau) \left[\mathcal{F}_{\alpha\beta}^{(2S)}(1_\sigma, 2_\tau) - \mathcal{F}_{\alpha\beta}^{(2S)}(1_\sigma|2_\tau) \right] =$$

$$\frac{8\pi}{15} \sum_{\sigma,\tau=1}^{p} n_\sigma n_\tau \int_{a_\sigma+a_\tau}^{\infty} R^2\, dR \Big[\alpha_{1_\sigma 1_\sigma}^{dd}(R)$$

$$+ \alpha_{1_\sigma 2_\tau}^{dd}(R) + \frac{3}{2}\left(\beta_{1_\sigma 1_\sigma}^{dd}(R) + \beta_{1_\sigma 2_\tau}^{dd}(R)\right) + 3\left(\gamma_{1_\sigma 1_\sigma}^{dd}(R) + \gamma_{1_\sigma 2_\tau}^{dd}(R)\right) \Big] g_{\alpha\beta} \quad .$$
$$(6)$$

Here $A_{20}(1_\sigma)$ is a hydrodynamic scattering coefficient [12 - 14], one of an infinite number of such coefficients which describe the flow disturbance produced when a single spherical particle of species σ is subject to a general $\boldsymbol{v}^{\text{inc}}$. These coefficients depend on the size of the particle and upon its structure (hard particle, droplet, porous particle). The evaluation of the overlap term depends on an integral theorem for the Oseen tensor as explained in [11]. The quantities $\alpha_{ij}^{dd}(R)$, $\beta_{ij}^{dd}(R)$, $\gamma_{ij}^{dd}(R)$, where $R = |R_2 - R_1|$, are scalar two-body mobility functions defined in [12 - 14] where it is shown how they can be computed numerically from the one-body scattering coefficents to arbitrary accuracy in series of powers of $1/R$.

From the expressions in (6) we may finally express the effective viscosity of a polydisperse suspension to second order in the total volume fraction ϕ of suspended particles in the form

$$\eta_\infty^{\text{eff}} = \eta \left[1 + [\eta]^{\text{poly}} \phi + k_{\text{H}}^{\text{poly}} ([\eta]^{\text{poly}} \phi)^2 \right] \quad , \qquad (7)$$

$$\phi = \sum_{\sigma=1}^{p} \phi_\sigma \quad , \quad k_{\mathrm{H}}^{\mathrm{poly}} = \frac{2}{5} + \Delta k_{\mathrm{H}}^{\mathrm{poly}} \quad , \tag{8}$$

with $\phi_\sigma = 4\pi a_\sigma^3 n_\sigma/3$, and $[\eta]^{\mathrm{poly}}$ is defined by the volume fraction weighted average of the intrinsic viscosities of the different species,

$$[\eta]^{\mathrm{poly}} = \frac{1}{\phi} \sum_{\sigma=1}^{p} [\eta]_\sigma \phi_\sigma \quad , \quad [\eta]_\sigma = \frac{A_{20}(1\sigma)}{(a_\sigma)^3} \quad . \tag{9}$$

The contribution of $2/5$ to $k_{\mathrm{H}}^{\mathrm{poly}}$ arises from the virtual overlap contribution and is independent of the polydispersity while the part denoted $\Delta k_{\mathrm{H}}^{\mathrm{poly}}$ is sensitive to the polydispersity and is expressed in terms of the two-body hydrodynamic interaction functions as

$$\Delta k_{\mathrm{H}}^{\mathrm{poly}} = \frac{\sum_{\sigma,\tau=1}^{p} \phi_\sigma M_{\sigma\tau} \phi_\tau}{\left(\sum_{\sigma=1}^{p} [\eta]_\sigma \phi_\sigma \right)^2} \quad , \quad M_{\sigma\tau} = \int_0^1 dt\, G_{\sigma\tau}(t)\, t^{-4} \quad , \tag{10}$$

with the function $G_{\sigma\tau}(t)$ defined by

$$\begin{aligned}
G_{\sigma\tau}(t) = \frac{3\,(a_\sigma + a_\tau)^3}{20\pi\eta a_\sigma^3 a_\tau^3} \Big[&\alpha_{1\sigma1\sigma}^{dd} + \alpha_{1\sigma2\tau}^{dd} \\
&+ \frac{3}{2}(\beta_{1\sigma1\sigma}^{dd} + \beta_{1\sigma2\tau}^{dd}) + 3(\gamma_{1\sigma1\sigma}^{dd} + \gamma_{1\sigma2\tau}^{dd}) \Big] \quad .
\end{aligned} \tag{11}$$

The dimensionless variable t is defined by $t = (a_\sigma + a_\tau)/R$, with $R = |R_{2\tau} - R_{1\sigma}|$.

3. Droplets

We now specialize to a suspension of spherical droplets all made of the same material, characterized by a shear viscosity η', but with a distribution of droplet sizes. The different species introduced above correspond to droplets of different radii. In such a case the intrinisic viscosity is species independent so that

$$[\eta]_\sigma = [\eta] = \frac{5 + 2\nu}{2(1 + \nu)} \quad , \quad \nu = \eta/\eta' \quad , \tag{12}$$

where we have used the scattering coefficient A_{20} appropriate to spherical droplets [13]. From (9) we see that the polydisperse intrinsic viscosity is identical with that for a monodisperse suspension of the same total volume fraction., $[\eta]^{\mathrm{poly}} = [\eta]$. The matrix $M_{\sigma\tau}$ is a function of the viscosity ratio ν and the size ratio $f = a_\tau/a_\sigma$, $M_{\sigma\tau} = M_{\sigma\tau}(f,\nu) = M_{\tau\sigma}(1/f,\nu)$. In the limits $\nu \to 0$ and $\nu \to \infty$ $M_{\sigma\tau}$ reduces to the value appropriate to a hard sphere with stick or slip boundary conditions respectively. As explained elsewhere [15] we have computed the quantity $M_{\sigma\tau}(f,\nu)$ for a wide range of values of f and ν by a high order series expansion of $G_{\sigma\tau}(t)$ (through terms of order t^{150}) coupled with an extrapolation procedure [6]. However, it is possible to give explicit results for the lowest order terms which contribute to $M_{\sigma\tau}$. For $G_{\sigma\tau}$ we have

$$G_{\sigma\tau} = \frac{3}{2}\frac{(1+f)^6}{f^3}\left(c_6^{\sigma\tau}t^6 + c_8^{\sigma\tau}t^8 + c_9^{\sigma\tau}t^9 + c_{10}^{\sigma\tau}t^{10} + \cdots\right) \quad , \tag{13}$$

which gives a corresponding expression for $M_{\sigma\tau}$

$$M_{\sigma\tau} = M_{\sigma\tau}^{(6)} + M_{\sigma\tau}^{(8)} + M_{\sigma\tau}^{(9)} + M_{\sigma\tau}^{(10)} + \cdots \quad . \tag{14}$$

The explicit forms of the low order $M_{\sigma\tau}^{(n)}$ are

$$M_{\sigma\tau}^{(6)} = \frac{3}{20}\left[\frac{5+2\nu}{1+\nu}\right]^3 \frac{1}{(1+f)^3} \quad ,$$

$$M_{\sigma\tau}^{(8)} = \frac{3}{40}\left[\frac{5+2\nu}{1+\nu}\right]^2 \frac{1}{(1+f)^5}\left[-16\frac{1+f^2}{1+\nu} + \frac{3f^2}{5}\frac{3-2\nu}{1+\nu} + \frac{59f^2}{15}\frac{7+2\nu}{1+\nu} + \frac{8f^2}{3}\frac{1-\nu}{1+4\nu} - \frac{3f^2}{(1+\nu)(3+2\nu)}\right] \quad ,$$

$$M_{\sigma\tau}^{(9)} = \frac{3}{80}\left[\frac{5+2\nu}{1+\nu}\right]^4 \frac{f^3}{(1+f)^6} \quad ,$$

$$M_{\sigma\tau}^{(10)} = \frac{3}{28}\left[\frac{5+2\nu}{1+\nu}\right]^2 \frac{1}{(1+f)^7}\left[\frac{28+56f^2}{(1+\nu)(5+2\nu)} - \frac{35(7+2\nu)f^2}{(1+\nu)(5+2\nu)} - \frac{35f^4}{1+\nu} + \frac{14f^4}{3}\frac{1-\nu}{2+5\nu} + \frac{23f^4}{5}\frac{9+2\nu}{1+\nu} + \frac{4f^4}{3}\frac{5-2\nu}{1+\nu}\right] \quad . \tag{15}$$

4. Results and Discussion

The effect of polydispersity is contained within the quantity Δk_H^{poly} given in (10) as a ratio of quadratic forms in the partial volume fractions. For a fixed total volume fraction we can vary these partial volume fractions to ensure that Δk_H^{poly} takes an extremum value for a given total volume fraction ϕ. In [15] we have studied this extremum problem in detail, particularly for the limit $\nu \to 0$ which corresponds to stick boundary condition hard spheres. We found that, for bimodal distributions with only two sizes of droplet, the extremum occurs when the total volume fraction is split evenly between the two sizes. In the limit $\nu \to 0$ (stick hard spheres) Δk_H^{poly} is reduced relative to its value in a monodisperse suspension while for the inviscid limit $\nu \to \infty$ (slip hard spheres) it is increased. In other words, polydispersity reduces the viscosity for extremely viscous droplets and increases the viscosity for inviscid droplets as compared with a monodisperse suspension of the same total volume fraction. Our results for the limit $\nu \to 0$ agree quantitatively with those calculated independently for stick hard spheres by a different method [16]. Some numerical results are shown in Fig. 1 where we plot, as a function of ν, values of $R = \Delta k_H^{\text{poly}}/\Delta k_H^{\text{mono}}$, which is the ratio of the value of Δk_H in a bimodal suspension (droplets of two sizes with equal partial volume fractions) to its value in a monodisperse suspension of the same total volume fraction.

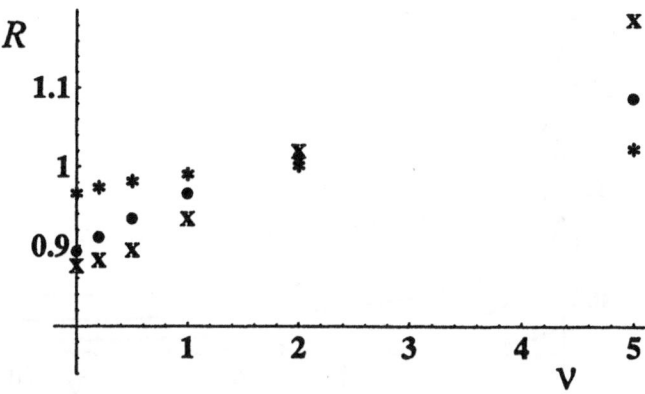

Fig. 1. Values of the ratio $R = \Delta k_{\mathrm{H}}^{\mathrm{poly}} / \Delta k_{\mathrm{H}}^{\mathrm{mono}}$ for $\nu = 0.0, 0.2, 0.5, 1.0, 2.0, 5.0$ at three different size ratios, $f = a_2/a_1 = 0.5(*), 0.2(\bullet), 0.0(X)$.

Since $k_{\mathrm{H}} = 2/5 + \Delta k_{\mathrm{H}}$, the change of k_{H} is not as great as the change in Δk_{H}, but if one species is much smaller in size than the other there are significant changes to k_{H} . For very viscous droplets, $\nu \approx 0$, k_{H} is reduced by 6% while in the inviscid limit, $\nu \to \infty$, k_{H} is increased by 20%. We also studied cases with three and four different sizes of droplet. The polydispersity effects are enhanced by having more than two species present. Thus for stick hard spheres, the monodisperse value of Δk_{H} is 0.400, while for two, three and four species it can fall to 0.350, 0.331 and 0.329 respectively. In these extremal cases the smallest droplets must have a radius less than a few percent of the radius of the largest particles. In addition to requiring a species of quite small particles to be present, the viscosity ratio ν must be reasonably large, $(\nu \gtrsim 5.0)$, or small ,$(\nu \lesssim 0.5)$, to get a significant change in the Huggins coefficient due to polydispersity. If these conditions are not met, the predicted change in the Huggins coefficient is so small as to be within the uncertainties of experimental measurement.

The qualitative difference between suspensions of very viscous or almost inviscid droplets arises from the details of how the droplets couple to an incident flow field. The scattering coefficient $A_{\ell 2}(\sigma)$, which describes how strongly an outgoing pressure field is excited by a droplet, has the form [12 - 14]

$$A_{\ell 2}(\sigma) = \frac{2\ell + 1}{2(1 + \nu)} a_\sigma^{2\ell+1} \quad , \tag{16}$$

so that it vanishes for inviscid droplets. This strong dependence on ν seems to make the main difference between the two limiting cases. In particular, A_{22} contributes to $M_{\sigma\tau}^{(8)}$ and $M_{\sigma\tau}^{(10)}$ in the small particle limit ($f \approx 0$) leading to a marked dependence of M_{12} upon ν as shown in Fig. 2.

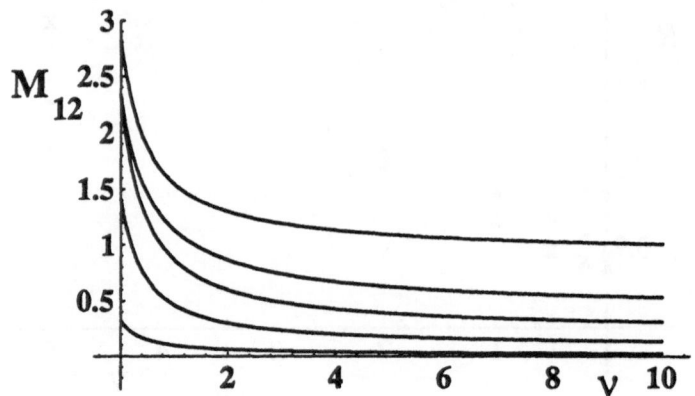

Fig. 2. Values of $M_{12}(f, \nu)$ as given approximately by (15). The f values increase monotonically from the top curve to the bottom curve, $f = 0.1, 0.5, 1.0, 2.0, 5.0$.

References

1.   A. Einstein: Ann. Physik **19** 289 (1906), Ann. Physik **34** 591 (1911)
2.   J. M. Peterson and M. Fixman: J. Chem. Phys. **39** 2516 (1963)
3.   G. K. Batchelor and J. T. Green: J. Fluid Mech. **56** 401 (1972)
4.   G. K. Batchelor: J. Fluid Mech. **83** 97 (1977)
5.   B. U. Felderhof and R. B. Jones: Physica **146A** 417 (1987)
6.   B. Cichocki and B. U. Felderhof: J. Chem. Phys. **89** 1049 (1988), J. Chem. Phys. **89** 3705 (1988)
7.   J. M. Rallison: Ann. Rev. Fluid Mech. **16** 45 (1984)
8.   A. T. J. M. Woutersen and C. G. de Kruif: J. Rheol. **37** 681 (1993)
9.   R. G. Cox and H. Brenner: J. Fluid Mech. **28** 391 (1967), Chem. Eng. Sci. **26** 65 (1971) ; P. Mazur and D. Bedeaux, Physica **76** 235 (1974) ; B. U. Felderhof, Physica **84A** 557, 569 (1976)
10.   B. U. Felderhof, G. W. Ford and E. G. D. Cohen: J. Stat. Phys. **28** 135 (1982)
11.   B. Cichocki, B. U. Felderhof and R. Schmitz: Physica **A154** 233 (1989)
12.   R. Schmitz and B. U. Felderhof: Physica **116A** 163 (1982)
13.   R. B. Jones and R. Schmitz: Physica **149A** 373 (1988)
14.   B. Cichocki, B. U. Felderhof and R. Schmitz: PhysicoChem. Hyd. **10** 383 (1988)
15.   R. B. Jones: Physica A, in press (1994).
16.   Norman J. Wagner and A. T. J. M. Woutersen: J. Fluid Mech., in press (1994)

Mesoscopic Theory of Liquid Crystals

W. Muschik
H. Ehrentraut, C. Papenfuß, S. Blenk

Institut für Theoretische Physik
Technische Universität Berlin
Hardenbergstraße 36
10623 BERLIN
Germany

The structural roots of the mesoscopic theory of liquid crystals are reviewed and shortly discussed.

1 Introduction

Liquid crystals are phases showing an orientational order of molecules which are of elongated or of plane shape, so that an orientational order of them can be defined. The molecules themselves may be uniaxial (i.e. their alignment can be described by one direction) or biaxial (for which two axes are necessary to fix their orientation). Besides the orientational order an additional spatial order of the centers of mass of the molecules is possible which causes a great variety of liquid crystal phases, such as nematics, smectics, cholesterics and a lot of other phases of various different structures. Because of thermal fluctuations the molecules are not totally aligned, but they have a certain distribution around a "mean orientation" which can be described by a normalized *macroscopic director field*. The name "macroscopic" originates from the fact, that the macroscopic director field belongs to all molecules of a volume element, whereas a special molecule may be aligned differently. So it is obvious to introduce a *microscopic director* which describes the alignment of a single molecule and which is different from the local macroscopic director (if it exists). Because the microscopic director is defined on a molecular level it is not a macroscopic field, but a so-called *mesoscopic variable*. Here "mesoscopic" means that the level of description is finer than the macroscopic level is, but that no microscopic concept such as molecular interaction or potentials are used.

Besides the microscopic and the macroscopic director other descriptions for alignment are in use which can be found in the following synopsis:

Up to now there are five different phenomenological concepts suitable to describe liquid crystals: The first one is the well known Ericksen-Leslie theory [1],[2] whose balance equations are formulated by use of the macroscopic director field mentioned above. But in fact this theory is not able to represent a change in the degree of orientational order. In general we need at each position and time a distribution function for describing the macroscopic orientation of the fluid. Therefore the macroscopic director has to be redefined statistically.

The second concept describes liquid crystals as micropolar media in the frame of a 3-director theory [3]. Instead of a balance equation for the macroscopic director the spin balance is taken into account, but no microscopic concepts are used.

The third concept [4] introduces besides the balance equations of a micropolar media an additional field, called microinertia tensor field. This field satisfying its own balance equation is coupled to the spin balance. The form of this coupling causes that all needle-shaped molecules of a volume element have always the same angular velocity. Therefore the degree of alignment cannot change (in a co-moving frame) as it happens in the phase transition from nematic to isotropic phase. Consequently the microinertia tensor field – although the molecules are not totally aligned – is not appropriate to describe dynamical situations in which the degree of alignment changes.

The fourth concept describing liquid crystals uses the alignment tensor [5] [6], a symmetric, traceless tensor which represents the first anisotropy moment of a multipole expansion of the orientation distribution function. This of course is a more general method for describing alignment as using a director theory because the alignment tensor is determined by five, the director by two variables.

The fifth concept describing liquid crystals introduces the above mentioned microscopic director related to the orientation of one rigid particle only. The anisotropic fluid is formally treated as a mixture by regarding all particles of a volume element of the same orientation as one component of the fluid [7]. Thus the orientation distribution function results as the fraction of the mass density of one component over the mass density of the mixture. Because mixture theories are well developed [8], [9] balance equations for liquid crystals can be written down very easily by use of this method. The domain on which these balances are defined is different from the usual one because it contains the microscopic director as an additional variable.

The aim of this paper is to sketch the mesoscopic theory of liquid crystals which is based on the concepts of the microscopic director and of the orientation distribution function.

2 Mesoscopic Concept

As discussed above a liquid crystal is composed of differently orientated particles for which we presuppose, that their orientation can be described individually by the direction of their microscopic directors. Thus the microscopic director n is defined as a unit vector pointing into the temporary direction of a needle-shaped rigid particle, or, if the particle is of a plane shape, the microscopic director is perpendicular to the particle. Because the microscopic director is normalized

$$n^2 = 1, \quad n \in S^2, \tag{1}$$

the microscopic alignment has two degrees of freedom. Since the particles may rotate, the microscopic director changes in time and we define the microscopic orientation change velocity u by

$$u := \frac{\mathrm{d}}{\mathrm{d}t}\, n, \quad \text{with} \quad u \cdot n = 0. \tag{2}$$

If we consider molecules of a volume element the alignment of their axes is specified by a distribution function f on the unit sphere, called *orientation distribution function* (ODF). The ODF describes the density generated by intersection points on the unit sphere S^2 between the molecule axes and the sphere. Because in general the alignment of the molecules is a function of position and time the ODF is defined on a six-dimensional space $S^2 \times \mathbb{R}^3 \times \mathbb{R}^1$

$$f(\, n,\, x,t) \equiv f(\cdot), \quad (\cdot) \equiv (\, n,\, x,t) \in S^2 \times \mathbb{R}^3 \times \mathbb{R}^1. \tag{3}$$

The 5-dimensional subspace $S^2 \times \mathbb{R}^3$ consisting of the orientation- and of the position-part is called *nematic space*. Because there are always two points of intersection opposite to each other on the S^2 the ODF shows the so-called head-tail symmetry

$$f(-\, n,\, x,t) = f(\, n,\, x,t). \tag{4}$$

By this ODF a mesoscopic classification of different types of orientation, so-called *phases*, becomes possible.

3 Orientational Balances

In the nematic space Ω each particle is represented by its five coordinates ($\, x,\, n$) at time t. We define fields on the nematic space, such as the field of orientational mass density which is defined by

$$\rho(\cdot) := \rho(\, x,t) f(\cdot)\,, \tag{5}$$

where $\rho(\, x, t)$ is the macroscopic mass density (25) of the considered one-component liquid crystal. $\rho(\cdot)$ describes the mass density of those molecules having the orientation n at position x at time t. In this interpretation $v(\cdot)$ is the material velocity of these molecules. Other such orientational fields (i.e. fields defined on the nematic space) are e.g. the external acceleration $k(\cdot)$, the stress tensor $T(\cdot)$, and the heat flux density $q(\cdot)$.

3.1 Global Balances

Using these orientational fields global balances on the nematic space can be formulated [10] which all have the following shape

$$\frac{\mathrm{d}}{\mathrm{d}t} \int_{G_t} \rho(\cdot)[\phi(\cdot)\, v(\cdot) + \, \Phi(\cdot)]\, d^2n\, d^3x = \, Z^{tot}, \tag{6}$$

$$Z^{tot} = \int_{G_t} [\rho(\cdot)\, K(\cdot) + \nabla_x \cdot\, S(\cdot)]\, d^2n\, d^3x. \tag{7}$$

Here $\phi(\cdot)$ is the convective part of the balance, $K(\cdot)$ the volume part, and $S(\cdot)$ the surface part of the external quantity Z^{tot}.

We obtain for the special balances the following identities [11]:

Mass

$$\phi(\cdot) \equiv 0, \quad \Phi(\cdot) \equiv 1, \quad Z^{tot} \equiv o, \tag{8}$$
$$K(\cdot) \equiv o, \quad S(\cdot) \equiv o, \tag{9}$$

Momentum

$$\phi(\cdot) \equiv 1, \quad \Phi(\cdot) \equiv o, \quad Z^{tot} = \text{total external force}, \tag{10}$$
$$K(\cdot) \equiv k(\cdot) = \text{external acceleration}, \tag{11}$$
$$S(\cdot) \equiv T^{\mathsf{T}}(\cdot) = \text{transposed stress tensor}, \tag{12}$$

Angular Momentum

$$\phi(\cdot) \equiv x\times, \quad \Phi(\cdot) \equiv I\, n \times u(\cdot), \quad Z^{tot} = \text{total external moment}, \tag{13}$$
$$K(\cdot) \equiv x \times k(\cdot) + n \times g(\cdot), \tag{14}$$
$$S(\cdot) \equiv [\, x \times T(\cdot)]^{\mathsf{T}} + n \times \pi(\cdot), \tag{15}$$

Total Energy

$$\phi(\cdot) \equiv (1/2)\, v(\cdot), \quad \Phi(\cdot) \equiv (1/2)I[\, n \times u(\cdot)]^2 + \varepsilon(\cdot), \tag{16}$$
$$Z^{tot} = \text{energy supply}, \tag{17}$$
$$K(\cdot) \equiv k(\cdot) \cdot v(\cdot) + g(\cdot) \cdot u(\cdot) + r(\cdot), \tag{18}$$
$$S(\cdot) \equiv v(\cdot) \cdot T(\cdot) + u(\cdot) \cdot \pi(\cdot) - q(\cdot). \tag{19}$$

Here $u(\cdot)$ is the field of the orientational change velocity, $g(\cdot)$ the orientational couple force density, $\pi(\cdot)$ the orientational couple stress tensor, $\varepsilon(\cdot)$ the density of orientational internal energy, and $r(\cdot)$ the orientational radiation supply.

Having taken into account the balances of mass, momentum, angular momentum, and energy we got all balances which are relevant for liquid crystals outside electrodynamics. These global balance equations formulated on the nematic space can be transformed into so-called orientational balance equations which are local in position, time, and the microscopic director. For this purpose we need a generalized Reynolds' transport theorem [10].

$$\frac{\mathrm{d}}{\mathrm{d}t} \int_{G_t} \Phi(\cdot) d^2 n\, d^3 x =$$
$$= \int_{G_t} \left[\frac{\partial}{\partial t} \Phi(\cdot) + \nabla_x \cdot [\, v(\cdot)\Phi(\cdot)] + \nabla_n \cdot [\, u(\cdot)\Phi(\cdot)] \right] d^2 n\, d^3 x. \tag{20}$$

By use of (20) we can transform the global balance equations (6) as usual into local ones.

3.2 Local Orientational Balances

Presupposing the global balances we obtain [12] by use of (20) local balance equations which are defined on the nematic space and which are therefore denoted as *local orientational balances*. The general shape of these balances is

$$\frac{\partial}{\partial t}\{\rho(\cdot)[\phi(\cdot)\,v(\cdot) + \,\boldsymbol{\Phi}(\cdot)]\} \; + \; \nabla_x \cdot \{\rho(\cdot)\,v(\cdot)[\phi(\cdot)\,v(\cdot) + \,\boldsymbol{\Phi}(\cdot)]\} +$$
$$+\nabla_n \cdot \{\rho(\cdot)\,u(\cdot)[\phi(\cdot)\,v(\cdot) + \,\boldsymbol{\Phi}(\cdot)]\} \; = \; \rho(\cdot)\,K(\cdot) + \nabla_x \cdot \,S(\cdot). \qquad (21)$$

Besides the local balances of mass, momentum, angular momentum, and energy we obtain from the balance of momentum by subtracting the balance of angular momentum the *orientational spin balance*:

$$s(\cdot) := I\,n \times \,u(\cdot) = \text{orientational spin density}, \qquad (22)$$

$$\frac{\partial}{\partial t}[\rho(\cdot)\,s(\cdot)] + \nabla_x \cdot [\rho(\cdot)\,v(\cdot)\,s(\cdot) - (\,n \times \,\pi(\cdot))^T] +$$
$$+\nabla_n \cdot [\rho(\cdot)\,u(\cdot)\,s(\cdot)] = \,\epsilon : \,T(\cdot) + \rho(\cdot)\,n \times \,g(\cdot) \qquad (23)$$

(ϵ is the Levi-Civita tensor). The right-hand side of this equation indicates that one part of the spin supply is caused by the antisymmetric part of the stress tensor.

According to the definition of the orientational mass density (5) we obtain from the orientational mass balance an additional balance of the ODF by inserting the definition of $f(\cdot)$:

$$\frac{\partial}{\partial t}f(\cdot) + \nabla_x \cdot (\,v(\cdot)f(\cdot)) + \,n \cdot \nabla_n \times (\,\omega(\cdot)f(\cdot)) +$$
$$+f(\cdot)[\frac{\partial}{\partial t} + \,v(\cdot) \cdot \nabla_x]\ln\rho(\,x,t) = 0 \; . \qquad (24)$$

Because this balance equation of the orientation distribution function includes the macroscopic field of the mass density

$$\rho(\,x,t) := \int_{S^2} \rho(\cdot)d^2 n \; , \qquad (25)$$

it is not independent of the macroscopic mass balance defined on $\mathbb{R}^3$. As can be seen from (25) macroscopic quantities are defined by integration over the orientational part.

3.3 Balances of Micropolar Media

Introducing the macroscopic fields by mesoscopic definitions the orientational balance equations transform into the balances for micropolar media by integrating over the unit sphere [10]. So the definition of the macroscopic material velocity

$$\rho(\,x,t)\,v(\,x,t) := \int_{S^2} \rho(\cdot)\,v(\cdot)d^2 n \; , \qquad (26)$$

and (25) results in the macroscopic mass balance. The shape of the macroscopic balances by integrating (21) is

$$\frac{\partial}{\partial t}\{\rho(\ x,t)[\phi(\ x,t)\ v(\ x,t)+\ \Phi(\ x,t)]\}+ \qquad (27)$$

$$+ \nabla_x \cdot \{\rho(\ x,t)\ v(\ x,t)[\phi(\ x,t)\ v(\ x,t)+\ \Phi(\ x,t)]\}=$$
$$= \rho(\ x,t)\ K(\ x,t)+\nabla_x \cdot\ S(\ x,t). \qquad (28)$$

Here the following mesoscopic definitions of macroscopic quantities are introduced:

$$\rho(\ x,t)\phi(\ x,t) := \int_{S^2} \rho(\cdot)\phi(\cdot)d^2n\ , \qquad (29)$$

$$\rho(\ x,t)\ v(\ x,t) := \int_{S^2} \rho(\cdot)\ v(\cdot)d^2n\ , \qquad (30)$$

$$\rho(\ x,t)\ K(\ x,t) := \int_{S^2} \rho(\cdot)\ K(\cdot)d^2n\ , \qquad (31)$$

$$\rho(\ x,t)\ \Phi(\ x,t) := \int_{S^2} \rho(\cdot)[\phi(\cdot)\ v(\cdot)+\ \Phi(\cdot)]d^2n\ +$$
$$+\ \rho(\ x,t)\phi(\ x,t)\ v(\ x,t), \qquad (32)$$

$$S(\ x,t) := \int_{S^2} \{S(\cdot)-\rho(\cdot)\ v(\cdot)[\phi(\cdot)\ v(\cdot)+\ \Phi(\cdot)]\}d^2n\ -$$
$$-\ \rho(\ x,t)\ v(\ x,t)[\phi(\ x,t)\ v(\ x,t)+\ \Phi(\ x,t)]. \qquad (33)$$

If the ϕ, Φ, K, and S are specified according to (8)-(19) we obtain the macroscopic balances of mass, momentum, spin, and energy. Besides these balances two more ones are essential: the entropy and the alignment tensor balance. The entropy balance reads

$$\frac{\partial}{\partial t}[\rho(\ x,t)\eta(\ x,t)] + \nabla_x \cdot [\rho(\ x,t)\eta(\ x,t)\ v(\ x,t)-\ \phi(\ x,t)]=$$
$$= \rho(\ x,t)\sigma(\ x,t) \qquad (34)$$

(η = field of specific entropy, ϕ = entropy flux density, σ = entropy production density). The second law is expressed by

$$\sigma(\ x,t) \geq 0. \qquad (35)$$

The alignment tensor balance is sketched in the next section.

4 Alignment Tensors

The mesoscopic description of liquid crystals offers the advantage to introduce macroscopic fields, denoted as alignment tensor fields, which describe orientation more precise than the field of the macroscopic director. The main idea of this approach is to expand the ODF $f(\cdot)$ into a series of multipoles [10] [13]. The coefficients of that expansion are tensors of even order. The balance equation (24) of the ODF induces alignment tensor balance equations of the following structure

$$\left(\frac{\partial}{\partial t} + v(\,x,t)\cdot\nabla_x\right)a_{\nu_1\cdots\nu_k} - k\left(\overline{\omega(\,x,t)\times a}\right)_{\nu_1\cdots\nu_k} \equiv$$

$$\equiv D_t^{(k)}\,a = \left\{\text{ all orders }\right\}. \tag{36}$$

Here the *alignment tensor of order l* is defined by

$$a_{\mu_1\ldots\mu_l}(\,x,t) := \frac{2l+1}{l!}\int_{S^2}f(\cdot)\,\overline{n_{\mu_1}\cdots n_{\mu_l}}\,d^2n \tag{37}$$

(n_j is the j-th Cartesian component of the microscopic director, and $\overline{n_{\mu_1}\cdots n_{\mu_l}}$ the total symmetric traceless product of them). The right-hand side of (36) has the typical form

$$\left\{\text{ all orders }\right\} = \mathcal{F}[\,v(\cdot) - v(\,x,t),\,\omega(\cdot) - \omega(\,x,t)] \tag{38}$$

with the property

$$\mathcal{F}[\,o,\,o] = 0\,. \tag{39}$$

Therefore (36) can easily be interpreted: Its left hand side is Jaumann's total time derivative for symmetric irreducible tensors. Consequently that side of the equation describes the changes of the alignment tensors which can be observed from a frame translated with $v(\,x,t)$ and rotated with $\omega(\,x,t)$. Thus the right hand side of (36) is due to the production of alignment. This production of course will vanish, if there are no deviations in the velocity distributions of $v(\cdot)$ and $\omega(\cdot)$ from their averages $v(\,x,t)$ and $\omega(\,x,t)$. The special form of the alignment production involves generally *all* the alignment tensors of any order. A subdynamic even of the alignment tensor of second order will not exist in general. Approximations of (36) are widespread [6].

5 Remark on Constitutive Equations

Because the balance equations include quantities which are determined by the considered material we need constitutive equations for solving the balances. In general there are two different methods to proceed [14]: We can make an ansatz for the constitutive equations, or we can look for a class of materials which is characterized by a chosen state space. In both cases the dissipation inequality (35) has to be exploited. After having inserted the energy balance equation into (34) the expressions for the spin, the stress tensor, and couple force are replaced by their mesoscopic definitions. A straightforward calculation results in a macroscopic dissipation inequality [15] whose shape is different from that of (34). In that dissipation inequality of different shape new macroscopic fields appear which are mesoscopically defined, e.g. the alignment tensor field of second order. After having chosen the state space the transformed dissipation inequality is exploited by the Coleman-Noll procedure [16]. The essential results are [15]:

310

- the coefficients of the free energy are restricted in comparison to Ericksen-Leslie theory of liquid crystals,
- the entropy flux density is different from heat flux density over temperature,
- the couple stress tensor can be calculated explicitly,
- the Landau condition for equilibrium in liquid crystals can be derived and extended to inhomogeneous alignment.

References

1. J. L. Ericksen. Anisotropic fluids. *Arch. Rat. Mech. Anal.*, 4 (1960) 231–237.
2. F. M. Leslie. Some constitutive equations for liquid crystals. *Arch. Rat. Mech. Anal.*, 28 (1968) 265–283.
3. A. C. Eringen and J. D. Lee. Relations of two continuum theories of liquid crystals. In J. F. Johnson and R. S. Porter, editors, *Liquid Crystals and Ordered Fluids, Vol. 2*, pages 315–330. Plenum Press, New York, 1974.
4. A. C. Eringen. Micropolar theory of liquid crystals. In J. F. Johnson and R. S. Porter, editors, *Liquid Crystals and Ordered Fluids, Vol. 3*, pages 443–474. Plenum Press, New York, 1978.
5. S. Hess. Irreversible thermodynamics of nonequilibrium alignment phenomena in molecular liquids and in liquid crystals. *Z. Naturforsch.*, 30a (1975) 728–733.
6. I. Pardowitz and S. Hess. On the theory of irreversible processes in molecular liquids and liquid crystals, nonequilibrium phenomena assosiated with the second and fourth rank alignment-tensors. *Physica*, 100A (1980) 540–562.
7. S. Blenk. Eine statistische Begründung phänomenologischer Bilanzgleichungen für Flüssigkristalle im Rahmen einer mikroskopischen Direktortheorie. Diplomarbeit, Institut für Theoretische Physik, Technische Universität Berlin, 1989.
8. W.H. Müller and W. Muschik. Bilanzgleichungen offener mehrkomponentiger Systeme
I. Massen- und Impulsbilanzen. *J. Non-Equilib. Thermodyn.*, 8 (1983) 29–46.
9. W. Muschik and W.H. Müller. Bilanzgleichungen offener mehrkomponentiger Systeme
II. Energie- und Entropiebilanz. *J. Non-Equilib. Thermodyn.*, 8 (1983) 47–66.
10. S. Blenk, H. Ehrentraut, and W. Muschik. Statistical foundation of macroscopic balances for liquid crystals in alignment tensor formulation. *Physica A*, 174 (1991) 119–138.
11. S. Blenk and W. Muschik. Orientational balances for nematic liquid crystals. *J. Non-Equilib. Thermodyn.*, 16 (1991) 67–87.
12. S. Blenk, H. Ehrentraut, and W. Muschik. Orientation balances for liquid crystals and their representation by alignment tensors. *Mol. Cryst. Liqu. Cryst.*, 204 (1991) 133–141.
13. H. Ehrentraut. Herleitung der Bilanzgleichungen für Alignment-Tensoren aus der mikroskopischen Orientierungsverteilung der Einzelmoleküle. Diplomarbeit, Institut für Theoretische Physik, Technische Universität Berlin, 1989.

14. W. Muschik. *Aspects of Non-Equilibrium Thermodynamics.* World Scientific, Singapore, 1990, sect. 6.4

15. S. Blenk, H. Ehrentraut, and W. Muschik. Macroscopic constitutive equations for liquid crystals induced by their mesoscopic orientation distribution. *Int. J. Engng. Sci.*, 30(9) (1992) 1127–1143.

16. B. D. Colemann and W. Noll. The thermodynamics of elastic materials with heat conduction and viscosity. *Arch. Rat. Mech. Anal.*, 13 (1963) 167–168.

Financial support by grant ERBCHRXCT920007 of the European Union is duly acknowledged.

On The Microscopic Theory of Phase Coexistence

Salvador Miracle Solé

Centre de Physique Théorique, CNRS Luminy, Case 907, F-13288
Marseille Cedex 9, France

1 Introduction

It is known that the equilibrium shape of a crystal is obtained, according to
the Gibbs thermodynamic theory, by minimizing the total surface free energy
associated to the crystal-medium interface, and that this shape is given by the
Wulff construction, provided one knows the anisotropic surface tension (or interfacial free energy per unit area). It is therefore important, even if a complete
microscopic derivation of the Wulff construction within statistical mechanics has
been proved only for some two-dimensional lattice models (see the recent work
by Dobrushin *et al.* [1,2] and also Ref. 3), to study the properties of the surface
tension $\tau(\mathbf{n})$, as a function of the unit vector $\mathbf{n}$ which specifies the orientation
of the interface with respect to the crystal axes. In the first approximation the
crystal can be modelled by a lattice gas. In these notes, we shall present some
new rigorous results on this subject which relate, in particular, to the problem
of the appearance of plane facets in the Wulff equilibrium shape. For this purpose several aspects of the microscopic theory of interfaces are analysed, and
another important quantity in this theory, the step free energy, is investigated
(a complete version of this work will appear later [4]). In the last Section we shall
report on some recent developments related to the theory of crystal growth.

2 Gibbs states and interfaces

First we recall some classical results about the Gibbs states of the Ising model (to
some extent these results were already discussed in Ref. 5). The model is defined
on the cubic lattice $\mathcal{L} = \mathbb{Z}^3$, with configuration space $\Omega = \{-1,1\}^{\mathcal{L}}$. The value
$\sigma(i)$ is the spin at the site i. The energy of a configuration $\sigma_\Lambda = \{\sigma(i), i \in \Lambda\}$,
in a finite subset $\Lambda \subset \mathcal{L}$, under the boundary conditions $\bar{\sigma} \in \Omega$, is

$$H_\Lambda(\sigma_\Lambda \mid \bar{\sigma}) = - \sum_{(i,j) \cap \Lambda \neq \emptyset} \sigma(i)\sigma(j)$$

where $\langle i, j \rangle$ are pairs of nearest neighbour sites and $\sigma(i) = \bar{\sigma}(i)$ if $i \notin \Lambda$. The partition function, at the inverse temperature $\beta = 1/kT$, is given by

$$Z^{\bar{\sigma}}(\Lambda) = \sum_{\sigma_\Lambda} \exp\left(-\beta H_\Lambda(\sigma_\Lambda \mid \bar{\sigma})\right)$$

It is known that this model presents, at low temperatures $T < T_c$ (where T_c is the critical temperature), two distinct thermodynamic pure phases, a positively and a negatively magnetized phase. This means two extremal translation invariant Gibbs states, which correspond to the limits, when $\Lambda \to \infty$, of the finite volume Gibbs measures $Z^{\bar{\sigma}}(\Lambda)^{-1} \exp\left(-H_\Lambda(\sigma_\Lambda \mid \bar{\sigma})\right)$, with boundary conditions $\bar{\sigma}$ equal to the ground configurations $(+)$ and $(-)$ (such that $\bar{\sigma}(i) = 1$ and $\bar{\sigma}(i) = -1$ for all $i \in \mathcal{L}$), respectively. On the other side, if $T \geq T_c$, the Gibbs state is unique.

Each configuration inside Λ can be geometrically described by specifying the *Peierls contours*, i. e., the boundaries between the spin 1 and spin -1 regions, which, under the above boundary conditions, are closed surfaces. The energy of the configuration is equal to twice the total area of the contours. The contours can be viewed as defects, or excitations, with respect to the ground states of the system, and are a basic tool for the investigation of the model at low temperatures.

In order to study the interface between the two pure phases one needs to construct a state describing the coexistence of these phases. Let Λ be a parallelepiped of sides L_1, L_2, L_3, parallel to the axes, and centred at the origin of $\mathcal{L}$, and let $\mathbf{n} = (n_1, n_2, n_3)$ be a unit vector in $\mathbb{R}^3$, such that $n_3 \neq 0$. Introduce the mixed boundary conditions $(\pm, \mathbf{n})$, for which $\bar{\sigma}(i) = 1$ if $i \cdot \mathbf{n} \geq 0$, and $\bar{\sigma}(i) = -1$ if $i \cdot \mathbf{n} < 0$. These boundary conditions force the system to produce a defect going transversely through the box Λ, a big Peierls contour that can be interpreted as a *microscopic interface*. The other defects that appear above and below can be described by closed contours inside the pure phases.

Consider now the microscopic interface orthogonal to the direction $\mathbf{n}_0 = (0, 0, 1)$. At low temperatures $T > 0$, we expect this interface, which at $T = 0$ coincides with the plane $i_3 = -1/2$, to be modified by deformations. It can be described by means of its defects, or excitations, with respect to the interface at $T = 0$. These defects, called *walls*, form the boundaries (which may have some width), between the smooth plane portions of the interface. In this way the interface structure, with its probability distribution in the corresponding Gibbs state, may then be interpreted as a "gas of walls" on a two-dimensional lattice.

Using the Peierls method, Dobrushin [6] proved the dilute character of this gas at low temperatures, which means that the interface is essentially flat (or rigid). The considered boundary conditions yield indeed a non translation invariant Gibbs state. Furthermore, cluster expansion techniques have been applied by Bricmont *et al.* [7,8], to study the interface structure in this case (see also Ref. 9).

The same analysis applied to the two-dimensional model shows a different behaviour at low temperatures. In this case the walls belong to a one-dimensional lattice, and Gallavotti [10] proved that the microscopic interface undergoes large fluctuations of order $\sqrt{L_1}$. The interface does not survive in the thermodynamic limit, $\Lambda \to \infty$, and the corresponding Gibbs state is translation invariant. Moreover, the interface structure can be studied by means of a cluster expansion for any orientation of the interface (see also Ref. 11). Such a problem in the three-dimensional case leads to very difficult problems of random surfaces. This is one of the serious difficulties which face the attempts to generalise the work by Dobrushin *et al.* [1] to the three-dimensional Ising model, since a very accurate description of the microscopic interface for any orientation **n** is needed in this work.

3 The surface tension

The free energy, per unit area, due to the presence of the interface, is the surface tension. It can be defined by

$$\tau(\mathbf{n}) = \lim_{L_1, L_2 \to \infty} \lim_{L_3 \to \infty} -\frac{n_1}{\beta L_1 L_2} \ln \frac{Z^{(\pm, \mathbf{n})}(\Lambda)}{Z^{(+)}(\Lambda)}$$

Notice that in this expression the volume contributions proportional to the free energy of the coexisting phases, as well as the boundary effects, cancel, and only the contributions to the free energy of the interface are left.

*Theorem 1. The thermodynamic limit $\tau(\mathbf{n})$, of the interfacial free energy per unit area, exists, and, as a function of **n**, extends by positive homogeneity to a convex function $f(\mathbf{x}) = |\mathbf{x}| \tau(\mathbf{x}/|\mathbf{x}|)$ defined for any vector $\mathbf{x} \in \mathbb{R}^3$.*

A proof of these statements was given in Ref. 12 using correlation inequalities (this being the reason for their general validity). Moreover, we know (from Refs. 13 and 14 and the convexity condition) that $\tau(\mathbf{n})$ is strictly positive for $T < T_c$ and that it vanishes if $T \geq T_c$.

The convexity of $f(\mathbf{x})$ may be interpreted as a thermodynamic stability condition. It is equivalent (as shown in Ref. 12) to the pyramidal inequality for the function $\tau(\mathbf{n})$. This condition, introduced in Ref. 1 for the two-dimensional Ising model (triangular inequality), was conjectured to hold true in general situations in Ref. 15.

According to the Wulff construction, the equilibrium shape of a crystal is given by

$$\mathcal{W} = \left\{ \mathbf{x} \in \mathbb{R}^3 \mid \mathbf{x} \cdot \mathbf{n} \leq \tau(\mathbf{n}) \text{ for every } \mathbf{n} \right\}$$

where $\tau(\mathbf{n})$ is the surface tension of the interface orthogonal to **n**. One obtains in this way the shape which has the minimum surface free energy for a given volume. Defined as the intersection of closed half-spaces, $\mathcal{W}$ is a closed bounded convex set, i.e., a convex body.

It turns out that if $f(\mathbf{x})$ is a convex function, then it is also the Minkowski's support function of the convex body $\mathcal{W}$ (i. e., $f(\mathbf{x}) = \sup_{\mathbf{y} \in \mathcal{W}} \mathbf{x} \cdot \mathbf{y}$). As a consequence of this fact the following *macroscopic properties* can be proved.

Theorem 2. Assume that the convexity condition is satisfied. A facet orthogonal to the direction $\mathbf{n}_0$ appears in the Wulff equilibrium crystal shape if, and only if, the derivative $\partial\tau(\theta, \phi)/\partial\theta$ is discontinuous at the point $\theta = 0$, for all ϕ. Moreover, the one-sided derivatives $\partial\tau(\theta, \phi)/\partial\theta$, at $\theta = 0^+$ and $\theta = 0^-$, exist, and determine the shape of the facet.

Here, the function $\tau(\mathbf{n}) = \tau(\theta, \phi)$ is expressed in terms of the spherical coordinates $0 \le \theta \le \pi$, $0 \le \phi \le 2\pi$ of $\mathbf{n}$, the vector $\mathbf{n}_0$ being taken as the polar axis. Actually, the shape of the facet is given by

$$\mathcal{F} = \left\{ \mathbf{x} \in I\!R^2 \mid \mathbf{x} \cdot \mathbf{m} \le \mu(\mathbf{m}) \text{ for every } \mathbf{m} \right\}$$

where $\mathbf{m} = (\cos\phi, \sin\phi) \in I\!R^2$ and $\mu(\phi) = (\partial/\partial\theta)_{\theta=0^+}\tau(\theta, \phi)$.

4 The step free energy

The step free energy plays, also, an important role in the problem under consideration. It is defined, again using appropriate boundary conditions, as the free energy associated with the introduction of a step of height 1 on the interface. This quantity can be regarded as an order parameter for the roughening transition, analogous, in some sense, to the surface tension in the case of a phase transition.

Let us consider again the interface orthogonal to a lattice axis, which, as we know from Section 2, is rigid at low temperatures. It is believed, that at higher temperatures, but before reaching the critical temperature T_c, the fluctuations of the considered interface become unbounded when the volume tends to infinity, so that the corresponding Gibbs state in the thermodynamic limit is translation invariant. The interface undergoes a roughening phase transition at a temperature $T = T_R$.

Approximate methods, used by Weeks *et al.* [16], suggest $T_R \sim 0.53\ T_c$, a temperature slightly higher then $T_c^{d=2}$ (the critical temperature of the two-dimensional Ising model), and actually van Beijeren [17] proved, using correlation inequalities, that $T_R \ge T_c^{d=2}$. The analogous result for the step free energy, i. e., that $\tau^{\text{step}} > 0$ if $T < T_c^{d=2}$, was proved in Ref. 18, as well as that $\tau^{\text{step}} = 0$ if $T \ge T_c$. Since then, however, it appears to be no proof of the fact that $T_R < T_c$.

At present one is able to study rigorously the roughening transition only for some simplified models of the microscopic interface. Thus, Fröhlich and Spencer [19] have proved this transition for the SOS (solid-on-solid) model, and several restricted SOS models, which are exactly solvable, have also been studied in this context (these models are reviewed in Refs. 20 and 21).

In order to define the step free energy we consider the box Λ as above and and introduce the (step,**m**) boundary conditions, associated to the unit vectors $\mathbf{m} = (\cos\phi, \sin\phi) \in I\!\!R^2$, by

$$\bar{\sigma}(i) = \begin{cases} 1 & \text{if } i > 0 \text{ or if } i_3 = 0 \text{ and } i_1 m_1 + i_2 m_2 \geq 0 \\ -1 & \text{otherwise} \end{cases}$$

Then, the *step free energy*, for a step orthogonal to **m** (such that $m_2 \neq 0$), is

$$\tau^{\text{step}}(\phi) = \lim_{L_1 \to \infty} \lim_{L_2 \to \infty} \lim_{L_3 \to \infty} -\frac{\cos\phi}{\beta L_1} \ln \frac{Z^{(\text{step},\mathbf{m})}(\Lambda)}{Z^{(\pm,\mathbf{n}_0)}(\Lambda)}$$

Clearly, this expression represents the residual free energy due to the considered step, per unit length.

When considering the configurations under the (step,**m**) boundary conditions, the step may be viewed as a defect on the rigid interface. It is in fact, a long wall going from one side to the other of the box Λ. A more careful description of it can be obtained as follows. At $T = 0$, the step parallel to the axis (i. e., for $\mathbf{m} = (0,1)$) is a perfectly straight step of height 1. At a low temperature $T > 0$, some deformations appear, connected by straight portions of height 1. The step structure, with its probability distribution in the corresponding Gibbs state, can then be described as a "gas" of these defects (to be called *step-jumps*), on a one-dimensional lattice. This description, somehow similar to the description of the interface of the two-dimensional Ising model used by Gallavotti [10], is valid, in fact, for any orientation **m** of the step. It can be shown that the gas of step-jumps is a dilute gas at low temperature and, as a consequence of this fact, cluster expansion techniques can be applied in order to study the step structure. Actually, the step-jumps are not independent since the rest of the system produces an effective interaction between them. Nevertheless, this interaction can be treated by applying the low temperature expansion, in terms of walls, for the rigid interface, to the regions of the interface lying at both sides of the step. ¿From this analysis one gets the following result.

Theorem 3. If the temperature is low enough (i.e., if $T \leq T_0$ where $T_0 > 0$ is a given constant), then the step free energy τ^{step}, exists in the thermodynamic limit, and extends by positive homogeneity to a strictly convex function. Moreover, it can be expressed in terms of an analytic function of T, for which a convergent power series expansion can be obtained from the above mentioned cluster expansion.

In fact,

$$\tau^{\text{step}}(\mathbf{m}) = 2J(|m_1| + |m_2|) - (1/\beta)((|m_1| + |m_2|)\ln(|m_1| + |m_2|)$$
$$- |m_1|\ln|m_1| - |m_2|\ln|m_2|) - (1/\beta)\varphi_{\mathbf{m}}(\beta)$$

where $\varphi_{\mathbf{m}}$ is an analytic function of $z = e^{-2\beta}$, for $|z| \leq e^{-2\beta_0}$. The first two terms in this expression, which represent the main contributions for $T \to 0$, come from the ground state of the system under the considered boundary conditions. The first term can be recognised as the residual energy of the step at zero temperature and, the second term, as $-(1/\beta)$ times the entropy of this ground state. The same two terms occur in the surface tension of the two-dimensional Ising model (see Ref. 22). By considering the lowest energy excitations, it can be seen that $\varphi_{\mathbf{m}}$ is $O(e^{-4\beta})$, and also, that the first term in which this series differs from the series associated to the surface tension of the two-dimensional Ising model, is $O(e^{-12\beta})$.

5 Facets in the equilibrium crystal

The roughness of an interface should be apparent when considering the shape of the equilibrium crystal associated with the system. One knows that a typical equilibrium crystal at low temperatures has smooth plane facets linked by rounded edges and corners. The area of a particular facet decreases as the temperature is raised and the facet finally disappears at a temperature characteristic of its orientation. The reader will find information and references on equilibrium crystals in the review articles of Refs. 20, 21, 23 and 24.

It can be argued, as discussed below, that the roughening transition corresponds to the disappearance of the facet whose orientation is the same as that of the considered interface. The exactly solvable SOS models mentioned above, for which the function $\tau(\mathbf{n})$ has been computed, are interesting examples of this behaviour (this subject has been reviewed in Ref. 12, Chapter VII). For the three-dimensional Ising model, Bricmont et al. [25] have proved a correlation inequality which establish τ^{step} as a lower bound to the one-sided derivative $\partial\tau(\theta)/\partial\theta|_{\theta=0^+}$ (here $\tau^{\text{step}} = \tau^{\text{step}}(0,1)$ and $\tau(\theta) = \tau(0,\sin\theta,\cos\theta)$). Thus $\tau^{\text{step}} > 0$ implies a kink in $\tau(\theta)$ at $\theta = 0$ and, according to the Wulff construction, a facet is expected.

In fact, τ^{step} should be equal to this derivative. This is reasonable, since the increment in surface tension of an interface tilted by an angle θ, with respect to the surface tension of the rigid interface, can be approximately identified, for θ small, with the free energy of a "gas of steps" (the density of the steps being proportional to θ). And, again, if the interaction between the steps can be neglected, the free energy of this gas can be approximated by the sum of the individual free energies of the steps.

As a result of the methods described in Section 3, it is possible to study the statistical mechanics of this "gas of steps", and to derive the following result.

Theorem 4. For $T < T_0$, we have

$$\partial\tau(\theta,\phi)/\partial\theta|_{\theta=0^+} = \tau^{\text{step}}(\phi)$$

i. e., the step free energy equals the one-sided angular derivative of the surface tension.

It is natural to expect that this equality is true for any T less than T_R, and that for $T \geq T_R$, both sides in the equality vanish, and thus, the disappearance of the facet is involved (these facts can be proved for certain SOS models of interfaces using correlation inequalities [26]). However, the condition that the temperature is low enough is important here. Only when it is fulfilled we have the full control on the equilibrium probabilities that is needed in the proofs.

The above relation, together with the discussion in Section 3, implies that one obtains the shape of the facet by means of the two-dimensional Wulff construction applied to the step free energy $\tau^{\text{step}}(\mathbf{m})$. Namely,

$$\mathcal{F} = \left\{ \mathbf{x} \in \mathbb{R}^2 \mid \mathbf{x} \cdot \mathbf{m} \leq \tau^{\text{step}}(\mathbf{m}) \text{ for every } \mathbf{m} \right\}$$

Then, from the properties of τ^{step}, it follows that the facet has a smooth boundary without straight segments and, therefore, that the crystal shape presents rounded edges and corners.

6 Nucleation and growing crystals

The phenomenon of nucleation takes place when a thermodynamic system, instead of undergoing a phase transition, stays in a metastable phase. The stable phase emerges via the formation of a suitable, sufficiently large droplet (*nucleus*). In fact, due to the competition, already at a microscopic level, between the volume free energy and the surface tension, small droplets have a tendency to shrink whereas large ones prefer to grow. The activation energy necessary for the formation of a critical nucleus, and the time which takes to overcome this energy barrier, become larger and larger, together with the size of the nucleus, as the parameters tend to their values at the coexistence point. This explains the very long life of a metastable state.

Different aspects of this important subject have recently been discussed from a rigorous point of view. We present here a brief account of the work by Kotecký and Olivieri [27] concerning the droplet dynamics in the Ising model (see also Ref. 28, by the same authors, where a more complex case is discussed). The subject was previously developped by Neves and Schonmann [29,30], following the approach introduced by Cassandro *et al* [31].

The anisotropic two-dimensional Ising model, with vertical and horizontal coupling constants $J_1 > J_2 > 0$, in the presence of a very small magnetic field $h > 0$, is studied. Let Λ be a square box of side L, with periodic boundary conditions, and let $H(\sigma)$ denote the energy of a configuration $\sigma \in \Omega_\Lambda = \{-1, 1\}^\Lambda$. We suppose that the volume is sufficiently large, $L > (2J_1/h)^3$.

A discrete time *stochastic dynamics* is then considered for this model. Namely, the Metropolis dynamics defined by the following updating rule: Given a configuration σ at time t one first chooses randomly a site $i \in \Lambda$ with uniform probability $1/|\Lambda|$. Then one flips the spin at site i with probability

$$\exp(-\beta \max\{H(\sigma^{(i)}) - H(\sigma), 0\})$$

where $\sigma^{(i)}(j) = \sigma(j)$, whenever $j \neq i$, and $\sigma^{(i)}(j) = -\sigma(j)$, for $j = i$. This dynamics is *reversible* with respect to the Gibbs measure.

The nucleation from a metastable state is studied for this model in the limit of very low temperatures (h fixed). It turns out that the critical nucleus, as well as the configurations on a typical path to it, *differ from the Wulff shape* of an equilibrium droplet. The critical droplet is in fact a square of side $\ell^* = [2J_2/h] + 1$ ($[\cdot]$ denotes the integer part), while the Wulff shape is a rectangle of sides proportional to J_1, J_2 (agreement could be expected, however, in the more customary region T fixed, small, and $h \to 0$).

A *path* of the process is a sequence $\omega = \sigma_0, \sigma_1, \ldots, \sigma_t, \ldots$ of configurations in Ω_Λ. We suppose that the process starts at the configuration $\sigma_0 = (-)$ (all spins $\sigma_0(i)$ in Λ equal to -1). We are interested in the first passage from the configuration $(-)$ to the configuration $(+)$, which takes place between the moments $\tau_{(-)} = \max\{t < \tau_{(+)} \mid \sigma_t = (-)\}$ and $\tau_{(+)} = \min\{t \mid \sigma_t = (+)\}$.

The configurations $r(\ell_1, \ell_2)$, which have a rectangle of sides ℓ_1, ℓ_2 as unique Peierls contour, play a particular role in the process. They correspond to the local minima of the energy (in the sense that one spin flip increases the energy). Now, the probability that starting from a given local minimum Q the system goes to a neighbouring local minimum Q', is determined by the energy barrier $H(S) - H(Q)$, where S is any configuration at which the energy on a path from Q to Q' reaches its maximum, but with the path chosen to minimalise it. In other words, the configurations in S are the *local saddle points* for which the minimax

$$\min_{\omega: Q \to Q'} \max_{\sigma \in \omega} H(\sigma)$$

is attained (here $\omega : Q \to Q'$ denotes a generic path with successive spin flips starting from a configuration in Q and ending at Q'). The considered probability is proportional to $\exp[-\beta(H(S) - H(Q))]$. On the other side, the system in the local minimum Q is likely to "stay" in its basin of attraction for a time of order $\exp[\beta(H(\bar{S}) - H(Q))]$, where $\bar{S}$ is the local saddle point with lowest energy through which it can escape from the local minimum Q, not necessarily in "the direction" of Q'. These are the basic mechanisms which determine the local dynamics. The task is then to find the class of paths which describe the most probable evolution.

Let us consider the probability of reaching a *global saddle point*, defined by the same minimax condition extended to all paths from the configuration $(-)$ to the configuration $(+)$. These configurations give rise to the *critical nucleus*.

It can be seen that the set of all global saddle points coincides with the set $\mathcal{P}$ of all configurations having as unique contour a rectangle, of sides $\ell^*, \ell^* - 1$, or $\ell^* - 1, \ell^*$, with a unit square attached to one of its longer sides. The relative energy of any $\bar{\sigma} \in \mathcal{P}$ is

$$\Gamma = H(\bar{\sigma}) - H((-)) = 2(J_1 + J_2)\ell^* - h((\ell^*)^2 - \ell^* + 1)$$

It is proved that the first excursion from $(-)$ to $(+)$ passes through a configuration from $\mathcal{P}$ and the time needed for this to happen is of the order $\exp(\beta\Gamma)$. Introducing the time $\tau_{\mathcal{P}} = \min\{t > \tau_{(-)} \mid \sigma_t \in \mathcal{P}\}$, the precise statement can be formulated as follows.

Theorem 5. We have

$$\lim_{T \to 0} \text{Prob} \left[\tau_P < \tau_{(+)} \right] = 1$$

and, moreover, for any $\epsilon > 0$,

$$\lim_{T \to 0} \text{Prob} \left[\exp(\beta(\Gamma - \epsilon)) < \tau_P < \exp(\beta(\Gamma + \epsilon)) \right] = 1$$

In addition, from the arguments in the proof of this result, one is getting very detailed information about a typical path followed by the process σ_t during its first excursion from $(-)$ to $(+)$.

   a) *First it passes through a monotonously growing sequence of subcritical rectangles $r(\ell_1, \ell_2)$, such that $|\ell_1 - \ell_2| = 0$ or 1, up to the critical square $r(\ell^*, \ell^*)$.*

   b) *After the vertical edge stays constant at the value ℓ^* while the horizontal edge grows up to L. Finally the vertical edge grows from ℓ^* to L.*

The precise statements involve the notion of ϵ-typical path, that is determined not only in terms of geometrical properties, but also with specified times of passage (by means of bounds analogous to those used above for τ_P) through certain configurations. The path is an ϵ-typical path (for any given $\epsilon > 0$) with a probability which tends to 1 when $T \to 0$.

Finally, let us mention that Schonmann [32] has recently discussed the regime in which the temperature is kept fixed and the field $h > 0$ is scaled to zero. As conjectured in Ref. 33, for the Ising model in any dimension $d \geq 2$, if the temperature is low enough, the relaxation time goes in this regime as an exponential of $1/h^{d-1}$. Moreover, before a time which grows also as an exponential of $1/h^{d-1}$ the system stays in a metastable situation.

Acknowledgements

It is a pleasure to thank Roman Kotecký for very valuable discussions.

References

1. R.L. Dobrushin, R. Kotecký and S.B. Shlosman: The Wulff construction: a global shape from local interactions. Providence: Amer. Math. Soc. 1992.
2. R.L. Dobrushin, R. Kotecký and S.B. Shlosman: J. Stat. Phys. **72**, 1 (1993)
3. C.E. Pfister: Helv. Phys. Acta **64**, 953 (1991)
4. S. Miracle-Sole: Surface tension, step free energy and facets in the equilibrium crystal. J. Stat. Phys. (to appear).
5. S. Miracle-Sole: In: Critical phenomena, Sitges international school. J. Brey and R.B. Jones eds. Berlin: Springer, 1976.
6. R.L. Dobrushin: Theory Prob. Appl. **17**, 582 (1972)
7. J. Bricmont, J.L. Lebowitz, C.E. Pfister and E. Olivieri: Commun. Math. Phys. **66**, 1 (1979).

8. J. Bricmont, J.L. Lebowitz and C.E. Pfister: Commun. Math. Phys. **66**, 21 (1979); Commun. Math. Phys. **69**, 267 (1979)
9. P. Holický, R. Kotecký and M. Zahradník: J. Stat. Phys. **50**, 755 (1988)
10. G. Gallavotti: Commun. Math. Phys. **27**, 103 (1972)
11. J. Bricmont, J.L. Lebowitz and C.E. Pfister: J. Stat. Phys. **26**, 313 (1981)
12. A. Messager, S. Miracle-Sole and J. Ruiz: J. Stat. Phys. **67**, 449-470 (1992)
13. J. Bricmont, J.L. Lebowitz and C.E. Pfister: Ann. Acad. Sci. New York **337**, 214 (1980)
14. J.L. Lebowitz and C.E. Pfister: Phys. Rev. Lett. **46**, 1031 (1981)
15. R.L. Dobrushin and S.B. Shlosman: In: Ideas and methods in mathematical analysis, stochastics and applications. S. Albeverio, S.E. Fenstad, H. Holden and T. Lindstrom eds. Cambridge: University Press, 1991
16. J.D. Weeks, G.H. Gilmer and H.J. Leamy: Phys. Rev. Lett. **31**, 549 (1973)
17. H. van Beijeren: Commun. Math. Phys. **40**, 1 (1975)
18. J. Bricmont, J.R. Fontaine and J.L. Lebowitz: J. Stat. Phys. **29**, 193 (1982)
19. J. Fröhlich and T. Spencer: Commun. Math. Phys. **81**, 527 (1981)
20. H. van Beijeren and I. Nolden: In: Topics in current Physics, Vol. 43. W. Schommers and P. von Blackenhagen eds. Berlin: Springer 1987.
21. D.B. Abraham: In: Phase transitions and critical phenomena, Vol. 10. C. Domb and J.L. Lebowitz eds. London: Academic Press 1986.
22. J.E. Avron, H. van Beijeren, L.S. Shulman and R.K.P. Zia: J. Phys. A: Mat. Gen. **15**, L81 (1982)
23. C. Rottman and M. Wortis: Phys. Rep. **103**, 59 (1984)
24. R. Kotecký: In: IX international congress of mathematical Physics. B. Simon A. Truman and I.M. Davies eds. Bristol: Adam Hilger 1889.
25. J. Bricmont, A. El Mellouki and J. Fröhlich: J. Stat. Phys. **42**, 743 (1986)
26. S. Miracle-Sole: In preparation
27. R. Kotecký and E. Olivieri: J. Stat. Phys. **70**, 1121 (1993).
28. R. Kotecký and E. Olivieri: J. Stat. Phys. **75**, 409 (1994)
29. E.J. Neves and R.H. Schonmann: Commun. Math. Phys. **137**, 209 (1991)
30. R.H. Schonmann: Commun. Math. Phys. **147**, 231 (1991)
31. M. Cassandro, A. Galves, E. Olivieri and M.E. Vares: J. Stat. Phys. **35**, 603 (1984)
32. R.H. Schonmann: In: Probability and phase transitions. G. Grimmet, ed. Dordrecht: Kluwer Academic Publishers, 1994.
33. M. Aizenman and J.L. Lebowitz: J. Phys. A: Math. Gen. **21**, 3801 (1988)

A Nonequilibrium Phase Transition Induced by Multiplicative Noise

Christian Van den Broeck[1], *Juan M.R. Parrondo*[2], *and Raul Toral*[3]

[1] LUC, B-3590 Diepenbeek, Belgium
[2] Dep. Física Aplicada I, Universidad Complutense de Madrid,
28040-Madrid, Spain
[3] Departament de Física, Universitat de les Illes Balears,
07071-Palma de Mallorca, Spain

1 Introduction

Over the past 15 years, the effect of noise on the behaviour of nonlinear systems has been a major theme of investigation. An important example of a noise-induced phenomenon is the so-called noise-induced transition [1], referring to the situation in which the form of a probability density, describing the steady state properties of a noisy nonlinear system, undergoes a qualitative change. In its simplest version, one considers a system described by a scalar variable obeying the following nonlinear stochastic differential equation:

$$\dot{x} = f(x) + g(x)\xi \ , \tag{1}$$

where ξ is a Gaussian white noise with intensity σ^2, interpreted in the Stratonovich sense. The steady state probability corresponding to (1) reads

$$P^{\mathrm{st}}(x) \sim \exp\left\{ \int^x dy \frac{f(y) - \dfrac{\sigma^2}{2}g(y)g'(y)}{\dfrac{\sigma^2}{2}g^2(y)} \right\} \ . \tag{2}$$

The extrema $\bar{x}$ of this probability density obey the following equation:

$$f(\bar{x}) - \frac{\sigma^2}{2}g(\bar{x})g'(\bar{x}) = 0 \ . \tag{3}$$

Note the appearance of an additional term, resulting from the noise, which can change the type or degree of nonlinearity of the steady state equation. For example, for

$$\dot{x} = -x + \lambda(1 - x^2) + (1 - x^2)\xi \tag{4}$$

with $|x| < 1$, it is found that the probability density is unimodal, with a maximum at $\bar{x} = 0$ (which corresponds to the deterministic steady state) for $\lambda = 0$ and an intensity of the noise $\sigma^2 < 1$. However the density becomes bimodal for $\sigma^2 > 1$ (with new solutions $|\bar{x}| \neq 0$ appearing, cf. (3)). This phenomenon has been called noise-induced bistability. Nevertheless, one has to keep in mind that this change in the form of the probability density does not correspond to a genuine bifurcation or phase transition with breaking of ergodicity or of symmetry. In the above example, transitions between the $x > 0$ and $x < 0$ "phases" occur constantly, fluctuations in the value of x are very large and one cannot talk about different macroscopic phases. Furthermore the $x \leftrightarrow -x$ symmetry of the model is not destroyed. One can reduce the frequency of the transitions by playing on time scales in more complicated models [2], but this is only a quantitative effect and there is no phase transition in the traditional sense of the word. Our purpose here is to investigate if a genuine phase transition can occur when considering spatially distributed systems. The answer turns out to be "yes", but the conditions for and characteristics of the transition are somewhat unexpected. In particular it is found that model (4) does not undergo a phase transition.

2 Mean Field Model

By spatially coupling units i that are described by scalar variables x_i with local dynamics identical to (1), one is led to the following set of stochastic differential equations:

$$\dot{x}_i = f(x_i) + g(x_i)\xi_i - \frac{D}{2d} \sum_{j \in n(i)} (x_i - x_j) , \tag{5}$$

where ξ_i are uncorrelated Gaussian white noises with strength σ^2 and $n(i)$ represents the neigbourhood of unit i. We will be considering a cubic lattice so that there are exactly $2d$ such neighbours. The multivariate steady state probability associated to (5) is only known for the case of additive noise $g \equiv 1$. To make progress in the multiplicative noise case, we introduce the Weiss mean field approximation and assume that $\sum_{j \in n(i)} x_j = 2d\langle x \rangle$ where $\langle x \rangle$ is the average value which uniform throughout the system [3]. In this way, the equations for all the units decouple, and (5) takes on a form similar to (1), but with f replaced by $f - D(x - \langle x \rangle)$. The solution for the single unit steady state probability thus reads (cf. (2), we dropped the subscripts i for simplicity of notation):

$$P^{\mathrm{st}}(x) \sim \exp \left\{ \int^x dy \frac{f(y) - \frac{\sigma^2}{2} g(y)g'(y) - D(y - \langle x \rangle)}{\frac{\sigma^2}{2} g^2(y)} \right\} . \tag{6}$$

The value of $\langle x \rangle$ follows from the self-consistent requirement that

$$\langle x \rangle = \int dx \; x \; P^{\mathrm{st}}(x) = F(\langle x \rangle) . \tag{7}$$

Whenever this nonlinear equation in $\langle x \rangle$ has multiple solutions, the mean field theory predicts symmetry breaking associated to the occurrence of a phase transition. To get an idea of what kind of results to expect, we consider the limit

$D \to \infty$, in which case (7) reduces to the following simple form:

$$f(\langle x \rangle) + \frac{\sigma^2}{2} g(\langle x \rangle) g'(\langle x \rangle) = 0 \ . \tag{8}$$

This equation should be compared with (3). First note the difference in interpretation. The solutions to (8) correspond to the various macroscopic phases of our system, not to extrema of some probability density. Secondly, we note the surprising difference in the sign between (3) and (8). One of the consequences is that models that exhibit noise induced bistability do not present, in their spatially extended version, a phase transition to an ordered phase. We now turn to a model which does exhibit such a phase transition [4].

3 Noise-induced Phase Transition

Consider the following model:

$$f(x) = -x(1+x^2)^2 \qquad g(x) = 1 + x^2 \ . \tag{9}$$

f and g have been chosen such that the system displays a perfect $x \leftrightarrow -x$ symmetry. Yet, for an intensity of the noise larger then some critical value $\sigma^2 > \sigma_c^2$, where the value of σ_c^2 depends on D, the mean field theory predicts the appearance of ordered phases with $\langle x \rangle \neq 0$. This is already apparent from (8). By Taylor expansion around $\langle x \rangle = 0$, one finds

$$f(\langle x \rangle) + \frac{\sigma^2}{2} g(\langle x \rangle) g'(\langle x \rangle) = -\langle x \rangle + \sigma^2 \langle x \rangle - 2\langle x \rangle^3 + \sigma^2 \langle x \rangle^3 + \dots$$

so that two new symmetry breaking solutions $\langle x \rangle \sim \pm\sqrt{\sigma^2 - 1}$ appear for $\sigma^2 \geq 1$.

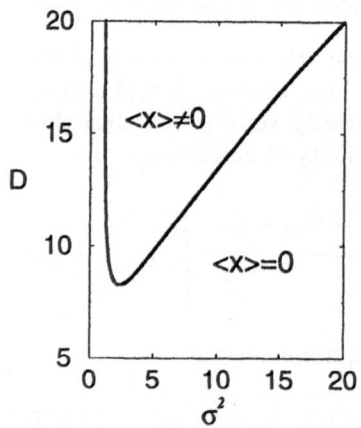

Fig. 1. Phase diagram for the noise-induced phase transition as predicted by the mean field theory.

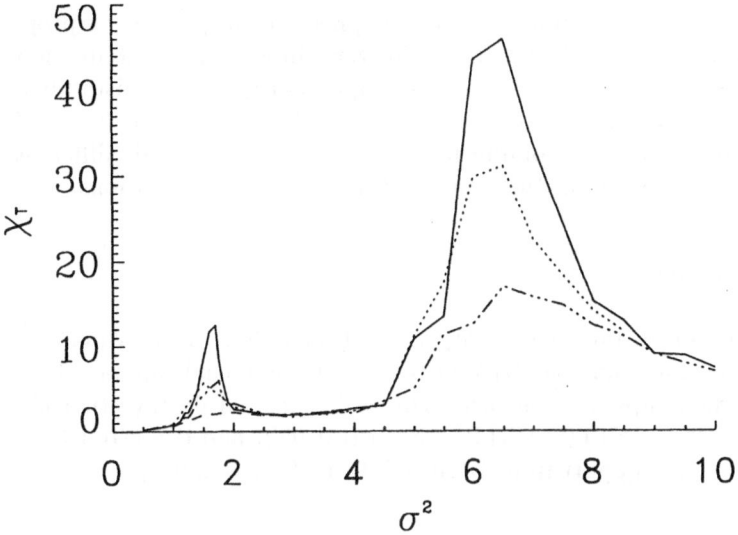

Fig. 2. Susceptibility in function of the intensity σ^2 for the value of $D = 20$ and for system sizes 10×10 (dashed line), 20×20 (dotted–dashed), 30×30 (dotted) and 40×40 (solid).

A numerical analysis of (7) confirms this result, and allows one to determine the region in parameter space (D, σ^2) where the ordered phase appears (see Fig. 1). One concludes from the mean field analysis that the ordered phase appears for a sufficiently strong spatial coupling, and in a window of intermediate noise strengths. In other words, the transition is reentrant. These qualitative features are confirmed by extensive simulations of a 2-dimensional system. Furthermore, these simulations give convincing evidence of the fact that the appearance or disappearance of the ordered state takes place through a genuine second order phase transition with all the properties normally associated to equilibrium phase transitions such as scaling, divergence of the susceptibility and of the temporal and spatial correlations, finite size effects, etc. As an illustration, we have plotted in Fig. 2 the susceptibility in function of the intensity σ^2 for the value of $D = 20$ and for system sizes up to 40×40. One clearly sees the development of divergencies at the first and reentrant location of the phase transition. Note that the mean field prediction overestimates the location of the reentrant transition (for more details see [4]). Further and more accurate simulations will be needed to determine the value of the critical exponents or to determine the universality class of this new type of nonequilibrium transition.

4 Discussion

The present work ends the speculation about whether multiplicative noise can be an essential ingredient for the formation of structure. The answer is affirmative, but our theoretical results also indicate that this phenomenon cannot be properly discussed within the context of a theory for zero-dimensional systems. For example, the model defined in (9) does not undergo a noise-induced bistability. We therefore believe that the physical ingredients leading to the formation of ordered structures under influence of multiplicative noise are essentially different from those needed or discussed in the context of noise-induced transitions.

Acknowledgements

We thank the Program on Inter-University Attraction Poles, Prime Minister's Office, Belgian Government and the NFWO Belgium for financial support. J.M. R.P. also acknowledges support from from Dirección General de Investigación Científica y Técnica (DGICYT) (Spain) Pro. No. PB91-0222 and Pro. No. PB91-0378. R.T. acknowledges support from DGICYT Pro. No. PB92-0046.

References

1. W. Horsthemke and R. Lefever, *Noise-Induced Transitions* (Springer-Verlag, Berlin, 1984).
2. R. Lefever, in *Spatial Inhomogeneities and Transient Behavior in Chemical Kinetics*, edited by. P. Gray et al. (Manchester University Press, New York, 1990).
3. Y. Onodera, Prog. Theor. Phys. **44**, 1477 (1970); M. Malek Mansour and G. Nicolis, J. Stat. Phys. **13**, 197 (1975); R. Desai and R. Zwanzig, J. Stat. Phys. **19**, 1 (1978); A.D. Bruce, Adv. Phys. **29**, 111 (1980); R.A. Dawson, J. Stat. Phys. **31**, 29 (1983); M.O. Hongler and R. Desai, J. Stat. Phys. **32**, 585 (1983); O.T. Valls and G.F. Mazenko, Phys. Rev. B **34**, 7941 (1986); M. Shiino, Phys. Rev. A **36**, 2393 (1987); K. Kaneko, Phys. Rev. Lett. **65**, 1391 (1990); P. Jung, U. Behn, E. Pantazelou, and F. Moss, Phys. Rev. A **46**, 1709 (1992); C. Van den Broeck, J.M. Parrondo, J. Armero, and A. Hernandez-Machado, Phys. Rev. E **49**, 2639 (1994).
4. C. Van den Broeck, J.M.R. Parrondo, and R. Toral, *Noise-induced phase transitions*, preprint.

Electric field domains in superlattices: Dynamics

L. L. Bonilla[1], J. A. Cuesta[1], J. Galán[1], F. C. Martínez[1] and J. M. Molera[1]

Universidad Carlos III de Madrid,
Escuela Politécnica Superior, Butarque 15
28911 Leganés, Spain.

1 Introduction

In this paper we will discuss several nonequilibrium phenomena in semiconductor superlattices (SL): formation and evolution of electric field domains. This is an old subject that goes back to Esaki and Chang in 1974 [1]. A SL is a succession of N quantum wells made out of alternating slabs of two (doped or undoped) semiconductors. The canonical example is the identical repetition of one superperiod composed of several tens of Å of GaAs (the "quantum valley") and several tens of Å of AlAs (the "quantum barrier"). When one such SL is subject to a large dc voltage bias across its growth direction, electric field domains [2] and time-dependent oscillations of the current may appear [3]. These phenomena are of potential importance in understanding the physics and technology of nanometer devices [4].

To understand the mechanism behind the formation of electric field domains in SL, let us consider the following oversimplified view of the Kazarinov and Suris's tight-binding results for conduction in a SL [5]. At zero applied electric field the dominant transport mechanism is miniband conduction: the energy levels of all the quantum wells (QW) are aligned along the growth direction of the SL and the electrons tunnel coherently from well to well across the SL. In reality the energy levels of the QWs are subbands (consider wells that extend infinitely on the transversal direction) and have a finite width, but we will continue speaking of QW levels and thinking in terms of one space dimension for simplicity. When an electric field is applied, the levels of different wells cease to be aligned and there appears a localized state (Wannier-Stark localization). As a consequence the electric current through the structure drops to a small value. When the applied electric field E is large enough, a new conduction mechanism appears: sequential resonant tunneling (SRT). Let

$$E \sim \frac{\mathcal{E}_2 - \mathcal{E}_1}{e\tilde{l}}, \tag{1}$$

where $\mathcal{E}_2$ and $\mathcal{E}_1$ are the energies of the first (e_1) and the second (e_2) levels of one well, $\tilde{l}$ the SL period and e the electron charge. Then the level $e1$ of one QW is aligned with the level e_2 of the following QW, and one electron may tunnel coherently betweem them. After this, scattering events (mainly scattering with phonons) send down the electron to the level e_1 again where it can proceed to the next QW in the same way. This SRT mechanism (which we denote by $e_1 \rightarrow e_2$) gives rise to a peak in the current-voltage diagram of the SL. For applied voltages (minus built-in potentials) which are not smaller than $N(\mathcal{E}_2 - \mathcal{E}_1)/e$, electric field domains may form: part of the SL is at at zero field (domain I) and the rest (domain II) at the field (1), so that the voltage across the SL equals the applied bias. For larger applied voltages other SRT mechanisms (between levels e_1 and e_3 for example) may give rise to domains with higher electric field. In doped SL domain formation was already experimentally demonstrated by Esaki and Chang [1]. Recent theories of domain formation include those of Laikhtman [6] [7], Prengel et al [8] (for doped SL) and ours [9] (for undoped SL). We now explain the relevant experimental results in undoped SL under laser illumination by Kwok et al. [3], our model based upon them and the corresponding results.

In a typical experiment, an undoped 40 period 90ÅGaAs/ 40ÅAlAs SL was mounted in a *p-i-n* diode and continuously illuminated by laser light at 4K [3]. When the laser power was in a certain interval and the applied *dc* voltage bias was large enough, damped time-dependent oscillations of the photocurrent (PC) and the peaks in the photoluminescence (PL) spectrum were observed. The applied fields in the experiment [3] are high, the potential barriers are wide, and the minibands correspondingly narrow so that the coherence length is comparable or smaller than the width of one quantum well. Then the quantum wells in the superlattice are weakly coupled and formation of electric field domains may appear [2, 10]. In Kwok et al. experiments, the time-dependent oscillations of the PC were observed in regions of the current-voltage diagram where domains type II and III (with electric fields corresponding to $e_1 \rightarrow e_2$ and $e_1 \rightarrow e_3$ SRTs) coexist [3]. Our theory aims to explain this behavior while other theories considered coexistence of zero field and $e1 \rightarrow e2$ domains [8, 7]. In these SL with wide QW, there are several important times that we have to consider in order to construct a reasonable model. First of all, the characteristic time scale of the PC oscillations in Ref. [3] is of 10-100 ns (the order of the recombination time). The time scale for carrier thermalization is of 0.1 ps while the carriers reach thermal equilibrium with the lattice after a time that ranges from 1 to 100 ps, smaller than the typical tunneling time of 500 ps (see [9] and references cited therein). This means that in time scales of the order of nanoseconds (the experimental time scale), we may consider the holes and electrons to be at local equilibrium within each QW j at the lattice temperature, and with given values of their densities, $\tilde{p}_j$ and $\tilde{n}_j$. The process of reaching a stationary state might be seen as the attempt of reaching a "global equilibrium" starting from "local equilibrium" through tunneling processes that communicate different QWs, self-consistency of the electric field and scattering and interband processes. In this spirit we consider the QWs as entities characterized by average values of the electric field, $\tilde{E}_j$ for the j-th well, and of the densities of holes and electrons, $\tilde{p}_j$ and $\tilde{n}_j$, respectively. We then propose the following transport equations to describe the dynamics of the SL [9]:

$$\tilde{E}_j - \tilde{E}_{j-1} = \frac{e\,\tilde{l}}{\epsilon}\,(\tilde{n}_j - \tilde{p}_j), \tag{2}$$

$$\epsilon\frac{d\tilde{E}_j}{d\tilde{t}} + e\,\tilde{v}(\tilde{E}_j)\,\tilde{n}_j = \tilde{J}, \tag{3}$$

$$\frac{d\tilde{p}_j}{d\tilde{t}} = \tilde{\gamma} - \tilde{r}\,\tilde{n}_j\,\tilde{p}_j, \tag{4}$$

where $j = 1, \ldots, N$. In this model Eq. (2) and Eq. (3) are, respectively, Poisson equation (averaged over one SL period) and Ampère's law; Eq. (4) is the hole rate equation containing the photogeneration rate (proportional to the laser power), $\tilde{\gamma}$ and the recombination constant $\tilde{r}$. The average permittivity is denoted by ϵ. $\tilde{v}(\tilde{E})$ is an effective electron velocity and $\tilde{J}$ is the total current density; the contribution of the heavier holes to the current is ignored.

In the calculation, we take $\tilde{v}(\tilde{E})$ as a datum and assume it does not vary with density. Qualitatively our results do not depend on the precise shape of $\tilde{v}(\tilde{E})$ provided it exhibits maxima at the resonant fields; the results reported here are based on the curve shown in the inset of Fig. 1. In the $3N$ Equations (2)- (4) there are $3N + 2$ unknowns, $\tilde{J}, \tilde{E}_0, \tilde{E}_j, \tilde{n}_j, \tilde{p}_j$, with $j = 1, \ldots, N$. One additional equation is the bias condition:

$$\frac{1}{N}\sum_{j=1}^{N}\tilde{E}_j = \frac{\tilde{\Phi}}{N\,\tilde{l}}. \tag{5}$$

Here $\tilde{\Phi}$ is the difference between the applied voltage and the built-in potential due to the doped regions outside the SL (1.5 Volts in Ref. [3]). $\frac{\tilde{\Phi}}{N\tilde{l}}$ is the average applied electric field on the SL, which we will henceforth call the *bias*. The missing condition is a boundary condition for the field at the zeroth QW, $\tilde{E}_0$, (*before* the SL). We do not have direct experimental evidence for what $\tilde{E}_0$ should be. Thus our choice for $\tilde{E}_0$ has to be validated *a posteriori* by comparing the results of our analysis with experiments and with the consequences of a different choice. We shall use throughout this paper the following boundary condition:

$$\tilde{E}_0(\tilde{t}) = \tilde{E}_1(\tilde{t}). \tag{6}$$

This condition does not allow a charge build-up at the first well and it is compatible with the existence of a stationary solution of model with the same constant values of the field and the carrier densities for all the QWs:

$$\tilde{v}\left(\frac{\tilde{\Phi}}{N\tilde{l}}\right) = \frac{\tilde{J}\,\tilde{r}^{1/2}}{e\,\tilde{\gamma}^{1/2}}, \quad \tilde{E}_j = \frac{\tilde{\Phi}}{N\tilde{l}}, \tag{7}$$

for all j. This equation says that the effective electron velocity is (except for constant scale factors) the same as the static I-V characteristic curve provided that *the field profile of the SL is static and uniform*. From experiments we know that such electric field profile is observed at low laser power [3], so that we can

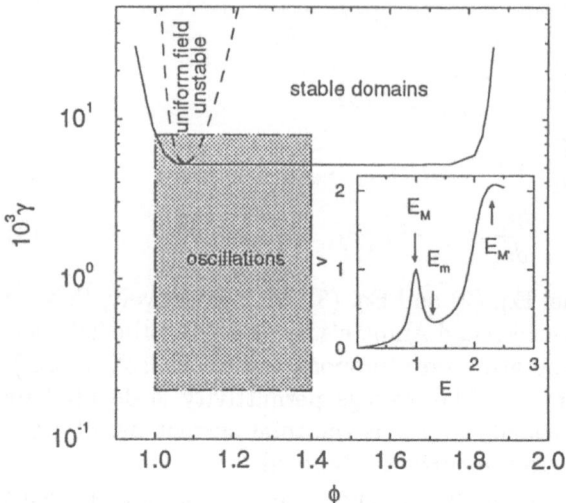

Fig. 1. Phase diagram of the model: Photogeneration rate (γ) vs. applied voltage bias (ϕ) in units of, respectively, $\tilde{n}/\tau_p$ and $\tilde{E}_M N \tilde{l}$; see text. The shaded region denotes the range of domain wall oscillations with $2 \times 10^{-4} < \gamma < 8 \times 10^{-3}$. The continuous (dotted) curve denotes the region for which the domain (uniform-field) solution is stable (unstable). Inset: The normalized field dependence of the carrier velocity used in the calculations.

infer the form of $\tilde{v}(\tilde{E})$ from experimental data. Later we shall see when the solution with uniform field profile ceases to be stable at high laser power and domains form. It is reassuring to remark that Prengel et al. used independently the same boundary condition (6) for similar reasons [8]. We could have also obtained the velocity curve from microscopic theories of static I-V curves for heterostructures (see Ref. [8]), but we find our identification with experimental data sufficient.

It is convenient to render the equations dimensionless by adopting as the units of electric field and velocity the values at the first maximum of the velocity curve, $\tilde{v}(\tilde{E})$, $\tilde{E}_M \simeq 10^5$ V/cm and $\tilde{v}_M$. Further we express the carrier density and bias in terms of

$$\tilde{n} = \frac{\epsilon \tilde{E}_M}{e \tilde{l}} \simeq 10^{17} - 10^{18} \, \text{cm}^{-3}, \tag{8}$$

and $\tilde{E}_M N \tilde{l}$ for $\tilde{l} = 130$ Å. This yields the time and current units, $\tau_p = 1/(\tilde{r}\,\tilde{n})$ and $\tilde{J}_M = e\tilde{v}_M \tilde{n}$; $\tau_p = 10$ ns determines the time scale of the oscillations [9]. The dimensionless equations then are:

$$E_j - E_{j-1} = n_j - p_j, \tag{9}$$

$$\beta \frac{dE_j}{dt} + v(E_j)\, n_j = J, \tag{10}$$

$$\frac{dp_j}{dt} = \gamma - n_j\, p_j, \tag{11}$$

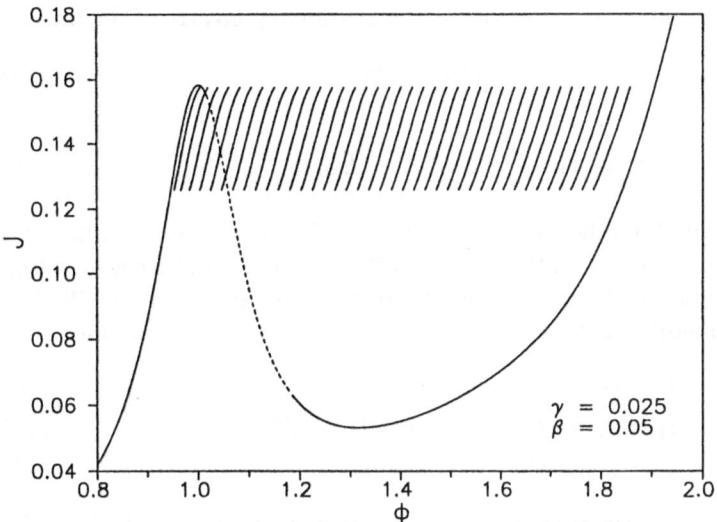

Fig. 2. Static characteristic curve: current (J) versus applied bias (ϕ) for laser power $\gamma = 0.025$ and $\beta = 0.05$. The solid line having the same shape as $v(E)$ corresponds to the stable uniform solution. In the part of the curve with negative slope there is a region where this solution becomes unstable (dashed line). The remaining 39 curves correspond to linearly stable two-domain solutions [9]. Important features to stress from this figure are: first, the coexistence, for a given ϕ, not only of several domain solutions but also of domain solutions and the uniform solution (see also Fig. 1), something relevant to the appearance of hysteresis and memory effects, and second, the average "flatness" of J in the bias region between the two local maxima of $v(E)$.

$$\frac{1}{N} \sum_{j=1}^{N} E_j = \phi, \tag{12}$$

$$E_0 = E_1. \tag{13}$$

Here $\beta = \tilde{r}\tilde{l}\tilde{n}/(\tilde{v}(\tilde{E}_M))$ goes from 0.01 to 1 and ϕ and $\gamma = \tilde{\gamma}/(\tilde{r}\tilde{n}^2)$ are the dimensionless control parameters. These equations are to be solved with initial conditions for the fields, $E_j(0)$, and the hole concentrations, $p_j(0)$, compatible with the bias (12) and the boundary condition (13). The initial conditions for the electron density, $n_j(0)$, then follow from (9).

2 Steady states

For given values of γ and J there are stationary solutions of the equations (9–11) compatible with the boundary condition (13). Using equation (12), we find the corresponding value of ϕ. From this we can obtain the curves $J = J(\phi)$ appearing in Fig. 2. We have studied two different types of stationary solutions, namely uniform solutions and two-domain non-uniform solutions.

The stationary versions of equations (10) and (11) may be solved for n_j and p_j in terms of the electric field E_j. When the result is inserted in (9), we find the following discrete mapping for the stationary electric field profiles $\{E_j, \ j = 1, \ldots, N\}$:

$$E_{j-1} = f(E_j; \gamma, J) \tag{14}$$

with

$$f(E; \gamma, J) \equiv E - \frac{J}{v(E)} + \frac{\gamma}{J} v(E), \tag{15}$$

which should obey the boundary condition $E_0 = E_1$. From these profiles and the bias condition (12), we obtain the current–voltage characteristic curve of Fig. 2, where only the stable stationary solution branches have been depicted [9]. The uniform stationary solution corresponds to the fixed points of the mapping (14):

$$E = \phi \tag{16}$$

$$n_j = p_j = \sqrt{\gamma} \qquad (j = 1, \ldots, N) \tag{17}$$

$$J = \sqrt{\gamma} v(\phi) \tag{18}$$

Equation (18) yield a static current-voltage characteristic curve which is the same as $v(E)$ except for a scale factor. See Fig. 2. The mapping (14) may have one or three fixed points. The latter is a necessary condition for non-uniform stationary field profiles to exist. For biases ϕ between the two maxima in the inset of Fig. 1, we find the linearly stable non-uniform solution branches depicted in Fig. 2, with two domains separated by a wall. They correspond to trajectories of the discrete mapping (14) that leave the fixed point with lower field on the first branch of $f(E; \gamma, J)$, E_L, after $j = k$ QWs and jump to the third branch of $f(E; \gamma, J)$ for $j > k$. See Fig. 3. These profiles exist only for

$$\gamma > \gamma_1 \equiv \frac{1}{4} \left(\min_{1 < \phi < E_m} \frac{v(\phi)}{|v'(\phi)|} \right)^2 \simeq 5.2 \times 10^{-3} \tag{19}$$

and J on an appropriate interval [9]. In general there are several stable stationary solutions for a given bias value, thereby yielding multistability and the possibility of hysteresis, in agreement with experiments [2] (see [11] for multistability and hysteresis with domain I/domain II stationary branches). There also exist other stationary solution branches which are unstable and have not been depicted in Fig. 2. They can be constructed with the help of the mapping (14) (they typically correspond to jumps to or from the second branch of $f(E; \gamma, J)$, see Fig. 3), and connect the stable branches of Fig. 2, [9].

3 Phase diagram and PC time-dependent oscillations

With the aim of interpreting the experimental results, we have to delimit the intervals of the parameters γ and ϕ where stable stationary solutions may be found. The $\gamma - \phi$ "phase diagram" of Figure 1 displays the regions where the uniform stationary solution is stable or unstable, the regions where linearly stable solutions with domains exist, and the regions where damped time-dependent oscillations of the photocurrent (PC) may be observed [9]. The information on

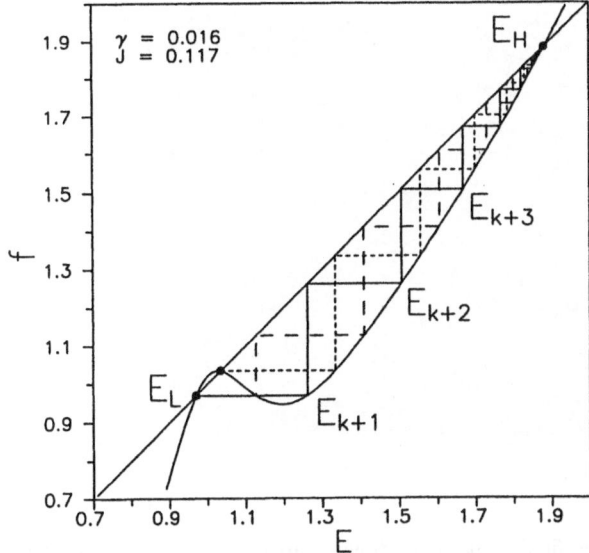

Fig. 3. The discrete mapping $f(E; \gamma, J)$ and E versus E for $\gamma = 0.016$ and $J = 0.117$. Jumps $1 \rightarrow 3$ and $2 \rightarrow 3$ between the branches of $f(E; \gamma, J)$ are allowed. The continuous lines show the discrete mapping process used to construct a $(1 \rightarrow 3)$ non-uniform stationary solution with two domains: After remaining at the low field domain value E_L, the field jumps at the $(m+1)$-th well to a different solution of $f(E; \gamma, J) = E_L$. In successive iterations the electric field tends to its high-field value E_H. The short-dashed lines show the same construction for a $2 \rightarrow 3$ solution, whereas the long-dashed lines correspond to a different (unstable) type of $2 \rightarrow 3$ solution explained in [9].

stationary solutions in the phase diagram was obtained with the help of the discrete mapping $f(E; \gamma, J)$, whereas that on time-dependent oscillations was found by direct numerical simulation of the model (9)-(13) with an initial condition that diferred from the uniform solution in that there was an excess positive charge in the first QW. This corresponds to turning on the laser at $t = 0$ for a fixed value of voltage and laser power [9].

The analysis of the steady states in Section 2 reveals that at low laser power $\gamma < \gamma_1$ ($\simeq 5.2 \times 10^{-3}$ in our rescaled parameters), there is only one steady solution: the uniform one, (17-16), which is linearly stable (to be more precise, the same stands for the whole region outside the full line of the $\gamma - \phi$ phase diagram shown in Fig. 1). Inside the full line in Fig. 1, multiple stable non-uniform steady states exist. From a dynamical point of view, the simulations show damped time-dependent oscillations of the current in the range $\gamma_0 < \gamma < \gamma_2$ ($\gamma_0 \simeq 2 \times 10^{-4}$, $\gamma_2 \simeq 8 \times 10^{-3}$) for ϕ on a subinterval of $(E_M, E_{M'})$ (the shaded rectangle of Fig. 1). In the region of the phase diagram where these oscillations occur we may distinguish two subregions, according to the situation reached after the oscillations stop. For $\gamma_0 < \gamma < \gamma_1$ the oscillation of the PC stops and the electric field profile resembles a two-domain non-uniform stationary solution. However, no such stationary profile exists as indicated in the previous section.

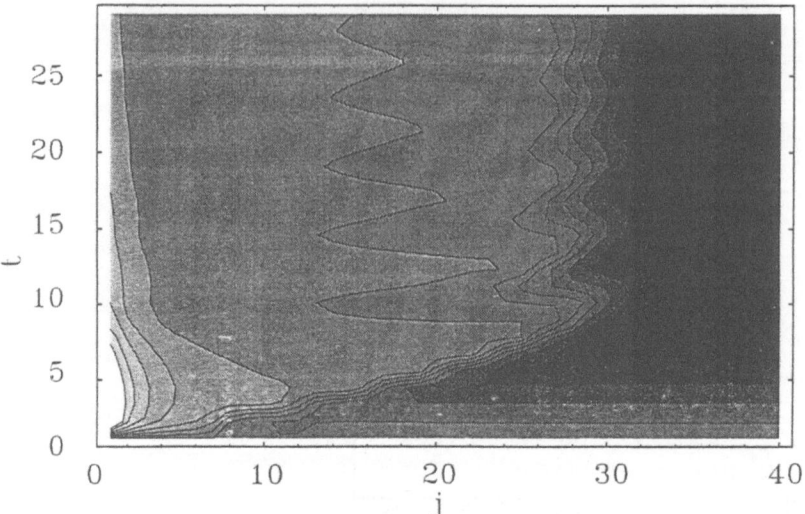

Fig. 4. Contour plot of the electric field (with time flowing in the y-axis and the QW number in the x-axis) showing the evolution of the domain wall in time for $\gamma = 2 \times 10^{-3}$ and $\phi = 1.2$. At $t = 0$ the electric field corresponds to the uniform solution except for a small perturbation at the first QW. At $t = 5$ the two domains are already defined (the darker the figure the higher the electric field) and the domain wall begins to oscillate. Most of the wells have an electric field centered around the first maximum of $v(E)$ (low field domain).

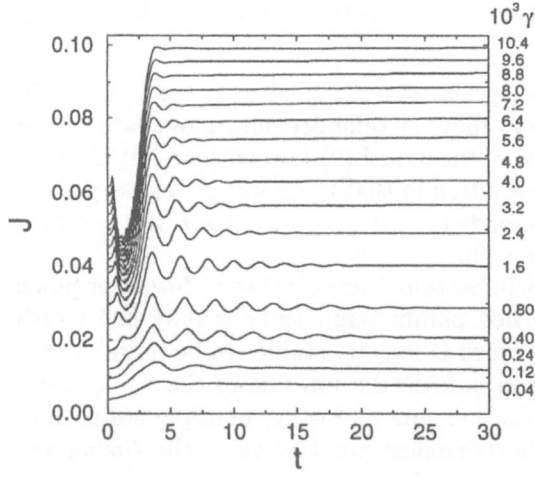

Fig. 5. Evolution of the PC, J, for different values of the photogeneration, γ (in units of 10^{-3}), which increases from bottom to top of the figure, as indicated at the right margin. The voltage ($\phi = 1.1$) is the same for all the curves.

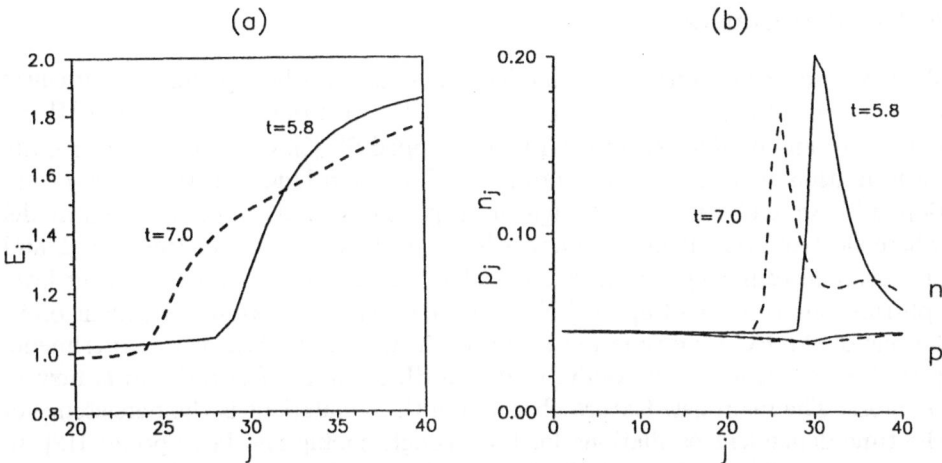

Fig. 6. (a) Electric field profile and (b) charge at each well corresponding to a maximum (full line) and a minimum (dashed line) of the PC. Parameter values are as in Fig. 4.

Thus there is a slow drift of the PC and of the field profile that ends at the values corresponding to the uniform stationary solution, after a time much longer than that required for the PC oscillations to stop. We may say that the non-uniform field profile with domains is "metastable" in this case [9]. For values of the photogeneration $\gamma_1 < \gamma < \gamma_2$, there are damped oscillations that end with an electric field profile corresponding to the true non-uniform stationary solution with domains. In all the cases, the numerical simulation shows that if the disturbance is large enough, a non-steady step-like electric field profile is formed after a short time. The domain wall of this profile oscillates back and forth in time about a fixed value of the position. As it oscillates, the domain wall changes its width, so that it spreads out over several wells, which makes easier the possibility of experimental detection. The electric field at the two domains in the profile also oscillates in time for a few periods before the oscillations stop (see Fig. 4 to visualize this process). Finally, outside the shaded rectangle of Fig. 1, the damping is so strong that there are no oscillations of the PC: the step-like electric field profile ($\gamma > \gamma_2$) or the uniform solution ($\gamma < \gamma_1$) are reached in a monotone fashion. These results are visualized in Fig. 5 in which the evolution of the current with time is represented for increasing values of γ in a range that covers the three situations just described.

These features of the solution of our model agree with experimental observations [3]. In Fig. 6 we have plotted the profiles of the electric field and of the electron and hole densities corresponding to a maximum and a minimum of the PC during one oscillation. Notice that the electronic charge oscillates from one side of the domain wall to the other during a period of the PC oscillation [3]. These results provide a clear picture of domain formation and posterior time evolution accounting for the time-dependent oscillations of the PC.

4 Final remarks

We have analyzed a very simple model that satisfactorily explains experimental observations of domain formation (field profiles with domain II/domain III coexistence) and oscillations of the PC in undoped SL [3]. Complementary results on formation of stationary field profiles with domain I/domain II coexistence in doped SL were obtained by Prengel et al. [8]. They used a more detailed model where electrons belonging to different levels of the QWs were distinguished, and the resonant tunneling current was calculated by perturbation theory in the level splitting due to a small applied electric potential [8]. No time-dependent oscillations of the PC were reported in [8] nor in the comparison with experiments [11]. An extension of our results to doped SL is straightforward and is now in progress. The main new feature observed in the simulations is the persistence of the time-dependent oscillations for low enough doping and laser power [12], in agreement with experimental observations by the Grahn group [13].

Among interesting open problems related to our work, we outline that of deriving our model from microscopic ones (see [8] for a perturbative derivation of a related model for doped SL) and that of the continuum limit of (9)-(13). In fact, our model resembles known drift-diffusion models in the continuum limit (such as the one in Ref. [14]) for which the uniqueness of the steady state can be proved easily, at least in the limit of small diffusion coefficient [15] (these continuum drift-diffusion models are also known to have Gunn effect oscillations [16, 17] among their possible solutions [18], which is not the case for Eqs. (9–13)). Thus it seems that the discreteness of our model is a crucial ingredient in reproducing both important static and dynamical features of transport in a SL made out of weakly coupled QWs. Work on derivation of discrete drift models from microscopic ones is now in progress.

L.L.B. thanks the organizing committee for the invitation to lecture in the XIII Sitges School. We thank Dr. O. M. Bulashenko, Prof. M. Kindelán, Dr. S.H. Kwok, Prof. R. Merlin, Mr. M. Moscoso, Dr. G. Platero and Prof. S. W. Teitsworth for fruitful discussions and collaboration on related topics and Dr. H. T. Grahn and Mr. J. Kastrup for showing us their laboratory and experimental results on doped superlattices. This work has been supported by the DGICYT grant PB92-0248, and by the EC Human Capital and Mobility Programme contract ERBCHRXCT930413. One of us (J.A.C.) also acknowledges financial support of the DGICYT grant PB91-0378.

References

1. L. Esaki and L. L. Chang, Phys. Rev. Lett. **33**, 495 (1974).
2. H. T. Grahn, H. Schneider and K. von Klitzing, Phys. Rev. B **41**, 2890 (1990).
3. S.-H. Kwok, Ph.D. Thesis. U. of Michigan, Ann Arbor, January, 1994. See also S.-H. Kwok, T. C. Norris, L. L. Bonilla, J. Galán, J. A. Cuesta, F. C. Martínez and J. M. Molera, H. T. Grahn, K. Ploog and R. Merlin, unpublished (1994).

4. See for instance the special issue of Physics Today, June (1993).

5. R. F. Kazarinov and R. A. Suris, Sov. Phys.–Semicond. **6**, 120 (1972).

6. B. Laikhtman, Phys. Rev. B **44**, 11260 (1991).

7. B. Laikhtman and D. Miller, Phys. Rev. B **48**, 5395 (1993) and preprint (1994).

8. F. Prengel, A. Wacker and E. Schöll, Phys. Rev. B **50**, (1994) issue of 15 July.

9. L. L. Bonilla, J. Galán, J. Cuesta, F. C. Martínez and J. M. Molera, Phys. Rev. B **50**, (1994) issue of 15 September.

10. H. T. Grahn et al, Surface Sci. **267**, 579 (1992).

11. J. Kastrup, H. T. Grahn, K. Ploog, F. Prengel, A. Wacker and E. Schöll, Appl. Phys. Lett. (1994), to appear.

12. L. L. Bonilla, J. Galán, M. Kindelán and M. Moscoso, unpublished (1994).

13. H. T. Grahn, private communication, June 1994.

14. S. W. Teitsworth, Appl. Phys. A **48**, 127 (1989). L. L. Bonilla and S. W. Teitsworth, Physica D **50**, 545 (1991).

15. L. L. Bonilla, Phys. Rev. B **45**, 11642 (1992).

16. M. P. Shaw, H. L. Grubin and P. R. Solomon, *The Gunn-Hilsum Effect*. Academic P., N. Y. 1979. M. P. Shaw, V. V. Mitin, E. Schöll and H. L. Grubin, *The Physics of Instabilities in Solid State Electron Devices*. Plenum P., N. Y. 1992.

17. F. J. Higuera and L. L. Bonilla, Physica D **57**, 161 (1992).

18. I. R. Cantalapiedra et al, Phys. Rev. B **48**, 12278 (1993). L. L. Bonilla et al, Semicond. Sci. Technol. **9**, 599 (1994).

Molecular dynamics simulation of phase separation and domain growth

Søren Toxvaerd

Department of Chemistry, H. C. Ørsted Institute
University of Copenhagen, DK-2100 Copenhagen Ø, Denmark

1 Non-Equilibrium Molecular Dynamics

The time evolution of a complex N-body system with analytic potentials can be simulated using the molecular dynamics (MD) technique, where the classical mechanical equations of motion are integrated using a finite difference algorithm. One criterion an MD algorithm must fulfill is that it scans the phase space *dynamically* correctly, leading to the correct equilibrium and non-equilibrium properties. Whereas the equilibrium behavior of complex systems is well explored by MD and by statistical methods such as the Monte Carlo (MC) technique, there has not been nearly the same effort in exploring the non-equilibrium dynamics, especially far from equilibrium. Perhaps this reservation is due to the fact that any finite difference algorithm must fail in giving the correct, individual (analytic) trajectories already after a short time interval due to the chaotic nature of the system. On the other hand it has been known for decades that MD of these systems does give the right (linear) response dynamics with the correct transport coefficients and functional form of the velocity auto correlation function [1]. Although this paradox of the long time behavior of complex systems being obtained correctly despite the fact that the individual "trajectory" points certainly deviate from an analytic solution is not fully understood, a crucial point is the nature of the algorithm which has to be time symmetrical and (mean) symplectic [2]. The simplest- and most commonly used example of such an algorithm is the time- centered formulation of Newton's equations of motion

$$\mathbf{q}_i(t + h) = 2\mathbf{q}_i(t) - \mathbf{q}_i(t - h) + h^2\mathbf{f}_i(t) \qquad (1)$$

for the i'th particle position $\mathbf{q}_i(t)$'s discrete time development, given by the force $\mathbf{f}_i(t)$ (mass included in the time unit). It appears under a variety of names in the literature (Verlet,Stoermer,Beeman, Leapfrog) and the Leap-frog reformulation of Eq. 1

$$(\mathbf{p}_i(t + h/2) - \mathbf{p}_i(t - h/2))/h = \mathbf{f}_i(t); \; (\mathbf{q}_i(t + h) - \mathbf{q}_i(t))/h = \mathbf{p}_i(t + h/2) \quad (2)$$

corresponds to Hamilton's formulation of classical dynamics.

MD on systems at equilibrium corresponds to microcanonical ensemble averaging. The velocity- or momentum, **p** in Eq. 1 and Eq. 2 plays no role for the dynamics and is not well defined because of the time shift between $q(t)$ and $p(t - h/2)$, and thus the "energy" is not well defined. However, in the case of dynamics of (coupled) harmonic systems the Verlet algorithm integrates exactly and with a conserved "shadow Hamiltonian" and there are indications that it is also the case for more complex systems [3], which can explain the paradox mentioned above. It is also possible to formulate constrained dynamics which mimic other ensembles, e.g. the canonical ensemble and where the dynamics is still time- reversible and (mean) symplectic by using a Nosé-Hoover (NH) thermostat [4]. The corresponding algorithms for time reversible constant temperature MD and constant pressure MD are given in [5] and [2].

The dynamics of phase separation is an example of non-equilibrium dynamics which can be explored by MD [6]. The analogous real experiment is obtained by quenching the system into a region of phase separation by a rapid change of one of its thermodynamical variables, typically the temperature T. The domain growth is then obtained e.g. from a measure of the growth of the corresponding peak in the structure factor $S(\mathbf{k}, t)$. As the domain(s) grow and coarsen in time all domain sizes will reach a state where they are much larger than all microscopic lengths. The growth law for the *average* domain size then has a power law of the form [7] $R(t) \propto t^n$, where n is the growth exponent, and the (radial) correlation function shows a scaled form $g(r/R(t), t) = G(x)$, with $x = r/R(t)$. Much effort has gone into classifying different systems by distinct universality classes [8], characterized by specific exponents and scaled functions. It is believed that the conservation of the order parameter is a relevant criterion for such a classification, and the systems examined by MD all belong to the class with conserved order parameters, e.g. spinodal decomposition of mixtures.

Thus the scenario is as follows: The MD makes it possible to perform a continuous space computer experiment of the non-equilibrium phase separation dynamics to be compared with corresponding experimentally obtained data and hydrodynamical models for phase separation dynamics. But in order to reach the late state with a dominating domain size much larger than microscopic (particle-) sizes, the number of particles, N, must be very "big", measured by what one usually deals with in today's MD and what can be handled in (big) computers. For a binary mixture, and in two dimensions N must be at least of the order $10^4 - 10^5$ [9]-[12] and the time increment h in the algorithm is of the order $10^{-14} s$ or less. With today's computers we can follow such systems in nano seconds. If we shall be able to reach the late state within so short a time interval, it means that the systems must be deeply quenched and to a degree which is not possible experimentally, but else it presents no problem. This is due to the fact that the speed of separation is driven by the excess of mixing energy, so by adjusting the pair potentials behind what is found in nature one can provocate the separation to take place much faster than else and to reach the scaling regime(s) within nano seconds.

2 MD of phase separations in 2 dimensional mixtures

The molecular dynamics simulations of 2-dimensional (2D) mixtures [9],[11] are performed by taking particles with different pair interactions. We have used Lennard-Jones -like potentials of the form

$$u_{AA}(r) = u_{BB}(r) = 4\epsilon((\sigma/r)^{12} - (\sigma/r)^m) \tag{3}$$
$$u_{AB}(r) = 4\epsilon((\sigma/r)^{12} + (\sigma/r)^m) \tag{4}$$

where all potentials are truncated at some distance. The particles which belong to the same type, A,B.., interact in the usual way whereas particles of different type repel each other according to Eq. 4. For simplicity all particles have the same size, σ. We can control the strengh of excess potential energy by varying the exponent m, and we have used $m = 3$ and $m = 6$, [9] and $m = 6$, [10], [12]- [14]. The phase diagram for a 2D LJ system is wellknown [15], it has a critical temperature $kT/\epsilon \approx .50$ so by quenching the system to the temperature $kT/\epsilon = 2$ we are far above the states where a phase separation between the components could be interfered by e.g. an intrinsic condensation. The phase diagram for $m = 6$ and $kT/\epsilon = 2$ is given in Figure 1. At constant temperature the (x_B, P) points

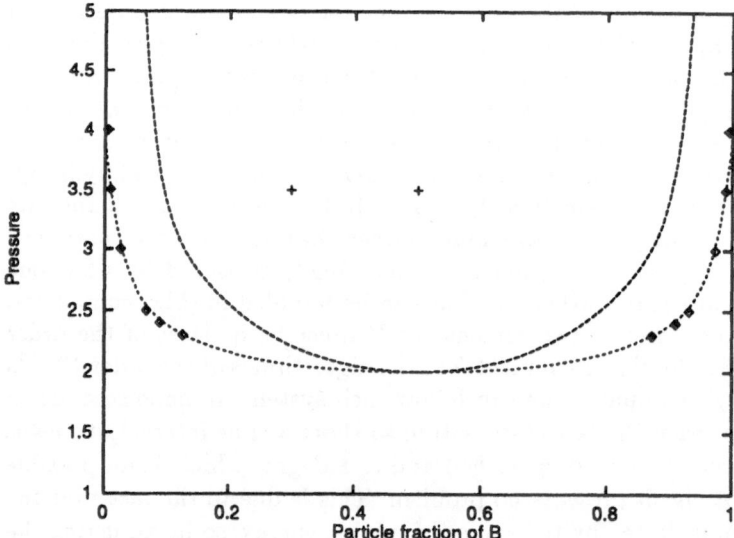

Fig. 1. Phase diagram for a binary mixture of Lennard-Jones particles with mixing energy given by Eq. 4. The diagram gives the pressure, $P\sigma^2/\epsilon$ as a function of the particle fraction x_B and is for the temperature $kT/\epsilon = 2$. The dashed line through the points, $\diamond$, is the coexisting line obtained from a Gibbs Ensemble MC simulation. The dashed line inside the immicibility region is the spinodal line and is obtained from a mean field model for binary mixtures. The two points, +, are the off-critical point $(x_B, P\sigma^2/\epsilon) = (.3, 3.5)$ and the critical concentration point $(x_B, P\sigma^2/\epsilon) = (.5, 3.5)$, respectively, for which growth exponents have been obtained.

of state are separated by the line of coexisting particle fraction, obtained by a Gibbs ensemble MC simulation of the pressure, P, in the coexisting phases [10]. The dotted line is the "spinodals", obtained by a mean field model.

The quenching is performed by running an ordinary one-component 2D-MD at $kT/\epsilon = 2$ and then, at time $t = 0$ to provide the particle with an identity according to eqs. 3 and 4, by which the homogeneous particle distribution rapidly changes and the phase separation starts. The latent heat of mixing is removed from the system by the NH-MD technique, but also other constant temperature MD thermostates have been applied.

A series of MD on such quenched systems have been performed recently, the results are collected in Table 1. Let us first discuss the spinodal decomposition

Table 1. Growth exponents for two dimensional mixtures

Num. of components	Particle fraction x_i	$N * 10^{-3}$	Method	Exponents
2 [9]	1/2	5-20	MD	$1/2 \to (1/4 \to)2/3$
2 [11]	1/2	17	MD	1/2
2 [12]	1/2	80	MD	$1/2 \to 2/3$
2 [9]	1/2	5	MC	$0.24 (\approx 1/4)$
2 [10]	0.30, 0.70	20	MD	0.37
2+surfactant [13]	$N_A = N_B, N_s$	10.5	MD	$\simeq 1/2$
3 [14]	1/3	10.5	MD	1/3

in a *two* component 2D mixture. Due to the perfect symmetry we know the exact x- location of the critical point, it must be at $x_c = .5$ and [9] and [11], [12] contains exponent(s) for spinodal decomposition at the critical particle concentration. It certainly has a scaling regime and with an exponent $\alpha = 1/2$ which is clearly different from 1/4 obtained by a corresponding MC quench [9]. This demonstrates the importance of the hydrodynamic modes which are missing in the MC-"dynamics". In [9] we concluded that the growth in 2D binary MD systems at critical concentrations was given by :"1/2, and a subsequent crossover to a faster growth, possibly mediated by a 1/4 transient". Later, more extensive *ensemble* simulations where the quenching was repeated from different starting positions demonstrated, however, that the observed 1/4 transient regime in the MD system in fact did not survive an averaging process over the independent realizations [11]; but recent [12] investigations based on ensemble averaging over a bigger system of 80 000 particles confirm the crossover to the faster growth with $\alpha = 2/3$ at late times in accordance with the prediction by Farrell and Valls [16]. They studied a Langevin model and concluded that it is the presence of heat diffusion which makes the growth follow a law with an exponent 2/3.

At critical quenches the domains exhibit a ramified cluster shape and where both phases only percolate the area in one, and the same (random) direction.

When the quench is performed (sufficiently) off-critically, the growth slows down with an exponent of the order $1/3 \approx .37$ [10]. This is to be expected because the solute no longer percolates the area in any directions, but performs droplets in the solvent. It seems plausible that for non-percolating mixtures the separation takes place partly by the coalescence of droplets and partly by an evaporation-condensation mechanism. The coalescence mechanism should give a growth proportional to $t^{1/2}$ [17] whereas the evaporation-condensation mechanism [18], in which particles evaporate from small droplets and condense into bigger ones , gives a $t^{1/3}$ growth. The exponent $\approx .37$ in [10] is obtained from one single off-critical quench in a mixture of 20 000 particles. The growth is certainly slowed down, but it is likely that both growth mechanisms still are present in the system which could explain a value of the growth exponent slightly bigger than $1/3$.

A quench of equally concentrated solutes in surfactants of amphifilic AB dimers nicely agrees with corresponding Langevin simulations and other theories [19]. In contrast to pure binary systems, characterized by an algebraic time dependence of the average domain size, the systems with surfactants exhibit non-algebraic, slow dynamics. The average domain size eventually saturates at a value inversely proportional to the surfactant concentration, and with a dynamical scaling which is independent of the surfactant concentration, all in agreement with a previous Langevin simulation [19].

Finally we have performed ensemble quenches for ternary and quarternary mixtures which have a completely different domain topology than that of binary mixtures, and with an exponent $\approx 1/3$ [14]. The growth exponent was consistently calculated from three different definitions of the domain size, ((1) from the first moment of the structure factor, (2) from the first zero of the correlation function, and (3) from the average linear domain size) and obtained as an ensemble average of twenty independent quenches. The domains are compact and the late time reaches a dynamical scaling regime during which the domains grow in agreement with the classical theory of Lifshitz and Slyozov [18]. This finding implies that hydrodynamic flow does not control the phase-separation process at late times in ternary and quaternary fluid system. The finding of the Lifshitz-Slyozov growth law in a multi-component system is in agreement with a recent MC simulation of spinodal decomposition in a three-state Potts model [20] and demonstrates the universality in ordering dynamics.

In summary: Molecular dynamics simulation of non-equilibrium systems offers a well founded technique for obtaining the real time dynamics. For systems driven by hydrodynamic modes, it is the only simulation technique which can be applied. Its limitation is that at present one can only follow up to of the order 10^5 particles in nano seconds and this is one of the reasons for that we have concentrated the investigations to two- dimensional systems.

Molecular dynamics simulation of phase separation and domain growth in two dimensional systems demonstrates the existence of scaling regimes with simple power-law growth in time: t^n. In order to obtain accurate estimates of the growth exponents it is , however, found to be necessary to perform ensemble averages over independent quench [11],[12] [14]. The MD simulations nicely confirm otherwise obtained results and serve as detailed experiments to be used for futher investigations of growth mechanisms.

Acknowledgement

Grant No. 11-0065-1 from the Danish Natural Science Research Council is gratefully acknowledged.

References

1. D. Levesque and W. T. Ashurst, Phys. Rev. Lett. **33**, 277 (1974).
2. S. Toxvaerd, Phys. Rev. E **47**, 343 (1993).
3. S. Toxvaerd, *to appear in* Phys. Rev. E (1994).
4. S. Nose', Mol. Phys. **52**, 255 (1984); W. G. Hoover, Phys. Rev. **A31**,1695 (1985).
5. S. Toxvaerd, Mol. Phys.(1992).
6. S. W. Koch, R. C. Desai, and F. F. Abraham, Phys. Rev. A **27**, 252 (1983).
7. K. Binder and D. Stauffer, Phys. Rev. Lett. **33**, 1006 (1974).
8. See e.g. O. G. Mouritsen, in *Kinetics of Ordering and Growth at Surfaces*, edited by M. G. Lagally (Plenum Publ. Co., New York,1990), p. 1.
9. E. Velasco and S. Toxvaerd, Phys. Rev. Lett. **71**, 388 (1993).
10. E. Velasco and S. Toxvaerd, J. Phys. C **6** A205 (1994).
11. P. Ossadnik, M. F. Gyure, H. E. Stanley and S. C. Glotzer, Phys. Rev. Lett. **72**, 2498 (1994).
12. E. Velasco and S. Toxvaerd, *submitted to* Phys. Rev. Lett. 1994.
13. M.Laradji, O. G. Mouritsen, S. Toxvaerd and M. J. Zuckermann, *to appear in* Phys. Rev. E 1994.
14. M. Laradji, O.G. Mouritsen and S. Toxvaerd, *submitted to Europhys. Lett.*.
15. B. Smit and D. Frenkel, J. Chem. Phys. **94** ,5663 (1991).
16. J.E. Farrell and O.T. Valls, Phys. Rev. B **40**,7027 (1989); B **42**,2353 (1990).
17. M. San Miguel, M. Grant and J.D. Gunton, Phys. Rev. A **31**, 1001 (1985).
18. I. L. Lifshitz and V. V. Slyozov, J. Phys. Chem. Solids **19**, 35 (1962).
19. M. Laradji, H. Guo, M. Grant, and M. J. Zuckermann, J. Phys. A **24**, L629 (1991).
20. C. Jeppesen and O. G. Mouritsen, Phys. Rev. B **47**, 14724 (1993).

Finite-Size Effects in the Kardar-Parisi-Zhang Equation

Raul Toral [1], Bruce Forrest [2]

[1]Departament de Fisica, Universitat de les Illes Balears, 07071-Palma de Mallorca, Spain,
[2]Institut für Polymere, E.T.H. Zürich, CH-8092 Zürich, Switzerland

Abstract: On the basis of a perturbative solution we study the numerical importance of finite–size effects in the Kardar–Parisi–Zhang equation for surface growth. The crossover behaviour between linear and non–linear regimes is studied numerically using convenient finite size scaling expresssions.

In the exciting field of non–equilibrium phenomena, much attention has been drawn recently to the problem of growth of random surfaces[1]. It has become clear that many different growth models share similar features such as scaling exponents and scaling functions and can thus be considered as belonging to the same *universality class*. The Kardar–Parisi–Zhang (hereafter referred to as KPZ) equation[2] is a prototype model for those systems in which the interface growth is driven by an external flux of particles. In the KPZ model, the surface height $h(r,t)$ on top of location r of a d–dimensional substrate satisfies a stochastic random equation:

$$\frac{\partial h(r,t)}{\partial t} = \nu \nabla^2 h(r,t) + \frac{\bar{\lambda}}{2}(\nabla h)^2 + \eta(r,t) \tag{1}$$

Every term in this equation models a physical phenomenon contributing to the surface evolution: ν, $\bar{\lambda}$ and D are parameters describing, respectively, surface relaxation, lateral growth and the effect of noise. This noise term aims to describe the random fluctuations in the incident flux of particles and is assumed to be a Gaussian random process of mean zero and correlations:

$$\langle \eta(r,t)\eta(r',t') \rangle = 2D\delta(r-r')\delta(t-t') \tag{2}$$

A convenient measure of the surface roughness is given by averaging the spatial fluctuations over different realizations of the noise:

$$w(t) = \sqrt{\langle \overline{h^2} - \overline{h}^2 \rangle} \tag{3}$$

Where the bar and the brackets denote the spatial and noise averages, respectively. Several time regimes can be found in the time evolution of the surface roughness. They can be summarized as follows:

(i) For very early times, the noise term dominates since its contribution to the equation grows as the square root of time. It is easy to find that the surface roughness grows in this time regime as $w(t) \sim t^{1/2}$.

(ii) For intermediate times, the linear term is the main contribution. The linear case ($\bar{\lambda} = 0$) is the Edwards–Wilkinson model[3] for which one can find easily that the surface roughness behaves as $w(t) = t^{\beta_0}$. The value for β_0 depends on the dimension of the substrate: $\beta_0 = 1/4$ for one-dimensional surfaces, $\beta_0 = 0$ (logarithmic growth) for the two-dimensional case.

(iii) For late times, the contribution of the relevant non-linear term becomes the dominant one and the surface roughness growth is characterized by a behaviour $w(t) = t^{\beta}$.

(iv) For very late times and finite substrate length L, the roughness saturates to a value $w(t \to \infty, L) \sim L^{\zeta}$.

Of course, in an experiment or in a numerical simulation, the transition between the different regimes is not sharp and different crossover behaviours can be observed. For the transition between non–linear (iii) and saturation (iv) regimes, a scaling law has been derived[4]:

$$w(t, L) = L^{\zeta} F(tL^{-z}) \tag{4}$$

ζ and z are the *roughness* and *dynamic* exponents, respectively. In order to recover the known limiting behaviours, the scaling function $F(x)$ behaves as $F(x) \sim x^{z/\zeta}$ (hence $\beta = z/\zeta$) for $x \ll 1$ and approaches a constant for large x. Galilean invariance implies the exact relation $z + \zeta = 2$ independently of dimension[5]. In one–dimensional substrates, a fluctuation–dissipation theorem[6] yields the exact values for the exponents $z = 3/2$, $\zeta = 1/2$, $\beta = 1/3$. It is interesting to notice that in the absence of non–linear terms a similar scaling relation holds but with different values for the exponents, namely $w(t, L) = L^{\zeta_0} F_0(tL^{-z_0})$, $\zeta_0 = 1/2$, $z_0 = 2$. So, in order to measure the correct crossover exponents and functions for non–linear to saturation behaviour, it is important to make sure that non–linear effects have fully developed before the saturation regime has started.

The precise knowledge of the dynamical exponents is very important since it allows a detailed characterization of the universality classes. Many numerical studies have been devoted to checking the scaling relations and to computing as accurately as possible the values for the scaling exponents[7]. Since a numerical simulation will deal necessarily with a system of finite-size, it is very important to analyze carefully the effect of finite-size effects in the different time regimes specified above and also on the crossover behaviour from one regime to another[8]. In this paper, we present a detailed study of the relevance that finite-size effects have in a numerical simulation of the KPZ equation and we show how it is possible, using convenient finite-size scaling forms, to compute the dynamical exponents that characterize the crossover form linear (ii) to non linear behaviour (iii).

It is possible to reparametrize the surface field $h \to (\nu/2D)^{1/2}h$ and time $t \to \nu t$ to obtain a somewhat simpler version of the KPZ equation:

$$\frac{\partial h(r,t)}{\partial t} = \nabla^2 h(r,t) + \frac{\lambda}{2}(\nabla h)^2 + \eta(r,t) \tag{5}$$

where $\lambda = (2D/\nu^3)^{1/2}\bar{\lambda}$ and the noise term satisfies $\langle \eta(r,t)\eta(r',t') \rangle = \delta(r - r')\delta(t - t')$. This reparametrization of the equation is achieved by simply setting $2D = \nu = 1$ in the original KPZ equation. In the numerical studies, typically one introduces a lattice discretization of the space variable $r_j = r_0 + ja_0$ (we restrict ourselves to one–dimensional systems from now on), so introducing a set of discrete variables $h_j(t) = h(r_j, t)$ in terms of which we write:

$$\frac{\partial h_j(t)}{\partial t} = (h_{j+1} + h_{j-1} - 2h_j) + \frac{\lambda}{2}\left(\frac{h_{j+1} - h_{j-1}}{2a_0}\right)^2 + \eta_j(t), \tag{6}$$

Periodic boundary conditions are usually assumed in order to avoid edge effects and the linear spacing a_0 is set to the unit length, $a_0 = 1$. It is possible to obtain the solution of the corresponding linear model ($\lambda = 0$) as formulated on the lattice to find that the surface roughness $w_{(0)}$ in this case is given by:

$$w_{(0)}^2(t, L) = \frac{1}{L}\sum_{k=1}^{L-1}\frac{1 - \exp(-2\alpha_k t)}{2\alpha_k}. \tag{7}$$

where $\alpha_k \equiv 4\sin^2(\pi k/L)$. For infinite size, this linear solution behaves asymptotically as:

$$w_{(0)}^2(t, L \to \infty) = \left(\frac{t}{2\pi}\right)^{1/2}\left(1 - \frac{1}{32t}\right) \tag{8}$$

(this expression has an error of less that 0.1% for $t \geq 1$) and, consequently, the dynamical exponent is $\beta_0 = 1/4$ as anticipated. This linear solution has a strong L-dependence as it is shown in figure (1) where we plot, in a double logarithmic scale, the time evolution of the surface roughness versus time for different system sizes. According to the above asymptotic solution this plot should yield a straight line of slope $1/2$. However, finite values of L have the effect of bending the curves so producing *effective* exponents which are smaller than the true exponents.

Similar conclusions can be drawn in the non–linear case. Here we do not know the exact solution but we can make use of a perturbative expansion in λ to find the leading correction (which is second order in λ) to the surface roughness as[8]:

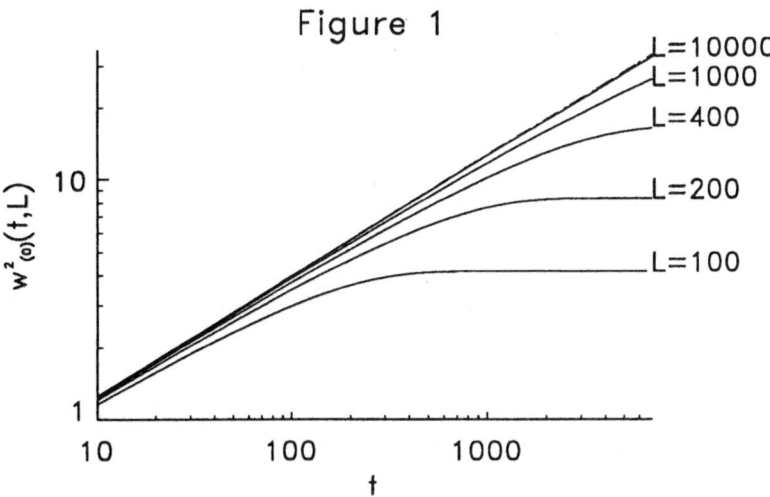

Fig. 1. Square of the surface roughness, $w_{(0)}^2$ for the linear solution (equation 9), as a function of system size L. For long times, the roughness saturates to a value that depends on L. For intermediate times, the slopes of the curves yield effective dynamical exponents less than the true asymptotic value $\beta_0 = 1/4$. The dashed line is the corresponding solution in the limit $L \to \infty$.

$$
\begin{aligned}
w^2(t, L)^{(2)} =\; & w_{(0)}^2(t, L) \\
& + \frac{\lambda^2}{16L^2} \sum_{k=1}^{L-1} e^{-2\alpha_k t} \sum_{k_1=0}^{L-1} \cos^2\left(\pi(k - k_1)/L\right) \times \left\{ \cos^2\left(\pi k_1/L\right) \right. \\
& \times [f(a, a, t) - f(a, b, t) - f(a, c, t) + f(a, d, t)] \\
& + \cot\left(\pi k/L\right) \sin\left(2\pi k_1/L\right) \\
& \left. \times [f(a, -a, t) - f(a, -b, t) + f(a, c, t) - f(a, d, t)] \right\} \\
& + o(\lambda^4)
\end{aligned}
\tag{9}
$$

with $a = -\alpha_{k_1} - \alpha_{k-k_1} + \alpha_k$, $b = -\alpha_{k_1} + \alpha_{k-k_1} + \alpha_k$, $c = \alpha_{k_1} - \alpha_{k-k_1} + \alpha_k$, $d = \alpha_{k_1} + \alpha_{k-k_1} + \alpha_k$ and $f(x, y, t) = \frac{xe^{(x+y)t} - (x+y)e^{xt} + y}{xy(x+y)}$ (and appropriate limits assumed for the case $xy(x + y) = 0$). Comparing with numerical simulations we see that the perturbative solution offers an accurate approximation for $\lambda = 0.5$ up to $t \approx 6000$ and for $\lambda = 1$ for $t \lesssim 1000$ (see figure (2)).

We can use the perturbative solution at $\lambda = 0.5$ to examine the effect of finite L on the solution in a similar fashion to the linear case studied above. In figure (3) we have plotted in a double logarithmic scale the time evolution of the roughness $w^2(t, L)^{(2)}$ as given by the second-order perturbation, equation (9), as a function of system size L. It is obvious from this figure that when dealing

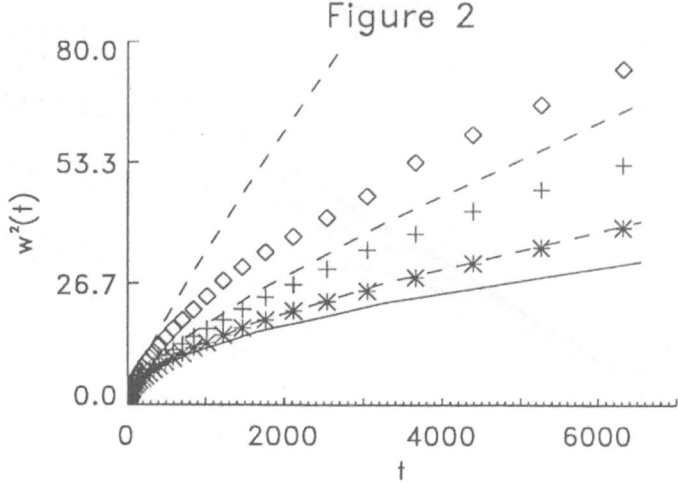

Fig. 2. Comparison of a numerical integration of the KPZ equation (symbols) for a system size $L = 10^4$ with the perturbative solution given in equation (9) (dashed lines) for the cases $\lambda = 2, 1, 0.5$ from top to bottom, respectively. The solid line is the exact linear solution

with finite system sizes the exponents obtained from these log-log plots should be considered only as effective exponents. Only for $L \gtrsim 10^4$ the relative error in the roughness is less than 1% up to times $t \simeq 6000$

One can obviously question the validity of the perturbation expansion to obtain dynamical exponents, but it is important to point out that that the finite-size effects worsen with increasing λ. This is obvious from figure (4) where we have plotted as a function of t the relative weight of the second-order term in the expansion at finite L compared to its asymptotic value ($L \to \infty$) as measured by the ratio $R \equiv w^2(t, L)^{(2)}/w^2(t, \infty)^{(2)}$ for $\lambda = 0$ and 0.5 at $L = 1000$ and 10000. Upon switching from $\lambda = 0$ to $\lambda = 0.5$ the ratio R is clearly seen to fall even further away from unity ($R = 1$ implies no finite-size effects). One can conclude that, at least for $d = 1$ and within the range of validity of the perturbative solution, the minimum value of L required in the linear case to avoid finite-size effects only provides a lower bound for $\lambda \neq 0$.

We turn now to the effect of finite-size effects in the crossover from linear to non-linear behaviour (regimes (ii) to (iii) above). For this crossover, two different scaling forms have been advocated. In a one–loop renormalization–group (RG) calculation Natterman and Tang (NT) found[9]:

$$w^2(t, L) = L^{2\zeta_0} f(\frac{t}{t_c}, \frac{L}{\xi_c}). \tag{10}$$

where the crossover time satisfies $t_c \sim \xi_c^{z_0}$. In the limit of small λ, one has the dependence: $\xi_c \sim \lambda^{-2}$ and $t_c \sim \lambda^{-4}$. On the other hand, Grossmann, Guo, Grant

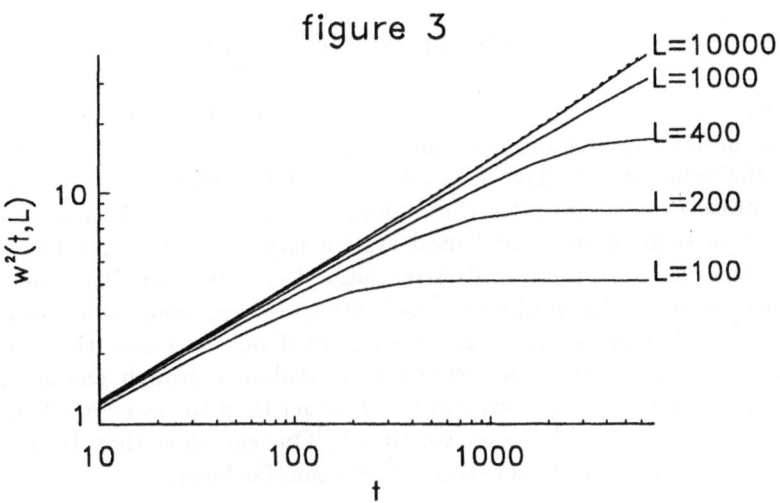

Fig. 3. Same as figure 1, but using the perturbative solution given in equation (9). Notice that again finite size effects show up as effective exponents in this log-log plot.

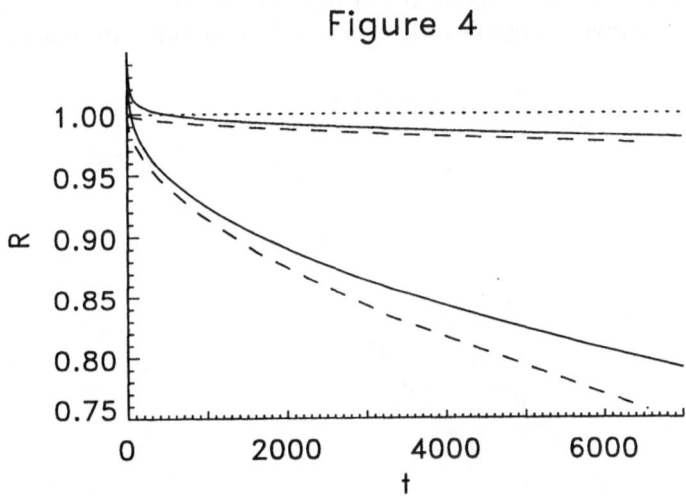

Fig. 4. Ratio $R = w^2(t, L)^{(2)}/w^2(t, \infty)^{(2)}$ from the second-order pertubation expression (equation (9)) as a function of time t for $\lambda = 0$ (solid lines) and $\lambda = 0.5$ (dashed lines) for two different values of $L = 1000$ (two lower lines), and $L = 10000$ (upper lines). If there were no finite size effects, R would take the constant value $R = 1$ (dotted line).

(GGG) proposed the form[10]:

$$w^2(t, L) = t^{2\beta_0} f\left(\frac{t}{L^{z_0}}, \frac{t}{\xi_c^{z_0}}\right). \tag{11}$$

and they made the ansatz that the crossover length scales as $\xi_c \sim \lambda^{-\phi}$ and on the basis of their numerical findings predicted the value $\phi = 3$, at variance with the value of NT. The data analysis of GGG that led to $\phi = 3$ used the asymptotic $L \to \infty$ form for the scaling relations, although in their numerical simulation these authors had used system sizes $L = 10^3$ and times $t = 10^4$ for which there are clear finite-size effects (see figure 3). It seems necessary, then, to reanalyze the problem using finite-size expressions. One can reduce the two proposed expressions to a similar form if one considers the nonlinear-to-saturation regime, when the crossover to nonlinear growth has already taken place. This limit requires a system size L larger than the crossover length ξ_c and also times larger than the crossover time t_c. One can show that, in this case, the expressions of NT and GGG reduce to the similar form:

$$w^2(t, L) = L f(t \lambda^{\frac{\phi}{4}} L^{-3/2}) \tag{12}$$

where $\phi = 3$ according to GGG and $\phi = 4$ according to NT. In order to check this expression, we have plotted in figure (5) $w^2(t, L)/L$ vs. $t \lambda^{\phi/4} L^{-3/2}$. It appears that choice of $\phi = 3$ gives a significantly superior data–collapse to that of $\phi = 4$.

Figure 5

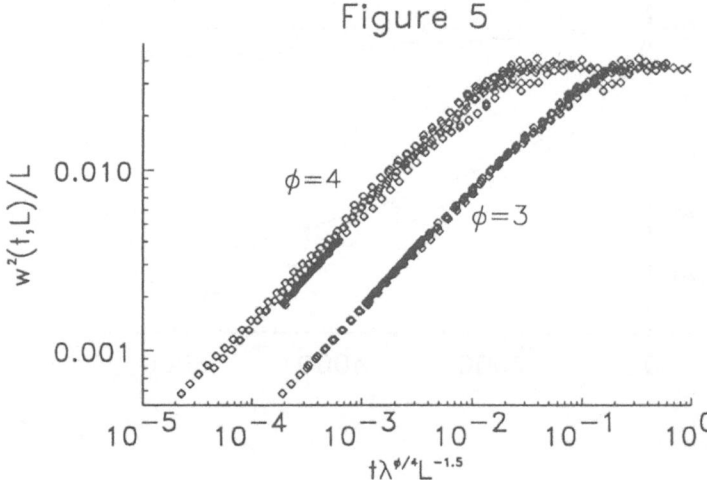

Fig. 5. Check of the finite-size scaling form, equation (11), with two different values for the exponent ϕ. Several simulations have been performed for system sizes $L = 100$, 500, 1000 and 10000 and values of λ ranging from $\lambda = 2$ to $\lambda = 10$. For the purpose of clarity, the data set with $\phi = 4$ has been shifted parallel to the horizontal axis.

To explain the failure of the RG calculation of Natterman and Tang to describe the simulation data one can argue that the specific RG expressions are oly valid asymptotically in the limit $\lambda \to 0$, $L \to \infty$ and that our simulations have not entered yet this asymptotic regime. Also, we note that the discretization in equation (5) can only be expected to reproduce the continuum behaviour used in the RG calculations if the mesh size a_0 is much less than the basic length scale $\nu^3/\lambda^2 D$ inherent in (1). In our parametrization, this corresponds to $\lambda < \sqrt{2}$. However, our results for the finite-size analysis (which favoured $\phi = 3$) were obtained with $2 \leq \lambda \leq 10$. Simulations at smaller values of λ (and therefore extended to much longer times in order to allow the development of non–linear effects) would be useful to resolve this issue.

References

[1]   Reviews on kinetic roughening in surface growth models can be found in: J. Krug and H. Spohn, in *Solids Far From Equilibrium: Growth, Morphology and Defects*, edited by C. Godrèche, (Cambridge Univ. Press, Cambridge, England, 1991); *Kinetics of Ordering and Growth at Surfaces*, edited by M. Lagally (Plenum, New York, 1990), and references therein; T. R. Thomas, *Rough Surfaces* (Longman, London, 1982).

[2]   M. Kardar, G. Parisi, and Y.-C. Zhang, Phys. Rev. Lett. **56**, 889 (1986); E. Medina, T. Hwa, M. Kardar, and Y.-C. Zhang, Phys. Rev. A **39**, 3053 (1989).

[3]   S. F. Edwards and D. R. Wilkinsons, Proc. R. Soc. Lon. A **381**:17 (1982).

[4]   F. Family and T. Vicsek, J. Phys. A **18**, L75 (1985). R. Jullien and R. Botet, J. Phys. A **18**, 2279 (1985).

[5]   P. Meakin, P. Ramanlal, L. M. Sander, and R. C. Ball, Phys. Rev. A **34**, 5091 (1986); J. Krug, Phys. Rev. A **36** 5465 (1987).

[6]   U. Dekker and F. Haake, Phys. Rev. A **11**, 2043 (1975).

[7]   K. Moser, J. Kertész, and D. E. Wolf, Physica A **178**, 215 (1991) and references therein.

[8]   B. Forrest and R. Toral, J. Stat. Phys. **70**, 703 (1993).

[9]   T. Nattermann and L-H. Tang, Phys. Rev. A **45**, 7156 (1992).

[10] B. Grossmann, H. Guo and M. Grant, Phys. Rev. A **43**, 1727 (1991);

COMPETING SCALING BEHAVIOURS IN THE LATE STAGE OF GROWTH KINETICS

A.Coniglio[a], P.Ruggiero[a] and M. Zannetti[b]

[a]*Dipartimento di Scienze Fisiche, Università di Napoli*
Mostra d'Oltremare, Padiglione 19, 80125 Napoli, Italy

[b]*Dipartimento di Fisica, Università di Salerno*
84081 Baronissi (SA), Italy

Abstract

The dependence of the scaling properties of the structure factor on space dimensionality, initial and final conditions, presence or absence of a conservation law is analysed in the framework of the large-N model for growth kinetics. The variety of asymptotic behaviours is quite rich, including standard scaling, multiscaling and a mixture of the two. A crossover structure due to the existence of competing fixed points is obtained.

In growth kinetics one deals with the relaxation to equilibrium of a system quenched from high to low temperature[1]. The processes of interest are those which exhibit scaling[2] in the asymptotic time regime. Denoting with T_I and T_F the initial and final temperatures, these processes can be grouped into three classes characterized by $(T_I > T_c, T_F < T_c)$, $(T_I > T_c, T_F = T_c)$ and $(T_I = T_c, T_F < T_c)$ where T_c is the critical temperature. In terms of the structure factor (Fourier transform of the equal time order parameter correlation function) the initial condition is of the form

$$C(\vec{k}, 0) = \frac{\Delta}{k^{\theta_1}} \tag{1}$$

where Δ is a constant and the value of θ selects the initial state: $\theta = 0$ corresponds to an uncorrelated initial state at infinite temperature ($T_I = \infty$) while $\theta = 2$ corresponds to the critical point ($T_I = T_c$). The most general form of the asymptotic scaling behaviour is

$$C(\vec{k}, t) \sim L^{\alpha(x)}(t) F(kL(t)) \tag{2}$$

where $x = kL(t)$ and $L(t)$ is a characteristic length which grows in time with a power law

$$L(t) \sim t^{1/z}. \tag{3}$$

If the exponent α does not depend on x scaling is standard, otherwise if there is a dependence of α on x the structure factor obeys multiscaling[3]. A scaling pattern of this type is completely characterized by the set of exponents z, $\alpha(x)$ and by the scaling function $F(x)$. These quantities depend to a different extent on the various elements entering in the specification of the process[4] which, in addition to (T_I, T_F), include the space dimensionality of the system, the vector dimensionality of the order parameter and the presence or absence of a conservation law.

In this talk the dependence of the scaling properties on the totality of these elements will be explored in detail in the framework of the large-N model[5]. This model is particularly important for two reasons i) at the moment it is the only available non trivial soluble model with a structure sufficiently rich to be adequate for this kind of investigation and ii) in principle corrections can be computed systematically via the $1/N$-expansion[6].

The dynamics of the model is governed by the Langevin equation

$$\frac{\partial \vec{\phi}(\vec{x}, t)}{\partial t} = -(i\nabla)^p \frac{\delta \mathcal{H}[\vec{\phi}, \mu]}{\delta \vec{\phi}(\vec{x}, t)} + \vec{\eta}(\vec{x}, t) \tag{4}$$

where $\vec{\phi}(\vec{x}) = (\phi_1(\vec{x}), ..., \phi_N(\vec{x}))$ is an N-component vector order parameter, $\mathcal{H}[\vec{\phi}, \mu]$ is a free energy functional of the Ginzburg-Landau type

$$\mathcal{H}[\vec{\phi}, \mu] = \frac{1}{2} \int d^d x \left[(\nabla \vec{\phi})^2 + r\vec{\phi}^2 + \frac{g}{2N} (\vec{\phi}^2)^2 \right] \tag{5}$$

and $\vec{\eta}$ is a gaussian white noise with expectations

$$< \vec{\eta}(\vec{x}, t) > \;=\; 0 \tag{6}$$

$$< \eta_\alpha(\vec{x}, t) \eta_\beta(\vec{x}', t') > \;=\; 2\Gamma T_F (i\nabla)^p \delta_{\alpha\beta} \delta(\vec{x} + \vec{x}') \delta(t - t'). \tag{7}$$

Here $p = 0$ for non conserved order parameter (NCOP), $p = 2$ for conserved order parameter (COP) and $\mu = (r, g)$ represents the pair of parameters which characterize the local potential.

The Gibbs equilibrium states $P_{eq}[\vec{\phi}, \mu] \sim \exp(-\frac{1}{T_F} \mathcal{H}[\vec{\phi}, \mu])$ are parametrized by the temperature T_F and by $\mu = (r, g)$. In the large-N limit ($N \to \infty$) there is a critical temperature $T_c(\mu) \sim -r/g$ and a phase diagram (Fig.1) in the three dimensional parameter space (T, μ) with a surface of critical points separating ordered states below it from disordered states above it[7]. The interesting portion of this phase diagram is the ($r \leq 0, g \geq 0$) sector at or below the critical surface where scaling is to be expected in a quench process. In particular we shall consider quenches to

1) $[T_F = T_c = 0, \mu_1 = (r = 0, g = 0)]$ trivial critical state at zero temperature

2) $[T_F = T_c > 0, \mu_1 = (r = 0, g = 0)]$ trivial critical states at finite temperature (T-axis)

3) $[T_F = T_c = 0, \mu_2 = (r = 0, g > 0)]$ non trivial critical states at zero temperature (g-axis)

4) $[T_F = T_c > 0, \mu_3 = (r < 0, g > 0)]$ non trivial critical states at finite temperature (critical surface)

5) $[T_F < T_c, \mu_3 = (r < 0, g > 0)]$ phase ordering region.

From the solution of the model one finds that the asymptotic scaling properties $[z, \alpha(x), F(x)]$ depend on (T_F, μ). There is a universality class, under each heading NCOP or COP, for each of the five regions (T_F, μ) listed above. In renormalization group language this means that there are five fixed points and that the extension of the universality classes depends on the relative stability of these fixed points. This in turn is regulated by the existence of critical dimensionalities which may depend on the initial condition.

After taking the large-N limit from (4) one derives[5] the equation of motion for the structure factor

$$\frac{\partial C(\vec{k}, t)}{\partial t} = -2\Gamma[k^{p+2} + k^p R(t)]C(\vec{k}, t) + 2\Gamma k^p T_F \tag{8}$$

where $R(t) = r + g \int \frac{d^d k}{(2\pi)^d} C(\vec{k}, t)$. Integrating with the initial condition (1), one finds

$$C(\vec{k}, t) = \frac{\Delta}{k^\theta} e^{-2\Gamma[k^{p+2}t + k^p Q(t)]} + 2\Gamma T_F k^p \int_0^t dt' e^{-2\Gamma[k^{p+2}(t-t') + k^p(Q(t) - Q(t'))]} \tag{9}$$

with the self-consistency condition $Q(t) = \int_0^t dt' R(t')$.

Analysing the above expression in the asymptotic time regime with NCOP ($p=0$) the following scaling behaviours are found[7]

1) $[T_F = 0, \mu_1]$
$$C(\vec{k}, t) \sim L^\theta(t) F(x) \tag{10}$$

2) $[T_F = T_c > 0, \mu_1]$
$$C(\vec{k}, t) \sim L^2(t) T_c F_0(x) \tag{11}$$

3) $[T_F = 0, \mu_2]$
$$C(\vec{k}, t) \sim \begin{cases} L^\theta(t) F(x) & \text{for } d > d_c \\ \frac{L^\theta(t)}{\log L(t)} F(x) & \text{for } d = d_c = \theta + 2 \\ L^{d-2}(t) F(x) & \text{for } d < d_c \end{cases} \tag{12}$$

4) $[T_F = T_c > 0, \mu_3]$
$$C(\vec{k}, t) \sim \begin{cases} L^2(t) T_c F_0(x) & \text{for } d \geq 4 \\ L^2(t) T_c F_e(x) & \text{for } d < 4 \end{cases} \tag{13}$$

5) $[T_F < T_c, \mu_3]$
$$C(\vec{k}, t) \sim L^d(t) F(x) \tag{14}$$

where $L(t) \sim t^{1/2}$, $x = kL(t)$ and

$$F(x) = e^{-x^2}/x^{\theta} \tag{15}$$

$$F_0(x) = (1 - e^{-x^2})/x^2 \tag{16}$$

$$F_{\epsilon}(x) = \int_0^1 dy (1-y)^{-\epsilon/2} e^{-x^2 y} \tag{17}$$

with $\epsilon = 4 - d$.

Let us now comment these results. The first observation is that with NCOP $L(t) \sim t^{1/2}$ is the only length in the problem implying that scaling is standard and $z = 2$ for any process. Instead, the exponent α and the scaling function may depend on the parameters of the final state. Specifically, there are two trivial fixed points at $T_F = 0$ and $T_F > 0$ with distinct asymptotic behaviours (10) and (11). Equations (12) and (13) show that these fixed points are attractive for quenches on the critical surface, respectively with $T_F = 0$ and $T_F > 0$, if the dimensionality is higher than a critical dimensionality. Otherwise a new exponent α or a new scaling function appears. In other words, for quenches on the critical surface there exist upper critical dimensionalities above which the non linerity in the problem becomes irrelevant, making available all the machinery of renormalized perturbation theory[8]. Totally different is the case of quenches below the critical surface in the phase ordering region. Then (14) shows that there does not exist an upper critical dimensionality, namely the non linearity of the problem is always relevant. Furthermore, the asymptotic behaviour with $0 < T_F < T_c$ is the same as for a quench to $T_F = 0$ implying that thermal fluctuations are irrelevant for quenches below the critical surface.

The variety of asymptotic properties is more complex when quenches with COP are considered. From (9) with $p = 2$ follows[7]

1) $[T_F = 0, \mu_1]$

$$C(\vec{k}, t) \sim L^{\theta}(t) \hat{F}_>(x) \tag{18}$$

2) $[T_F = T_c, \mu_1]$

$$C(\vec{k}, t) \sim L^2(t) T_c \hat{F}_0(x) \tag{19}$$

3) $[T_F = 0, \mu_2]$

$$C(\vec{k}, t) \sim \begin{cases} L^\theta(t)\hat{F}_>(x) & \text{for } d > d_c \\ L^\theta(t)\hat{F}(x) & \text{for } d = d_c = \theta + 2 \\ \lambda^\theta(t)\hat{F}_<(x') & \text{for } d < d_c \end{cases} \qquad (20)$$

where $x = kL(t)$ and $x' = k\lambda(t)$

4) $[T_F = T_c > 0, \mu_3]$

$$C(\vec{k}, t) \sim \begin{cases} L^2(t)T_c\hat{F}_0(x) & \text{for } d \geq 4 \\ L^2(t)T_c\hat{F}_\epsilon(x) & \text{for } d < 4 \end{cases} \qquad (21)$$

5) $[0 < T_F < T_c, \mu_3]$

$$C(\vec{k}, t) = T_F \frac{L^{\alpha(x)}(t)}{x^2} \qquad (22)$$

6) $[T_F = 0, \mu_3]$

$$C(\vec{k}, t) \sim L^{\alpha_0(x)}(t)\frac{1}{x^\theta} \qquad (23)$$

with

$$L(t) \sim t^{1/4} \qquad (24)$$

$$\lambda(t) \sim t^{1/(d+2-\theta)} \qquad (25)$$

$$\hat{F}_>(x) = \frac{e^{-x^4}}{x^\theta} \qquad (26)$$

$$\hat{F}(x) = \frac{1}{x^\theta}e^{-[x^4+cx^2]} \qquad (27)$$

$$\hat{F}_<(x') = \frac{e^{-x'^2}}{x'^\theta} \qquad (28)$$

$$\hat{F}_0(x) = \frac{1}{x^2}(1 - e^{-x^4}) \qquad (29)$$

$$\hat{F}_\epsilon(x) = \frac{1}{x^2}\int_0^x dx'x'^3 e^{-[(x^4-x'^4)+b(x^2-x'^2)]} \qquad (30)$$

$$\alpha_0(x) = (d - \theta)\varphi(x) + \theta. \qquad (31)$$

$$\alpha(x) = \begin{cases} 2 + (d - 2)\varphi(x) & \text{for } x < x^* \\ 2 & \text{for } x > x^* \end{cases} \qquad (32)$$

$$\varphi(x) = 1 - (x^2 - 1)^2 \qquad (33)$$

and where x^* is the non trivial zero of $\varphi(x)$.

The general discussion of fixed points and the relative stability goes along the same lines as for NCOP with important modifications due to the existence of *two* divergent lengths $L(t)$ and $\lambda(t)$. Therefore, the exponent z depends on which of the two is the dominant one. For quenches on the critical surface $L(t)$ prevails yielding $z = 4$ except for the quench to $[T_F = 0, \mu_2]$ with $d < d_c = 2 + \theta$. In that case $\lambda(t)$ prevails yielding $z = d + 2 - \theta$. In the phase ordering region $[T_F < T_c, \mu_3]$, instead, $\lambda(t)$ and $L(t)$ diverge in the same way up to a logarithmic factor $\lambda \sim L(\log L)^{1/4}$ yielding the multiscaling behaviour of Eq.(22) and (23). The exponents $\alpha_0(x)$ and $\alpha(x)$ are plotted in Fig.3. The behaviour of $\alpha(x)$ exhibits multiscaling for $x < x^*$ and standard scaling for $x > x^*$ with the same value of $\alpha = 2$ as for quenches on the critical surface. In other words the asymptotic behaviour of the structure factor for quenches in the phase ordering region with $0 < T_F < T_c$ is dominated by the zero temperature fixed point for small k, while is dominated by the critical fixed point for large k. Therefore, contrary to what one finds with NCOP, when

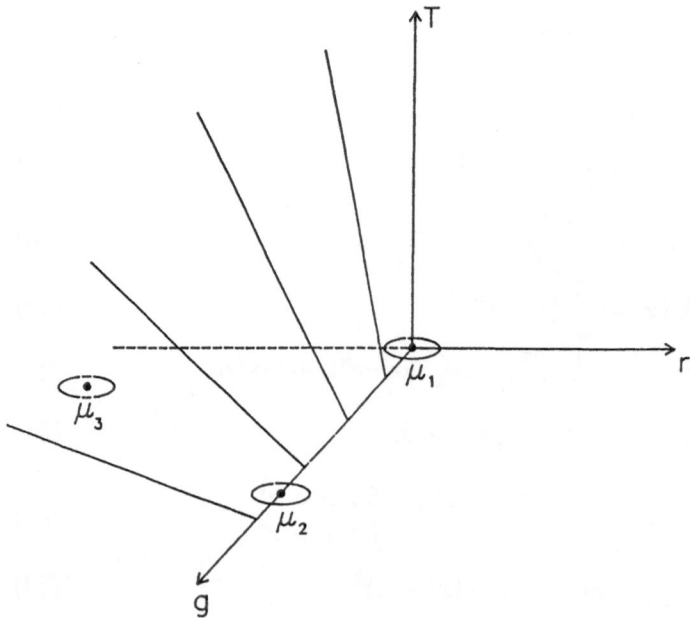

Fig.1 - Manifold of final equilibrium states with the critical surface separating disordered states (above) from ordered states (below).

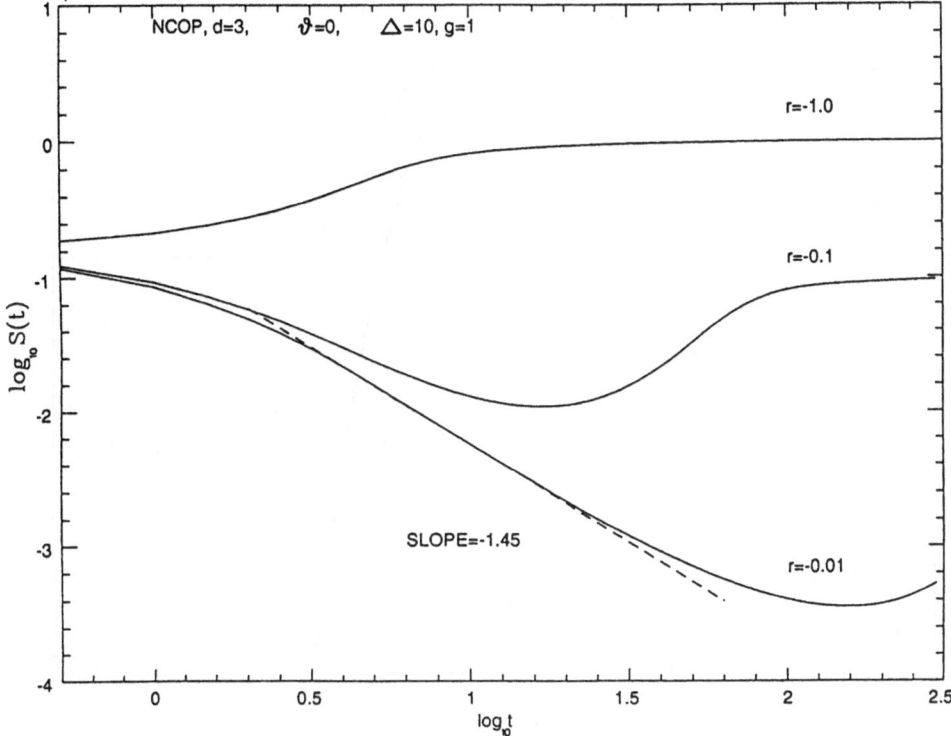

Fig.2 - Behaviour of $S(t)$ in a quench to μ_3 for NCOP with $\sigma = 2, \Delta = 10, d = 3, \theta = 0$ at fixed $g = 1$ as r approaches the μ_2-axis. The straight dashed line has slope -1.45.

the order parameter is conserved temperature fluctuations are not irrelevant for quenches below the critical point.

From the existence of distinct fixed points it is natural to expect crossovers relating the different asymptotic behaviours. For simplicity we address this question with NCOP. In that case, as remarked above, the nature of the final state affects only the exponent α. This can be obtained from the study of the quantity

$$S(t) = \int \frac{d^d k}{(2\pi)^d} C(\vec{k}, t) \sim L^{\alpha - d}. \qquad (34)$$

The behaviour of $S(t)$ computed numerically for $d = 3, \theta = 0, g = 1$ and decreasing values of $|r|$ is displayed in Fig.2. The curves clearly show that while away from μ_2, e.g. at $r = -1.0$, the behaviour of $S(t)$ shows only the

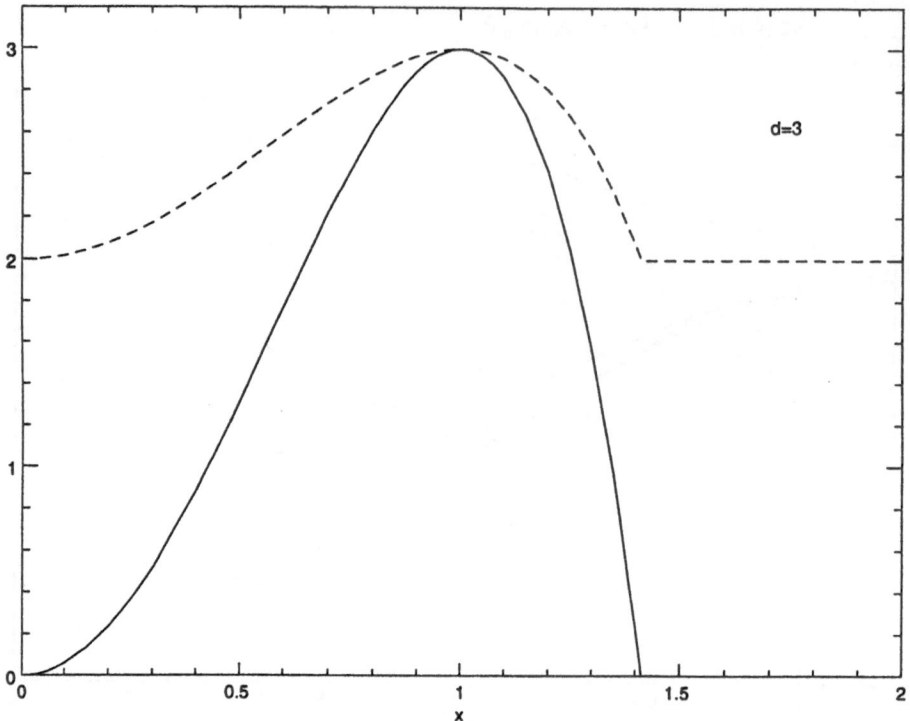

Fig.3 - Behaviour of $\alpha_0(x)$ (continous line) and $\alpha(x)$ (dashed line).

asymptotic scaling regime $S(t) \sim -r/g = 1$, as $|r|$ is decreased and μ_2 is approached for r sufficiently small, e.g. $r = -0.01$, $S(t)$ indeed displays at intermediate times the power law $L^{-d} \sim t^{-3/2}$ which holds as the asymptotic behaviour in quenches to μ_2.

In summary, what the solution of the large-N model shows is that the scaling properties are quite different with and without conservation law. This is due to the existence of only one divergent length $L(t)$ for NCOP and of two divergent lengths $L(t)$ and $\lambda(t)$ for COP. It is the interplay between these two lengths which leads to phenomena not observed with NCOP such as the dependence of the growth exponent z on the final state and multiscaling for quenches inside the phase ordering region. This latter feature might be a peculiarity of the $N \to \infty$ limit as it has been recently argued[9]. Instead, the crossover structure which emerges as the parameters of the quench are moved over the manifold of final equilibrium states is a generic feature which is expected to hold beyond the large-N model. In particular, the picture illustrated above suggests the possibility of observing a crossover in the growth law in a quench of a system with scalar ($N = 1$) COP. In that case asymptotically $L(t)$ grows according to (3) with $z = 3$. On the other hand in the trivial theory ($r = 0, g = 0$) one has $z = 4$ for COP, irrespective of the order parameter being a scalar or a vector. Thus, for a quench to ($T_F = 0, \mu_3$) sufficiently close to ($T_F = 0, \mu_1$) with $N = 1$ it should be possible to observe the influence of the trivial fixed point at early time producing a crossover from $z = 4$ to $z = 3$.

References

1. Reviews can be found in J.D.Gunton, M. San Miguel and P.S.Sahni in *Phase transitions and critical phenomena*, edited by C.Domb and J.L.Lebowitz (Academic Press, New York, 1983), Vol. 8, p.267;

   H.Furukawa, *Adv. Phys.* **34**, 703, (1985); K.Binder, *Rep. Progr. Phys.* **50**, 783, (1987); A.J.Bray, Lectures given at NATO Advanced Study Institute on *"Phase Transitions and Relaxation in Systems with Competing Energy Scales"*, 1993 Geilo, Norway.

2. K.Binder and D.Stauffer, *Phys. Rev. Lett.* **33**, 1006, (1974); J.Marro, J.L.Lebowitz and M.H.Kalos, *ibid.* **43**, 282, (1979); H.Furukawa, *Progr. Theor. Phys.* **59**, 1072, (1978); *Phys. Rev. Lett.* **43**, 136, (1979)

3. A.Coniglio and M.Zannetti, *Europhys. Lett.* **10**, 575, (1989)

4. The problem of the definition of universality classes for growth kinetics as been addressed in Z.W.Lai, G.F.Mazenko and O.T.Valls *Phys. Rev.* B **37**, 9481, (1988)

5. G.F.Mazenko and M.Zannetti, *Phys. Rev.* B **32**, 4565, (1985); Z.Rácz and T.Tél, *Phys. Lett.* **60A**, 3, (1977); G.F.Mazenko and M.Zannetti

, *Phys. Rev. Lett* **32**, 4565, (1984); F.de Pasquale and P.Tartaglia, *Phys. Rev. B* **33**, 2081, (1986); A.Coniglio and M.Zannetti, in *From phase transitions to cahos* edited by G.Gyorgyi, I.Kondor, L.Sasvári and T.Tél, World Scientific (1992), p. 100

6. T.J.Newman and A.J.Bray *J. Phys. A: Math. Gen.* **23** 4491, (1990)

7. A.Coniglio, P.Ruggiero and M.Zannetti, *Scaling and Crossover in the Large-N Model for Growth Kinetics, Phys. Rev. E*, July 1994

8. H.K.Janssen, B.Schaub and B.Schmittmann, *Z.Phys.* **73**, 539, (1989)

9. A.J.Bray and K.Humayun, *Phys. Rev. Lett.* **68**, 1559, (1992)

Stretched-Exponential Relaxation of Transient Electric Birefringence in Polymer-Like Reverse Micelles

G.J.M. Koper[1], C. Cavaco[2], and P. Schurtenberger[2]

[1] Dept. of Physical and Macromolecular Chemistry, Leiden University, Gorlaeus Laboratories, PO Box 9502, 2300 RA Leiden, The Netherlands
[2] Institut für Polymere, ETH-Zentrum, CH-8092 Zürich, Switzerland

1 Introduction

Quite a number of amphiflic molecules are known to form reverse micelles in organic solvents [1]. The addition of a small amount of water to such micellar solutions usually induces a spherical growth of the micelles with no significant changes in physical properties such as viscosity. However, a completely different behaviour can be found for lecithin reverse micelles in some organic solvents [2]. These solutions can be transformed into transparant, highly viscous, and thermodynamically stable viscoelastic systems by adding very small quantities of water. With isooctane for example the viscosity increases by as much as a factor of 10^6 upon the addition of three molecules of water per lecithin molecule [3, 4]. It was postulated that the addition of water to lecithin reverse micellar solutions induces one-dimensional micellar growth into long cylindrical micelles [5]. Above the crossover lecithin volume fraction, ϕ^*, these micelles are believed to entangle and form a transient network similar to that found in semidilute polymer solutions. This structural model was subsequently tested for lecithin/isooctane solutions by means of small-angle neutron scattering (SANS), quasielastic light-scattering (QLS), rheological measurements and fluorescence recovery after fringe-pattern photobleaching measurements (FRAP) [3, 4, 5, 6]. The postulated model was confirmed and on the basis of the obtained data it was concluded that these micellar solutions may serve as an ideal model system for structural and dynamical studies of *living polymer* systems [8, 7], where the term living (or equilibrium) polymer is used for linear macromolecules that can break and recombine.

Most surfactant systems in which worm-like micelles are demonstrated to exist are aquaous [9, 10, 11, 12, 13, 14, 15] where there are only a few systems reported with organic solvents [16, 17]. The oil-continuous microemulsions are the preferred systems because no complicating effects arise due to additional contributions from electrostatic interactions or salt effects. Also for dielectric studies and electric birefringence experiments these systems are easier to handle due to the much lower conductivity compared to the aquaous systems.

Recently, results from conformation space renormalization group theory for semi-dilute polymer solutions have been succesfully applied to the lecithin microemulsions using data from static and dynamic light scattering experiments [18, 20, 21]. There are two important aspects resulting from these analyses that need to be mentioned here: (1) the exponent ν appearing in the scaling relation between the radius of gyration and the mass of the polymer, $R_g \sim M^\nu$, has for these systems the value 0.588 which is the good solvent value for polymers, and (2) the size distribution of the equilibrium polymers is exponential where the mean value scales with lecithin concentration as $\overline{M} \sim c^\alpha$ with the value of $\alpha \approx 1.2$. This value is at least twice as large as theoretical predictions that give $\alpha = 0.5$ from law of mass-action or Flory-Huggins lattice model calculations and $\alpha \approx 0.6$ from a scaling theory approach for semi-dilute polymer solutions [9].

Transient electric birefringence is a tool to investigate the internal dynamics of polymers [22, chapter 5]. Within the Rouse/Zimm picture of polymers [23] internal relaxation is described in terms of modes, each mode decaying exponentially. Because the timescale, common to all modes, depends on the size of the polymer one expects for a living polymer system with an exponential size distribution stretched-exponential relaxation of the electric birefringence [24]. Wu et al. [25] have performed transient electric birefringence experiments on an aquaous living polymer system and only find approximately a stretched exponential decay. In this paper we perform the same experiments on the water/lecithin/oil microemulsion and demonstrate that in these systems the transient electric birefringence does show stretched exponential decay. Moreover we show that the associated time scales and amplitudes demonstrate the theoretically predicted scaling behaviour for semidilute flexible polymer systems.

The paper is organized as follows: First we shall derive the equations that govern the static and dynamic electric birefringence response for living polymer systems. Then we describe the experiments and discuss the obtained results.

2 Theory

The dynamics of flexible polymers (of fixed size) in *dilute* suspensions is usually described in terms of the Rouse-Zimm model [23, chapter 4]. The polymer is pictured as a chain of beads and the beads are held together by springs so that the segment length is b. In the same spirit one can attribute a polarizability tensor to each segment as [26]

$$\mathbf{A}_n = \frac{\gamma}{b^2} r_n^2 \mathbf{1} + \frac{\Delta\gamma}{b^2} (\mathbf{r}_n \mathbf{r}_n - \frac{1}{3} r_n^2 \mathbf{1}) \tag{1}$$

with segment vector $\mathbf{r}_n$. The polarizabilities γ and $\Delta\gamma$ depend on the detailed chemical structure of the polymer. In the prescence of an electric field a dipole moment will be induced in each segment which gives rise to an additional potential energy. The electric field causes anisotropy in the segment vector distribution which is reflected by the segment averages. For a polymer coil the anisotropy in segment vector distribution causes *form birefringence* due to the spheroidal deformation of the coil. The effect is usually quite small because of the small difference in effective refractive index of the coil and the solvent. The effective index of refraction of the coil is related to the isotropic part of the polarizability in eq.(1) through the Clausius-Mossotti relation. The second contribution originates from the anisotropy in the polarizability of the segments. The orientations of the segment vectors give rise to what is called the *intrinsic birefringence*. To relate the polarizability of the individual segments to macroscopic birefringence we use the Clausius-Mossotti relation from which the refractive index tensor can be calculated by summing up the anisotropic contributions from the individual segments [23, section 4.7]. In sufficiently small fields the static birefringence Δn indeed obeys the Kerr law, i.e. the birefringence is linear in the electric field E squared, and the Kerr constant is given by

$$K = \frac{\Delta n}{E^2} = \frac{2\pi(n^2+2)^2}{9n}\rho\Delta\gamma N\frac{\Delta\gamma_E}{3k_BT} \tag{2}$$

with n the refractive index of the solvent, $\Delta\gamma_E$ is the polarizability anisotropy of the segments at low frequencies. The important aspects of the Kerr constant for flexible polymers is that it is *linear* in both the polymer number density ρ and the number of segments per polymer N and hence in the segment density c.

The field free decay of the birefringence, after switching off the electric field, can be calculated using the Rouse-Zimm formalism. The result is a sum over exponentially decaying modes with time constants $\tau_p = \tau_1 p^{-3\nu}$ with $p = 1, 2, \ldots$ and with

$$\tau_1 = \frac{\eta N^{3\nu}b^3}{k_BT} \tag{3}$$

where η is the viscosity of the solvent. For sufficiently long times only the slowest mode will contribute and hence

$$\Delta n(t) \simeq e^{-2t/\tau_1} \qquad (t \to \infty). \tag{4}$$

The important aspect is that the time constant scales with the number of segments as $\tau_1 \sim N^{3\nu}$.

For the *semidilute* regime we can use scaling arguments to arrive at the segment density dependence of the Kerr constant and the birefringence decay. We begin by noting, that

$$K = c\tilde{F}(N, cb^3) \tag{5}$$

with $\tilde{F}$ a scaling function. Scaling theory considers how physical quantities change when λ segments are grouped into one. Under such a transformation $N \to N/\lambda$, $b \to b\lambda^\nu$ and $c \to c/\lambda$, see [23, section 5.3]. For the Kerr constant to remain invariant under this transformation it must be written as

$$K = \frac{c}{N}F(c/c^*) \tag{6}$$

with $F(x) = \tilde{F}(1, x)$ and with the overlap density

$$c^* \simeq N^{1-3\nu}/b^3. \tag{7}$$

This expression is valid in both the dilute and the semidilute regime. In the semi-dilute regime the Kerr constant should not depend on N but only on the segment density c, so that

$$K \sim c^{3\nu/3\nu-1}. \tag{8}$$

Taking for ν the value 0.588 [18, 20, 21] the exponent of the segment concentration dependence of the Kerr constant is 2.3. In the semidilute regime the relaxation time is Rouselike and becomes concentration dependent. Effective medium theory [23, section 5.7] gives

$$\tau_1 \simeq cN^2 \tag{9}$$

To extend this result to equilibrium polymers the exponential decay eq.(4) has to be averaged over the exponential size distribution. For the field-free decay we then find (omitting algebraic factors)

$$\overline{\Delta n(t)} \sim \exp\left\{-\left(\frac{t}{\overline{\tau}}\right)^{1/p}\right\}. \tag{10}$$

The exponent p and the time constant $\overline{\tau}$ are given in table 1.

	dilute regime	semidilute regime
K	c	$c^{3\nu/3\nu-1}$
p	$3\nu + 1$	3
$\overline{\tau}$	$c^{3\nu\alpha}$	$c^{1+2\alpha}$

Table 1. Summary of expressions for the static Kerr constant, the exponent p and the time constant $\overline{\tau}$ for flexible polymers in the dilute and the semidilute regime.

3 Materials and Methods

Soybean lecithin was obtained from Lucas Meyer (Epikuron 200) and used without further purification. Cyclohexane and isooctane (spectroscopic grade) were purchased from Fluka. Samples were prepared as described elsewhere [19, 21].

Electrooptic birefringence measurements were performed with a conventional setup [27] using a rectangular quartz cuvette (Hellma, Germany) with two platinum electrodes in teflon separators. The transient electrooptic birefringence experiment consists of applying a rectangular electric field pulse to the suspension and recording both the electric field and the induced birefringence response. The response time of the measurement system is better than 0.1 μs and the pulse rise and decay times are better than 40 ns, both much faster than the response of the suspension.

The (static) Kerr constant was measured as follows: For each electric field pulse the induced birefringence and the electric field were determined from the plateau regions in the recorded traces. This was repeated for typically 15 values of the electric field and the Kerr constant was subsequently found from the slope of the induced birefringence versus the electric field squared, see eq.(2). The obtained Kerr constants were corrected for the contribution from the solvent.

The field free decay curves of the birefringence were obtained by repeating the pulse experiments for typically 8 times and by averaging the obtained traces. The start of the birefringence decay was marked by the trailing side of the (simultaneously recorded) rectangular field pulse. Both for the static and the transient birefringence experiments the pulse duration was chosen such that a clear plateau value was reached.

4 Results and Discussion

Fig. 1 summarizes the results for the static Kerr constant as a function of polymer volume fraction. The polymer volume fraction is linearly related to the segment concentration c. Clearly the Kerr constant does not depend linearly on volume fraction and fits better the expression for the semidilute regime. In table 2 the obtained values for the exponents are summarized. For both the cyclohexane and the isooctane system we obtain 2.1 ± 0.3 which is in accordance with the value of 2.3 obtained using the value of $\nu = 0.588$ from static and dynamic light scattering experiments on the same systems [18, 20, 21], see table 1.

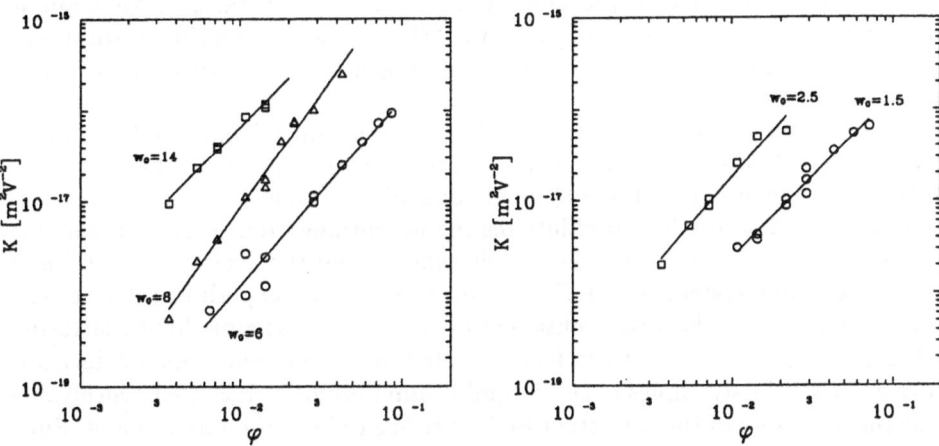

Fig. 1. Static Kerr constant versus volume fraction for various values of the water to surfactant ratio w_0. The left graph shows the values for the cyclohexane systems and the right graph for the isooctane systems.

The measured decay curves were fit to the expression (10) for the field-free decay of dilute systems. The exponent ν was taken to be 0.588 so that the exponent p becomes $3\nu + 1 = 2.8$. Experimentally, this value cannot be distinguished from the value 3 for semidilute systems and therefore we kept the

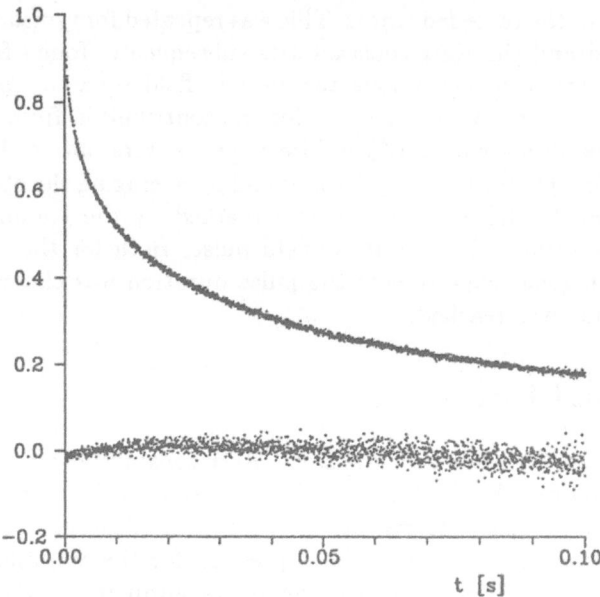

Fig. 2. Normalized field-free birefringence decay curve (above) and residues (below) from a fit to eq.(10) for an isooctane system of $w_0 = 2.5$ and volume fraction $\phi = 0.0144$.

value of p fixed at 3. The fit procedure yielded the time constants $\bar{\tau}$. An example of a birefringence decay curve together with the residues[3] from the fit are shown in fig. 2. Including more modes in the birefringence decay (10) does not yield better fits to the data.

Fig. 3 summarizes the results for the time constant. Using for the relation between the time constant $\bar{\tau}$ and the volume fraction ϕ for flexible polymers in the dilute regime (see table 1) yields unrealistic values for the exponent α. From the relation for the semidilute regime we obtained much more reasonable values for the exponent α, for the cyclohexane system the average is $\alpha=1.3$ and for the isooctane system $\alpha=1.1$. These values are, within experimental accuracy, in accordance with the values obtained by static and dynamic light scattering [18, 20, 21]. The discrepancies in fact indicate that the presented model does not fully describe the dynamics of these equilibrium polymers. It is well conceivable that the dynamics of the surfactant molecules has to be taken into account. Such processes have been completely neglected in the present model.

Acknowledgements

It is a pleasure to thank D. Bedeaux and P. Cirkel for many illuminating discussions on polymer theory. We also thank J. v.d. Ploeg for technical support.

---

[3] The residue of a fit function $y(t)$ to experimental data points y_n at t_n is given by
$$\Delta y_n \equiv (y_n - y(t_n))/y(t_n).$$

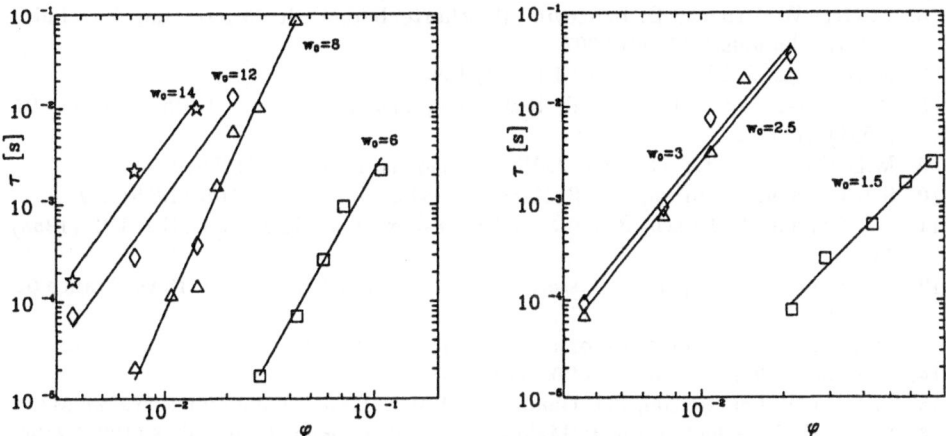

Fig. 3. Relaxation time constant versus volume fraction for various values of the water to surfactant ratio w_0. The left graph shows the values for the cyclohexane systems and the right graph for the isooctane systems.

w_0	Kerr constant exponent	time constant exponent
cyclohexane		
6	2.0	4.0
8	2.4	4.8
12		3.0
14	1.8	3.0
isooctane		
1.5	1.8	2.8
2.5	2.0	3.4
3.0		3.4

Table 2. Summary of obtained exponents for the volume fraction dependence of the Kerr constant K and the relaxation time constant $\bar{\tau}$ (see table 1) as obtained from the static and transtient birefringence for different systems.

The authors are grateful for a visitors grant from the Dutch Science Foundation (NWO) for one of us (CC). Part of this work was supported by the Portuguese National Fund for Technological and Scientific Research through the grant BD/1256/91-IC.

References

1. P.L. Luisi, and L.J. Magid, CRC Crit. Rev. Biochem. **20** (1986) 409.
2. R. Scartazinni and P.L. Luisi, J. Phys. Chem. **20** (1988) 829.
3. P. Schurtenberger, R. Scartazzinni, and P.L. Luisi, Rheol. Acta **28** (1989) 372.
4. P. Schurtenberger, R. Scartazzinni, L.J. Magid, M.E. Leser, and P.L. Luisi, J. Phys. Chem. **94** (1990) 3695.
5. P.L. Luisi, R. Scartazzini, G. Haering, and P. Schurtenberger, Colloid Polym. Sci. **268** (1990) 356.

6. A. Ott, W. Urbach, D. Langevin, P. Schurtenberger, R. Scartazzini, P.L. Luisi, J. Phys. Liquids **2** (1990) 5907.
7. M.E. Cates, J. Phys. France **49** (1988) 1593.
8. P. Schurtenberger, L.J. Magid, S.M. King, and P. Lindner, J. Phys. Chem. **95** (1991) 4173.
9. M.E. Cates and S.J. Candau, J. Phys: Condens. Matter **2** (1990) 6869.
10. S.J. Candau, E. Hirsch, and R. Zana, J. Colloid Interf. Sci. **105** (1985) 521.
11. S.J. Candau, E. Hirsch, R. Zana, and M. Adam, J. Colloid Interf. Sci. **122** (1988) 430.
12. J. Marignan, J. Appell, P. Bassereau, G. Porte, and R.P. May, J. Phys. France **50** (1989) 3553.
13. J. Appell and G. Porte, Europhys. Lett. **12** (1990) 185.
14. T. Imae, Colloid Polym. Sci. **267** (1989) 707.
15. F. Kern, P. Lemarechal, S.J. Candau, and M.E. Cates, Langmuir **8** (1992) 437.
16. P. Terech, V. Schaffhauser, P. Maldivi, and J.M. Guenet, Langmuir **8** (1992) 2104.
17. Z. Zhou, Y. Georgalis, W. Liang, J. Li, R. Xu, and B. Chu, J. Colloid Interf. Sci. **116** (1987) 473.
18. P. Schurtenberger and C. Cavaco, J. Phys. II France **3** (1993) 1279.
19. P. Schurtenberger, Q. Peng, M. Leser, and P.L. Luisi, J. Colloid Interface Sci. **156** (1993) 43.
20. P. Schurtenberger and C. Cavaco, J. Phys. II France **4** (1994) 305.
21. P. Schurtenberger and C. Cavaco, Langmuir **10** (1994) 100.
22. R.L. Jernigan and D.S. Thompson, In: *Molecular Electro-optics. Part I: Theory and Methods*, Ed. C.T. O'Konski, Marcel Dekker Inc., New York 1976.
23. M. Doi and S.F. Edwards, *The Theory of Polymer Dynamics*, Clarendon Press, Oxford 1986.
24. T. Bellini, F. Mantegazza, R. Piazza, and V. Degiorgio, Europhys. Lett. **10** (1989) 499.
25. X-l. Wu, C. Yeung, M.W. Kim, J.S. Huang, and D. Ou-Yang, Phys. Rev. Lett. **68** (1992) 1426.
26. J.S. Bowers and R.K. Prud'homme, J. Chem. Phys. **96** (1992) 7135.
27. S.P. Stoylov, *Colloid Electro-Optics: Theory, Techniques, Applications*, Academic Press, London 1991.

LIST OF POSTERS

R. Villanova, *Ising model on 2D random lattices*

M. Mayorga and R.M. Velasco, *The Rayleigh peak in a molecular fluid in the presence of a temperature gradient*

A. M. Somoza, C. Sagui and R. C. Desai, *Spinodal decomposition in an order-disorder phase transition*

X.F.Yuan, R.C.Ball and S.F.Edwards, *Computer Simulations of Complex Fluid Flows*

U. Ritschel, *The Role of Initial Conditions in Critical Relaxation*

J. Vitting Andersen, Y. Brechet, and H. Jeldtoft Jensen, *Fracturing described by a spring-block model*

N. Brilliantov, *Formation kinetics and fractal properties of random space filling tiling*

P. García-Fernández and A. Sánchez-Díaz, *Squeezing in the self-pulsing domain of lasers with a two-photon saturable absorber*

E.G. Petrov, *Kinetic equations for quantum-dynamic system interacting with strong external fields and thermal bath*

V.I. Kalikmanov and M.E.H. van Dongen, *Semi-phenomenological theory of binary nucleation*

O. Pla, V.G. Benza, and F. Nori, *Theory of Motion and Avalanches in Rotating Drums*

J. Martinez-Linares, C. Mirasso, P. Garcia-Fernandez, and F.J. Bermejo, *Analytical and Numerical Evaluation of Transmission Characteristics of Chirped DFB Lasers in Dispersive Optical Fibers*

A. Santos, V. Garzó and J. M. Montanero, *Absence of a normal solution to the Boltzmann equation far from equilibrium*

H. Pascal and J. Stephenson, *Some Non-Linear Diffusion Equations and Fractal Diffusion*

P. Neu and R. Speicher, *Spectra of hamiltonians with generalized dynamical disorder*

J.A. Cuesta, F.C. Martínez, J.M. Molera, R. Brito, *A theoretical approach* conttitlea*to two dimensional traffic flow models*

S. N. Laughton, *Separable Stochastic Neural Netorks with and without Detailed Balance*

H.M. Polatoglou, *Theory for the disordering near the [001] twist boundaries in Cu_3Au*

M. Zimmer, *Supersymmetry breaking and fluctuation effects in a nonequilibrium model*

M. Zapotocky, P. M. Goldbart and N. Goldenfeld, *Kinetics of phase-ordering in uniaxial and biaxial nematic films*

M. J. de Oliveira and T. Tomé, *Inhomogeneous Random Sequential Adsorption of Particles on Bipartite Lattices*

A D Rutenberg and A J Bray, *Growth Laws and Scaling in Phase Ordering Kinetics*

R. Montagne, A. Amengual, E. Hernández-García, and M. San Miguel, *Transient pattern formation by multiple front propagation*

I. Graham, E. Hernández-García, and M. Grant, *Damage spreading during domain growth*

O. Alarcón-Waess and E. González-Tovar, *Time-dependent electrolyte friction on a non-spherical charged tracer*

H. Huffel, *Stochastic Quantization of Gauge Theories*

T. Tome and J. E. Satulovsky, *Nonequilibrium Phase Transitions in a Lattice Model for Predator-Prey Systems*

N. G. Almarza, E. Enciso, V. del Prado, F.J. Bermejo, *Dynamical behaviour of Simple Fluids by Molecular Dynamics Simulation. Departure from*

Linearized Behavior of the Dynamical Structure Factor

G. A. Perez Alcazar, *Theoretical interpretation of the antiferro- paramagnetic transition for the FeMnAl system in the FCC disordered phase*

L. E. Zamora Alfonso, *Theoretical magnetic phase diagram for the (Fe.65Ni.35)1-xMnx system*

A.K. Evans, *Subdynamics in a simple system*

A. Diaz-Guilera, C.J. Perez Vicente, A. Corral and A. Arenas, *From synchronization to self-organized criticality in a model of biological oscillators*

A. R. Kolovsky, *How many degrees of freedom should have a thermostat?*

V. Špička, P. Lipavský, *Quasiparticle Boltzmann equation in semiconductors*

N. Nakazawa, *Stochastic Quantization of Dissipative Systems*

M. C. Miguel and J.M. Rubí, *Relaxation dynamics of ferromagnetic particles in suspension*

I. Pagonabarraga and J.M. Rubí, *Influence of hydrodynamic interactions on the adsorption of large particles.*

J. Mach, P.P. Trigueros, J. Claret, F. Mas and F. Sagués, *Fractal electrodeposition*

J.M. Lopez, M.A. Rodriguez, L. Pesquera, *Analysis of self-averaging properties in the transport of particles through random media*

J.M. Noriega, M.A. Rodriguez, and L. Pesquera, *Propagation of trigger waves in a disordered medium*

P. L. Taylor, *Spinodal - the curve that never exists. What, never? Hardly ever!*

J. Gronewald and D. Bedeaux, *The surface tension of a curved interface*

A.S. Cukrowski and S. Fritzsche, *Thermal effects in a system composed of simple reacting spheres*

List of Participants

AGUDA, B.D., Dept. of Chemistry, Univ. of Waterloo, Waterloo, Ontario, Canada N2L 3G1, bdaguda@watsci.uwaterloo.ca

AHLERS, G., Dept. of Physics, Univ. of California, Sta. Barbara,CA 93106, USA

ALARCON-WAESS, O. , D. de Física, U. Américas-Puebla, A.P.100 Hacienda Sta.Catarina M., Cholula M72820 Puebla, México, alarcon@udlapvms.pue.udlap.mx

ALDER, B., Lawrence Livermore Nat. Lab., Livermore, CA 94550, USA

ALONSO PEREDA, J.J., P.M.M.H.,E.S.P.C.I., 10 rue Vauquelin, 75231 Paris cedex 05, France, jjalonso@pmmh.espci.fr

ALVES, D., Fac.de Ciencias, Dept. de Física Teorica, Pza. San Francisco sn., 50009 Zaragoza, Spain fteorica@cc.unizar.es

AMI, S., Fachbereich Physik,WE2, Freie Universitat Berlin, Arnomallee 14, D-14195 Berlin, Germany

ARENAS, A., Dept. Física Fonamental, Univ. of Barcelona, Diagonal 647, 08028 Barcelona, Spain, alex@ulyses.ffn.ub.es

ATTINGER, S. Institut Theoretische Physik, Univ. Heidelberg, Philosophenweg 19,D-69120 Heidelberg, Germany, attinger@hybrid.tphys.uni-heidelberg.de

AZCOITI, V., Dept. Física Teorica, Univ. Zaragoza, Campus Universitario, 50009 Zaragoza, Spain, vicente@cc.unizar.es

BAFALUY, J., Dept. de Física, Univ. Autónoma de Barcelona, Edifici C, 08193 Bellaterra, Barcelona, Spain, javier@ulises.uab.es

BALSERA, M., Univ. of Illinois at Urbana Champaign, Loomis Lab. of Physics, 1110W Green St., Urbana, IL61801 USA, manel@kolmogorov.physics.uiuc.edu

BEDEAUX, D., Gorlaeus Laboratories, Einsteinweg 55, P.O. Box 9502, 2300 RA Leiden The Netherlands, bedeaux@rulfc1.LeidenUniv.nl

BETHGE, A., Inst. TheoretischePhysik, Univ. of Heidelberg, Philosophenweg 19, D-69117 Heidelberg, Germany, bethge@hybrid.tphys.uni-heidelberg.de

BIEL, J., Fac. Ciencias, Univ. of Granada, 18071 Granada, Spain

BINDER, K., Inst. Physik, Univ. of Mainz, Staudinger Weg 7, D-55099, Mainz, Germany

BONET-AVALOS, J., Inst. Charles Sadron, 6 rue Boussingault, 67083 Strasbourg, France, bonet@janus.u-strasbg.fr

BONILLA, L.L., Univ. Carlos III, Butarque 15, 28911 Leganes, Madrid, Spain bonilla@ing.uc3m.es

BOYA, L.J. , Dep. Física Teorica, Univ. of Zaragoza, Pz. San Francisco sn. 50009 Zaragoza, Spain

BREY, J., Física Teorica, Fac. de Física, Apdo. 1065, 41080 Sevilla, Spain, brey@cica.es

BRITO, R., Dpto. Física Aplicada I, Fac. Ciencias Físicas, UCM, 28040 Madrid, Spain, brito@seneca.fis.ucm.es

CASAS-VAZQUEZ, J., Dept. de Física, Univ. Autónoma de Barcelona, Edifici C, 08193 Bellaterra, Barcelona, Spain

CHEN, YI-HONG, CSIC, Facultat de Ciencias C-III, Universidad Autónoma, 28049 Madrid ,Spain, chen@queso.icm.uam.es

COHEN, E.G.D., The Rockefeller Univ., 1230 York Avenue, 10021 New York, USA

COOLEN, A.C.C., Dept.of Physics,Theoretical Physics, Univ. of Oxford, 1 Keble road, Oxford OX1 3NP, England, coolen@thphys.ox.ac.uk

CORRAL, A., Dept. Física Fonamental, Univ. of Barcelona, Diagonal 647, 08028 Barcelona, Spain, alvaro@ulyses.ffn.ub.es

CUESTA, J.A., Univ. Carlos III, Butarque 15, 28911 Leganes , Madrid, Spain cuesta@ing.uc3m.es

CUKROWSKI, A. S., Institute of Physical Chemistry, Polish Acad. of Sci., ul. Kasprzaka 44/52, Pl-01224 Warsaw, Poland, cukrowsk@ichf.edu.pl

De CASTRO, T.T.M., Instituto de Física da Universidade de Sao Paulo, Caixa Postal 20516 01452-990 Sao Paulo Brazil tcastro@uspif.if.usp.br

De OLIVEIRA, M.J., Instituto de Física, Univ.de Sao Paulo, Caixa Postal 20516, 01452-990 Sao Paulo, SP, Brazil mjoliveira@uspif.if.usp.br

DE SIENA, S., Dip. de Física, Univ. de Salerno, I-84100 Salerno, Italy, desiena@vaxsa.dia.unisa.it

DE LA RUBIA, J., Dept. Física Fundamental, UNED, Apto. de Correos 60.141, 28080 Madrid, Spain, jrubia@uned.es

DEL RIO PORTILLA, J. A, Lab. de Energia Solar IIM, Univ. Nacional Autónoma de México A.P.34, 62580 Temixco, Morelos, México,ntonio@redvax1.dgsca.unam.mx

DELGADO, J., Dept. Llenguatges: Sistemes Informatics (FIB,UPC), Pau Gargallo 5, 08028 Barcelona, Spain

DIAZ-GUILERA, A., Dept. Física Fonamental, Univ. of Barcelona, Diagonal 647, 08028 Barcelona, Spain, albert@ulyses.ffn.ub.es

DITTRICH, T., Dept. of Physics, Univ. Augsburg, Memmingerstr. 6, D-86135 Augsburg, Germany, dittrich@physik.uni-augsburg.de

DUFTY, J.W., Dept. of Physics, Univ. of Florida, Gainesville, FL 32611, USA dufty@phys.ufl.edu

ERNST, M.H., Institute for Theoretical Physics, Univ. of Utrecht, P.O. Box 80.006 3508 TA Utrecht, Netherlands,ernst@fys.ruu.nl

EVANS, A., TCM, Cavendish Lab., University of Cambridge,Madingley Road, Cambridge, CB3 OHG, England, ake10@cus.cam.ac.uk

FEIGENBAUM, M., The Rockefeller Univ., 1230 York Avenue, 10021 New York, USA

FELDERHOF, B.U., Institut fur Theoretische Physik, RWTH Aachen, Templergraben 55D-52056 Aachen, Germany, ufelder@thphys.physik.rwth-aachen.de

FERNANDEZ TEJERO, C., Fac. de Ciencias Físicas, Univ. Complutense, Ciudad Universitaria,28040 Madrid, Spain, tejero@seneca.fis.ucm.es

FISCHER, E.W., Max-Planck Institut fur Polymerforschung, Ackermannweg 10, Postfach 3148, 55021 Mainz, Germany

FOLLANA ADIN, E., Dep. Fisica Teorica, Univ. Zaragoza, Campus Universitario, 50009 Zaragoza, Spain, follana@cc.unizar.es

FRENKEL, D., FOM Institute for Atomic and Molecular Physics, Kruislaan 407, 1098SJ, Amsterdam, The Netherlands, frenkel@amolf.amolf.nl

GABARRO VALLES, J., Dept. LLenguatges: Sistemes Informatics (FIB,UPC), Pau Gargallo 5, 08028 Barcelona, Spain

GARCIA ALMAZARA, N., Dept. Quimica-Física I, Fac. Ciencias Quimicas, Univ. Complutense, E-28040 Madrid, Spain, almarza@quim.ucm.es

GARCIA FERNANDEZ, P., Instituto de Estructura de la Materia, CSIC, Serrano 123, 28006 Madrid, Spain, imtpg39@cc.csic.es

GARCIA LADONA, E., Dept. de Física, Univ. Autónoma de Barcelona, Edifici C, 08193 Bellaterra, Barcelona, Spain, emilio@ulises.uab.es

GARRIDO, P.L., Fac. de Ciencias, Univ. Granada, 18071 Granada, Spain garrido@ugr.es

GILLES, L., CSIC, Instituto de Estructura de la Materia, Serrano 123, 28006, Madrid, Spain, emluc@roca.csic.es

GOMILA, G., Dept. Física Fonamental, Univ. of Barcelona, Diagonal 647, 08028 Barcelona, Spain, gabriel@ulyses.ffn.ub.es

GONZALEZ MIRANDA, J., Dept. Física Fonamental, Univ. of Barcelona, Diagonal 647, 08028 Barcelona, Spain, jgm@hermes.ffn.ub.es

GRAHAM, R., Facbereich Physik Univ., GHS, D-4300 Essen, Germany

GROENEWOLD, J., Univ. of Leiden, Gorlaeus Labs, PO Box 9502, 2300 RA Leiden, The Netherlands, jang@rulfc1.leidenuniv.nl

GUINEA, F., S. Fernando de Jarana 14, 28002 Madrid, Spain

HERNANDEZ-GARCIA, E., Dept. de Física, Univ. de les Illes Balears, 07071 Palma de Mallorca , Spain, dfsehg4@ps.uib.es

HUEFFEL, H., Inst. of Theoretical Physics, Univ. of Vienna, Austria, a8241dag@helios.edvz.univie.ac.at

JONES, R.B., Dept. of Physics, Queen Mary and Westfield College, Mile End Road, London E1 4NS, United Kingdom, rbj@v2.ph.qmw.ac.uk

JOU, D., Dept. de Física, Univ. Autónoma de Barcelona, Edifici C, 08193 Bellaterra, Barcelona, Spain

KALIKMANOV, V.I. , Physics Department, W&S, Eindhoven University, P.O.Box 513, 5600MB, Eindhoven, The·Netherlands, tnnttu1@urc.tue.nl

KAWASAKI, K., Physikalisches Institut, Univ. Bayreuth, Helmut Brand, D-95440 Bayreuth, Germany

KOLOVSKY, A.R., Fachbereich Physik, Universitat Essen, D-45117 Essen, Germany, phy1b0@vm.hrz.uni-essen.de

KOPER, G., Univ. Leiden, Gorlaeus Laboratories, PO Box 9502, 2300 RA Leiden, The Netherlands, koper@rulfc1.leidenuniv.nl

LAUGHTON, S., Theoretical Physics, Oxford Univ., 1 Keble Road, Oxford OX1 3NP England, stephen@thphys.ox.ac.uk

LEBOWITZ, J., Dept. Math. and Physics, Rutgers Univ., New Brunswick, NJ 08903 USA, lebowitz@math.rutgers.edu

LOZANO FERNANDEZ, C., Instituto de estructura de la materia, CSIC, Serrano 123, 28006 Madrid, Spain, imtcl11@cc.csic.es

MACH I DROUHIM, J., Dept. Quimica-Física, Univ. of Barcelona, Diagonal 647, 08028 Barcelona, Spain, jordi@daphne.qf.ub.es

MANRUBIA, S.C., Dept. Física i Enginyeria Nuclear,UPC, Sor Eulalia de Anzizu, Campus Nord B4,O8028 Barcelona, Spain, susanna@f_731.upc.es

MARRO, J., Inst. Carlos I de Física Teorica y Computacional, Univ. of Granada, 18071 Granada, Spain

MARTINEZ, F.C., Univ. Carlos III, Butarque 15, 28911 Leganes, Madrid, Spain dopico@ing.uc3m.es

MASOLIVER, J., Dept. Física Fonamental, Univ. of Barcelona, Diagonal 647, 08028 Barcelona, Spain, jaume@hermes.ffn.ub.es

MAYORGA, M., Univ. Autónoma Metropolitana, Iztapalapa D.F., 09340 México, ines@xanum.uam.mx

MAZUR, P., Inst. Lorentz, Leiden Univ., PO Box 9502, 2300 RA Leiden, The Netherlands

MIGUEL, M.C., Dept. Física Fonamental, Univ. of Barcelona, Diagonal 647, 08028 Barcelona, Spain, carmen@ulyses.ffn.ub.es

MIRACLE, S., Centre de Physique Théorique, CNRS, Luminy, Case 907, F-13288 Marseille cedex 9, France

MOLERA, J.M., Univ. Carlos III, Butarque 15, 28911 Leganes, Madrid, Spain molera@ing.uc3m.es

MONTAGNE, R., Dept. de Física, Univ. de les Illes Balears, 07071 Palma de Mallorca, Spain, dfsrhm4@ps.uib.es

MORKEL, C., Physics Department, TU Munchen E21 D-85747 Garching, Germany

MUSCHIK, W., Ins. f. Theoretische Physik, Technische Uni. Berlin, Sekr. PN 7-1 Hardenbergstr.36, D-10623 Berlin, womu0433@w421zrz.physik.tu-berlin.de

MUÑOZ, M.A., Depto. Física Aplicada, Univ. Granada, Campus Fuentenueva, 8071 Granada, Spain, mamartinez@ugr.es

NAKAZAWA, N., Dept. of Physics, Shimane University, Matsue 690, Japan nakazawa@nbivax.nbi.dk

NEU, P., Institut fur Theoretische Physik, Uni. Heidelberg, Philosophenweg 19, D-69120 Heidelberg, Germany, bz3@vm.urz.uni-heidelberg.de

PAGONABARRAGA, I., Dept. Física Fonamental, Univ. of Barcelona, Diagonal 647, 08028 Barcelona, Spain, ignacio@ulyses.ffn.ub.es

PARISI, G., Dip. Física, Univ. di Roma I, Pia Aldo Moro 4, Roma, Italy, parisi@roma1.infn.it

PASTOR, R., Dept. Física Fonamental, Univ. de Barcelona, Diagonal 647, 08028 Barcelona, Spain, romu@hermes.ffn.ub.es

PEREZ ALCAZAR, G. A. , Inst. de Ciencia de Materiales, CSIC, Serrano 144, 28006 Madrid, Spain materiales@cc.csic.es

PEREZ-VICENTE, C., Dept. Física Fonamental, Univ. of Barcelona, Diagonal 647, 08028 Barcelona, Spain, conrad@ulyses.ffn.ub.es

PEREZ-MADRID, A. Dept. Física Fonamental, Univ. of Barcelona, Diagonal 647, 08028 Barcelona, Spain, agusti@ulyses.ffn.ub.es

PESQUERA, L., Dpto. Física Moderna, Univ. de Cantabria, 39005 Santander, Spain, pesquera@ccucvx.unican.es

PFLUEGL, W., Institut fur Theoretische Physik, Altenbergerstr. 69, 4040 Linz, Austria , pfl@convex.alijku11.edvz.uni-linz.ac.at

PLA, O., Ins.de Ciencia de Materiales, CSIC, Fac. de Ciencias C-III, Univ. Autónoma de Madrid, 28049 Madrid, oscar@queso.icm.uam.es

POLATOGLOU, H.M., Aristotle Univ. Thessaloniki, Physics Dep., GR-54006 Thessaloniki, Greece, polatoglou@olymp.ccf.auth.gr

POMEAU, Y., Ecole Normale Superieure, 24 rue Lhomod, 75231 Paris cedex 05

PRADOS, A., Física Teorica, Fac. Física, Univ. Sevilla, Apdo. de correos 1065, 41080 Sevilla, Spain, prados@cica.es

REISS, H., Gorlaeus Laboratories, Einsteinweg 55, P.O. Box 9502, 2300 RA Leiden The Netherlands

RITSCHEL, U., Fachbereich Physik, Univ. Essen, D-45117 Essen, Germany, phy350@vm.hrz.uni-essen.de

ROBINSON, A., Dept. Física Fonamental, Univ. of Barcelona, Diagonal 647, 08028 Barcelona, Spain, crusoe@hermes.ffn.ub.es

RODRIGUEZ, M.A.,m Dept. de Física Moderna, Univ. Cantabria, 39005 Santander, Spain, rodriguez@ccucvx.unican.es

RUBI, M., Dept. Física Fonamental, Univ. of Barcelona, Diagonal 647, 08028 Barcelona, Spain, miguel@ulyses.ffn.ub.es

RUIZ MONTERO, M. J. , Física Teorica, Fac.de Física, Apdo. Correos 1065, 41080 Sevilla, Spain, majose@cica.es

RUTENBERG, A., Department of Physics and Astronomy, University of Manchester, Manchester UK M13 9PL, ar@v2.ph.man.uk

SALUEÑA, C., Dept. Física Fonamental, Univ. of Barcelona, Diagonal 647, 08028 Barcelona, Spain, clara@ulyses.ffn.ub.es

SAN MIGUEL, M., Dept. Física, Univ. de les Illes Balears, 07071 Palma de Mallorca, Spain, dfsmsm0@ps.uib.es

SANCHEZ DIAZ, A., Instituto de Estructura de la Materia, CSIC, Serrano,123, 28006 Madrid, Spain, emangel@roca.csic.es

SANCHEZ REY, B., Física Teorica, Fac. Física, Apdo. de Correos 1065, 41080 Sevilla, Spain, bernardo@cica.es

SANCHO, J.M., Dept. Estruc. y Constit. de la Materia, Univ. of Barcelona, Diagonal 647, 08028 Barcelona, Spain

SANTOS, A., Dpt. de Física, Univ. Extremadura, E-06071 Badajoz, Spain, andres@ba.unex.es

SCHAAF, P, Institut Charles Sadron, 6 rue Boussingault, 67083 Strasbourg, France, wojta@janus.u-strasbg.fr

SCHREIBER, G. , Fak. f. Physik, Theor. Vielteilchenphysik, Philosophenweg 19, D-69120 Heidelberg, Germany, georg@hybrid.tphys.uni-heidelberg.de

SHERRINGTON, D., Theoretical Physics, Univ. of Oxford, 1 Keble Rd., Oxford OX1 3NP, England, sherr@thphys.ox.ac.uk

SOLE, R.V., Dept. Física i Enginyeria Nuclear, UPC, Sor Eulalia de Anzizu, Campus Nord B5,08028 Barcelona, Spain

SOMOZA, A.M., Ins. Ciencias de Materiales, CSIC, Univ. Autónoma de Madrid (C-XII), E-28049 Madrid, Spain, andres@fluid1.fmc.uam.es

SPICKA, V., Solid state theory group, Physics Dep., Blackett lab., Imperial College, London SW7 2BZ, UK, vaclav@sst.ph.ic.ac.uk

STEPHENSON, J. Physics U. of Alberta, Edmonton Alberta, T6G 2J1 Canada jsjs@phys.ualberta.ca

TAYLOR, P.L., Dept. of Physics, Case Western Reserve Univ., Cleveland OH 44106-7079 USA, plt@po.cwru.edu

TORAL, R., Dept. de Física, Univ. de les Illes Balears, 07071 Palma de Mallorca, Spain, dfsrtg0@ps.uib.es

TORRENT SERRA, M.C ., Dept. Estruc. y Constit. de la Materia, Univ. of Barcelona, Diagonal 647, 08028 Barcelona, Spain

TOXVAERD, S., Dept. of Chemistry, H.C. Oersted Institute, Universitetsparken 5, Copenhagen 2100 Oe, Denmark, tox@st.ki.ku.dk

VAN KAMPEN, N., Inst. Theor. Phys. Rijksuniverste, Utrecht Princetonplein 5, Utrecht , The Netherlands

VAN LEEUWEN, J.M.J., Instituut-Lorentz, Nieuwsteeg 8, 2311 SB Leiden, The Netherlands, jmjvanl@rulgm0.LeidenUniv.nl

VAN DER BROECK, Ch., Limburgs Universitair Centrum, B-3590 Diepenbeek, Belgium, chris@luc.ac.be

VERBERG, R., Interfaculty Reactor Institute, Delft Univ. of Technology, Mekelweg 15,, 2629JB Delft, The Netherland, rolf@iri.tudelft.nl

VERDASCA, J., Univ. libre de Bruxelles, Campus Plaine CP231, Boulevard du Triomphe, B-1050 Bruxelles, Belgique, jalves@cenoli.ulb.ac.be

VILLANOVA, R., Univ. Pompeu Fabra, Escola de Ciencies Empresarials, Rambla de Santa Monica 32, 08002 Barcelona, villanova@ifae.es

VITTING ANDERSEN, J., Dept. of Physics, McGill Univ., 3600 Univ. Str. Montreal, Quebec, Canada H3A 2T8, vitting@frodo.physics.mcgill.ca

VOLLMER, D., Inst. fur Physical Chemistry, Klingelbergstr 80, CH-4056 Basel Switzerland, kammann@urz.unibas.ch

WIO, H. S., Centro Atomico Bariloche, Division Física Teorica, 8400 San Carlos de Bariloche, Argentina, wio@cab.edu.ar

YUAN, XUE-FENG, Polymer and Colloid Science Group, Cavendish Laboratory, Madingley Road, Cambridge CB30HE, England, xy100@phy.cam.ac.uk

ZAMORA ALFONSO, L.E., Instituto de Ciencia de Materiales, CSIC, Serrano 144, 28006 Madrid, Spain, materiales@cc.csic.es

ZANNETTI, M., Dipartimento di Física, Univ. di Salerno I-84081 Baronissi, Salerno, Italy , zannetti@napoli.infn.it

ZAPOTOCKY, M., Dept. Physics, Uni. Illinois at Urbana-Champaign,, 1110 West Green St. Urbana, IL 61801, USA, zapotock@uiuc.edu

ZIMMER, M., Forschungszentrum Julich, D-52425 Julich, Germany m.zimmer@kfa-juelich.de

CONTRIBUTIONS TO THE SITGES CONFERENCES

MALLORCA INTERNATIONAL SCHOOL OF PHYSICS

"The Many Body Problem" (1969)

ENZ,C.P. "Green Functions Applied to Phonon Problems"
VERBOVEN, E.J. "Kinematical Properties of Equilibrium Gas"
THIRRING, W. "The Mathematical Structure of the BCS-Model and Related Models"
FUJIWARA, I. "Functional Integration Methods in Quantum Mechanics"
CAIANIELLO, E.R.C. "Field Equations and Form-Invariant Renormalization"
BOYA, L.J. "Introduction to Brückner Theory to Nuclear Matter"
DOVER, C.B. "Coupled Boson-Fermion Systems"
HORING, N.J. "Plasmon Resonances in the Quantum Strong Field Limit"

II SITGES INTERNATIONAL SCHOOL OF PHYSICS

"Irreversibility in the Many-Body Problem" (1972)

ROSENFELD, L. "General Introduction to Irreversibility"
FARQUHAR, I.E. "Ergodicity and Related Topics"
WERGELAND, H. "Irreversibility in Many-Body Systems"
MIRACLE, S. "Infinite Dynamical Systems and Time Evolution: Rigorous Results"
BALESCU, R. "Non-Equilibrium Statiscal Mechanics"
RESIBOIS, P. "Hydrodynamical Concepts in Statistical Physics"
VAN KAMPEN, N. "Fluctuations"
ENZ, C.P. "Hydrodynamics of Magnetic Crystals"
VELARDE, M.G. "Symmetry Breaking Instability"
SANTOS, E. "Brownian Motion and the Stochastic Theory of Quantum Mechanics"

III SITGES INTERNATIONAL SCHOOL OF STATISTICAL MECHANICS

"Transport Phenomena" (1974)

PEIERLS, R. "Some Simpe Remarks on the Basis of Transport Theory"
PRIGOGINE, I. and MAYNE, F. "Entropy, Dynamics and Scattering
    Theory"
KUBO, R. "Response, Relaxation and Fluctuation"
MAZUR, P. "Fluctuating Hydrodynamics and Renormalization of
    Susceptibilities and Transport Coefficients"
BIEL, J. "Irreversibility of the Transport Equations"
LEBOWITZ, J.L. "Ergodic Theory and Statistical Mechanics"
DE LEENER, M. "Correlation Functions in Heisenberg Magnets"
VELARDE, M.G. "On the Enskog Hard-Sphere Kinetic Equation and
    the Transport Phenomena of Dense Simple Gases"
HAUGE, E.H. "What Can We Learn from Lorentz Models?"
STINCHCOMBE, R.B. "Conductivity in a Magnetic Field"
BEENAKKER, J. "Transport Properties in Gases in the Presence
    of External Fields"
SNIDER, R.F. "Transport Properties of Dilute Gases with Internal
    Structure"

IV SITGES CONFERENCE

"Critical Phenomena" (1976)

WEGNER, F.J. "Critical Phenomena and Scale Invariance"
GREEN, M.S. "Invariant Properties of the Renormalization Group"
MA, S.-K. "Scale Transformations in Dynamic Models"
ENZ, C.P. "Critical Dynamics in Fokker-Planck Formalism"
SZEPFALUSY, P. "Dynamic Critical Phenomena and
    the Renormalization-Group Application to a Lattice Dynamic Model"
HAAG, R. "The Algebraic Approach to Quantum Statistical Mechanics:
    Equilibrium States and Hierarchy of Stability"
MIRACLE-SOLE, S. "Theorems on Phase Transitions with a
    Treatment for the Ising Model"
LEBOWITZ, J.L. "Statistical Mechanics of Equilibrium Systems:
    Some Rigurous Results"
GALLAVOTTI, G. "Probabilistic Aspects of Critical Fluctuations"
KADANOFF, L.P. "The Application of Renormalization Group Techniques
    to Quarks and Strings"
BROUT, R. "The Role of Spontaneous Broken Symmetry"

V SITGES CONFERENCE

"Stochastic Processes in Nonequilibrium Systems" (1978)

VAN KAMPEN, N. "An Introduction to Stochastic Processes for Physicists"
SANTOS, E. "Stochastic Differetial Equations with Non-Markov Processes"
MAZO, R.M. "Aspects of the Theory of Brownian Motion"
GRAHAM, R. "Path-Integral Methods in Nonequilibrium Themodynamics and Statistics"
HAKEN, H. "Synergetics - A Field Beyond Irreversible Thermodynamics"
ALDER, B.J. "Computer Results on Transport Properties"
PENROSE, O. "Kinetics of Phase Transitions"
POMEAU, Y. "Stochastic Behavior of Simple Dynamical Systems"

VI SITGES CONFERENCE

"Systems Far From Equilibrium" (1980)

GARRIDO, L. and LLOSA, J. "Prologue: The Intrinsic Fokker-Planck Equation"
LANGER, J.S. "Kinetics of Metastable States"
SUZUKI, M. "Instability, Fluctuations and Critical Slowing Down"
BINDER, K. "Spinodal Decomposition"
NICOLIS, G. "Bifurcations and Symmtry-Breaking in Far From Equilibrium Systems"
KAWASAKI, K., ONUKI, A. and OHTA, T. "Some Topics in Nonequilibrium Critical Phenomena"
AHLERS, G. "Onset of Convection and Turbulence in a Cylindrical Container"
GOLLUB, J.P. "The Onset of Turbulence: Convection, Surface Waves, and Oscillators"
MORI, H. and FUJISAKA, F. "Statistical Dynamics of Turbulence"
ZWANZIG, R. "Problems in Nonlinear Transport Theory"
GARCIA-COLIN, L.S. "Non-Linear Transport Theory"
BREY, J.J. "On Non-Linear Fluctuations From Statistical Mechanics"
DORFMAN, J.R. and KIRKPATRICK, T. "Kinetic Theory of Dense Gases Not in Equilibrium"
LURIE, D. and WAGENSBERG, J. "Information Theory and Ecological Diversity"

VII SITGES CONFERENCE

"Dynamical Systems and Chaos" (1982)

GARRIDO, L. and SIMO, C. "Prologue: Some Ideas about Strange
Attractors"
CHIRIKOV, B.V. "Chaotic Dynamics in Hamiltoniann Systems with
Divided Phase Space"
KATOK, A. "Periodic and Quasi-Periodic Orbits for Twist Maps"
ROSSLER, O.E. "Macroscopic Behavior in a Simple Chaotic Hamiltonian
System"
HUBERMAN, B.A. "Quantum Dynamics"
SIGGIA E.D. "A Universal Transition from Quasi-Periodicity to Chaos -
Abstract"
GEISEL, T. and NIERWETBERG, J. "Self-generated Diffusion
and Universal Critical Properties in Chaotic Systems"
RUDNICK, J. "Subharmonics and the Transition to Chaos"
FEIGENBAUM, M. "Low Dimensional Dynamics and
the Period Doubling Scenario"
GUCKENHEIMER, J. "Strange Attractors in Fluid Mechanics"
LIBCHABER, A. "Experimental Aspects of the Period Doubling Scenario"
NEWHOUSE, S.E. "Entropy and Smooth Dynamics"

VIII SITGES CONFERENCE

"Applications of Field Theory to Statistical Mechanics" (1984)

GARRIDO, L. and SAGUES, F. "Prologue: A Functional Perturbative
Approach to the Classical Statistical Mechanics"
NELSON, D.R. "The Structure and Satistical Mechanics of Glass"
FROLICH, J. "The Statistical Mechanics of Surfaces"
WAGNER, H. "Surface Effects in Phase Transitions"
DE DOMINICIS, C. "On the Ising Spin Glass, I. Mean Field"
DE DOMINICIS, C. and KONDOR, I. "On the Ising Spin Glass, II.
Fluctuations"
BREZIN, E. "The Wetting Transition"
BREZIN, E. "Grassmann Variables and Supersymmetry in the
Theory of Disordered Systems"
AIZENMAN, M. "Rigurous Studies of Critical Behavior"
WEGNER, F. "Anderson Transition and Non-Linear σ-Model"
WALLACE, D.J. "Non-Perturbative Renormalization in Field Theory"
GONZALEZ-ARROYO, A. "Stochastic Quantization:
Regularization and Renormalization"
BRYDGES, D.C. and SPENCER, T. "Self Avoiding Random Walk
and the Renormalization Group"

IX SITGES CONFERENCE

"Fluctuations and Stochastic Phenomena in Condensed Matter" (1986)

GRAHAM, R. "Macroscopic Potentials, Bifurcations
and Noise in Dissipative Systems"

GUNTON, J.D. "Dynamics of Topological Defects in
First Order Phase Transitions"

HOHENBERG, P.C. and CROSS M.C. "An Introduction to
Pattern Formation in Nonequilibrium Systems"

EDWARDS, S.F. "The Statistical Mechanics of Polymer Melts and Glasses"

DE DOMINICIS, C. and MOTTISHAW, P. "On the Replica Symmetric
Ising Spin Glasses"

ITZYKSON, C. "Conformal Invariance and Finite Size Effects in
Critical Two Dimensional Statistical Models"

CASTELLANI, C., DI CASTRO, and C. STRITANI, G.
"Generalized Non-Linear σ-Model and Effective Landau Theory
for Disordered Interacting Electron Systems"

SCHNEIDER, T. "Relationship Between D-Dimensional Models with
Langevin Dynamics, Associated Quantum Systems and
(D+1)-Dimensional Classical and Static Models"

MARRO, J. "Phase Transitions and Stationary Nonequilibrium States"

HAAKE, F., KUS, M. and SCHARF, R. "Quantum Mechanical Chaos
Criteria for a Kicked Top"

BOVIER, A. "Short Range Spin Glasses at Low Temperatures"

GROSSMAN, S. "Diffusion in Fully Developed Turbulence.
A Random Walk in a Random Structure"

CASTELLANI, C. "Multifractal Wavefunction at the
Localization Threshold"

TEN BOSCH, A. and MAISSA, P. "Effects of Screening in
Liquid Crystal Polymers"

KRAMER, B. and SCHREIBER, M. "Localization, Quantum Interference
and Transport in Disordered Solids"

DRESS, A.W.M. "On the Computational Complexity
of Composite Systems"

GRABERT, H. "Dissipative Quantum Tunneling"

X SITGES CONFERENCE

"Far From Equilibrium Phase Transitions" (1988)

LINDEMBERG, K., BROWN, and D. WANG, X. "A Review
of Current Issues in the Quantum Theory of Envelope Solitons"

SAN MIGUEL, M. "Fluctuations in the Transient Dynamics of
Nonlinear Optical Systems"

CROSS, M.C. "Theoretical Methods in Pattern Formation in
  Physics, Chemistry and Biology"
BONILLA, L.L. "Two Nonequilibrium Phase Transitions:
  Stochastic Hopf Bifurcation and Onset of Relaxation Oscillations
  in the Diffusive Sine-Gordon Model"
JAUSLIN, H.R. "Exactly Solvable Multistable Fokker-Planck Models
  with Arbitrarily Prescribed N Lowest Eigenvalues"
ABRAHAM, N.B. "Phase and Frequency Dynamics in Laser Instabilities"
MIKHAILOV, A.S. "Fluctuations and Critical Phenomena in
  Reaction-Diffusion Systems"
LE BERRE, M., POMEAU, Y., RESSAYRE, E., TALLET,
A., GIBBS, H.M., KAPLAN, D.L. and ROSE, M.J.
  "From Deterministic Chaos to Noise in Retarded Feedback Systems"
GUYON, E., ROUX, S. and HANSEN, A. "Non-Local and
  Non-Linear Problems in the Physics of Disordered Media"
LUCKE, M. "Convection in Binary Mixtures:
  Propagating and Standing Patterns"
MANDEL, P., ZEGHLACHE, H. and ERNEUX, T.
  "Time-Dependent Phase Transitions"
RISKEN, H. and VOGEL, K.
  "Quantum Treatment of Dispersive Optical Bistability"
LUGLIATO, L.A., OLDANO, C., SARTIRANA, L., KAIGE, W.,
NARDUCCI, L.M., OPPO, G.-L., PERNIGO, M.A., TREDICCE, J.R.,
PRATI, F. and BROGGI, G. "Spontaneous Symmetry Breaking
  and Spatial Structures in Optical Systems"
JASNOW, D. "Scaling for an Interfacial Instability"
MAZENKO, G.F. "Fied Theory for Growth Kinetics"

XI SITGES CONFERENCE

"Statistical Mechanics of Neural Networks" (1990)

GARRIDO, L. and RUBI, M. "Introduction:
  On the Statistical-Mechanical Formulation of Neural Networks"
ABBOT L.F. and KEPLER, T.B. "Model Neurons:
  From Hodgkin-Huxley to Hopfield"
KUHN, R. "Statistical Mechanics for Networks of Analog Neurons"
RIEGER, H. "Properties of Neural Networks with Multi-State Neurons"
SCHURMANN, B., HOLLATZ, J. and RAMACHER, U.
  "Adaptive Recurrent Neural Networks and Dynamic Stability"
TORRAS I GENIS, C. "Neural Oscillators: Experiments and Models"
TREVES, A. and ROLLS, E.T. "Neural Networks in the
  Hippocampus Involved in Memory"
VIANA, L., COTA, E. and MARTINEZ, C. "Basins of Attraction
  and Spurious States in Neural Networks"

WONG, K.Y.M. and SHERRINGTON, D. "Tailoring the Performance
of Attractor Neural Networks"
BERNASCONI, J. "Learning and Optimization"
HERTZ, J.A. "Statistical Dynamics of Learning"
NICOLIS, S. "Learning and Retrieving Marked Patterns"
PEREZ-VICENTE, C.J. "A Learning Algorithm for Binaty Synapses"
KINZEL, W. "Statistical Mechanics of the Perceptron
with Maximal Stability"
PRIETO, A., MARTIN-SMITH, P., MERELO, J.J., PELAYO,
F.J., ORTEGA, J., FERNANDEZ, F.J.,and PINO, B. "Simulation
and Hardware Implementation of Competitive Learning Neural
Networks"
RUJAN, P. "Learning in Multilayer Networks:
A Geometric Computational Approach"
BOUTEN, M. "Storage Capacity of Diluted Neural Networks"
CAMPBEL, C. and WONG, K.Y.M. "Dynamics and Storage Capacity
of Neural Networks with Sign-Constrained Weights"
ERDOS, P. and NIEBUR, E."The Neural Basis
of the Locomotion of Nematodes"
GUSTAFSON, K. "Reversibility in Neural Processing Systems"
ZHAOPING, L. and HERZ, A.V.M. "Lyapunov Functional
for Neural Networks with Delayed Interactions
and Statistical Mechanics of Temporal Associations"
NOEST, A.J. "Semi-Local Signal Processing in the Visual System"
SOURLAS, N. "Statistical Mechanics and Error-Correcting Codes"
HAKEN, H. "Synergetic Computers- An Alternative to Neurocomputers"
GESZTI, T., CSABAI, I., CZAKO, F., SZAKACS, T., SERNEELS, R.
and VATTAY, G. "Dynamics of the Cohonen Map"
BOCHEREAU, L., BOURGINE, P. and DEFFUANT, G. "Equivalence
Between Connectionist Classifiers and Logical Classifiers"
BOLLE, D. and DUPONT, P. "On Potts-Glass Neural Networks
with Biassed Patterns"
COOLEN, A.C.C. "Ising-Spin Neural Networks with Spatial Structure"
MARRO, J. and GARRIDO, P.L. "Kinematically Disordered
Lattice Systems"
BAVAN, A.S. "A Programming System for Implementing Neural Networks"
AHUJA, S.B. and WOO-YOUNG S. "An Auto-Augmenting Neural-Network
Architecture for Diagnostic Reasoning"
DORRONSORO, J.R. and LOPEZ, V. "Formal Integrators
and Neural Networks"
VIRASORO, M.A. "Disordered Models of Acquired Dyslexia"
KURTEN, K.E. "Higher-Order Memories in Optimally Structured
Neural Networks"
SHERRINGTON, D. and WONG, K.Y.M. "Random Boolean Networks
for Autoassociative Memory: Optimization and Sequential Learning"

XII SITGES CONFERENCE

"Complex Fluids" (1992)

LEKKERKERKER, H.N.W. and STROOBANTS, A. "The Phase
Behavior of Colloid-Polymer and Colloid-Colloid Mixtures"

ACKERSON, B.J. and SCHATZEL, K. "Dynamics of Crystallization
in Model Hard Sphere Suspensions"

LIVOLANT, F., LEFORESTIER, A., DURAND, D. and DOUCET, J.
"Structure of DNA Mesophases"

JANNINK, G., KUNZ, W., VAN DER MAAREL, J.R.C., CALMETTES, P.
and COTTON, J.P. "Charge Structure in Electrolytes and Polyelectrolytes.
Experimental Evidence and Interpretation"

RUBI, J.M., SALUEÑA, C. and PEREZ-MADRID, A. "Ferrofluids.
Hydrodynamical and Statistical Aspects"

BACRI, J.-C. and PERZYNSKI, R. "Ferrofluids. Magneto-Optic Effects
in Time Dependent Magnetic Fields"

BACRI, J.-C. and PERZYNSKI, R. "Colloidal Stability Influence
on Rehology of Magnetic Fluids"

COHEN-ADDAD, J.P. "NMR and Dynamics in Polymeric Systems.
Melts, Gels and Blends"

MOUSSAID, A., SCHOSSELER, F., MUNCH, J.P. and CANDAU, S.J.
"Weakly Charged Polyelectrolyte Solutions"

FRENKEL, D. "Order Through Disorder:
Entropy-Driven Phase Transitions"

D'AGUANNO, B., KLEIN, R., MENDEZ-ALCARAZ, J.M.
and NAGELE, G. "Polydisperse Complex Fluids"

ALMAZARA, N.G. "Monte Carlo Simulation of Liquid n-Alkanes"

TURQ, P., BERNARD, O., KUNZ, W. and BLUM, L. "Transport in
Electrolytes Using the Mean Spherical Approximation:
Electrical Conductance and Self-Diffusion Coefficient as a Function
of Concentration in Solutions"

PADRO, J.A. "Computer Simulations of Macromolecules in Solution:
Modelling of Solvent Effects on Ions in Water"

CUESTA, J.A. "Orientational Freezing Within
the Effective Liquid Approach"

DOI, M. "Rheology of Textured Materials"

SKJELTORP, A.T. "Physical Modelling Using Microparticles"

JOHNER, A. "Bridging in Grafted Layers: Statics and Kinetics"

NEMIROVSKY, A.M. and WITTEN, T.A. "Stress Relaxation
in Diblock Copolymers"

KROGER, M. and SELLERS, S. "A Molecular Theory for Spatially
Inhomogeneous, Concentrated Solutions

of Rod-Like Liquid Crystal Polymers"
BOTET, R. and PZOSZAJCZAK M. "Intermittency Patterns
of Fluctuations in Disaggregating Systems"
HAYAKAWA, H. and SHIN-ICHI, S.
"Void Fraction Dynamics in Fluidization"
LANGEVIN, D. "Micelles and Microemulsions"
WEAIRE, D. "Froths and Foams"
KOPER, G.J.M., SMEETS, J. and BEDEAUX, D.
"Clustering and Relaxation in Oil-Continuous Microemulsions"
KUNZ, W., CALMETTES, P., BELLISSENT-FUNEL, M.-C.,
JANNINK, G., CARTAILLER, T., and TURQ, P. "Neutron Scattering
Experiments on Nonaqueous Electrolyte Solutions"
GOLDBART, P.M. and OLMSTED, P.D. "Nematogenic Fluids
in Shear Flow: A Laboratory for Nonequilibrium Physics"
GRANEK, R. "Dynamic Structure Factor of Sponge Phases"
TERAMOTO, A. and SATO, T. "Liquid Crystal Formation in Semiflexible
Polymer Solutions: Effects of Chain Stiffness, Electrostatic Interaction,
and Polydispersity"

Lecture Notes in Physics

For information about Vols. 1–419
please contact your bookseller or Springer-Verlag

Vol. 420: F. Ehlotzky (Ed.), Fundamentals of Quantum Optics III. Proceedings, 1993. XII, 346 pages. 1993.

Vol. 421: H.-J. Röser, K. Meisenheimer (Eds.), Jets in Extragalactic Radio Sources. XX, 301 pages. 1993.

Vol. 422: L. Päivärinta, E. Somersalo (Eds.), Inverse Problems in Mathematical Physics. Proceedings, 1992. XVIII, 256 pages. 1993.

Vol. 423: F. J. Chinea, L. M. González-Romero (Eds.), Rotating Objects and Relativistic Physics. Proceedings, 1992. XII, 304 pages. 1993.

Vol. 424: G. F. Helminck (Ed.), Geometric and Quantum Aspects of Integrable Systems. Proceedings, 1992. IX, 224 pages. 1993.

Vol. 425: M. Dienes, M. Month, B. Strasser, S. Turner (Eds.), Frontiers of Particle Beams: Factories with e+ e- Rings. Proceedings, 1992. IX, 414 pages. 1994.

Vol. 426: L. Mathelitsch, W. Plessas (Eds.), Substructures of Matter as Revealed with Electroweak Probes. Proceedings, 1993. XIV, 441 pages. 1994

Vol. 427: H. V. von Geramb (Ed.), Quantum Inversion Theory and Applications. Proceedings, 1993. VIII, 481 pages. 1994.

Vol. 428: U. G. Jørgensen (Ed.), Molecules in the Stellar Environment. Proceedings, 1993. VIII, 440 pages. 1994.

Vol. 429: J. L. Sanz, E. Martínez-González, L. Cayón (Eds.), Present and Future of the Cosmic Microwave Background. Proceedings, 1993. VIII, 233 pages. 1994.

Vol. 430: V. G. Gurzadyan, D. Pfenniger (Eds.), Ergodic Concepts in Stellar Dynamics. Proceedings, 1993. XVI, 302 pages. 1994.

Vol. 431: T. P. Ray, S. Beckwith (Eds.), Star Formation Techniques in Infrared and mm-Wave Astronomy. Proceedings, 1992. XIV, 314 pages. 1994.

Vol. 432: G. Belvedere, M. Rodonò, G. M. Simnett (Eds.), Advances in Solar Physics. Proceedings, 1993. XVII, 335 pages. 1994.

Vol. 433: G. Contopoulos, N. Spyrou, L. Vlahos (Eds.), Galactic Dynamics and N-Body Simulations. Proceedings, 1993. XIV, 417 pages. 1994.

Vol. 434: J. Ehlers, H. Friedrich (Eds.), Canonical Gravity: From Classical to Quantum. Proceedings, 1993. X, 267 pages. 1994.

Vol. 435: E. Maruyama, H. Watanabe (Eds.), Physics and Industry. Proceedings, 1993. VII, 108 pages. 1994.

Vol. 436: A. Alekseev, A. Hietamäki, K. Huitu, A. Morozov, A. Niemi (Eds.), Integrable Models and Strings. Proceedings, 1993. VII, 280 pages. 1994.

Vol. 437: K. K. Bardhan, B. K. Chakrabarti, A. Hansen (Eds.), Non-Linearity and Breakdown in Soft Condensed Matter. Proceedings, 1993. XI, 340 pages. 1994.

Vol. 438: A. Pękalski (Ed.), Diffusion Processes: Experiment, Theory, Simulations. Proceedings, 1994. VIII, 312 pages. 1994.

Vol. 439: T. L. Wilson, K. J. Johnston (Eds.), The Structure and Content of Molecular Clouds. 25 Years of Molecular Radioastronomy. Proceedings, 1993. XIII, 308 pages. 1994.

Vol. 440: H. Latal, W. Schweiger (Eds.), Matter Under Extreme Conditions. Proceedings, 1994. IX, 243 pages. 1994.

Vol. 441: J. M. Arias, M. I. Gallardo, M. Lozano (Eds.), Response of the Nuclear System to External Forces. Proceedings, 1994. VIII. 293 pages. 1995.

Vol. 442: P. A. Bois, E. Dériat, R. Gatignol, A. Rigolot (Eds.), Asymptotic Modelling in Fluid Mechanics. Proceedings, 1994. XII, 307 pages. 1995.

Vol. 443: D. Koester, K. Werner (Eds.), White Dwarfs. Proceedings, 1994. XII, 348 pages. 1995.

Vol. 444: A. O. Benz, A. Krüger (Eds.), Coronal Magnetic Energy Releases. Proceedings, 1994. X, 293 pages. 1995.

Vol. 445: J. Brey, J. Marro, J. M. Rubí, M. San Miguel (Eds.), 25 Years of Non-Equilibrium Statistical Mechanics. Proceedings, 1994. XVII, 387 pages. 1995.

Vol. 446: V. Rivasseau (Ed.), Constructive Physics. Results in Field Theory, Statistical Mechanics and Condensed Matter Physics. Proceedings, 1994. X, 337 pages. 1995.

Vol. 447: G. Aktaş, C. Saçlıoğlu, M. Serdaroğlu (Eds.), Strings and Symmetries. Proceedings, 1994. XIV, 389 pages. 1995.

Vol. 448: P. L. Garrido, J. Marro (Eds.), Third Granada Lectures in Computational Physics. Proceedings, 1994. XIV, 346 pages. 1995.

Vol. 449: J. Buckmaster, T. Takeno (Eds.), Modeling in Combustion Science. Proceedings, 1994. X, 369 pages. 1995.

Vol. 450: M. F. Shlesinger, G. M. Zaslavsky, U. Frisch (Eds.), Lévy Flights and Related Topics in Physics. Proceedigs, 1994. XIV, 347 pages. 1995.

Vol. 452: A. M. Bernstein, B. R. Holstein (Eds.), Chiral Dynamics: Theory and Experiment. Proceedings, 1994. VIII, 351 pages. 1995.

Vol. 453: S. M. Deshpande, S. S. Desai, R. Narasimha (Eds.), Fourteenth International Conference on Numerical Methods in Fluid Dynamics. Proceedings, 1994. XIII, 588 pages. 1995.

Vol. 454: J. Greiner, H. W. Duerbeck, R. E. Gershberg (Eds.), Flares and Flashes, Germany 1994. XXII, 477 pages. 1995.

Vol. 455: F. Occhionero (Ed.), Birth of the Universe and Fundamental Physics. Proceedings, 1994. XV, 387 pages. 1995.

Vol. 456: H. B. Geyer (Ed.), Field Theory, Topology and Condensed Matter Physics. Proceedings, 1994. XII, 206 pages. 1995.

New Series m: Monographs

Vol. m 1: H. Hora, Plasmas at High Temperature and Density. VIII, 442 pages. 1991.

Vol. m 2: P. Busch, P. J. Lahti, P. Mittelstaedt, The Quantum Theory of Measurement. XIII, 165 pages. 1991.

Vol. m 3: A. Heck, J. M. Perdang (Eds.), Applying Fractals in Astronomy. IX, 210 pages. 1991.

Vol. m 4: R. K. Zeytounian, Mécanique des fluides fondamentale. XV, 615 pages, 1991.

Vol. m 5: R. K. Zeytounian, Meteorological Fluid Dynamics. XI, 346 pages. 1991.

Vol. m 6: N. M. J. Woodhouse, Special Relativity. VIII, 86 pages. 1992.

Vol. m 7: G. Morandi, The Role of Topology in Classical and Quantum Physics. XIII, 239 pages. 1992.

Vol. m 8: D. Funaro, Polynomial Approximation of Differential Equations. X, 305 pages. 1992.

Vol. m 9: M. Namiki, Stochastic Quantization. X, 217 pages. 1992.

Vol. m 10: J. Hoppe, Lectures on Integrable Systems. VII, 111 pages. 1992.

Vol. m 11: A. D. Yaghjian, Relativistic Dynamics of a Charged Sphere. XII, 115 pages. 1992.

Vol. m 12: G. Esposito, Quantum Gravity, Quantum Cosmology and Lorentzian Geometries. Second Corrected and Enlarged Edition. XVIII, 349 pages. 1994.

Vol. m 13: M. Klein, A. Knauf, Classical Planar Scattering by Coulombic Potentials. V, 142 pages. 1992.

Vol. m 14: A. Lerda, Anyons. XI, 138 pages. 1992.

Vol. m 15: N. Peters, B. Rogg (Eds.), Reduced Kinetic Mechanisms for Applications in Combustion Systems. X, 360 pages. 1993.

Vol. m 16: P. Christe, M. Henkel, Introduction to Conformal Invariance and Its Applications to Critical Phenomena. XV, 260 pages. 1993.

Vol. m 17: M. Schoen, Computer Simulation of Condensed Phases in Complex Geometries. X, 136 pages. 1993.

Vol. m 18: H. Carmichael, An Open Systems Approach to Quantum Optics. X, 179 pages. 1993.

Vol. m 19: S. D. Bogan, M. K. Hinders, Interface Effects in Elastic Wave Scattering. XII, 182 pages. 1994.

Vol. m 20: E. Abdalla, M. C. B. Abdalla, D. Dalmazi, A. Zadra, 2D-Gravity in Non-Critical Strings. IX, 319 pages. 1994.

Vol. m 21: G. P. Berman, E. N. Bulgakov, D. D. Holm, Crossover-Time in Quantum Boson and Spin Systems. XI, 268 pages. 1994.

Vol. m 22: M.-O. Hongler, Chaotic and Stochastic Behaviour in Automatic Production Lines. V, 85 pages. 1994.

Vol. m 23: V. S. Viswanath, G. Müller, The Recursion Method. X, 259 pages. 1994.

Vol. m 24: A. Ern, V. Giovangigli, Multicomponent Transport Algorithms. XIV, 427 pages. 1994.

Vol. m 25: A. V. Bogdanov, G. V. Dubrovskiy, M. P. Krutikov, D. V. Kulginov, V. M. Strelchenya, Interaction of Gases with Surfaces. XIV, 132 pages. 1995.

Vol. m 26: M. Dineykhan, G. V. Efimov, G. Ganbold, S. N. Nedelko, Oscillator Representation in Quantum Physics. IX, 279 pages. 1995.

Vol. m 27: J. T. Ottesen, Infinite Dimensional Groups and Algebras in Quantum Physics. IX, 218 pages. 1995.

Vol. m 28: O. Piguet, S. P. Sorella, Algebraic Renormalization. IX, 134 pages. 1995.

Vol. m 29: C. Bendjaballah, Introduction to Photon Communication. VII, 193 pages. 1995.

Vol. m 30: A. J. Greer, W. J. Kossler, Low Magnetic Fields in Anisotropic Superconductors. VII, 161 pages. 1995.

Vol. m 31: P. Busch, M. Grabowski, P. J. Lahti, Operational Quantum Physics. XI, 230 pages. 1995.